T0318301

Personnel Protection and Safety Equipment for the Oil and Gas Industries

Personnel Protection and Safety Equipment for the Oil and Gas Industries

Alireza Bahadori, PhD
School of Environment
Science and Engineering
Southern Cross University
Lismore, NSW
Australia

AMSTERDAM • BOSTON • HEIDELBERG • LONDON
NEW YORK • OXFORD • PARIS • SAN DIEGO • SAN FRANCISCO
SINGAPORE • SYDNEY • TOKYO
Gulf Professional Publishing is an imprint of Elsevier

Gulf Professional Publishing is an imprint of Elsevier
225 Wyman Street, Waltham, MA 02451, USA
The Boulevard, Langford Lane, Kidlington, Oxford, OX5 1GB, UK

Copyright © 2015 Elsevier Inc. All rights reserved.

No part of this publication may be reproduced or transmitted in any form or by any means,
electronic or mechanical, including photocopying, recording, or any information storage
and retrieval system, without permission in writing from the publisher. Details on how to
seek permission, further information about the Publisher's permissions policies and our
arrangements with organizations such as the Copyright Clearance Center and the Copyright
Licensing Agency, can be found at our website: www.elsevier.com/permissions.

This book and the individual contributions contained in it are protected under copyright by
the Publisher (other than as may be noted herein).

Notices
Knowledge and best practice in this field are constantly changing. As new research and
experience broaden our understanding, changes in research methods, professional practices,
or medical treatment may become necessary.

Practitioners and researchers must always rely on their own experience and knowledge in
evaluating and using any information, methods, compounds, or experiments described herein.
In using such information or methods they should be mindful of their own safety and the
safety of others, including parties for whom they have a professional responsibility.

To the fullest extent of the law, neither the Publisher nor the authors, contributors, or editors,
assume any liability for any injury and/or damage to persons or property as a matter of
products liability, negligence or otherwise, or from any use or operation of any methods,
products, instructions, or ideas contained in the material herein.

ISBN: 978-0-12-802814-8

British Library Cataloguing-in-Publication Data
A catalogue record for this book is available from the British Library

Library of Congress Cataloging-in-Publication Data
A catalog record for this book is available from the Library of Congress

For information on all Gulf Professional Publishing publications
visit our website at http://store.elsevier.com/

This book has been manufactured using Print On Demand technology.
Each copy is produced to order and is limited to black ink. The online version
of this book will show color figures where appropriate.

Working together
to grow libraries in
developing countries

www.elsevier.com • www.bookaid.org

Dedicated to the loving memory of my parents, grandparents, and to all who contributed so much to my work over the years.

Contents

Biography

Alireza Bahadori, PhD is a research staff member in the School of Environment, Science and Engineering at Southern Cross University, Lismore, NSW, Australia. He received his PhD from Curtin University, Perth, Western Australia.

During the past 20 years, Dr Bahadori has held various process and petroleum-engineering positions and has been involved in many large-scale projects at NIOC, Petroleum Development Oman (PDO), and Clough AMEC PTY LTD.

He is the author of around 250 articles and 12 books. His books have been published by many major publishers, including Elsevier.

Dr. Bahadori is the recipient of highly competitive and prestigious Australian government's international postgraduate research award as part of his research in oil and gas area. He also received Top-Up Award from the state government of Western Australia through the Western Australia Energy Research Alliance (WA:ERA) in 2009. Dr. Bahadori serves as a reviewer and member of editorial board and reviewer for a large number of journals.

Preface

The vast complexity and variety that nature presents within the oil, gas, and petrochemical industries requires specific books for personnel safety and protection measures. This book outlines the mentioned specific requirements.

Sanitation and first aid are two key factors, and the minimum requirements for keeping plants/machinery, workplaces, and personnel in healthy conditions are covered in this text.

The serious consequences of poor sanitary conditions and insufficient first-aid procedures in the oil, gas, and petrochemical industries can be briefly categorized as follows:

a. Unsafe working conditions
b. Malfunctioning machineries
c. Poor health of personnel

In the control of occupational diseases caused by breathing air contaminated with harmful dusts, fogs, fumes, mists, gases, smokes, sprays, or vapors, the primary objective should be to prevent atmospheric contamination. This should be accomplished as far as feasible by accepted engineering control measures (e.g., enclosure or confinement of the operation, general and local ventilation, and substitution of less toxic materials).

The focus of this book is on the minimum requirements for protection of the respiratory system from inhalation of particulate matter, noxious gases, and vapors, and oxygen deficiency. The factors affecting the choice of respiratory equipment are discussed. The equipment covered include:

a. Respirators for dusts, gases, and gases with dusts
b. Breathing apparatus, self-contained closed and open-circuits
c. Airline, fresh-air, and compressed types
d. Dust hoods and suits (positive-pressure, powered)
e. Underwater breathing apparatus
f. Ventilatory resuscitators

This book is designed to assist in the selection of respiratory protective devices for use against atmospheric contaminants. Atmospheres can be contaminated by dust or gas, or be deficient in oxygen. These hazards occur singly or in combination. Additionally, each contaminant may have special characteristics of its own that require protection. For instance, radio-active or corrosive contaminants require the use of special clothing. Some gases, liquids, and soluble solids absorb through the skin and these also require special protection.

Contaminated atmospheres are described generally as nuisance atmospheres that are not toxic or immediately dangerous to health; hazardous atmospheres that are of low toxicity or cause easily reversible biological changes; dangerous atmospheres of a high toxicity or where health hazards are more severe; and atmospheres immediately dangerous to life.

Respiratory protective devices should either filter the contaminated atmosphere to produce air suitable for respiration or supply such air from an alternative source. The air is supplied to the breathing area (the nose and mouth of the wearer) by one of the following: a mouthpiece and nose clip; a half-mask covering the nose and mouth; a full face piece covering the eyes, nose and mouth; a hood covering the head down to the shoulders; or a suit covering the head and body down to waist and wrists.

The use of sealed radioactive sources has become so widespread that a resource to is needed to help guide users. Safety is the prime consideration in establishing a standard for the use of sealed radioactive sources. However, as the application of sources becomes more diversified, a text is needed to specify the characteristics of a source and the essential performance and safety-testing methods for a particular application.

Safety belts and harnesses are means of protective equipment that are filled around the upper parts of the body protecting the user against falls and creating self-confidence when used in the correct manner. In designing and selecting a belt or harness for any particular work, care should be taken to ensure that the equipment gives the user, as far as it is compatible with safety, the maximum degree of comfort and freedom of movement, and also in the event of the user falling, the greatest possible security against injury. Self-locking anchorage, lanyards, and other component parts are safety protective devices used to protect against falls. In assessing the performance of safety belts and harnesses the focus is on maintenance, inspection, and storage of equipment.

This book specifies the minimum requirements for types, classes, materials, design, physical and performance details that afford protection to all workers in industrial plants. This book is divided into several parts and will provide a separate section for each category of protective clothing.

Acknowledgments

I would like to thank the Elsevier editorial and production team and Ms. Katie Hammon and Ms. Kattie Washington of Gulf Professional Publishing for their editorial assistance.

Breathing apparatus for personnel safety and protection

1

1.1 Introduction

This chapter is designed to assist in the selection of respiratory protective devices (RPDs) for use against atmospheric contaminants. Atmospheres can be contaminated by dust, gas, or by being deficient in oxygen. These hazards occur singly or in any combination. In this context, dust can include mist and fume and gas can include vapor.

Each contaminant can have special characteristics of its own that require protection in addition to those discussed in this section. For instance, radioactive or corrosive materials require the use of special clothing. Some gases, liquids, and soluble solids absorb through the skin, and these also require special protection.

Contaminated atmospheres are generally described as nuisance atmospheres that are not toxic or immediately dangerous to health, hazardous atmospheres that are of low toxicity or easily cause reversible biological changes, dangerous atmospheres of a high toxicity or where the health hazards are more serious, and atmospheres immediately dangerous to life.

RPDs should either filter the contaminated atmosphere to produce air suitable for respiration, or supply such air from an alternative source. The air is supplied to the breathing area (the nose and mouth of the user) by one of the following: a mouthpiece and nose clip; a half-mask covering the nose and mouth; a full facepiece covering the eyes, nose, and mouth; a hood covering the head down to the shoulders; or a suit covering the head and body down to the waist and wrists.

In this chapter, for each type of RPD a nominal protection factor is given. This factor is a guide to the effectiveness of the device when used correctly. It indicates the degree to which the atmospheric contaminant is reduced by the respirator within the breathing zone. Thus, a device that reduces the level of contamination 10 times will have a nominal protection factor of 10, while one that reduces it 1000 times will have a nominal protection factor of 1000. These figures should be used in conjunction with the maximum allowable concentration, or threshold limit value, of the contaminant and its actual concentration in the atmosphere. Generally, a substance with a threshold limit value of 10 parts per million, which has a concentration of 1000 parts per million in the atmosphere, will require the use of equipment with a nominal protection factor of at least 100.

Breathing apparatus sets allow firefighters to enter areas filled with smoke or other poisonous gases. These sets consist of a cylinder that contains compressed air, a mask that is worn over the whole face, a gauge to tell the firefighter the pressure in the cylinder, a distress signal unit that activates if the firefighter stops moving, and other

Personnel Protection and Safety Equipment for the Oil and Gas Industries.
DOI: http://dx.doi.org/10.1016/B978-0-12-802814-8.00001-8
© 2015 Elsevier Inc. All rights reserved.

safety equipment. The cylinder can supply up to 45 minutes of air, although hard work and other factors can reduce this duration. Breathing apparatus (BA) is also available in a twin-cylinder configuration for longer use.

The following requirements will be discussed in more detail in subsequent sections of this chapter.

In the control of occupational diseases caused by breathing air contaminated with harmful dusts, fogs, fumes, mists, gases, smokes, sprays, or vapors, the primary objective is to prevent atmospheric contamination. This is accomplished as far as feasible by accepted engineering control measures (e.g., enclosure or confinement of the operation, general and local ventilation, and substitution with less toxic materials).

Respirators should be provided by employers when such equipment is necessary to protect the health of employees. Employers should provide respirators that are applicable and suitable for the task, and should establish and maintain a respiratory protective program that covers these general requirements.

Employee should use the provided respiratory protection in accordance with the instructions and training received, and should guard against damage to the respirator. Employees should report any malfunction of the respirator.

1.1.1 Minimal acceptable program

Standard operating instructions governing the selection and use of respirators should be observed. Respirators should be selected on the basis of the hazards to which workers are exposed. The user should be instructed and trained in the proper use of respirators and their limitations.

All types of BA should be regularly cleaned and disinfected. Those issued for the exclusive use of one employee should be cleaned after each day's use, or more often if necessary. Those used by more than one employee should be thoroughly cleaned and disinfected after each use. Appropriate surveillance of work-area conditions and degree of employee exposure or stress should also be maintained. There should be regular checkups and evaluations to determine the continued effectiveness of the program.

The company's industrial hygiene, health physics, safety engineering, or fire department should administer the program in close liaison with the company's medical department. Responsibility for the program should be given to one individual to assure it is maintained. In small plants having no formal industrial hygiene, health physics, safety, fire, or medical department, the respirator program should be handled by an upper-level superintendent, foreman, or other qualified individual. The program's administrator should also have sufficient knowledge to properly supervise the program.

1.1.2 Medical limitations

A firefighter should be assigned tasks requiring use of RPDs only if it has been determined that he is able to perform these tasks while using the device(s). Firefighters with punctured eardrums should wear earplugs. The assigned physician should determine what health, physical, and psychological conditions are pertinent. The firefighter's medical status pertaining to use of RPDs should be reviewed at least annually.

1.1.3 Communication

Although full facepieces distort the human voice to some extent, the exhalation valve usually provides a pathway for speech transmission over short distances. Also, most types of full facepieces are available with speaking diaphragms to improve speech intelligibility. In addition, there are a variety of electronic communication units that utilize a microphone inside the full facepiece, connected directly to an amplifier and speaker, to a telephone, or to a radio transmitter. Connecting cables from microphones pass through the face piece. If the cables are removed for any reason, they should be carefully replaced or any hole in the facepiece should be carefully sealed.

1.1.4 Use of unapproved respiratory protective devices

Unapproved self-contained breathing devices are risky and should not be purchased or used.

Some of the problems associated with the use of full facepieces at low temperatures are poor visibility and freezing of exhalation valves. All full facepieces should be designed so that incoming fresh-air sweeps over the inside of the eyepieces to reduce misting. Anti-mist compounds should be used to coat the inside of eyepieces to reduce misting at room temperatures and down to temperatures approaching 0°C. Full facepieces are also available with inner masks that direct moist exhaled air through the exhalation valves, and when properly fitted are likely to provide adequate visibility at low temperatures.

At very low temperatures the exhalation valve collects moisture and may freeze open, allowing the user to breathe contaminated air, or freeze closed, preventing normal exhalation. Dry air suitable for respiration should be used with self-contained and compressed airline BA at low temperatures. The dew point of the breathed gas should be appropriate to the ambient temperature. High-pressure connections on self-contained BA will leak because of metal contraction at low temperatures, but the only problem is likely to be an outward leak.

A worker in areas of high ambient or radiant temperatures is under stress, and any additional stress caused by the use of respiratory protective devices should be minimized. This can be done by using devices having low weight and low breathing resistance. Supplied air respirators, hoods, and suits with an adequate supply of cool breathing air should be used.

1.2 Selection of respiratory protective equipment

A respiratory protective device (RPD), also known as a respirator, is a piece of safety equipment used for personal protection. Respirators are designed to prevent the inhalation of contaminated air and belong in two main categories:

- Air-purifying respirators: These respirators are designed to filter or clean contaminated air from the workplace before it is inhaled by the RPD user. They are available as either disposable respirators, or as nondisposable respirators with disposable filters.

Figure 1.1 Respiratory apparatus.

• Air-supplied respirators: These respirators deliver clean, breathable air from an independent source to the user. Air-supplied respirators are typically used for high-risk environments, such as oxygen-deficient atmospheres and confined spaces.

The multiplicity of hazards that may exist in a given operation requires careful evaluation followed by intelligent selection of protective equipment. This selection is made even more complex by the many types of equipment available, each having its limitations, areas of application, and operational and maintenance requirements. Figure 1.1 shows a respiratory apparatus.

The selection of correct respiratory protection equipment for any given situation requires consideration of the nature of the hazard, the severity of the hazard, work requirements and conditions, and the characteristics and limitations of the available equipment. Where there is any doubt about the suitability of respirator, BA should be used. Particular care is necessary in the case of odorless gases or fumes as these do not give a warning of their presence.

Some respirators and BA deliver air to the user at pressures slightly above ambient pressures. These can provide greater protection since any gas flow through leaks in the system will be outward. However, for this to happen the pressure bias needs to be sufficient to ensure that the pressure within the mouth and nose remains positive throughout the respiratory cycle. In the case of BA, should the leak rate be excessive, the duration will be reduced. Since the reduction of pressure on inhalation depend on respiratory resistance, the performance of positive-pressure devices should always be checked with all resistances to respiration attached.

1.3 Severity and location of the hazard

Only protective devices that arrange for the provision of an independent atmosphere suitable for respiration are appropriate for use in oxygen-deficient atmospheres, and self-contained, airline, or fresh-air hose apparatus should be used. The concentration of the contaminant and its location should be considered, and the length of time for which protection will be needed, entry and exit times, accessibility of a supply of air suitable for respiration, and the ability to use air lines or move about freely while wearing the protective device should also be considered. Where flammable or explosive atmospheres might arise, the equipment should be suitable for use in these circumstances, too.

1.3.1 Nature of the hazard

The chemical and physical properties, toxicity, and concentration of the hazardous materials should be considered in the selection of RPDs.

- Immediately dangerous atmospheres: It should be assumed that atmospheres immediately dangerous to life or health–either oxygen-deficient or containing high concentrations of dangerous gases and vapors–will be encountered by firefighters.
- Oxygen-deficient atmospheres: Self-contained BA of suitable service time should be used where oxygen deficiency exists or may exist.
- Unusual hazards: Unusual factors can add new dimensions to a hazardous situation and should be anticipated when selecting RPDs. Some examples include:
- Absorption through, or irritation of, the skin: Some airborne contaminants are extremely irritating to the skin (e.g., ammonia and hydrochloric acid), while others are capable of being absorbed through the skin and into the bloodstream with serious, possibly fatal results. Hydrocyanic-acid gas and many of the organic-phosphate pesticides, such as thiophosphate insecticides and tetraethyl pyrophosphate (TEPP), will penetrate unbroken skin. The RPD will not provide complete protection against these contaminants. If such materials are encountered, an effective full body suit of impermeable material should be worn with appropriate respiratory protection.
- Ionizing radiation of skin and whole body: The RPD will not protect the skin or whole body against ionizing radiation from airborne concentrations of certain radioactive materials.

1.3.2 Storage

RPDs should be stored in the original carrying or storage case or in a wall or apparatus rack specially designed for quick removal and for protection of the BA. Demand-type BA should be stored with the main-line regulator valve open and the main cylinder valve and regular bypass valve closed. The facepieces of all devices should be positioned carefully to avoid distortion of rubber parts during storage. Head-harness straps should be fully extended.

1.4 Special considerations

1.4.1 Corrective lenses with full facepieces

Providing respiratory protection for individuals wearing corrective lenses is a serious problem. A proper seal cannot be established if the temple bars of eyeglasses extend through the sealing edge of the full facepiece. As a temporary measure, glasses with short temple bars or without temple bars may be taped to the user's head. Wearing of contact lenses in contaminated atmospheres with a respirator should not be allowed.

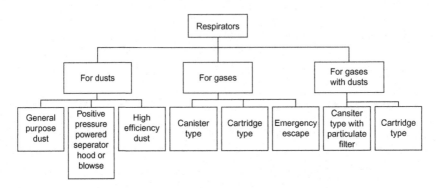

Figure 1.2 Different types of respirators.

There are systems for mounting corrective lenses inside full facepieces. When a worker must wear corrective lenses as part of the facepiece, the facepiece and lenses should be fitted by qualified individuals to provide good vision, comfort, and a gastight seal.

1.4.2 Eyewear with half-mask facepiece

If corrective glasses or goggles are required, they should be worn so as not to affect the fit of the facepiece. Proper selection of equipment will minimize or avoid this problem.

1.5 Classification of respiratory protective equipment

There are two distinct methods of providing personal protection against contaminated atmospheres, discussed below and also shown in Figure 1.2.

1.5.1 By purifying the air breathed

In this apparatus, the inhaled air is drawn through a medium that removes the harmful substances, which depends on the contaminating agent. Devices that achieve this are known as respirators.

Note: In addition to dust respirators, other devices are available that consist of a filtering medium held to the nose and mouth by simple means to remove coarse nuisance dusts from the inhaled air. They should never be used in the presence of harmful or toxic dusts.

1.5.2 By supplying air or oxygen from an uncontaminated source

In this apparatus, the inhaled air is conveyed by air line, or alternatively, air or oxygen is supplied from cylinders or other containers carried by the person at risk. Devices that achieve this are known as BA. The main types of BA are shown in Figure 1.3.

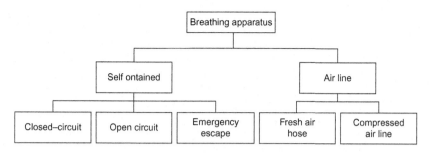

Figure 1.3 The main types of breathing apparatus.

A respirable atmosphere independent of the ambient air is supplied to the user by one of the following.

- Self-Contained Breathing Apparatus (SCBA). Supply of air, oxygen, or oxygen-generating material carried by user. Normally equipped with full facepiece, but some include a mouthpiece.
  1. Closed-Circuit SCBA (oxygen only)
     a. Compressed- or liquid-oxygen type: High-pressure O_2 from a gas cylinder passes through a high-pressure reducing valve and, in some designs, through a low-pressure admission valve to a breathing bag or container. Liquid oxygen is converted to low-pressure gaseous oxygen and delivered to the breathing bag. The user inhales from the bag through a corrugated tube connected to a mouthpiece or facepiece and a one-way check valve. Exhaled air passes through another check valve and tube into a container of carbon-dioxide removing chemical and re-enters the breathing bag. Make up O_2 enters the bag continuously or as the bag deflates sufficiently to actuate an admission valve. A pressure relief system is provided and a manual bypass system and saliva trap is also provided depending on the design.
     b. Oxygen-generating type: Water vapor in the exhaled breath reacts with a chemical in the canister to release O_2 to the breathing bag. The user inhales from the bag through a corrugated tube and one-way check valve at the facepiece. Exhaled air passes through a second check-valve breathing-tube assembly into the canister. The O_2 release rate is governed by the volume of exhaled air, and CO_2 is removed by the canister fill.
  2. Open-Circuit SCBA (compressed air, compressed oxygen, liquid air, or liquid oxygen)
     a. Demand type: The demand valve permits oxygen or airflow only during inhalation. Exhaled breath passes to the ambient atmosphere through a valve(s) in the facepiece, and a bypass system is provided in case of regulator failure except on escape-type units.
     b. Pressure-demand type (see note 2) below): Equipped with full facepiece only. Positive pressure is maintained in the facepiece at all times. The user usually has the option of selecting the demand or pressure-demand mode of operation.
- Hose Mask and Air-line Respirator
  1. Hose Mask: Equipped with a full facepiece, nonkinking breathing tube, rugged safety harness, and a large diameter heavy-duty nonkinking air supply hose. The breathing tube and hose are securely attached to the harness. A check valve allows airflow only toward the facepiece. The facepiece is fitted with an exhalation valve. The harness has provision for attaching a safety line.

   a. Hose mask with blower: Air is supplied by a motor-driven or hand-operated blower.

   b. Hose mask without blower: The user provides the force need to pull air through the hose. The hose inlet is anchored and fitted with a funnel or funnel-like object covered with a fine mesh screen to prevent entrance of coarse particulate matter. Up to 23 m of hose length is permissible.

2. Air-line Respirator: Respirable air is supplied through a small-diameter air line from a compressor or compressed-air cylinders. The air line is attached to the user by belt and can be detached rapidly in an emergency. A flow-control valve or orifice is provided to govern the rate of airflow to the user. Exhaled air passes to the ambient atmosphere through a valve(s) or opening in the enclosure (facepiece, hood, suit). Up to 76 m. of air line is permissible.

   a. Continuous flow of air: Equipped with a half-mask or full facepiece or a helmet (abrasive blasting) or hood covering the user's head and neck. At least 4 cubic feet of air per minute to tight-fitting facepieces and 6 cubic feet per minute to loose-fitting hoods and helmets should be required.

   b. Demand type: Equipped with a half-mask or full facepiece. The demand valve permits flow of air only during inhalation.

   c. Pressure-demand type: Equipped with a half-mask or full facepiece. A positive pressure is maintained in the facepiece at all times.

3. Supplied Air Suit: A form of continuous air-line respirator (see air-line respirator above). The suit is one or two pieces and made of leak-resistant material. Air is supplied to the suit through a system of internal tubes to the head, trunk, and extremities. Air exhausts through valves located in appropriate parts of the suit.

- Combination Self-Contained and Air-line Respirators: Normally, a demand or pressure-demand type air-line respirator with a full or half-mask facepiece, together with a small compressed-air cylinder to provide air if the normal supply fails. User immediately returns to a respirable atmosphere if the normal air supply fails.

Notes: 1) Equipped with a demand valve that is activated on initiation of inhalation and permits the flow of breathing atmosphere to the facepiece. On exhalation, pressure in the facepiece becomes positive and the demand valve is deactivated. 2) A small positive pressure is maintained at all times in the facepiece by a spring-loaded or balanced regulator and exhalation valve.

- AIR-PURIFYING RESPIRATORS: Half-mask, full facepiece, or mouthpiece respirator equipped with air-purifying units to remove gases, vapors, and particulate matter from the ambient air prior to its inhalation. Some air-purifying respirators are blower-operated and provide respirable air to the facepiece (or hood) under a slight positive pressure.

- Gas and Vapor-Removing Respirators: Packed sorbent beds (cartridge or canister) remove single gases or vapors (e.g., chlorine gas), a single class of gases or vapors (e.g., organic vapors), or a combination of two or more classes of gases and vapors (e.g., acid gases, organic vapors, ammonia, and carbon monoxide) by absorption, adsorption, chemical reaction, or catalysis or a combination of these methods.

1. Full Facepiece Respirator (Gas Mask): Equipped with a single, large chin canister or harness-mounted canister with breathing tube and inhalation and exhalation valves. Canisters come in the "super" size, "industrial" size (regular), and chin style. The service life is approximately proportional to the canister size for a given type of canister. Canisters are marked in bold letters with the contaminant against which they protect and are color coded for quick identification. The maximum concentration in which the canister can be safely used is indicated on the label. Figure 1.4. shows a full facepiece respirator (gas mask)

Figure 1.4 Full Face piece Respirator (Gas Mask).

Figure 1.5 A typical Half-Mask Respirator.

2. Half-Mask Respirator (Chemical-Cartridge Respirator): Equipped with one or more cartridge and exhalation and inhalation valves. Figure 1.5. shows a typical half-mask respirator.

3. Mouthpiece Respirator: A compact device designed for quick application when the atmosphere is unexpectedly contaminated with a hazardous material. Normally consists of a housing with a mouthpiece and a single cartridge, a nose clamp, exhalation and inhalation valves, and a neckband.

- Particulate-Removing Respirators: Filter media in pads, cartridges, or canisters remove dust, fog, fume, mist, smoke, or spray particles. Filters are designed to remove a single type of particle (silica dust) or classes of particles (dust and fumes). Filters may be replaceable or a permanent parts of the respirator. Some filters can be used only once, while others are reusable and should be cleaned according to the manufacturer's instructions.

1. Full Facepiece Respirator: Normally equipped with a high-efficiency filter canister designed to protect against hazardous particulates. Equipped with inhalation and exhalation valves.

2. Half-Mask Respirator: Normally equipped with one or two dust, mist, or fume filters designed to protect against nuisance and low-to-moderate toxicity dusts, fumes, and mists, and exhalation valves. A knitted fabric cover is sometimes worn on dust respirators to decrease discomfort.

3. Mouthpiece Respirator: Infrequently used as a particulate respirator.

- Combination Gas, Vapor, and Particulate-Removing Respirators: Some canisters and cartridges contain both filters and sorbents to provide protection against contaminants. Some filters are designed to be attached to a sorbent cartridge as a pre-filter (e.g., for a paint-spray operation).

- Combination Atmosphere-Supplying and Air-Purifying Respirators: These provide the user the option of using one of two different modes of operation. It may be an air-line respirator with an air-purifying attachment to provide protection in the event the air supply fails or an air-purifying respirator with a small air cylinder in case the atmosphere unexpectedly exceeds safe conditions for use of an air-purifying respirator.
- ATMOSPHERE-SUPPLYING RESPIRATORS: Atmosphere-supplying respirators provide against oxygen deficiency and most toxic atmospheres. The breathing atmosphere is independent of ambient atmospheric conditions.

General Limitations: Except for the supplied-air suit no protection is provided against skin irritation by materials such as ammonia and HCl, or against sorption of materials such as HCN, tritium, or organic phosphate pesticides through the skin. Facepieces present special problems to individuals required to wear prescription lenses.

- Self-Contained Breathing Apparatus (SCBA): The user carries his own breathing atmosphere. Use is permissible in atmospheres immediately dangerous to life or health.

Limitation: The period over which the device will provide protection is limited by the amount of air or oxygen in the apparatus, the ambient atmospheric pressure (service life is cut in half by a doubling of the atmospheric pressure), and work load. A warning device should be provided to indicate to the user when the service life has been reduced to a low level. Some SCBA devices have a short service life (a few minutes) and are suitable only for escape (self-rescue) from an irrespirable atmosphere. The chief limitations of SCBA devices are their weight, bulk, or both, limited service life, and the training required for their maintenance and safe use.

1. Closed-circuit SCBA: The closed-circuit operation conserves oxygen and permits longer service life.
2. Open-circuit SCBA-demand and Pressure-demand: The demand type produces a negative pressure in the facepiece on inhalation, whereas the pressure-demand type maintains a positive pressure in the facepiece and is less apt to permit inward leakage of contaminants.
   - Hose Mask or Air-line Respirator: The respirable air supply is not limited to the quantity the individual can carry, and the devices are lightweight and simple.
   - Limitations: The user is restricted in movement by the hose or air line and must return to a respirable atmosphere by retracing the route of entry. The hose or air line is subject to being severed or pinched off.

1. Hose Mask
   a. Hose mask with blower : If the blower fails, the unit still provides protection, although a negative pressure exists in the facepiece during inhalation. Use is permissible in atmospheres immediately dangerous to life or health.
   b. Hose mask without blower: Limited to use in atmospheres from which the user can escape unharmed without aid of the respirator.
2. Air-line Respirators (Continuous-Flow, Demand, and Pressure-Demand Types): The demand type produces a negative pressure in the facepiece on inhalation, whereas continuous-flow and pressure-demand types maintain a positive pressure in the facepiece at all times and are less apt to permit inward leakage of contaminants. Limitation: Air-line respirators are limited to use in atmospheres not immediately dangerous to life or health. Air-line respirators provide no protection if the air supply fails.
3. Supplied-Air Suit: These suits protect against atmospheres that affect the skin or mucous membranes. Limitations: Some contaminants, such as tritium, can penetrate the suit

material and limit its effectiveness. Other contaminants, such as fluorine, can react chemically with the suit material and damage it. These suits are limited in use to atmospheres not immediately dangerous to life or health.

- Combination Self-Contained and Air-line Respirators: Equipping an air-line respirator with a small cylinder of compressed air to provide an emergency air supply qualifies the respirator for use in immediately dangerous atmospheres.
- AIR-PURIFYING RESPIRATION: Air-purifying respirators do not protect against oxygen-deficient atmospheres nor against skin irritation by, or sorption through the skin of, airborne contaminants. The maximum contaminant concentration against which an air-purifying respirator will protect is determined by the designed efficiency and capacity of the cartridge, canister, or filter. For gases and vapors and for particles having a threshold limit value (TLV) of less than $0.1 \, mg/m^3$, the maximum concentration for which the air-purifying unit is designed is specified on the label. Respirators without a blower to maintain a constant positive pressure within the facepiece will not provide the maximum design protection specified unless the facepiece is carefully fitted to the user's face to prevent inward leakage. The time period over which protection is provided is dependent on the canister, cartridge, or filter type, concentration of contaminant, and the user's respiratory rate. The proper type of canister, cartridge, or filter should be selected for the particular atmosphere and conditions. Air-purifying respirators generally cause discomfort and objectionable resistance to breathing, although these problems are minimized in blower-operated units. Respirator facepieces present special problems to individuals required to wear prescription lenses. These devices do have the advantage of being small, light, and simple to use. Figure 1.6 shows a typical air-purifying respirator.
- Gas- and Vapor-Removing Respirators: A rise in canister or cartridge temperature indicates that a gas or vapor is being removed from the inspired air. This is not a reliable indicator of canister performance. An uncomfortably high temperature indicates a high concentration of gas or vapor and requires an immediate return to fresh air.

1. Full Facepiece Respirator (Gas Mask): Should avoid use in atmospheres immediately dangerous to life or health if the contaminant(s) lacks sufficient warning properties (i.e., odor or irritation).
2. Half-Mask Respirator (Chemical-Cartridge Respirator): Should not use in atmospheres immediately dangerous to life or health and should be limited to low concentrations of gases

Figure 1.6 A typical air-purifying respirator.

and vapors. A fabric covering should not be worn on the facepiece since it will permit gases and vapors to pass. No protection is provided to the eyes.
3. Mouthpiece Respirator (Chemical Cartridge): Should not be used in atmospheres immediately dangerous to life or health. Mouth breathing prevents detection of contaminants by odor. The nose clip should be securely in place to prevent nasal breathing. No protection is provided to the eyes.
4. Self-Rescue Mouthpiece Respirator: Designed for self-rescue from immediately dangerous atmospheres with gases and vapors. Mouth breathing prevents detection of contaminants by odor. The nose clip should be securely in place to prevent nasal breathing. No protection is provided to the eyes.
   - Particulate-Removing Respirators: protect against nonvolatile particles only. No protection against gases and vapors. The filter should be replaced or cleaned when breathing becomes difficult due to plugging by retained particles. These respirators should not be used during shot- and sand-blasting operations. Abrasive-blasting respirators should be used instead.
1. Full Facepiece Respirator: Should avoid use in atmospheres immediately dangerous to life or health if the contaminant(s) lacks sufficient warning properties (i.e., odor or irritation).
2. Half-Mask Respirator: Should not be used in atmospheres immediately dangerous to life or health. A fabric covering on the facepiece is permissible only in atmospheres of coarse dusts and mists of low toxicity. No protection is provided to the eyes.
3. Mouthpiece Respirator (Filter): Should not be used in atmospheres immediately dangerous to life or health. Mouth breathing prevents detection of contaminants by odor. The nose clip should be securely in place to prevent nasal breathing. No protection is provided to the eyes from irritating aerosols.
4. Self-Rescue Mouthpiece Respirator (Filter): Designed for self-rescue from atmosphere having immediately dangerous concentrations of toxic particles. Mouth breathing prevents detection of contaminants by odor. The nose clip should be securely in place to prevent nasal breathing. No protection is provided to the eyes from irritating aerosols.
   - Combination Particulate and Vapor and Gas-Removing Respirators: The advantages and disadvantages of the component parts of the combination respirator as described above apply.
   - Combination Atmosphere-Supplying and Air-Purifying Respirator: The advantages and disadvantages, expressed above, of the mode of operation being used will govern use. The mode with the greater limitations (air-purifying mode) will mainly determine the overall capabilities and limitations of the respirator since the user may for some reason fail to change the mode of operations, even though conditions may require such change.

1.5.3 Self-contained breathing apparatus (SCBA)

The following should be kept in mind in the use of BA.

Since additional weight can reduce a firefighter's ability to carry out assigned tasks, weight reduction is a key concern. SCBAs should be rated in 30-minute duration, the predominant SCBA used by the firefighters, and should be limited to a maximum composite weight of 11.4 kg. Purchasers are advised to specifically address weight in their purchase specifications regardless of the rated service time.

Many standards and regulations require covering the quality of air to be used in SCBA. However, it is most important to remind the user that the quality of the air in the cylinder is of great concern, and that it should have a dew point compatible with the ambient temperature to be encountered.

Figure 1.7 A typical Self-Contained Breathing Apparatus (SCBA).

SCBA that is certified by recognized organizations as positive pressure but capable of supplying air to the user in a negative pressure, demand-type mode may not meet the requirements of standard. Figure 1.7 shows a typical SCBA.

The SCBA manufacturer should provide, with each SCBA, instructions and information for maintenance, cleaning, disinfecting, storage, and inspection. The SCBA manufacturer should also provide, with each SCBA, specific instructions regarding the use, operation, and limitations of it, as well as training materials.

The use of SCBA by firefighters is always assumed to be in atmospheres immediately dangerous to life or health (IDLH). There is no way to predetermine hazardous conditions, concentrations of toxic materials, or percentages of oxygen in air in a fire environment, during overhaul (salvage) operations, or under other emergency conditions involving spills or releases of hazardous materials. Thus, SCBA are required at all times during any firefighting, hazardous materials, or overhaul operations.

Although SCBA that meet standards have been tested to more stringent requirements than required for approval, there is no inherent guarantee against SCBA failure or firefighter injury. Even the best-designed SCBA cannot compensate for either abuse or the lack of a respirator training and maintenance program. The severity of these tests should not encourage or condone abuse of SCBA in the field.

To assure proper utilization of equipment in actual situations, after training and instruction, users should actually use the SCBA in a series of tasks representing or approximating the physical demands likely to be encountered in order to gain confidence in its use.

In addition to the degree of user exertion, other factors that may affect the service time of the SCBA include:

- Physical condition of the user
- Emotional conditions, such as fear or excitement, which may increase the user's breathing rate
- Degree of training or experience the user has had with such equipment
- Whether or not the cylinder is fully charged at the beginning of use
- Facepiece fit
- Use in a pressurized tunnel or caisson (at two atmospheres the duration will be one-half the duration obtained at one atmosphere; at three atmospheres, the duration will be one-third the duration obtained at one atmosphere)
- Condition of the SCBA

Compressed oxygen should not be used in supplied-air respirators or in open-circuit self-contained BA that have previously used compressed air. Compressed air might contain low concentrations of oil. When high-pressure oxygen passes through an oil-or grease-coated orifice, an explosion or fire could occur.

Breathing air should be supplied to respirators from cylinders or air compressors. Cylinders should be tested and maintained in accordance with applicable standards.

Compressors should be designed so as to avoid entry of contaminated air into the system and suitable in-line air purifying sorbent beds and filters installed to further assure breathing air quality.

A receiver of sufficient capacity to enable the respirator user to escape from a contaminated atmosphere in event of compressor failure, and alarms to indicate compressor failure and over-heating, should be installed in the system. Air-line coupling should be incompatible with outlets for other gas systems to prevent inadvertent servicing of air-line respirators with nonrespirable gases or oxygen.

1.5.4 Open-circuit escape BA

This section specifies requirements for the design, construction, and performance of two types of open-circuit escape BA using a source of compressed air. Type 1 is for use under hard work conditions such as walking up flights of stairs and running. Type 2 is for use under less arduous conditions such as walking on level ground, walking down flights of stairs, and climbing a few stairs.

Note: Type 1 equipment is also suitable for type 2 applications. The apparatus should be designed and constructed to enable the user to breathe air on demand from a high-pressure air cylinder or other approved container via either a lung-governed demand valve or another device that adequately controls the air supply and is connected to a facepiece, hood, or mouthpiece. The exhaled air should pass from the facepiece, hood, or mouthpiece either via an exhalation valve or directly to the atmosphere.

- Demand valve: Without positive pressure – the opening pressure of the lung-governed supply mechanism, measured at a constant airflow rate of 10L/min., should not exceed 3.5 mbar* at all cylinder pressures above 50 bar. With positive pressure – the demand valve should operate such that the minimum facepiece or hood-cavity pressure should be not less than 0 mbar at a sinusoidal airflow rate of 80L/min (2.5L tidal volume × 32 respirations per minute) at all cylinder pressures above 50 bar. All apparatus should have a rated duration of no less than 5 min. (1 mbar $= 100 \text{N/m}^2 = 100$ pa.)

Note: A minute volume of 40L is appropriate to the work rate of a worker walking at a steady speed of 6.5 km/h for type 1 and a minute volume of 30L is appropriate to the work rate of a worker walking at a steady speed of 5.5 km/h for type 2. In practice, the time for which protection is afforded should be longer or shorter than the rated duration and will depend upon the work rate, stress level, and physical characteristics of the user.

- Condition of the inhaled air: Oxygen content – when the apparatus is tested the oxygen content of the inhaled air should not fall below 17% (by volume). Carbon dioxide content – the maximum carbon dioxide content of the inhaled air throughout the rated duration should not exceed the value corresponding to the rated duration of the apparatus.

• Resistance to breathing: Without positive pressure – Neither the inspiratory side nor the expiratory side of the circuit should have a dynamic resistance greater than 6 mbar. With positive pressure – the expiratory side of the circuit should have a dynamic resistance not greater than that indicated by a straight line jointing a value of 6 mbar at zero flow to a value of 9 mbar at a sinusoidal airflow rate of 80 L/min.

1.5.5 Closed-circuit escape breathing apparatus

This section specifies requirements for the design, construction, and performance of two types of closed-circuit escape BA using a source of compressed oxygen or chemically bound oxygen. Figure 1.8 shows a closed-circuit escape BA. Type 1 is for use under hard work conditions such as walking up flights of stairs and running. Type 2 is for use under less arduous conditions such as walking on level ground, walking down flights of stairs, and climbing a few stairs. Type 1 equipment is also suitable for type 2 applications. The apparatus should be designed and constructed so that exhaled air passes from a facepiece, hood, or mouthpiece, into the breathing circuit where it is purified, and fresh oxygen is added to the air and returned to the user via a breathing bag.

Notes: In the compressed-oxygen type chemicals in the canister absorb the exhaled carbon dioxide, and fresh oxygen is supplied to the breathing circuit. In the

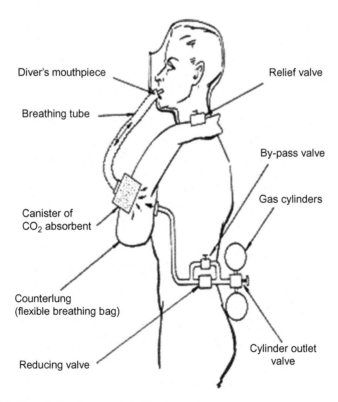

Figure 1.8 A Closed-circuit escape breathing apparatus.

chemical-oxygen type chemicals in the canister react with the exhaled carbon dioxide and moisture to produce oxygen.

• Rated duration: All apparatus should have a rated duration of no less than 5 min using a breathing simulator set at a flow rate of 40 L/min for type 1 and 30 L/min for type 2 apparatus.

Note: A minute volume of 40L is appropriate to the work rate of a worker walking at a steady speed of 6.5 km h for type 1 and a minute volume of 30L is appropriate to the work rate of a worker walking at a steady speed of 5.5 km/h for type 2. In practice, the time for which protections are provided should be longer or shorter than the rated duration and will depend upon the work rate, stress level, and the physical characteristics of the user.

• Condition of the inhaled air: Oxygen content – when the apparatus is tested the oxygen content of the inhaled air should not fall below 17% (by volume). Carbon dioxide content – the maximum carbon-dioxide content of the inhaled air throughout the rated duration should not exceed the value corresponding to the rated duration of the apparatus.
• Resistance to breathing: Neither the inspiratory side nor the expiratory side of the circuit should have a dynamic resistance greater than 6.5 mbar.

1.6   Fresh-air hose and compressed air-line breathing apparatus

This section covers types of fresh-air hose and compressed air-line BA designed to enable a worker to work in irrespirable or hazardous atmospheres for longer periods than are generally possible with open- and closed-circuit types of BA. It does not deal

Figure 1.9 A typical application of fresh air hose and compressed air line breathing apparatus.

with apparatus designed only for escape purposes. Figure 1.9 shows a typical application of fresh-air hose and compressed air-line BA.

For guidance on the type of respiratory protection that should be provided for particular conditions, and the provision of suitable air supplies for air-line BA, reference should be made to relevant standards. In addition, particular care should be taken in the choice of BA itself, when such equipment is to be used in very high or very low ambient temperatures.

Additionally, certain toxic substances that may occur in some atmospheres can be absorbed by the skin. Where these do occur, respiratory protection alone is not sufficient and the whole body should be protected. When this apparatus is being used in atmospheres immediately dangerous to life, a full facepiece should be worn. For conditions of very heavy work a flow in excess of 120 L/min is desirable.

This type of apparatus includes:

- Fresh-air hose apparatus
  Without blower (short distance)
  With hand blower
  With motor-operated blower
- Compressed air-line apparatus
  Constant flow type
  Demand valve type

1.6.1   General requirements

- Fresh-air hose apparatus (without blower): The apparatus consists of a full facepiece or mouthpiece with nose clip, with a valve system, connected by an air hose to uncontaminated air that is drawn through a hose of adequate diameter to enable a flow of 120 L/min., to be achieved by the breathing action of the user. The hose should not normally exceed 9 m in length.
- Fresh-air hose apparatus (with hand blower): The apparatus consists of a full facepiece or mouthpiece with nose clip, with a valve system, by which uncontaminated air is forced through a hose of adequate diameter by a hand-operated blower, at a minimum flow of 120 L/min., and through which the user can inhale in an emergency, whether the blower is operated or not. The hose should not exceed 36 m in length.
- Fresh-air hose apparatus (with motor-operated blower): The apparatus consists of a full facepiece air hood or suit or half-mask, with a valve system, by which uncontaminated air is forced through a hose of adequate diameter by a motor-operated blower at a flow of no less than 120 L/min and through which the user can inhale in an emergency whether or not the blower is operated. The hose should not exceed 36 m in length.
- Compressed-air-line apparatus (constant-flow type): The apparatus consists of a full facepiece, a half-mask, or an air hood or suit connected to a supply of breathable air fed continuously to the user. The airflow is regulated by a flow-control valve from a source of compressed air. An air line connects the user to a supply of fresh air that is fed to him at a flow of at least 120 L/min at the stated operating pressure.

1.6.2   Compressed-air-line apparatus (Demand-Valve Type)

The apparatus consists of a full facepiece connected to a demand valve that admits breathable air to the user when he inhales and closes when he exhales. An air line connects the user to a supply of compressed air. The airflow available to each user should meet the requirements of the demand valve (pressure between 345 KN/m^2 and 1035 KN/m^2).

1.6.3   Resistance to breathing

With the air-supply system working at any flow chosen by the testing authority but within its designed range of pressures and airflow, or with a blower operated in such a way that the operator would not become unduly fatigued after 30 min, or with the fresh-air hose alone (if not supplied with blower or bellows), then with the maximum length of tube for which the apparatus has been submitted for approval, half of it coiled to an inside diameter of 500 mm, or, for compressed air apparatus, with half the maximum length of the air line coiled either on the drum supplied by the manufacturer or, if a drum is not supplied, on a drum not exceeding 300 mm in diameter neither the inspiratory nor the expiratory side of the apparatus should have a dynamic resistance greater than 50 mm H_2O (1 mm $H_2O = 10 N/M^2$).

If any of the air-supply systems detailed ceases to operate, the user should still be able to inhale through the tube without undue distress. This provision should be satisfied if the total inspiratory resistance, with the air-supply system inoperative but not disconnected and with the maximum length of the tube for which the apparatus has been approved, is not greater than 125 mm H_2O at a continuous airflow of 85 L/min.

1.6.4   Requirements for fresh-air hose apparatus

For fresh-air hose supply systems with blower, hand-operated blowers should be capable of being operated by one worker without undue fatigue for at least 30 min. Rotary-type blowers should be capable of maintaining a positive air pressure with either direction of rotation, or else be made to operate in one direction only. In the former case, the direction of operation in which the blower delivers the least volume of air against the designed working pressures should be used in testing. When motor-operated blowers are used where flammable surroundings may arise it is essential that the suitability of the equipment for use in these environments be considered.

Note: An airflow indicator should be on the blower to indicate the flow rate.

Without a blower the hose should be fitted with a strainer at the free end to exclude debris. Provision should be made for securely anchoring the free end of the hose and strainer so that it cannot be dragged into the contaminated atmosphere.

1.6.5   Requirements for compressed-air-line apparatus

For compressed-air-line supply systems, the air supply should be in the range of 138 kN/m2 to 1035 kN/m2, and a pressure regulator should be fitted if necessary. When the supply of air is from high-pressure cylinders the flow from a pressure regulator of constant-flow type must remain constant to within 10% of the preset flow at all pressures above 1000 kN/m2. The pressure regulator should not be capable of adjustment without the use of tools. In addition, where the air is supplied from cylinders the apparatus should be provided with an alarm signal on the high-pressure side to indicate the approach of the depletion of the air supply. This device should not substantially deplete the remaining air supply. Pressure gauges should be provided on the high- and low-pressure sides if cylinders are used.

1.6.6 High-efficiency dust respirators

This section covers respirators for use in areas where highly toxic particulate materials (including radioactive substances) are handled.

The best personal respiratory protection is provided by positive pressure, supplied-air equipment. In many cases, however, it is not practical to use such equipment, and high-efficiency respirators provide an alternative form of protection. These respirators have definite limitations and should be issued and fitted only by persons who are competent to do so and who are aware of the conditions surrounding their use. They afford no protection against oxygen deficiency.

When a respirator is used in an atmosphere containing radioactive particles these will be trapped in the filter and will themselves constitute a hazard and should emit ionizing radiations that, over a period of time, could effectively reduce the protecting properties of the filter. In the absence of specialist advice, a respirator should not be re-used in such conditions without changing the filter after each period of use and monitoring the inside of the facepiece. Since the health of the user depends upon the condition of the apparatus, it is essential that adequate provision is made for cleaning and maintenance as necessary, for regular examination, and for supervision to ensure proper storage and use.

The design should be such that the respirator provides protection against solid particles or, in certain cases, water-based mists, where no hazard from toxic gas or vapor exists. Respirators should consist of a facepiece covering the eyes, nose, mouth, and chin, held securely in position by a head harness.

The design should also include a filter, or filters, through which the inhaled air passes. Filters should be readily replaceable and should be capable of being fitted without edge leakage or other loss of efficiency; they should also be encapsulated. For some applications it is desirable that readily replaceable pre-filters be available.

A valve system such that air inhaled by the user passes through the filter(s) and exhaled air passes direct to the surrounding atmosphere through a nonreturn valve is also needed.

The weight of the assembled respirator should be as low as practicable and preferably symmetrically balanced to ensure the maximum retention of the face seal and to minimize muscular strain, particularly when worn in circumstances involving vigorous movement.

1.7 Positive-pressure, powered dust respirators

This section covers situations in which the face-seal leakage of typical half-masks and of some full-face dust respirators is unacceptably high.

Positive-pressure, powered dust respirators provide increased comfort and reduce the respiratory load to negligible proportions while providing an overall standard of protection equivalent to that specified of high-efficiency dust respirators (if high-efficiency filters are used), or to respirators for use against harmful dusts and gases (if standard filters are used).

This section discusses two types of such respirators: one for use with high-efficiency filters and giving a standard of performance equal to that of the high-efficiency dust respirator, and the other designed for use with standard filters and giving a standard of performance equal to that of the dust masks specified in widely accepted standards. Should the positive pressure supply fail, the degree of protection is reduced, and the user should leave the hazardous atmosphere as quickly as possible. These respirators should not be used in atmospheres immediately dangerous to life, and filters should be checked periodically for clogging.

Such devices afford no protection against oxygen deficiency. Respirators, moreover, have very definite limitations and should only be issued and fitted under the supervision of a competent person aware of the conditions in which they are to be used. When determining the type of protection that should be provided for any particular conditions, reference should be made to relevant standards.

This section covers a respirator comprising a mask supplied with filtered air from a power pack that provides protection in adverse environmental conditions (particulate hazards) while reducing respiratory load.

1.7.1 Design

The design should be such that the respirator provides protection against solid particles or, in certain cases, water-based mists, where no hazard from toxic gas or vapor exists. The respirator should consist of:

- Full facepiece covering the eyes, nose, and mouth, or a half-mask covering the nose and mouth, held securely in position by a head-harness.
- Power pack supplying filtered air directly to the facepiece by means of a flexible hose.
- Filter(s) through which all the air supplied to the facepiece passes. Filters should be readily replaceable and should be capable of being fitted without edge leakage or other loss of efficiency. For some applications it is desirable that a readily replaceable pre-filter(s) be available.
- A valve system such that both the exhaled and surplus air pass direct to the surrounding atmosphere through a nonreturn valve.

1.7.2 Power pack

The power pack should be capable of supplying air to the facepiece at a minimum rate of 120 liters/min ($4.24\,ft^3$/min) for a period of 4 h without replacement of the power source. If a rechargeable battery is used, it should be possible to recharge it completely within 14 h and it should be the nonspillable type. Unless the battery is a sealed type, a safe venting device should be incorporated in the pack.

1.8 Respirators for protection against harmful dust and gas

This section deals with respirators for protection against harmful dusts and gases.

When determining the type of protection that should be provided for any particular conditions, reference should be made to relevant standards. Respirators provide no protection against oxygen deficiency and have very definite limitations, therefore, they should only be issued and fitted under the supervision of a competent person aware of the conditions surrounding their use. Reference can be made to standards covering respirators for use against highly toxic materials (including radioactive substances) in particulate form.

This section covers dust respirators providing protection against dusts and other particulate matter; it also specifies requirements for gas respirators (canister type) providing protection against the limited concentration of the gases listed in Tables 1.3 and 1.4, and gas respirators (cartridge type) providing protection against low concentrations of certain relatively nontoxic gases.

For the purposes of this section, the term *dust* includes other particulates such as mists and fumes, and the term *gas* includes vapor.

1.9 Dust respirators

This section covers two types of dust respirator: Types A and B. Type A respirators are low-resistance respirators in that the test requirements impose a maximum inhalation resistance of 2 mbar* (20 mm H_2O). They are intended for use against dusts of low toxicity in work conditions where low breathing resistance is important.

Type B are higher resistance respirators, and the test requirements impose a maximum inhalation resistance of 3.2 mbar* (32 mm H_2O). They are required to be more efficient in stopping particles of fine dusts than are Type A respirators.

1.9.1 Design

The design of dust respirators should be such that they provide protection against solid particles or, in certain cases, mists, where no toxic gas or vapor is present. Dust respirators should consist of:

Facepiece held securely in position by a head harness.

Filter(s) through which all the inhaled air passes. Filters should be readily replaceable and capable of being fitted without edge leakage or other loss of efficiency; they also should be encapsulated. For some applications, it is desirable that readily replaceable pre-filters be available.

Valve system such that all air inhaled by the user passes through the filter(s). The exhaled air should pass directly to the surrounding atmosphere through a nonreturn valve.

1.10   Gas respirators, canister type

1.10.1   Design

Gas respirators (canister type) are designed to protect the user from the gases listed in Tables 1.1 and 1.2. The shelf life of some canisters is given in Table 1.2 in terms of the time of exposure at the maximum concentration; however, allowance must be made for facepiece leakage. Gas respirators should be one of the following types:

Full facepiece connected to a canister or canister-containing absorbent and/or adsorbent materials, by means of a breathing tube, and arranged with valves so that all air inhaled by the user passes through the canisters. The exhaled air should pass direct to the surrounding atmosphere through a nonreturn valve(s).

Similar to above but with the canister or canisters connected directly to the facepiece.

1.10.2   Canisters

The canister should be connected to the facepiece tightly so that when fitted all the inhaled air passes through it. It should be readily replaceable without the use of special tools. Metal canisters should be varnished internally and painted or varnished externally, or otherwise rendered corrosion-resistant and, where necessary, should be provided with a suitable carrying harness. The color marking of canisters should be visible when the canister is fitted in the harness. The particulate filter, if provided, should be integral with the canister and in such a position that the inhaled air passes through it first. The charcoal employed in this type of canister should be impregnated with no less than 0.01% w/w of silver (on the dry charcoal).

1.11   Gas respirators, cartridge type

1.11.1   Design

The design of gas respirators (cartridge type) should be such as to provide protection against low concentrations of certain relatively nontoxic gases. The respirator should consist of a facepiece held securely in position with a head harness and connected to a cartridge(s) containing absorbent or adsorbent material, and arranged with valves so that all air inhaled by the user passes through the cartridge. The exhaled air should pass direct to the surrounding atmosphere through a nonreturn valve(s). A particulate filter should be incorporated, and for some applications readily replaceable prefilters should be available. The cartridge should be readily replaceable without the use of special tools, and should be designed or marked to prevent incorrect assembly. Table 1.3 shows substances covered by cartridge-type respirators.

1.12   Positive-pressure, powered dust hoods and suits

For positive pressure, powered dust respirators are prepared, but the use of a hood or a hood integral with a suit in place of a respirator facepiece will, for certain applications,

Table 1.1 **List of substances showing application of canisters**

Substance	Canister type	Substance	Canister type
Acetaldehyde	C.C	Hydrogen sulphide	S.H.C
Acetone	C.C	Ketene	C.C
Acetone cyanohydrin	S.H.C	Mercury and compounds	
Acridine	C.C	on mercury	C.C
Acrylaldehyde (acrolein)	C.C	Methanol	C.C
Ammonia	A.	Methyl bromide	O.
Amyl acetate	C.C	Nitrous fumes	N.F.
Amyl alcohol	C.C	Particulate smokes	C.C
Aniline	C.C	Particulate smokes	S.H.C.
Arsine	C.C	Particulate smokes	N.F.C.
Benzene	C.C	Petroleum vapor	
Bromine	C.C	(see Note 1)	C.C
Bromomethane	O.	Phenol	C.C
Carbon disulphide	C.C	Phosgene	C.C
Carbon tetrachloride	C.C	Pyridine	C.C
Chlorine	C.C	Sulphur dioxide	C.C
Chloromethane	O.	Sulphur dioxide	S.H.C.
Cyanide dusts	C.C	Sulphur chloride	C.C
Cyanogen chloride	O.	Sulphur trioxide	C.C.
Diazomethane	C.C	Sulphur trioxide	S.H.C.
Dichloromethane	O.	Sulphuric acid	C.C
Diethyl ether	C.C	Sulphuryl chloride	C.C.
Diketene	C.C	Lead alkyl compounds	
Ethylene oxide	C.C	containing TEL and TML	C.C.
Formaldehyde	C.C	Thionyl chloride	C.C.
Hydrogen bromide	C.C. or S.H.C	Toluene	C.C.
Hydrogen chloride	C.C. or S.H.C	Trichloroethylene	C.C.
Hydrogen cyanide	D.	Xylene	C.C.
Hydrogen cyanide	S.H.C	Organic compounds	
Hydrogen fluoride	C.C. or S.H.C	boiling at temperatures	
Hydrogen sulphide	C.C	above 60°C	C.C.

Notes:
1) Breathing apparatus should be used in connection with the lighter petroleum hydrocarbons.
2) Canister respirators should not be used to protect against coal gas or any other gas containing carbon monoxide.
3) Type O canisters may also be used for protection against gases and vapors listed against C.C. canisters, with the exception of those cases where a particulate filter is required.
4) 'C' denotes particulate filter incorporated.

provide increased comfort. This part specifies such an apparatus for use either with high-efficiency filters and giving a standard of performance equal to that of the high-efficiency dust respirator or with standard filters and giving a performance equal to that of the dust mask.

Should the positive pressure air supply fail, the user should leave the hazardous atmosphere as quickly as possible. These devices afford no protection against oxygen deficiency; they should not be used in atmospheres immediately dangerous to

Table 1.2 List of canisters showing substances covered and color marking

Canister Type	British colour marking of canisters	Recommended for use against	Life		Absorption test		
			Maximum concentration by volume	Maximum exposure to maximum concentration	Test gas	Test Concentration by volume	Minimum exposure
			%	min		%	min
A	Blue	Ammonia	2	60	Ammonia	2	60
C.C.	Black with grey stripe	Organic compounds boiling above 60°C, acetaldehyde, acetone, acridine, acrylaldehyde (acrolein), amyl acetate, amyl alcohol, aniline, arsine, benzene bromine, carbon disulphide, carbon tetrachloride, chlorine, cyanide dusts, diazomenth- ane: diethyl ether, diketene ethylene oxide, formaldehyde. hydrazine, hydrogen bromide, hydrogen chlor- ide, hydrogen fluoride, hydrogen sulphide, isocyanates, ketene, merc-ury and compounds of mercury (organic and inorganic), methanol, particulate smokes and dusts, petrol- eum vapor, phenol, phosgene, pyridine, sulphur dioxide, sulphyr chloride, sulphur trioxide, sulphuric acid, sulphuryl chloride, sulphur monochloride, thionyl chloride, toluene, trichloroethylene, xylene	1	30	(a) Phosgene and (b) Carbon tetrachloride	1	30

	Colour	Gas/substance			Substance		
D	White	Hydrogen cyanide	1	30	Hydrogen cyanide	1	30
H	Half black Half blue	All under C.C. (except particulates), Ammonia	1	30	Ammonia Carbon tetrachloride	1 1	30 30
N.F.C.	Orange with grey stripe	Nitrous fumes, particulate smokes and dusts	1	20	Nitrous fumes (NO$_2$/NO)	1	20
O	Black with orange stripe	All under C.C. (except particulates). bromomethane (methyl bromide,) chloroethane, chloromethane, cyanogen chloride, vinyl chloride, vinylidene chloride	1	30	(a) either chloromethane or cyanogen chloride and (b) As for C.C.	1	30
S.H.C.	Red with white and grey stripes	2-cyanopropan-2-ol (acetone cyanhydrin), hydrogen chloride, hydrogen bromide, hydrogen fluoride sulphur dioxide, sulphur trioxide, hydrogen sulphide; acid gases including hydrogen cyanide, particulate smokes and dusts	1	30	Hydrogen cyanide	1	30

Table 1.3 **Substances covered by cartridge type respirators**

Type	Color	Recommended for use against	Life	
			Maximum concentration by volume %	Maximum exposure to maximum concentration min
CART-RIDGE(S)	BLACK	AS FOR C. C. CANISTERS PROVIDED THAT SUBSTANCES HAVE THRESHOLD LIMIT VALUES EXCEEDING ONE HUNDRED PARTS PER MILLION (0. 01 %)	0.1	20

life, and filters should be checked periodically for clogging. Hoods should only be issued and fitted under the supervision of a competent person aware of the conditions in which they are to be used. Where flammable surroundings may arise, the suitability of the equipment for use in such surroundings must be considered when determining the types of protection that should be provided for any particular conditions.

This section discusses requirements for hoods and suits supplied with filtered air, providing protection in adverse environmental conditions (particulate hazards) while reducing the respiratory load.

1.12.1   Design

The design should be such that the equipment provides protection against solid particles or, in certain cases, mists, where no hazard from a toxic gas or vapor exists. It should consist of the following:

- Hood or a hood integral with a sleeved suit that should be so designed that all the exhaled and surplus air passes from the inside of the hood or suit to the surrounding atmosphere at the lower extremities
- Power pack supplying filtered air directly to the hood
- Filter(s) through which all the air supplied to the hood or the suit passes. These filters should be readily replaced and should be capable of being fitted without edge leakage or other loss of efficiency. For some applications it is desirable that one or more readily replaceable prefilters should be available.

1.12.2   Hood and suit

The hood should preferably be made in a one-size-fits all for adult users. It should have a transparent area for viewing and be comfortable to wear. The suit, which should be either of integral construction or otherwise sealed airtight with the hood, should be made to fit adult users, with sleeve and waist openings elasticated or otherwise designed to constrict the openings.

1.12.3 Power pack

The power pack should be capable of supplying air to the hood at a minimum rate of 120 liters per minute for a period of 4 h without replacement of the power source. If a rechargeable battery is used, it should be possible to recharge it completely within 14 h and it should be the nonspillable type. Unless the battery is a sealed type, a safe-venting device should be incorporated in the pack.

1.13 Underwater breathing apparatus

Strict attention to the care and maintenance of all types of BA used underwater is of vital importance at all times. More damage is caused to underwater apparatus by mishandling out of water, by incorrect maintenance and during storage, than is incurred during actual diving operations. It is essential that the equipment be thoroughly examined for damage or defects before and after every occasion on which it is used. All defects should be fixed before the equipment is used again.

The recommendations made in the following sections concerning the handling of the various components should be carefully observed and the manufacturer's instructions with regard to assembly, adjustment, replacement of components, and maintenance should be obtained and followed. It should also be kept in mind that the lightweight construction of the self-contained BA renders it vulnerable to damage through mistreatment. Careless manipulation with the wrong tools may not only cause defects, but render further maintenance expensive or impossible.

The corrosive action of seawater and water-borne contaminants should also never be underestimated, and if precautions are not taken to clean the apparatus properly after use, serious damage may be caused to all parts of the apparatus while it is stowed away. It is worth as chemical and petroleum wastes that are not noticeable at the time, but which will start corrosive action if left in contact with the apparatus.

1.13.1 Cylinders

- Use of proper cylinder: Cylinders having a water capacity of approximately 5.4 liters or more to contain compressed air are subject to the regulations.
- Storage: The condition of the inside of the cylinder should be maintained by keeping it dry at all times. The inside of the cylinder should be filled with dry air and never completely discharged as this can lead to water getting back into the cylinder and causing contamination. Cylinders should be stored, preferably in the vertical position, in a cool, dry place adequately protected from the weather and away from excessive heat and direct exposure to the sun. Once a cylinder has been put into service, it should never be left completely discharged. A slight positive pressure should always register on the gauge.

- Care of cylinders: Accessories fitted to the cylinder, even if they are plated or are made of stainless steel, should be insulated from the cylinder by suitable means, i.e., a plastic or nylon coating or a rubber sleeve. After use, particularly in seawater, the cylinder should be removed from its harness and boot and then washed carefully in clean, fresh water to remove all traces of saltwater and dirt, especially from any cracks. The cylinder and valve should be thoroughly dried. To ensure the good service of the cylinder valve, care should be taken when opening and closing it to ensure that excessive force is not applied once the stop has been reached as this causes damage to the internal components of the valve and could result in a leak through the valve.
- Before storage, or when the cylinder has been completely discharged and seawater may have entered the cylinder, the cylinder valve should be removed and the cylinder washed internally and externally in clean fresh water and thoroughly dried. This operation should be done by a competent person. Additionally, the cylinder should not be stored with the valve downwards. Cylinders should also be retested periodically in accordance with current regulations.

1.13.2 Compressed air for human respiration

- Preparation of compressed air cylinders: Cylinders should be internally and externally clean and free of scales or other foreign matter.
- Compression of atmospheric air: Atmospheric air should be compressed by means of suitable compressors to attain the air purity and pressure desired. Precautions should be taken to ensure that only uncontaminated air is admitted into the compressor intake. Attention should be paid to the location of the compressor intake and to the provision of suitable intake screening or filtration. Where compressors are driven by an internal combustion engine, extra care should be taken, by extending the exhaust of the engine or the inlet of the compressor, to avoid the compressor drawing in the exhaust gases of the engine. The compressor manufacturer should be consulted concerning the maximum length and the minimum cross-sectional area of such an extension to avoid reducing the efficiency of the engine or compressor. When compressors are being used in the vicinity of other machinery, adequate precautions should be taken to avoid intake of fumes from these machines. The maintenance and operation of compressors should be carried out in accordance with the manufacturer's instructions, with particular attention paid to the condition of piston rings, driers, filters, and accessories. No lubricant other than that recommended by the compressor manufacturer should be used. The air discharged from the compressor should be subjected to the processes necessary to achieve the degree of purity specified in standards. At regular intervals, not exceeding 6 months and after a major overhaul, a sample of the compressed air delivered by the compressor should be analyzed to check that the standards of purity are being maintained.
- Purity of the breathing air: A sample of air taken from the cylinder should not contain impurities in excess of the limits given in Table 1.4. The air should be free from all odors and contamination by dust, dirt, or metallic particles, and should not contain any other toxic or irritating ingredients.

Note: Odor and cleanliness of compressed air is difficult to check accurately without special equipment. A rough check should be made by cracking open the cylinder valve and smelling the escaping air, and by noting any discoloration or moisture when the air is passed gently through a wad of tissue or filter paper. Filled cylinders whose contents have an objectionable odor or show signs of discoloration or moisture, as described above, should not be used.

Table 1.4 **Maximum limits of impurities in cylinder air**

Impurity	Limit
CARBON MONOXIDE	10 p.p.m. (V/V) (11 mg/m^3)
CARBON DIOXIDE	500 p.p.m. (V/V) (900 mg/m^3)
OIL	1 mg/m^3
WATER	0.5 g/m^3

Notes: At ambient temperature below 4°C there is a risk of freezing inside the apparatus and particular attention should be given to the dryness of the air under these circumstances. It is recommended that the air be dried chemically under extremely cold conditions. It is also acknowledged that the above limits are not in total agreement, but the values are considered to be obtainable in practice for everyday use for underwater BA.

1.14 Ventilatory resuscitators

This section discusses performance and safety requirements for ventilatory resuscitators (hereinafter referred to as resuscitators) intended for use of any size. It covers both operator-powered and gas-powered resuscitators. Electrically-powered resuscitators, automatic pressure-cycled gas-powered resuscitators, devices that have been designed only to deliver gases to a patient breathing adequately, or devices that are designed to assist or provide for the ventilation of a patient for an extended period of time are outside the scope of this book.

1.14.1 Classification

Resuscitators for use with patients up to 40 kg body mass should be classified by the body mass range for which they are suitable. This range should be derived on the basis that resuscitators should deliver a tidal volume of 15 mL/kg body mass. Resuscitators delivering a tidal volume of 600 mL and above, i.e., suitable for patients over 40 kg, should be classified as adult resuscitators.

Note: Resuscitators designed to deliver a tidal volume of 20 mL to 50 mL are usually suitable for use with neonates.

1.14.2 Physical requirements

- Size: The resuscitator, with a container, if provided, should pass through a rectangular opening 300 mm × 600 mm in size. Resuscitators are required where access to patients is difficult, such as in crawl spaces and through manholes.
- Resuscitator mass: Except for gas-powered resuscitators designed to be an integral part of a neonatal critical care system, the mass of the esuscitator container and contents (including any full gas cylinders) should not exceed 18 kg.
- Ease of operation: The resuscitator should be designed to facilitate effective operation by one person when used with a face mask to provide adequate ventilation of the patient's lungs (as defined by this specification). All performance characteristics in this section should be satisfied with resuscitation by one person, unless otherwise specified by the manufacturer in the instructions for use.

- Cleaning and sterilization: Components in contact with the patient breathing mixture should withstand sterilization or be labeled for single-use only (disposable). The suggested methods of cleaning and sterilization should be generally recognized as effective. Examine the manufacturer's instructions for sterilization or disinfection and determine if they meet commonly described methods.
- Disassembly and reassembly: A resuscitator designed to allow disassembly by the user (e.g., for cleaning, etc.) should be designed so as to minimize incorrect reassembly when all parts are put together. For resuscitators designed to be disassembled, the manufacturer should include resuscitator disassembly and assembly instructions that include a schematic diagram showing the correct assembly. After reassembly the operator should be instructed to perform the manufacturer's recommended test procedure to ensure proper functioning. Verify by inspection and testing for proper operation.
- Pressure-limiting system indication: If a resuscitator is equipped with an over-pressure limiting system, there should be an audible or visible warning to the operator when the pressure-limiting system is activated.
- Operator-powered resuscitators: In accordance with the requirements of its classification an operator-powered resuscitator should deliver a minimum oxygen concentration of at least 40% (V/V) when connected to an oxygen source supplying not more than 15 L/MINT., and should be capable of delivering at least 85% (V/V) (see note). The manufacturer should state the range of concentrations at representative flows, e.g. 2 L/MINT., 6 L/MINT., 8 L/MINT., etc. If the resuscitator is intended to be hand-operated, only one hand should be used to compress the compressible unit, and the hand of the person carrying out the test should not exceed the dimensions given in Figure 1.10. Although 40% (V/V) oxygen concentration is adequate under some circumstances, 85% (V/V) or higher oxygen concentrations are preferable for the treatment of severely hypoxemic patients during resuscitation. This concentration should be achievable at supplementary oxygen flows of 15 L/MINT. or less because specifiying greater than 15 L/MINT. would exceed the normal calibration

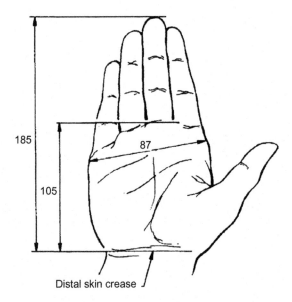

Figure 1.10 Maximum hand dimensions.

standard, clinically used for flowmeters for adult use, and could potentially lead to inaccurate control of oxygen flows and jamming of the patient valve in the inspiratory position.

Note: The 85% (V/V) requirement should be accomplished with the use of an attachment.

- Tidal volume: Resuscitators intended for use with infants and children up to 40 kg body mass should be classified according to the body mass range for which they are suitable. This body mass range should be derived from a requirement for a tidal volume of 15 ml/kg body mass. Resuscitators delivering a tidal volume of 600 ml and over should be classified as adult resuscitators. The tidal volumes specified should be delivered under the test conditions without the use of the override mechanism on any pressure-limiting system.

Note: Resuscitators designed to deliver a tidal volume of 20 ml to 50 ml are usually suitable for use with neonates. For adult ventilation a typical tidal volume is approximately 600 ml. The compliances and resistances given here are representative of the possible compliances and resistances found in adults and children needing resuscitation. The tidal volume requirements of 15 ml/kg are higher than normal and are commonly used during resuscitation to allow for mask leakage. The ventilatory frequencies are typical values used in pediatric and adult resuscitation.

- Pressure limitation (operator-powered resuscitators): Experience with infant resuscitation suggests that a maximum inspiratory pressure of 4.5 kPa ($\approx$45 cm H_2O) will not produce lung damage and will permit adequate tidal volume in most patients weighing under 10 kg. Pressure-limiting systems are not specified for operator-powered resuscitators designed for use with patients weighing over 10 kg. However, it is essential that resuscitators with such systems satisfy the tidal volume requirements specified here without the use of an override mechanism. When airway pressure is limited to below 6 kPa ($\approx$60 cm H_2O), an override mechanism is essential in order to ventilate those patients with low lung compliance and/or high airway resistance.
- Pressure-limiting system: It is essential that maximum delivery pressure is limited on all gas-powered resuscitators. Airway pressure at 4.5 kPa ($\approx$45 cm H_2O) is considered adequate for ventilation of the lungs, but unlikely to produce barotrauma. The selection of higher settings for difficult clinical problems necessitates risk of barotrauma.

1.14.3 Gas-Powered resuscitators

The approximate duration of a single (gas cylinder 457 mm long and 102 mm outside diameter size) containing 340 L of gas when the resuscitator is delivering a minute volume of 10 L (or the nearest setting to this) and a concentration of:

at least 85% V/V oxygen and;
the manufacturer's selected concentration less than 85% V/V oxygen, if the resuscitator will deliver such a concentration.

High oxygen concentrations are important for resuscitating patients who are extremely hypoxemic. Lower percentages of oxygen will lengthen the duration of oxygen supply. The performance of air-entrainment "mixing" devices is influenced by the flow settings of the resuscitator and the compliance and resistance of the patient.

- Pressure-limiting system: A pressure-limiting system should be incorporated in gas-powered resuscitators. When the resuscitator is supplied with gas at the range of pressures 270 kPa and 550 kPa the airway pressure should not exceed 6 kPa ($\approx$60 cm H_2O). An override mechanism should be provided to enable the operator to select a higher pressure. However, automatic, pressure-cycled, gas-powered resuscitators should not be equipped with any type of override mechanism. If provided with a locking mechanism, pressure override mechanisms should be so designed that the operating mode, i.e., on or off, is readily apparent to the user by obvious control position, flag, etc.

Notes:

1. A setting for the pressure-limiting system higher than 6 kPa ($\approx$60 cm H_2O) should be made available for certain patients, although the selection of such a setting requires medical advice.
2. There should be an audible or visible warning to the operator when the pressure-limiting system is operating.
   - Expiratory resistance: In the absence of positive end-expiratory devices, the pressure generated at the patient connection port should not exceed 0.5 kPa ($\approx$5 cm H_2O).: To facilitate exhalation, expiratory resistance should be minimized unless there are special clinical indications to impose such resistance.
   - Inspiratory resistance: The design of a resuscitator should be such that it is possible for the patient to breathe spontaneously without excessive subatmospheric pressure when the resuscitator is applied to the patient's airway but is not activated by the operator. The pressure at the patient connection port should not exceed 0.5 kPa ($\approx$5 cm H_2O) below atmospheric pressure.
   - Inspiratory flow: All gas-powered resuscitators should be capable of delivering 40 L/min $\pm$ 10% inspiratory flow against a back pressure of 2 kPa ($\approx$20 cm H_2O) when tested by the method described in relevant standards.

     Note: Devices with fixed flows should be set to this value. Devices with operator-adjustable flows should include this value in their range of adjustment. To minimize the risk of gastric distension, 40 l/min is considered to be the maximum flow that should be used when resuscitating with a mask. This is in accordance with the recommendations of the American Heart Association (AHA Standards for CPR and ECC), and these recommendations are generally accepted worldwide. Higher flows may be used with intubated patients because of the decreased risk of gastric distension.
   - Automatic pressure-cycled, gas-powered resuscitators: It is required that pressure-cycled resuscitators meet the performance requirements of standards, but they are of limited use on patients with poor lung compliance and/or high airways resistance because the cycling pressure is achieved without adequate ventilation. A "negative" pressure phase should not be used as it is associated with a fall in functional residual capacity (FRC) and arterial oxygen partial pressure (Po_2).
   - Pressure for initiation: It is important that the patient not have to generate large amounts of negative pressure to initiate gas flow from the demand valve in order to minimize respiratory work. A negative pressure of -0.2 kPa (≈ -2 cm H_2O) is physiologically acceptable.
   - Peak inspiratory flow: The minimum peak inspiratory flow should be 100 l/min for at least 10 s, at an outlet pressure of no more than 0.8 kPa ($\approx$8 cm H_2O). Peak flows as outlined above are necessary in order to satisfy the inspiratory needs of the typical patient; large amounts of negative pressure should not be required to generate these flows as it would cause fatigue in a spontaneously breathing patient.

- Termination pressure: Positive pressure indicates that adequate tidal volume has been delivered. When the pressure becomes slightly positive it indicates that the patient should be allowed to exhale and hence flow should stop.

1.14.4 Gas supply

- Gas cylinders, cylinder valves, and yoke connections: If provided, gas cylinders, cylinder valves, and yoke connections of the pin index type should meet the requirements given in International Organization for Standardization (ISO) 407, "small medical gas cylinders-yoke type, valve connections."
- Duration of gas supply: Small, easily portable cylinders are commonly used with resuscitators. It is important that the operator have some idea of size of oxygen supply under simulated conditions of use.

1.15 Nominal protection factor

There is now a range of standard specifications dealing with the design construction and performance of respirators and BA. In specifying equipment with different inward leakage characteristics, as determined by prescribed test procedures, the standards describe equipment offering different degrees of protection against inhalation of harmful substances. The selection of the most suitable type of equipment for particular circumstances requires an understanding of the limits of protection of the equipment available as well as an understanding of the hazard against which protection is required.

As an aid to the selection of respiratory protective equipment, the term "nominal protection factor" is introduced in this section for each type of equipment. It is derived from a "maximum allowed inward leakage" figure and is defined as the ratio of the concentration of contaminant present in the ambient atmosphere to the calculated concentration within the facepiece, at maximum allowed inward leakage, when the respirator or BA is being worn, for example.

Inward leakage of ambient atmosphere occurs at the face-seal of respirator or BA when the design and operation of the equipment is such that the pressure within the facepiece falls below atmospheric pressure during inhalation. At the same time, there is a small inward leakage through an exhalation valve and, in the case of a dust respirator, there will generally be a measurable penetration through the filter itself. An inward leakage can occur at the neck cord of a positive pressure hood or at the waistband and wrists of a positive-pressure suit.

The sum of the maximum leakages, or penetrations, allowed for each type of equipment is referred to by the expression "maximum allowed inward leakage," expressed as a percentage.

Note: Dust respirators are now available with face-seal leakage test figures significantly below 5%. Individual manufacturers should be consulted where the highest standard of fit is desirable. In other cases, maximum allowed inward leakages will be between 15% for a Type A dust respirator and 0.05% for a self-contained BA. Leakage into BA in which a positive pressure is continuously maintained should, of course, be negligible.

It should be noted that the face-seal leakage figures of facepieces and hence, in part, nominal protection factors, are based on a test from which subjects whom the equipment under test clearly does not adequately fit are excluded. Abnormal face-seal leakages will result from poorly fitting facepieces. When a respirator or BA is available in more than one size it is important that the best fitting size for the individual is worn.

For the estimation of the nominal protection factor of a dust respirator, the filter penetration figure to be taken is that of the mechanical type. In this way, allowance is made for deterioration in the performance of other types of filter during storage. Excessive loadings of dust filters can produce increased breathing resistance that tends to increase face-seal leakage, and in some cases, they should increase the leakage of dust through the filters, with a consequent decrease in the protection given.

In any potentially hazardous situation a prior assessment, based on the best available information, should always be made of the maximum concentration of the contaminant in air that a person can be expected to breathe without ill effect. It should be kept in mind that the estimated nominal protection factor can only be applied to respiratory protective equipment that has been properly maintained.

In order to make use of nominal protection factors in practice, it is necessary to know both the concentrations of harmful contaminants in air that are likely to be encountered and the maximum allowable concentration in the air inhaled by the user of the respiratory protective equipment. A comparison of these concentrations will indicate the protection factor required and hence the type of respiratory protective equipment.

A guide to the maximum allowable concentration of a substance in inhaled air over extended periods of time is provided by its published threshold limit value (TLV). The TLV of a substance is defined as the concentration of that substance in air to which it is believed nearly all persons can be exposed, without adverse effect, for a 7- or 8-hour work- day and a 40-hour week. Most TLVs are to be taken as time-weighted average concentrations, i.e., excursions above the limit are permissible provided that they are compensated by equivalent excursions below the limit during the workday. In certain cases, the TLV is a ceiling value that should not be exceeded and is annotated to that effect in the standards.

In the absence of a ceiling value, considerably higher concentrations of a substance in air is allowed when exposure is an isolated occurrence of short duration, e.g., an emergency situation, than when it is closely repetitive or of extended duration.

Masks and respiratory equipment materials

2

2.1 Introduction

Respiratory protective equipment will either filter the contaminated atmosphere or supply clean air from other sources. The following devices are used for breathing:

- Mouthpiece and nose clip
- Half-mask covering the nose and mouth
- Full facepiece covering the eye, nose, and mouth
- Hood covering the head down to the shoulders, or a suit covering the head and body down to waist and wrists

For selection of proper equipment the following points must be considered:

- Nature of hazardous operation of process
- Type of air contaminant including its physical properties, physiological effects on the body, and its concentration
- Period of time for that respiratory protection has to be used
- Location of hazard area to be considered with the distance of uncontaminated respirable area
- State of health of personnel involved
- Functional and physical characteristics of respiratory protective devices

Supervisors should be aware of the classes of hazards for which a given type of respiratory equipment is approved to be used and should not permit its use for protection against hazards for which it is not designed. Employees must realize that failure to wear the appropriate respiratory equipment will endanger their lives, and only those with a valid physical limitation should be prevented from using and entering the environment presenting respiratory hazards. Training is of vital importance in the operation and maintenance of equipment.

2.2 Masks and respiratory equipment (Breathing apparatus)

2.2.1 Classification of respiratory equipment

There are two distinct methods of providing personal protection against contaminated atmosphere:

- By purifying the air breathed
- By supplying air or oxygen from an uncontaminated source

Personnel Protection and Safety Equipment for the Oil and Gas Industries.
DOI: http://dx.doi.org/10.1016/B978-0-12-802814-8.00002-X
© 2015 Elsevier Inc. All rights reserved.

2.2.2 Classification of environment

The environment may be contaminated by particles and/or by gases and vapors. Oxygen deficiency may also occur. Temperature and humidity should also to be taken into consideration, as shown in Figure 2.1.

2.2.3 Classification of respiratory protective devices (see Figure 2.2)

There are two distinct methods of providing personal respiratory protection against contaminated atmospheres:

- By purifying the air (filtering device)
- By supplying air or oxygen from an uncontaminated source (breathing apparatus)
- Filtering devices (see Figure 2.3)

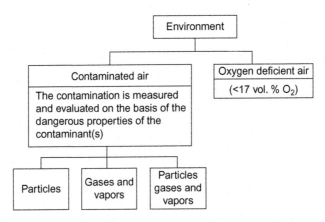

Figure 2.1 Classification of the environment.

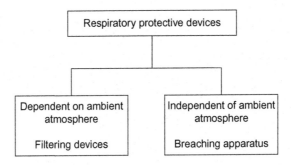

Figure 2.2 Classification of respiratory protective devices.

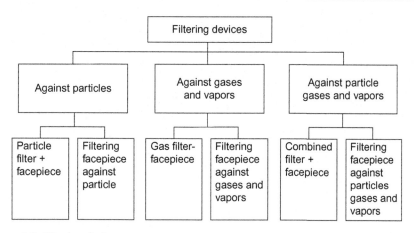

Figure 2.3 Filtering devices.

The inhaled air passes through a filter to remove contaminants. The filtering devices can be unassisted or power-assisted. Particle filters are divided into the following classes:

Low-efficiency filters
Medium-efficiency filters
High-efficiency filters

Medium- and high-efficiency filters are graded according to their ability to remove solid and liquid or solid particles only. Gas filters are divided into the following classes:

Low-capacity filters
Medium-capacity filters
High-capacity filters

2.2.4 Breathing apparatus

The main types of BA are represented in Figure 2.4.

2.3 Selection of breathing apparatus

Respiratory equipment must be regarded as emergency equipment or equipment for occasional use. Air contaminants range from relatively harmless to toxic dusts, vapors, mists, fumes, and gases that may be extremely harmful. Before respiratory equipment is purchases, the chemical and other offending substances must be determined and the extent of hazard evaluated to determine the best choice. Some types of respiratory protective equipment are illustrated in Figure 2.5.

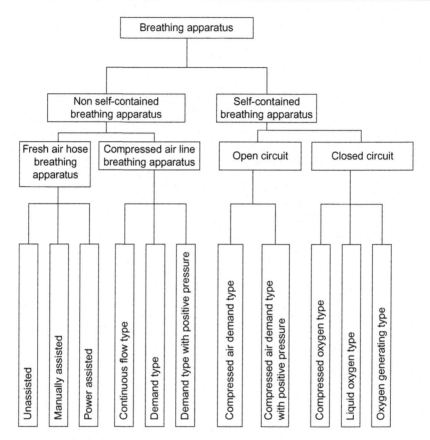

Figure 2.4 Breathing apparatus.

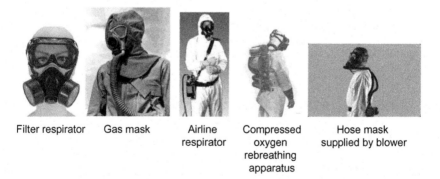

| Filter respirator | Gas mask | Airline respirator | Compressed oxygen rebreathing apparatus | Hose mask supplied by blower |

Figure 2.5 Sample types of respiratory protective equipment.

2.4 Respirators for dusts and gases

Respirators for dusts and gases are filtering devices that are dependent on ambient atmosphere. There are three types:

Filtering against particles (dust respirators)
Filtering against gases and vapors (gas masks)
Filtering against particles, gases, and vapors

2.4.1 Filtering facepiece dust respirators

A filtering dust respirator consists of the following:

Facepiece covering at least the nose and mouth and consisting wholly or partly of filter materials through which the exhaled air may be discharged through the same materials or through an exhalation valve
Head harness that may or may not be adjustable to hold the facepiece securely in position on the head
- Construction: Materials used should be suited to withstand handling and wear over the period for which the respirator is designed to be used. No materials of highly flammable nature of the same order such as cellulose nitrate should be used. Materials that may come into contact with the skin should be nonstaining, soft, and pliable, should not have harmful effects on the skin, and no known toxic substances (e.g., asbestos) should be used as filtering medium. When fitted in accordance with the manufacturer's instructions the complete respirator should allow the correct wearing of safety helmet, ear defenders, and glasses or eye protectors. The respirator should be light in weight, cause minimal interference with vision, allow freedom of movement, and be comfortable to wear.
- Perforworkerce requirements: Respirators should meet testing requirement in accordance with British Standard (BS) 6016.
- Marking: Respirators should be marked legibly with the following:
Name, trademark, or other means of identification
Number of official standard used
Size (if more than one size is available)
Type
Month and the year of manufacture and probable shelf life

2.4.2 High-efficiency dust respirators

- Performance requirement: Assembled respirator should be capable of passing the following tests:
Resistance test
    1. Inhalator: The respirator is connected in an airtight manner to a hallow former of suitable shape, and air is drawn from the atmosphere through the filter at a rate of 85 LPM. The resistance imposed by the respirator should not exceed 50 mm H_2O.
    2. Exhalation: The assembled respirator is connected in an airtight manner to a hallow former of suitable shape, and air is blown into the former at a rate of 85 LPM and passed through the respirator. The resistance imposed by the respirator should not exceed 12.5 mm H_2O.
    3. Valve leak test: The valve assembly when dry should not have a leak exceeding 30 ml/min when tested at constant suction head of 25 mm H_2O.

- Facepiece: The respirator should be marked legibly with the following:
Name, trademark, or other means of identification of the manufacturer
Number of official standard used
Year of manufacture

2.4.3 Positive-pressure dust respirators

- Construction: The respirator should be constructed of materials able to withstand normal use and exposure to extreme environmental temperature and humidity. The materials used should be such that every component (except a battery when used) has a portable effective shelf life of at least 5 years, if properly stored and maintained. No materials of highly flammable nature of the same order such as cellulose nitrate should be used. Materials that may come into contact with skin should be nonstaining, soft, and pliable, and should not contain known dermatitic substances. Except for filters, all external surfaces and finishes should be such that they will not as far as practical retain toxic and radioactive particles.
- Facepiece: The components of the facepiece should withstand a test underwater at an air pressure of 1.7kN/m^2 and be free from leaks. The facepiece should be made in a one-size-fits-all size for adult users. It should be designed to meet the requirements of normal operating conditions and tested in accordance with BS 2091. The facepiece should cause the least possible interference with vision and freedom of movement. The full facepiece should have a suitable and preferably replaceable eyeshield and should permit the wearing of glasses. It should be capable of withstanding cleaning and sterilizing by effective methods specified by the manufacturer. The weight should be as low as possible and symmetrically balanced to ensure the maximum retention of the face seal and minimize muscular strain, particularly when worn in circumstances involving vigorous head movement. It should reduce as far as practical the proportion of air exhaled that can be rebreathed from within the mask itself.

2.4.4 Filters

- High-efficiency filters: The effectiveness of the initial protection given by the filter system when assessed by the method described in BS 4400 or equivalent should be such that penetration at a flow rate of 120 LPM should not exceed 0.1%, and in the case of resin wool and resin felt filters, it should not exceed 0.05%.
- Standard filters: When similarly assessed, the effectiveness of the initial protection given by the filter system should be such that the penetration at a flow rate of 120 LPM should not exceed 5% in the case of resin wool and resin felt filters the penetration should not exceed 2%. Substances (e.g., asbestos) should not be used as a filtering medium. Filter should be designed to be irreversible. Filters should be readily replaceable and capable of being fitted without edge leak or other loss of efficiency. The exhaled air should pass directly to the surrounding atmosphere through a nonreturn valve.

Figure 2.6 and 2.7 show different types of filters.

2.4.5 Harness

The head harness should hold the facepiece firmly and comfortably in position. It should be simply fitted and adjusted and easily removed for cleaning and decontamination. The head harness should be adjustable and if consisting only of straps, these

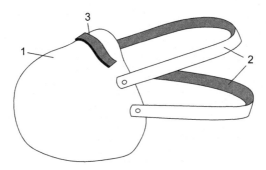

Figure 2.6 Filtering Facepiece. (1) Facepiece; (2) Head harness; (3) Nosepiece.

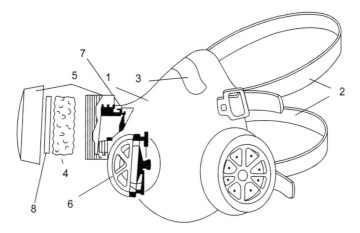

Figure 2.7 Filtering Devices. (1) Face blank; (2) Head harness; (3) Nosepiece; (4) Filter; (5) Filter housing; (6) Exhalation valve; (7) Inhalation valve; (8) Prefilter.

should be adjustable and no less than a 19 mm nominal width at the point of contact with the head and designed so that they have to be fully slackened on removal to ensure the user must readjust the straps before each use. The head harness of a half-mask, if consisting only of straps, should not have less than a 19 mm width at the point in contact with the head, with the adjustment in a position that will not interfere with the comfort or fit of safety helmet. After adjustment, the harness should permit the removal of the mask to a position round the mask and its replacement on the face, preferably without loss of adjustment.

2.4.6   Connecting fittings

The components of the respirator should be designed to connect together by simple means so that individual components can be replaced easily and without special tools. They should be designed and marked to prevent incorrect assembly.

2.4.7 Performance requirement

Respirators should be capable of meeting the requirement of BS 2091, for resistance of inhalation and exhalation, protection against dust, valve leak, resistance to clogging by dust and humidity, and resistance of filter to rough use. Full face masks should be tested in accordance with BS 7355 EN 136 testing. Half-masks and quarter-masks should be tested in accordance with BS 7356 EN 140 testing.

2.4.8 Marking

All units of the same model should be provided with a type identified by marking sub-assemblies and piece parts with considerable bearing on safety. The manufacturer should be identified by name, trademark, or other means of identification. Where the reliable performance of piece parts may be affected by aging, means of identifying the date (at least the year) of manufacture should be given.

The marking should be provided with the following:

- Serial number
- Year of manufacture
- The marking should be as clearly visible and as durable as possible. For parts that cannot be marked, the relevant information should be included in the instruction manual.

2.5   Positive-pressure, powered dust hood and suits

2.5.1 Construction

Hoods and suits should be constructed of suitable materials able to withstand normal use and exposure to extreme environmental temperatures and humidity. The material used should be such that every component (except a battery when used) has a probable effective shelf life of at least 5 years. No materials of highly flammable nature of the same order as cellulose nitrate should be used. Materials that may come into contact with the skin should be nonstaining, soft, and pliable, and should not contain known dermatetic substances. Except for filters all surfaces and finishes should be such that they will not as far as practical retain toxic or radioactive particles.

The equipment should be compact and should cause the least possible interference with vision and freedom of movement. The weight of equipment should be as low as practical and balanced to ensure comfort and minimize muscular strain, particularly when worn in circumstances involving vigorous movement.

2.5.2 Hood and suits

The hood should be made in a one-size-fits-all size for adult users. It should have a transparent viewing area and be comfortable to wear. The suit should be either of integral construction or sealed airtight with the hood and should be made in sizes to fit adult users, with sleeve and waist openings elasticated or otherwise designed to construct the openings. Hood and suit should be capable of withstanding, without appreciable deterioration, of repeated treatment by the method described in BS 2091 or any methods of equivalent effectiveness specified by manufacturer.

2.5.3   Power pack

The power pack should be capable of supplying air to the hood at a minimum rate of 120 Liter Per Minute (LPM) for a period of 4 hours, without replacement of the power source. If a rechargeable battery is used it should be possible to recharge it completely within 14 hours and it should be of nonspillable type; if the battery is of a sealed type, a safe venting device should be incorporated in the pack.

- Harness: Any harness fitted should hold the equipment firmly and comfortably in position. It should be simply fitted and adjusted and easily removed for cleaning. So far as is practical, it should be constructed in such a way that a contaminated garment may be removed easily without increasing local environmental contamination and without risk of transferring contamination to the user or other handler.
- Connecting fittings: The components should connect together by a simple means so that the individual component may be replaced by a simple operation without the use of special tools. They should be designed or marked to prevent incorrect assembly. Any hose providing connection from the power pack to the hood should be kink resistant.

2.5.4   Performance requirement

The equipment should be tested in accordance with BS 2091 or equivalent. Full, half-, and quarter-masks should be tested in accordance with BS 7355 EN 136 and BS 7356 EN 140.

2.5.5   Marking

The equipment should be marked as follows:

- Marking on the hood and suit
  Name, trademark, or other means of identification of manufacturer
  Number of official standard used
- Marking on supplying container
  Year of manufacture of the hood and suit
- Marking on the filter
  Name, trademark, or other means of identification of the manufacturer
  Number of official standard used
  Year of manufacture

2.5.6   Gas respirators, canister type

- Construction: Gas respirators (canister type) are designed to protect the user from certain gases. The shelf life of the canister is also given in terms of the time of exposure at the maximum concentration, and allowance must be made for facepiece leaks.
- Performance requirement: Gas respirators (canister type) should be tested in accordance with BS 2091, which consists of tests for facepiece, valve leak, resistance of canister to

rough use, and performance. Canisters should be tested with the test gases, test concentration, and minimum exposure.

- Marking: Respirators covered within this section of the standard should be marked with the following particulars:
- Marking on the facepiece

  Name, trademark, or other means of identification of manufacturer

  Number of official standard used
- Marking on supplying container

  The month and the year of manufacture should be marked on the supplying container.
- Marking on the breathing tube

  The number of official standard used should be marked on breathing tube.
- Marking on the canister

  Name and trademark or other means of identification of the manufacturer

  Number of official standard used

  Color showing relevant standards, with bands or stripes being at least 25 cm wide

  List of gases against which protection is given by the canister

  Month and the year on which the canister was filled

  Warning notice indicating limiting condition of use (i.e., maximum concentration percentage and maximum exposure in minutes) and the warning: "Not for use in static tanks, enclosed places, or any circumstances where a high concentration of gas is likely to be present or in atmosphere deficient in oxygen."

2.6 Gas respirators, cartridge type

2.6.1 Design and construction

The design and construction of gas respirators (cartridge type) should provide protection against low concentrations of certain relatively nontoxic gases. The respirator should consist of a facepiece held securely in position with a head harness and connected to a cartridge(s) containing absorbent or absorbent material, and arranged with valves so that all air inhaled by the user passes through the cartridges. The exhaled air should pass directly to the surrounding atmosphere through a nonreturn valve(s). A particulate filter should be incorporated, and for some applications it is desirable that readily replaceable prefilters be available.

The cartridge should be readily replaceable without the use of special tools, and should be designed or marked to prevent incorrect assembly. When testing, the fit of the respirator should ensure that leakage of the test contaminant between the facepiece and the user's face does not exceed a mean value of 5% for the 10 test subjects. Respirators should also be capable of meeting performance requirements in accordance with BSI 2091 or equivalent.

2.6.2 Tests and certification

- Respirators should be tested and certified by manufacturer. The tests should cover the following in accordance with BSI 2091:

  Static leak of outlet valves

Dust clogging test

Face seal leak test

Sterilization test for material and construction

Rough use test

- Marking: Respirators covered within this section should be marked with the following particulars:
- Marking on the facepiece

The facepiece should be marked with:

Name, trademark, or other means of identification of the manufacturer

Number of official standard used

- Marking on the supplying container: The month and year of manufacture should be marked on the supplying container.
- Markings on the cartridge: The cartridge should be marked with:

Name, trademark, or other means of identification of the manufacturer

Number of official standard used

Indication of the gases against which protection is given by the cartridge

Month and year in which the cartridge was filled

Warning notice indicating conditions of use, e.g., "for low concentration of nontoxic substances."

- Markings on the cartridge container: The cartridge container should be marked with the following warning: "Not for use in static tanks, enclosed places, or in any circumstances where a high concentration of gas is likely to be present, or in atmospheres deficient in oxygen."

2.7 Combination respirators

Combination chemical and mechanical filter respirators use dust, mist, or fume filters with chemical cartridge for dual or multiple exposure. Normally, the dust filter clogs up before the chemical cartridge is exhausted. It is therefore preferable to use respirators with independently replaceable filters. The combination respirator is well suited for spray painting and welding.

Note: It is recommended that chemical cartridges and canister respirators not be used in emergencies.

2.8 Fresh-air hose and compressed-air-line breathing apparatus

The following are five methods of operation:

- Fresh-air hose apparatus without blower
- Fresh-air hose apparatus with hand blower
- Fresh-air hose apparatus with motor-operated blower
- Compressed-air-line apparatus (constant-flow type)
- Compressed-air-line apparatus (demand type)

All the above apparatuses consist of a full facepiece or mouthpiece with nose clip.

2.8.1 Fresh-air hose apparatus

Fresh-air hose apparatus is provided with a valve system connected by an air hose to uncontaminated air that is drawn through a hose of adequate diameter by the breathing action of the user. The hose should not exceed 9 m in length and ¾ inch diameter.

- Fresh-air hose apparatus (with hand-operated blower): The apparatus is connected by an air hose and uncontaminated air is forced through a hose of adequate diameter by a hand-operated blower, and through which the user can inhale in and in the case of emergency whether or not the blower is operated the user has a chance to leave the contaminated area. The hose should not exceed 35 m in length (see Figure 2.8).

When the flow of air along the tube is mechanically assisted, the inward leak of contaminated atmosphere is greatly reduced. If the air hose is relatively short, restricts movement, and necessitates return to a respirable atmosphere along the route of entry. The size of the hose should be 9 m in length and DN-20 (¾ inch) diameter. Care is necessary to prevent damage to the hose.

- Fresh-air hose without blower: Air suitable for respiration is drawn by the breathing of the user. The hose is anchored and fitted with a suitable device to prevent entry of course particles.

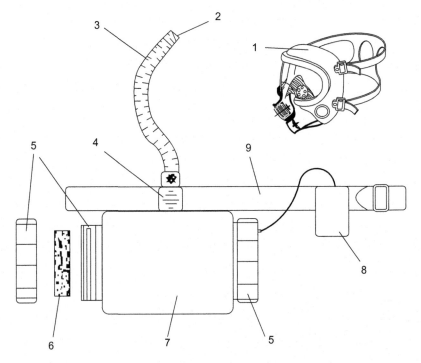

Figure 2.8 Powered filtering device. (1) Facepiece; (2) Equipment connector; (3) Breathing hose; (4) Coupling; (5) Filter housing; (6) Filter; (7) Blower; (8) Battery; (9) Belt or carrying strap.

2.8.2 Fresh-air hose apparatus (with motor-operated blower)

In this apparatus, uncontaminated air is forced through a hose of adequate diameter by a motor-operated blower at a flow rate of minimum 120 LPM and through which the user can inhale in case of an emergency. In this type of apparatus a continuous flow of air suitable for respiration is forced through a hose by a motor-operated blower. A full facepiece, a half-mask, or a mouthpiece and nose clip may be used. The hose length should not exceed 35 m. If a facepiece is used and the blower fails, negative pressure will exist in the facepiece during inhalation, although the apparatus will continue to provide some protection while the user leaves the hazardous area.

2.8.3 Air-line hose mask connected to a source of respirable air under pressure

This apparatus provides compressed air suitable for respiration through a flexible air hose attached to a compressed air line. A filter may be included in the air line to remove contaminants, and suitable valves are used to control air supply. The apparatus can be used only where a suitable continuous supply of clean compressed air is available either from a compressor system or from cylinders. The flexible hose should not exceed 90 m, and while this type allows more movement than the fresh-air hose apparatus, its restrictions are basically the same. Care should be taken in the choice of tubing to be used in very high or very low ambient temperatures. No protection is provided if the air supply fails, unless special provision is made.

2.8.4 Compressed-air-line apparatus (constant-flow type)

This apparatus may consist of an air hood or suit connected to a supply of breathable air fed continuously to the user. The airflow is regulated by a flow-control valve from a source of compressed air. The most extreme condition requiring respiratory equipment is that in which rescue and emergency repair work must be done in atmospheres extremely corrosive to the skin and mucous membranes, in addition to being actually poisonous and immediately hazardous to life. Where high ambient temperature may be encountered or where body heat may build up, the hose line supplying the air should be connected to the suit itself, as well as to the helmet, and personal air-conditioning devices utilizing a vortex tube should be used.

2.8.5 Compressed-air-line apparatus (demand-valve type)

The apparatus consists of a full facepiece connected to a demand valve that admits breathable air to the user when he inhales and closes when exhales. An air line connects the user to a supply of compressed air.

2.8.6 Materials

All materials used in the construction of fresh-air hose and compressed-air-line breathing apparatus should have adequate mechanical strength, durability, and resistance to deterioration by heat, and, where applicable, by contact with seawater. Such materials should be antistatic and fire resistant as far as is practical. Exposed parts of the apparatus should not be made of magnesium, titanium, aluminium, or alloys containing these metals since, when impacted, they can create frictional sparks capable of igniting flammable gas mixtures.

Materials that may come into contact with the skin should be nonstaining, soft, and pliable, and should not contain known dermatetic substances. The apparatus should be sufficiently robust to withstand the rough use it is likely to receive in service.

2.8.7   Separation of parts

The design and construction of the apparatus should permit its parts to be readily separated for cleaning, examination, and testing. The couplings required to achieve this should be readily connected and secured, where possible by hand, and any means of sealing should be done when the joints and couplings are disconnected during normal maintenance. All parts requiring manipulation by the user should be readily accessible and easily distinguishable from one another by touch. All adjustable parts and controls should be constructed so that their adjustment is not liable to accidental alteration during use.

2.8.8   Leak tightness

The apparatus should be designed and constructed to prevent the opening of the external atmosphere within the limits set out in this chapter at any place other than the fresh-air inlet.

2.8.9   Cleaning and decontamination

The design of the apparatus should be such facilitate cleaning. All exposed surfaces should be capable of withstanding treatment without appreciable deterioration.

2.9   Facepiece

Where facepieces are used, they should be designed to meet the following requirements. For leak tightness see BS 4667:

- Facepieces should cover the eyes, nose, mouth, and should provide adequate sealing on the face of the user of the BA against the outside gas, when the skin is dry or moist, when the head is moved, and when the user is speaking.
- Facepieces should fit against the contours of the face so that when tested the inward leak of the test contaminant between the facepiece and the user's face should not exceed a value of 0.05% of the inhaled air for any one of the 10 test subjects. It is unlikely that this requirement can be met by users with beards or with glasses.
- Facepieces should be light in weight and comfortable to wear for long periods. The weight should be symmetrically balanced to ensure the maximum retention of the face seal and

to minimize muscular strain, particularly when worn in circumstances involving vigorous movements.

- Facepieces should have suitable and, preferably, replaceable eyepieces or eyeshields.
- Facepieces should be secured to the face with an adjustable and replaceable head harness, and they should be fitted with a strap to support them when not being worn.
- Means for speech transmission should be incorporated.
- Manufacturer should provide a way to reduce misting of the eyepieces or eyeshields so that vision is not interfered with.

2.10 Half-masks

When half-masks are used they should comply with the requirements specified in BS 7356, except that they should not cover the eyes, and the inward leak test should be carried out when air is being supplied at a flow of 120 LPM.

2.11 Mouthpiece

If the apparatus is fitted with a mouthpiece it should be designed to provide a reliable seal with the mouth and should be secured against accidental displacement with an adjustable head harness. It is recommended that a plug or cover be provided to close the orifice of the mouthpiece when not in use.

2.12 Nose clip

A nose clip should be provided if a mouthpiece is used and should be designed to afford maximum security against accidental displacement. It should not slip when the nose becomes moist with perspiration, and suitable means should be provided for attaching it to the apparatus to prevent loss.

2.13 Head harness

The head harness should hold the facepiece, half-mask, or mouthpiece firmly and comfortably in position. It should be simply fitted and adjusted and should be able to be cleaned and decontaminated. Any fabric used in the construction of a head harness should be resistant to shrinkage. The head harness should also be adjustable and, if consisting only of straps, these should be adjustable and no less than 19 mm (nominal) in width at the points in contact with the head and designed to ensure that the user must readjust the straps before each use. The means of adjustment should be slip-proof.

2.14   Air hood or suit

The air hood or suit should be light in weight and comfortable to wear for long periods. It should have a transparent area that provides a clear view. A minimum air supply should be specified by the manufacturer (which should be no less than 120 LPM), and the inward leak of the external atmosphere into the hood or suit should not exceed a value of 0.1% (see Figures 2.9 and 2.10).

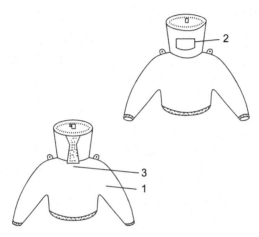

Figure 2.9 Schematic of Blouse. (1) Blouse; (2) Visor; (3) Connector.

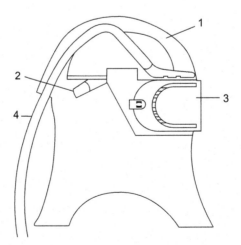

Figure 2.10 A schematic of hood. (1) Hood; (2) Head harness; (3) Visor; (4) Air supply hose.

2.15   Inhalation and exhalation valves

In fresh-air hose apparatus (lung-operated or with hand blower) an inhalation valve should be fitted in such a position as to minimize the rebreathing of expired air. Where a breathing bag or other flexible reservoir is fitted, the inhalation valve should be located between the bag or reservoir and the mouthpiece or facepiece.

All apparatus except air hoods and suits should be provided with an exhalation valve to allow the escape of exhaled air and any excess air delivered by the air supply, and it should be capable of being operated automatically by the pressure in the breathing circuit. The exhalation valve should be designed so that the inward leak of the external atmosphere does not exceed 0.0025% when the moist valve is tested. The valve should be protected against dirt and mechanical damage.

When it is possible in these types of apparatus for the pressure in the facepiece to fall below atmospheric in normal use, the exhalation valve should be shrouded or include an additional nonreturn valve or other device.

The design of valve assemblies should be such that valve discs or the assemblies can be readily replaced; it should not be possible to fit an inhalation valve assembly in the expiratory circuit or an exhalation valve assembly in the inspiratory circuit.

- Demand Valve: A demand valve, when fitted, should be connected directly or by a nonkink-able hose to the facepiece. When in any position the demand valve should operate at an inlet pressure between 345 kN/m² and 1035 kN/m² and should be capable of supplying air at a minimum flow of 120 LPM.
- Flow-Control Valve: The flow-control valve when fitted should be placed on the waistbelt or harness in a position where it can be easily adjusted. It should provide an adequate flow to the facepiece or hood at all stated supply pressures, and the valve in the fully closed position should pass at least 57 LPM at the minimum stated supply pressure.
- Breathing Tubes: If the air-supply hose is the low-pressure type, a flexible, nonkinking breathing tube (tubes) should be used to connect it to the mouthpiece of facepiece and allow free head movement.

2.16   Harness or belt

A harness or belt should be provided to prevent a pull on the breathing tube or on the mouthpiece, facepiece, or air hood. Buckles should be so constructed that once adjusted they will not slip. The attachment connecting the hose to the harness or belt should be able to to withstand a pull of 1000 N in all directions. Any fabric used in the construction of a body harness should be resistant to shrinkage.

2.17   Condition of the inhaled air (Carbon-Dioxide content)

When the apparatus using a mouthpiece is tested, the carbon-dioxide content of the inhaled air (including dead-space effects) should not exceed 1.0% (by volume). When the apparatus with an air flow of less than 120 LPM using a facepiece or half-mask is tested the carbon-dioxide content of the inhaled air (including dead-space effects) should not exceed 1.5% (by volume).

2.18 Resistance to breathing

With the air-supply system working at any flow chosen by the testing authority but within its designed range of pressures and air flow, or with a blower operated in such a way that the operator would not become unduly fatigued after 30 min, or with the fresh-air hose alone (if not supplied with blower or bellows), then with the maximum length of tube for which the apparatus has been submitted for approval, half of it coiled to an inside diameter of 500 mm, neither the inspiratory nor the expiratory side of the apparatus should have a dynamic resistance greater than 50 mm H_2O.

If any of the air-supply systems ceases to operate, the user should still be able to inhale through the tube without undue distress. This provision should be satisfied if the total inspiratory resistance, with the air-supply system inoperative but not discon-nected and with the maximum length of the tube for which the apparatus has been submitted for approval, is not greater than 125 mm H_2O at a continuous air flow of 85 LPM.

When tested the apparatus should be such that it may be worn without avoidable discomfort, so that users show no undue signs of strain attributable to wearing the apparatus, and that it impedes the user as little as possible when in a crouched position or when working in a confined space.

Apparatus intended for use in low temperatures should function satisfactorily.

2.19 Requirements for fresh-air hose apparatus

2.19.1 Fresh-air hose supply systems

*   With blower: Hand-operated blowers should be capable of being operated by one worker without undue fatigue for at least 30 min. Rotary-type blowers should be capable of main-taining positive air pressure in either direction of rotation, or else be made to operate in one direction only. In the former case, the direction of the operation in which the blower deliv-ers less air against the designed working pressures should be used in testing. When motor-operated blowers are used where flammable surroundings may arise it is essential that the suitability of the equipment for use in such surroundings be considered. Figure 2.11 shows a typical assisted fresh-air hose BA power-assisted type.

Note: It is recommended that an airflow indicator to be provided at the blower to indicate the flow rate.

*   Without blower: The hose should be fitted with a strainer at the free end to exclude debris. Provision should be made for securely anchoring the free end of the hose and strainer so that it cannot be dragged into the contaminated atmosphere.
*   Low-pressure hoses of fresh-air supply: The hose of low-pressure air supply should meet the requirement of BS 4667 or equivalent.

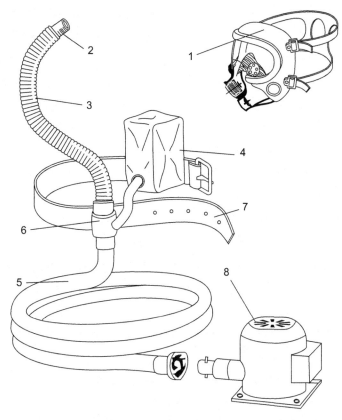

Figure 2.11 Assisted fresh air hose breathing apparatus power assisted type. (1) Facepiece; (2) Equipment connector; (3) Breathing hose; (4) Breathing bag; (5) Air supply hose; (6) Coupling; (7) Belt or body harness; (8) Blower (Motor-driven) or compressed air injector.

- High-pressure tubing: High-pressure tubing should meet the requirement of BS 4667 or equivalent.
- Marking: Marking in the facepiece, half mask, hood, and suit:
  Name, trademark or other identification of the manufacturer
  For hood and suit, the designed air flow in LPM
  Whether or not designed in low temperature
  - Marking on hose
    Name, trademark, or other means of identification of the manufacturer:
    Designed air flow in LPM
    Mark of heat resistant
    Working pressure of high-pressure hose
  - Marking on the flow control (maximum working pressure):
- Marking on blower
- Name, trademark, or other means of identification of the manufacturer
- Designed air flow in LPM
- Maximum length of air hose for which the blower is designed

2.20 Instruction

Breathing apparatus should be supplied and accompanied by updated instructions for maintenance and use. These instructions should include the following information:

- Size of the facepiece, half-mask, hood or suit (if more than one size is available)
- For hoods and suits, the designed air flow in LPM
- Whether or not designed for use in low temperatures
- Guidance on the fit of the facepiece, and adjustment of face seal where relevant
- For the hose of fresh-air hose apparatus, the designed minimum air flow in LPM
- For hoses, the words "heat resistant"
- Working pressure of high-pressure hose
- Maximum and minimum working pressures of the flow-control valve
- For the blower, the designed minimum air flow in LPM
- Maximum length of air hose for which the blower is designed
- Warning that adequate protection may not be provided by the apparatus in certain highly toxic atmospheres and that guidance is given in BS 4275
- Warning that allowance should be made for the fact that it is likely that faceseal fit will be adversely affected by glasses, sideburns, or beards
- Warning that at very high work rates the pressure in the facepiece may become negative at peak inhalations

The manufacturer should guarantee in writing that all fresh-air hose and compressed-air-line apparatus have been tested and certified by a recognized international organization.

2.21 Self-contained breathing apparatus (SCBA)

SCBA are classified as:

- Closed-circuit (recirculating)
- Open-circuit (demand)

There are two types of closed-circuit apparatus:

- Oxygen-generating type, in which oxygen-generating chemicals in a container are activated by the moisture in the user's expired breath
- Compressed air or liquid oxygen type, which employs a contained or compressed or liquid oxygen

2.21.1 Closed-circuit SCBA

Closed-circuit SCBA is designed to enable workers to work in an irrespirable atmosphere for longer periods than are generally possible with the open circuit type and with the greater freedom of movement than is allowed by the air-line type. The apparatus is designed and constructed so that exhaled air passes from the facepiece or mouthpiece through a breathing tube into a purifier containing chemicals that absorb the exhaled carbon dioxide. Oxygen is fed into breathing circuit from a cylinder of compressed oxygen or from a liquid oxygen/air container. The oxygen and purified gas mixture is fed to the user who inhales from a breathing bag, and any excess gas is released through a relief valve.

2.21.2   Materials and design

- Body harness: The body harness should be designed to allow the user to put it on quickly and easily without assistance and should be adjustable for fit. Buckles fitted to waist and shoulder harness should be constructed so that once adjusted they will not slip. Any fabric used in the construction of a body harness should be resistant to shrinkage. For certain applications, the body harness should be detachable to permit water-immersion testing, or the parts should be water-resistant. Where the body harness incorporates a way to attach a lifeline, the harness, together with the snap hook, should be capable of withstanding a drop test of 1 m when loaded to 75 kg.
- Inhalation and exhalation valves: The design of valve assemblies should be such that valve discs or the assemblies can be readily replaced; it should not be possible to fit an inhalation valve assembly in the expiratory circuit or an exhalation-valve assembly in the inspiratory circuit.

2.21.3   Relief valve

Breathing apparatus of the closed-circuit type should be provided with a relief valve-operated automatically by the pressure in the breathing circuit and designed so that inward leak of the external atmosphere does not exceed 0.0025% when the moist valve is tested. The relief valve, which should include an additional nonreturn valve, should be protected against dirt and mechanical damage. There should also be a way to seal the relief valve to permit leak testing.

- Performance characteristics of the relief valve:
  The opening pressure of the moist-relief valve measured at a constant flow of 1 LPM should be between 15 mm H_2O and 40 mm H_2O in any position of the valve.
  The flow resistance, at a constant flow of 300 LPM of that part of the expiratory breathing circuit between the relief valve and the breathing bag, should not be greater than the minimum opening pressure of the relief valve.
  In apparatus using liquid air or liquid oxygen, the resistance of the relief valve to an air flow of 100 LPM should not exceed 50 mm H_2O in any position of the valve.
  In apparatus using compressed oxygen, the resistance of the relief valve to an air flow of 50 LPM should not exceed 50 mm H_2O in any position of the valve.

2.21.4   Reducing valve or pressure reducer

In apparatus using a reducing valve or pressure reducer alone, i.e., without a supplementary lung-governed oxygen supply, the flow of oxygen should not be less than 2 LPM for the effective duration of the apparatus. Except for apparatus with a pressure reducer, the oxygen flow during the reserve period should not fall to less than 1.8 LPM. The flow of oxygen from a constant flow type of reducing valve should remain constant to within 10% of the preset flow at all cylinder pressures above 10 atmospheres. The reducing valve, if adjustable, should be provided with a suitable locking device to prevent accidental alteration of the oxygen supply.

2.21.5 Lung-governed oxygen supply (demand valve)

The opening pressure of the lung-governed supply mechanism measured at a constant flow of 10 LPM should not exceed 35 mm H_2O. Apparatus operating with a lung-governed constant supply of less than 2.0 LPM should have an automatic scavenge device with which sufficient "air" is removed from the circuit to the outside to maintain an oxygen content of no less than 21%.

2.21.6 By-pass valve

Apparatus equipped with a pressure reducer or a reducing valve and/or lung-governed valve should be provided with a manually operated self-closing type of by-pass valve, whereby the user can obtain a supply of oxygen at a flow of between 60 LPM and 300 LPM at all cylinder pressures above 5000 kN/m² independently of the reducing valve or lung-governed valve.

2.21.7 Pressure gauge

Apparatus using compressed oxygen should have a pressure gauge. The gauge should incorporate a suitable blow-out release so that in the event of an explosion or fracture of the pressure element of the gauge, the blast will be away from the front. The gauge's window should be made nonsplintering glass or of clear plastic material. An efficient valve should be provided to isolate the gauge and connections to it from the rest of the circuit. The pressure gauge should be placed to enable the gas cylinder pressure to be read conveniently by the user. The pressure gauge should also incorporate a way to indicate an adequate warning period. This will vary according to user requirements.

2.21.8 Warning device

Where apparatus using compressed oxygen has an audible warning device that sounds when the cylinder pressure drops to a predetermined level to warn the user that he must withdraw immediately to fresh air, the device should have the following characteristics:

- If operated by compressed oxygen, an average consumption of not more than 2 LPM
- Operation should begin when no more than 80% of the fully charged capacity of the cylinder has been used
- A frequency of between 2500 Hz and 4000 Hz
- Clear audibility to the user and those in the immediate area until the pressure gauge needle at least reaches the warning zone marked on the pressure gauge

2.21.9 Flexible tubes

Flexible tubes and fittings of the high-pressure system should be capable of withstanding (without damage) a test pressure of twice the maximum designed working pressure. It should not be possible to fit a low-pressure tube or hose into a higher-pressure part of the circuit.

2.21.10 Gas cylinder and valve

Gas cylinders should comply with specifications approved by the company. Such approval may involve restrictions in application. The main valve should be designed so that the full pressure in the gas cylinder cannot be applied rapidly to other parts of the apparatus. The valve should be designed so that the valve spindle can not be completely unscrewed from the assembly during normal operation of the valve. The valve should also be either lockable in the open position or designed so that it cannot be closed inadvertently by contact with a surface.

2.21.11 Oxygen supply

The total volume of oxygen available should be sufficient to meet an average consumption of no less than 2 LPM for the effective duration of the apparatus. In apparatus without a supplementary lung-governed oxygen supply, an additional 10% capacity should be provided to allow for the possible use of the by-pass valve.

2.21.12 Breathing bag

The breathing bag should be made of strong, flexible material and should be protected against collapse or damage by external agencies. The breathing bag should be reliably and tightly joined to the couplings. The coupling at the inhalation side should be shaped in such a way that its opening cannot be closed by the bag itself. In apparatus using compressed oxygen, the capacity of the breathing bag, when correctly fitted and with the casing closed, should be at least 5 liters.

2.22 Tests and certification

Closed-circuit BA should be tested and certified, and the manufacturer should supply testing certificates with the apparatus. The test should cover the following:

- Capacity and function of breathing bag
- Condition of inhaled air
- Resistance to breathing air
- Comfort
- Durability of materials when subjected to cleaning and decontamination
- Inward leak of facepiece
- Practical performance
- Laboratory performance tests
- Low-temperature tests
- Test for inward leak on relief valve

2.23 Marking

BA manufactured in compliance with standards should be marked with the following:

- Marking on the facepiece
  Name, trademark, or other means of identification of the manufacturer
  Number of official standard used
- Markings on the apparatus
  Name, trademark, and other means of identification of the manufacturer
  Number and part of official standard used
  Working duration:
  1. with warning device
  2. without warning device

2.24 Instructions

BA manufactured in compliance with standards should be supplied accompanied by dated instructions for maintenance and use that should include where appropriate:

- Working duration
- Guidance on fit of facepiece and adjustment of faceseal where relevant
- Warning that adequate protection may not be provided by the apparatus in certain highly toxic atmospheres and guidance
- Warning that allowance should be made for the fact that it is likely that faceseal fit will be adversely affected by glasses, sideburns, or beards
- Grain size of carbon dioxide absorbent

2.25 Open circuit – SCBA

A self-contained open-circuit compressed-air BA is an apparatus that has a portable supply of compressed air that is independent of the ambient atmosphere. Compressed-air BAs are designed and constructed to enable the user to breath air on demand from a high-pressure air cylinder (or cylinders) either via a pressure reducer and a lung-governed demand valve or a lung-governed demand valve connected to the facepiece. The exhaled air passes without recirculation from the facepiece via the exhalation valve to the ambient atmosphere. Compressed-air BAs are classified according to the following effective air volume at a pressure of 1 bar absolute and a temperature of 20°C:

at least	600 L
at least	800 L
at least	1200 L
at least	1600 L
at least	2000 L

2.25.1   Requirements

The apparatus should be of simple and reliable construction and as compact as possible. The design of the apparatus should allow its reliable inspection. The apparatus should be sufficiently robust to withstand the rough use it is likely to receive in service and designed so that it will continue to function satisfactorily while temporarily submerged in water at a maximum depth of one meter, and thereafter, until the air in the cylinder is exhausted. The apparatus is not designed for prolonged use underwater. The apparatus should be designed so there are no parts or sharp edges likely to be caught on projections in narrow passages.

The apparatus should also be designed so the user can remove it and, while still wearing the facepiece, continue to breath the air from the apparatus. The apparatus should be designed to ensure its full function in any orientation. The main valve(s) of the air cylinder(s) should be arranged so the user can operate them while wearing the apparatus.

2.25.2   Materials

All materials used in the construction should have adequate mechanical strength, durability, and resistance to deterioration, i.e., by heat or by contact with seawater. Such materials should be anti-static as far as it is practical. Exposed parts, excluding cylinders, i.e., those that may be subjected to impact during wearing, should not be made of magnesium, titanium, aluminum, or alloys containing such proportions of these metals as they can, on impact, create frictional sparks capable of igniting flammable gas mixtures.

2.25.3   Cleaning and disinfection

The materials used should withstand the cleaning and disinfecting agents recommended by the manufacturer, and the cleaning and disinfection process should be approved by the testing authority. The weight of the apparatus when ready for use with facepiece and fully charged cylinder should not exceed 18 kg.

2.25.4   Connection (Couplings)

The design and construction of the apparatus should permit its components to be easily separated for cleaning, examining, and testing. The demountable connections should be readily connected and secured, where possible by hand. Any means of sealing used should be retained in position when the connection(s) is (are) disconnected during normal maintenance. The connection between the facepiece and the apparatus may be achieved by a permanent or special type of connection or by a screw-thread connection.

2.25.5   Body harness

The body harness should be designed to allow the user to put on and take off the apparatus quickly and easily without assistance, and it should be adjustable. All adjusting devices should be constructed so that once adjusted they will not slip inadvertently. The harness should be constructed so that when tested in practical performance tests the apparatus can be worn without avoidable discomfort. The user should not be strained by wearing the apparatus, and the apparatus should impede the user as little as possible when in a crouched position or when working in confined space. In addition to the machine tests, the apparatus should also undergo practical performance tests under realistic conditions. The apparatus should operate trouble-free over the temperature range − 30°C to + 60°C. Apparatus specifically designed for temperatures beyond these limits should be tested and marked accordingly.

2.25.6   Protection against particulate matters

The component parts of the apparatus supplying compressed air should be reliably protected against the penetration of particulate matter that may be contained in the compressed air.

2.25.7   High-pressure parts

Metallic high-pressure tubes, valves, and couplings should be capable of withstanding a test pressure of 50% above the maximum filling pressure. Non-metallic parts should be capable of withstanding a test pressure twice the maximum filling pressure of the cylinder.

2.25.8   Cylinder valves

The design of the cylinder valve should be such as to ensure safe performance. The valve should be designed so that the valve spindle cannot be completely unscrewed from the assembly during normal operation of the valve. The valve should also be designed so that it cannot be closed inadvertently by contact with a surface by one of the following methods:

•   The valve should be designed so that a minimum of two turns of the handwheel are required to open fully the valve.
•   The valve should be lockable in the open position.

Apparatus fitted with more than one cylinder may be fitted with individual valves on each cylinder.

2.25.9   Cylinder valve connection (Valve outlet)

It should not be possible to connect cylinders with a higher maximum filling pressure, e.g., 300 bar (4350 psi) to an apparatus that is designed only for a lower maximum filling pressure, e.g., 200 bar (2900 psi).

Note: Only cylinders of equal maximum filling pressure should be connected to an apparatus with more than one cylinder.

2.25.10   Pressure reducer

Any adjustable medium-pressure stage should be reliably secured against accidental alteration and adequately sealed so that any unauthorized adjustment can be detected. A pressure reducer safety valve should be provided, if the apparatus cannot take the full cylinder pressure. If a pressure reducer safety valve is incorporated, it should be designed to operate within the manufacturer's design parameters.

At maximum operating pressure of the pressure reducer safety valve the apparatus still has to permit breathing. The maximum pressure built up at the inlet of the lung-governed demand valve should be such that the user can continue breathing. Where demand valves open with medium pressure, a pressure reducer safety valve need not be installed, provided the previous requirements are met.

2.25.11   Pressure gauge

The apparatus should be equipped with a reliable pressure indicator that will read the pressure in the cylinder(s) when opening the valve(s) to ensure that the individual or the equalized contents are measured, respectively. The pressure gauge should be placed to enable the pressure to be read conveniently. The pressure gauge tube should be sufficiently robust to withstand rough use. Where the tube is protected by sheathing, the enclosed space should be vented to the atmosphere.

The pressure gauge should also be resistant to dust and water and should withstand immersion in water at a depth of 1 meter for 24 hours. After testing, no water should be visible in the device. The pressure gauge should be graduated from the zero mark up to a value of at least 50 bar above the maximum filling pressure of the cylinder. The design of the gauge should allow the reading of the indicated pressure to within 10 bar. When the pressure gauge and connection hose are removed from the apparatus, the flow should not exceed 25 LPM at maximum filling pressure of the cylinder.

2.25.12   Warning device

The apparatus should have a suitable warning device that operates when the cylinder pressure drops to a predetermined level to warn the user. The warning device should respond at the latest when only one-fifth of the total breathable air volume is left (tolerance + 50 L) but at least 200 L are still available. After warning device sounds, the user should be able to continue to breathe without difficulty.

If there is an audible warning device the sound-pressure level should be a minimum of 90 dB(A) as a continuous intermittent warning. The frequency range should be between 2000 Hz and 4000 Hz. The air loss that might be caused by the warning signal should not exceed an average of 5 LPM from the response of a signal to a pressure of 10 bar or no more than 50 L for those warning devices not operating continuously. The duration of the warning at 90 dB(A) should be at least 15 s for a continuous signal and 60 s for an intermittent signal.

2.25.13   Flexible hoses and tubes

Hoses may be extensible or compressible but should not collapse, and the temporary elongation should be at least 20%. Tubes for the demand valves (connections included) should withstand for at least 15 min. twice the operating pressure of the pressure reducer safety valve or at least 30 bar, whichever is higher. Any hose or tube connected to the facepiece should permit free head movement and should not restrict or close off the air supply under chin or arm pressure during practical performance tests.

2.25.14   Lung-governed demand valve

- Breathable air supply: The breathable air supply should be capable of a flow rate of at least 300 LPM at all cylinder pressures above 20 bar and of at least 150 LPM at a cylinder pressure of 10 bar.
- Without positive pressure: The negative pressure for opening of the lung-governed supply demand valve should be between 0.5 m bar and 3.5 m bar when tested using a continuous flow of 10 LPM from maximum filling pressure down to 10 bar. A self-opening of the demand valve at negative pressure of less than 0.5 m bar should not occur. At a flow rate of 300 LPM the negative pressure should not exceed 10 m bar at all pressures down to 20 bar.
- Supplementary air supply: Apparatus without positive pressure should be provided with a manually operated means of providing a supply of air at a flow rate of at least 60 LPM at all cylinder pressures above 50 bar, independently of the normal operation of the demand valve. Apparatus with positive pressure can be provided with such a device.

All parts requiring manipulation by the user should be readily accessible and easily distinguishable from one another by touch. All adjustable parts and controls should be constructed so that their adjustment is not liable to accidental alteration during use.

2.26   Breathing resistance

2.26.1   Inhalation resistance

- Without positive pressure: The inhalation resistance of an apparatus without a facepiece should not exceed 4.5 m bar at all cylinder pressures from full to 10 bar. Where a lung-governed demand valve is permanently attached to a full face-mask the negative pressure should not exceed 7 m bar.
- With positive pressure: The apparatus should be designed in such a way that at a flow rate of 300 LPM positive pressure is maintained in the cavity of the mask adjacent to the face seal. This requirement should be valid at all cylinder pressures above 20 bar.

2.26.2   Exhalation resistance

- Without positive pressure: The exhalation resistance of an apparatus with facepiece should not exceed 3.0 m bar.
- With positive pressure: The exhalation valve should have an opening resistance not exceeding 6 m bar, a resistance not exceeding 7 m bar at a continuous flow of 160 LPM, and a resistance not exceeding 10 m bar at a continuous flow of 300 LPM. The static pressure in the mask cavity (inner mask, if applicable) under conditions of equilibrium should not exceed 5 m bar.

2.27 Tests and certification

Open-circuit BA should be tested and certified by testing authorities, and tests certificate should be supplied with the apparatus by the manufacturer. The tests should cover the following in accordance with EN 137 or equivalent, e.g., National Fire Protection Association (NFPA) Supplement 2 Code 1981 Chapter 1 to 4 (1991):

- Visual inspection
- Practical performance tests
- Resistance to temperature
- Pressure reducer test
- Warning device test
- Flexible hose test
- Lung-governed demand valve
- Breathing resistance demand

2.28 Instructions for maintenance and storage

1. On delivery, instructions should accompany every apparatus enabling trained and qualified persons to use it.
2. The instructions should comprise the technical data, the range of application, and instructions necessary for correct fitting, care, maintenance, and storage.
3. The instructions should state that the air supply should meet the requirements of breathable air.

Note: To assure reliable operation of the equipment, the data given for moisture content in Table 2.1 should not be exceeded.

4. Any other information the supplier may care to provide.
5. If the sub-assemblies are too small to be marked, those sub-assemblies should be given in the maintenance list. All units of the same model should be provided with a type of identifying marking. Sub-assemblies and components with considerable bearing on safety should be marked so they can be identified. The manufacturer must be identified by name, trademark, or other means of identification. Where the reliable performance of components may be affected by aging, the date (at least the year) of performance should be marked. The marking of the apparatus should be provided with the following:
- Serial number
- Year of manufacture
- The pressure reducer should be durably marked with a serial number. The marking should be such that the year of production can be determined. In addition, provision should be made to mark the date (year and month) and test marks of the last testing performed.

Table 2.1 Maximum moisture content

Filling pressure bar	Moisture mg/m^3
200	50
300	35

2.29 Open-circuit escape SCBA

Open-circuit SCBA is comprised of a compressed-air cylinder, a demand valve (or other device that adequately controls the air supply), and breathing tube to a face-piece, hood, mouthpiece, or nose clip. The exhale air passes to the atmosphere. The apparatus is intended for escape purposes only from irrespirable atmosphere.

2.29.1 Requirements

Materials, design, and construction should meet the specification of SCBA with the following exceptions:

- Time for which the apparatus will function when tested in accordance with BS 4667: Part 4 or equivalent using a breathing simulator set at a flow rate of 40 LPM and the rated duration should not be less than 5 minutes.
- If an air line is used with this type of equipment, the apparatus should be provided with a leak-tight check valve and connector and if more than one apparatus is connected to the air line, the air flow available to each demand valve through the maximum length of air line supplied should not be less than 40 LPM for each demand valve.

2.29.2 Instructions

The apparatus should be supplied with dated instructions for storage, maintenance, and use that should include:

- Rated duration
- Guidance on assembly and use
- Warning that adequate protection may not be provided by the apparatus in some circumstances within certain highly toxic atmospheres
- Warning that allowance should be made when the face-seal fit of the facepiece could be adversely affected if the user wears glasses or has sideburns or a beard. The apparatus should be visibly marked with the following:
  Name trademark or other means of identification of manufacturer
  Number of official standard
  Year of manufacture
  Rated duration

2.29.3 Test and Certification

The manufacturer should issue a certification from a testing authority stating that all tests in accordance with Appendices A to G of BS 4667: Part 4 have been carried out and reporting that no malfunctioning of the apparatus has occurred.

2.30 Ventilatory resuscitators

This section specifies performance and safety requirements of a ventilatory resuscitator intended for use in patients of any body mass. It covers both operator-powered and gas-powered resuscitators. It also covers equipment essential for the use of a resuscitator as a self-contained piece of portable equipment, e.g., oxygen-supplying system and carrying case.

2.30.1 Dimensions

The resuscitator including any carrying case or frame should pass through a rectangular opening with dimensions of 400 × 300 mm.

2.30.2 Mass

The mass of the resuscitator complete with its carrying case or frame and accessories including (for gas-powered resuscitators) the gas cylinder should not exceed 16 kg.

2.31 Performance requirement

2.31.1 Ventilation performance

- Tidal volume: When tested, the resuscitator should deliver the tidal volume range appropriate to its classification (patients up to 40 kg or over).
- Pressure limitation:
  Operator-powered resuscitator: If a pressure-limiting system is provided in an operator-powered resuscitator classified for use with patients with body mass exceeding 10 kg, pressure at the patient-connection port should not exceed 6 kPa.
  Gas-powered resuscitators: A pressure-limiting system should be provided in oxygen-powered resuscitators so that the pressure at the patient-connection port does not exceed 4.5 kPa. No mechanism to override the pressure-limiting system should be provided.

2.31.2 Supplementary oxygen and delivered oxygen concentration

- Operator-powered resuscitator: An operator-powered resuscitators should be provided with a connection for supplementary oxygen that accepts elastomeric tubing having an inside diameter of 6 mm. The operator-powered resuscitator should deliver an oxygen concentration of at least 40 vol.%. If fitted with an attachment to raise the delivered oxygen concentration, the resuscitator should deliver an oxygen concentration of at least 85 vol.%.
- Patient valve malfunctioning: An operator-powered resuscitator with the ability to provide supplementary oxygen should not fail to cycle from inspiration to expiration.

2.31.3 Resistance to spontaneous breathing

The pressure at the patient-connection port should not exceed 0–5 kPa for expiratory resistance and should not fall by more than 0–5 kPa below atmospheric pressure for inspiratory resistance.

2.31.4   Oxygen supply for oxygen-powered resuscitators

- Oxygen cylinders, cylinder valve, and connections: Oxygen cylinder connections should be noninterchangeable between oxygen services. A cylinder pressure gauge or content indicator should be provided for each cylinder connected to the resuscitator. If a detachable device is provided for opening the cylinder it should be closed with a retaining chain or similar attachment that can withstand a static load of no less than 200 N without breakage.
- Cylinder pressure regulator: Except when the resuscitator is designed to be controlled by adjusting the pressure regulator, the cylinder pressure regulator should be preset and should not be adjustable by the operator. The regulator should be fitted with a relief valve that opens at a pressure no more than double its delivery pressure.
- Filter: A sintered filter having a pore-size index no greater than 100 μm should be provided in the supply to the pressure regulator.
- Container capacity: The resuscitator carrying case or frame for oxygen-powered resuscitator should accommodate one or more oxygen cylinders that should enable the resuscitator to supply at least 180 L of oxygen containing 85% of oxygen or more. Oxygen-powered resuscitators should deliver an oxygen concentration of at least 85 vol.%.
- Testing and Certification: The manufacturer of the resuscitator should supply certification by an official testing authority with the equipment, stating that all tests in accordance with BS 6850 or ISO 8382 Appendices A to K have been carried out before shipment.
- Markings
  - Operating instructions: Basic operating instructions should be provided either on the resuscitator or on the resuscitator carrying case or frame.
  - Marking on the resuscitator: The following information should be marked on the resuscitator:
    Classification (body mass range of adult)
    Number and date of official standard
    Nominal pressure setting of the pressure-limiting system
    Identification reference to the batch or the date of manufacturer
    For oxygen-powered resuscitators, the recommended range of gas supply pressure
- Cylinder connections: Each cylinder connection should be clearly and permanently marked with the name and chemical symbol of the gas it accommodates.

2.31.5   Information to be provided by the manufacturer

The manufacturer should provide operating and maintenance instructions. The size and shape of these the instruction manual should be such that it can be enclosed within or attached to the carrying case or frame. The manual should include the following:

- Warning that the resuscitator should only be used by trained personnel
- Instructions on how to make the resuscitator operational in all intended modes of operation
- Specification dealing the following:
  Body mass range for which the resuscitator is suitable for use
  Range of frequency
  Attainable delivery pressure
  Operating environmental limit
  Storage environmental limit
  For operator-powered resuscitator, the delivered oxygen concentrations at stated supplementary flows

Total volume range

Expiratory and inspiratory resistance, excluding details of any such resistance imposed by special fittings that are supplied by the resuscitator

End expiratory pressure in normal use if greater than 0.2 kPa

For patient demand valves, the pressure of termination of flow if above atmospheric

Details about the pressure-limiting system and the mechanism to override it, if any

- Instructions for dismantling and reassembly of components, if applicable, including an illustration of the parts in their correct location
- Recommended methods of cleaning and disinfection or sterilization of the parts in their correct location
- Test of resuscitator function that may be performed by the operator at the point of use
- List of parts that may be replaced by the operator
- Recommendations for use of resuscitator in hazardous or explosive atmosphere, including a warning if it will entrain or permit the patient to inhale gases from atmosphere. The manufacturer should describe how to prevent entrainment or inhalation.
- Warning that in the presence of high oxygen concentration there operators should not smoke or keep the resuscitator near open open flames, and that oil should not be used with the resuscitator
- Fault-finding and correction procedures

2.31.6   Oxygen-powered resuscitator

In addition to the above information, the manual for an oxygen-powered resuscitator should include the following information:

- Approximate duration of a single oxygen cylinder 457 mm long and with a 102 mm outside diameter containing 340 L of oxygen when the resuscitator is delivering a minute volume of 10 L and a concentration of at least 85% oxygen or the manufacturer's selected concentration is less than 85% oxygen
- If the resuscitator will deliver concentration of oxygen less than 85 vol.% the concentration delivered at the maximum and minimum tidal volume settings
- Flow from the patient-connection port against leak pressure of 1.5 kPa and 3 kPa
- Recommended range of oxygen supply pressure and flows
- Duration of oxygen cylinder supplied with the resuscitator under given conditions

Selection, inspection, and maintenance of respiratory protective equipment

3.1 Introduction

The number of possible hazards that may exist in a given operation requires careful evaluation followed by intelligent selection of protective equipment. This selection is made even more complex by the many types of equipment available, each having limitations, areas of application, and operational and maintenance requirements.

The selection of correct respiratory protection equipment for any given situation requires consideration of the nature of the hazard, the severity of the hazard, work requirements and conditions, and the characteristics and limitations of available equipment. Where there is any doubt about the suitability of a respirator, breathing apparatus (BA) should be used. Particular care is necessary in the case of odorless gases or fumes since these do not give a warning of their presence.

Some respirators and BA deliver air to the user at pressures slightly above ambient pressure. These can provide greater protection since any gas flow through leaks in the system will be to the outside. However, for this to be so, the pressure bias needs to be sufficient to ensure that the pressure within the mouth and nose remain positive throughout the respiratory cycle. In the case of BA, should the leak rate be excessive, the duration will be reduced. Since the reduction in pressure on inhalation depends on respiratory resistance, the performance of positive-pressure devices should always be checked with the respiration attached.

3.2 Severity and location of the hazard

Only protective devices that arrange for the addition of an independent atmosphere suitable for respiration are appropriate for use in oxygen-deficient atmospheres and self-contained, air line, or fresh-air hose apparatus should be used. The concentration of the contaminant and its physical location should be considered and account should also be taken of the length of time for which protection will be needed, entry and exit times, accessibility of a supply of air suitable for respiration, and the ability to use air lines or move about freely while wearing the protective device. Where flammable or explosive atmospheres may arise, the equipment should be suitable for use in such circumstances.

Personnel Protection and Safety Equipment for the Oil and Gas Industries.
DOI: http://dx.doi.org/10.1016/B978-0-12-802814-8.00003-1
© 2015 Elsevier Inc. All rights reserved.

3.3 Work requirements and conditions

3.3.1 Working duration

The amount of time spent working usually determines the length of time for which respiratory protection is needed and includes the time taken to enter and leave the contaminated area. With self-contained BA and canister or cartridge respirators, the protection time is limited, whereas compressed air line and fresh-air hose apparatus provide protection for as long as the facepiece is supplied with adequate air suitable for respiration. Dust respirators normally provide protection until the filter loading becomes excessive. Some canister respirators have a means for indicating the remaining time left through a window in the canister. Canisters and cartridges should be changed according to the manufacturer's instructions. Warning devices are sometimes fitted to self-contained BA. The user should understand the operation and limitations of each type of warning device.

3.3.2 Activity of user

The work area to be covered, work rate, and mobility required of the user in carrying out his work should be considered in the selection of respiratory protective equipment. Canister, cartridge, and dust respirators minimally interfere with movements, but high resistance to breathing within respirators in heavy-work conditions can result in distressed breathing. Compressed air line and fresh-air hose apparatus severely restrict the area their users can cover and present a potential hazard where trailing lines and hoses can come into contact with machinery. Self-contained BA present a size and weight penalty that may restrict movement in confined spaces and when climbing.

The user's work rate determines his respiratory minute volume, maximum inspiratory-flow rate, and inhalation and exhalation breathing resistance. The respiratory minute volume is of great significance in self-contained and compressed air line apparatus operated from cylinders since it determines their working duration, which may, in moderate work conditions, be only one-third of that in a condition of rest.

All facepieces will restrict vision to some degree and the restriction should be taken into account when training users. Other problems include the wearing of prescription glasses and having focial hair.

The ability to withstand stress caused by temperature extremes is especially important in emergency situations when only immediately available protective devices can be used. Eye protection may be necessary when half-masks and mouthpieces and nose clips are worn. In such cases, the eye protectors should be compatible with the respiratory protective equipment.

3.3.3 User acceptability and facepiece fit

User acceptability and facepiece fit are factors are of prime importance in the selection of equipment. The user's acceptance of a particular device depends on the degree

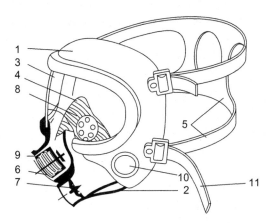

Figure 3.1 Full face mask. (1) Face blank; (2) Facepiece seal; (3) Visor; (4) Inner mask; (5) Head harness; (6) Connector; (7) Exhalation valve; (8) Check valve; (9) Inhalation valve; (10) Speech diaphragm; (11) Neck strap (Carrying strap).

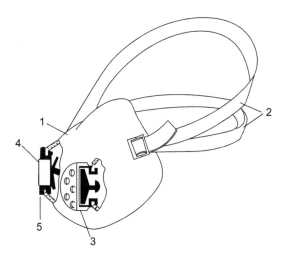

Figure 3.2 Half mask. (1) Face blank; (2) Head harness; (3) Exhalation valve; (4) Inhalation valve; (5) Connector.

of facepiece discomfort, interference with vision, its weight, breathing resistance, and individual physical condition and psychological factors. The fit depends on facepiece design, facial features, and hair, and is usually the most important factor in obtaining proper protection with a respirator, particularly with the half-mask type. Figures 3.1 and 3.2 show the half-mask and full-mask types, respectively.

3.4 Use of respiratory protective equipment

3.4.1 Operating procedures

Standard procedures should be developed that should take account of all the information and guidance given in these recommendations and in the manufacturer's instructions. All possible emergency and routine uses of respiratory protective equipment should be anticipated and procedures detailed. Written procedures should be prepared concerning the safe use of respiratory protective devices in dangerous atmospheres that might be encountered in normal operations or emergencies. Users should be familiar with these procedures and with the available devices.

3.4.2 Issue of equipment

The appropriate equipment should be specified for each job and should be specified in the work procedures by a qualified person supervising the respiratory protection program. The person responsible for issuing equipment should be adequately instructed to ensure that appropriate types are issued. Each device that is given to an individual should be durably marked with his name, but the marking should not affect performance. Dates of issue should also be recorded.

3.4.3 Work in confined spaces and toxic or oxygen-deficient atmospheres, e.g., tanks and vessels

When the equipment is worn in an atmosphere from which the user would be unable to escape without its protection, he should, where practicable, use a lifeline and be attended by another person. Where possible, he should be under the surveillance of a person in uncontaminated air to whom suitable respiratory equipment is immediately available, and who is instructed in the methods of resuscitation including the administration of oxygen and means for summoning assistance in an emergency.

3.4.4 Training in proper use

For safe use of any respiratory protective device, it is essential that the user be thoroughly instructed in its use and maintenance. Both supervisors and workers should be instructed by a competent person. Minimum training should include:

1. Instruction on the nature of the hazard and a close appraisal of what may happen if the device is not used
2. Instructions as to why the particular device is appropriate to the hazard
3. Explanation of the device's capabilities and limitations
4. Instruction and training in the actual use of the device and close supervision to ensure that it continues to be properly used
5. Disinfection

Users should be given regular practice in handling the equipment, fitting it properly, testing facepiece fit, and becoming familiar with it.

3.4.5   Test for fit of facepiece

The fit of the facepiece is important and users should receive instructions including demonstrations and practice in how it should be fitted and how to determine if it is on correctly. A satisfactory fit of a full facepiece cannot be expected when glasses are worn unless they are specially made for this purpose; beards and whiskers are also likely to affect fit adversely. To ensure proper protection, the facepiece fit should be checked by the user each time he puts it on. This may be done as follows:

• Negative pressure test: Close the inlet of the equipment. Inhale gently so that the facepiece collapses slightly, and hold the breath for 10 seconds. If the facepiece remains in its slightly collapsed condition and no inward leak of air is detected, the tightness of the facepiece is probably satisfactory. If the user detects leak, he should re-adjust the facepiece, and repeat the test. If leak is still noted, it can be concluded that this particular facepiece will not protect the user. The user should not continue to tighten the headband straps until they are uncomfortably tight to achieve a gas-tight face fit.
Note: It may not be possible to carry out this test with certain types of apparatus in which event the manufacturers should be consulted.

3.4.6   Use in low temperatures

Some of the major problems in the use of full facepieces at low temperatures are poor visibility and freezing of exhalation valves. All full facepieces should be designed so that the incoming fresh air sweeps over the inside of the eyepieces to reduce misting. Anti-mist compounds may be used to coat the inside of eyepieces to reduce misting at room temperatures and down to temperatures approaching 0°C. Full facepieces are available with inner masks that direct moist exhaled air through the exhalation valves, and when properly fitted are likely to provide adequate visibility at low temperatures.

At very low temperatures the exhalation valve may collect moisture and freeze open allowing the user to breathe contaminated air, or freeze closed, preventing normal exhalation. Dry air suitable for respiration should be used with self-contained and compressed air line BA at low temperatures. The dew point of the breathed gas should be appropriate to the ambient temperature. High-pressure connections on self-contained BA may leak because of metal contraction at low temperatures but the only risk is likely to be an outward leakage.

3.4.7   Use in high temperatures

A worker working in areas of high ambient or radiant temperatures is under stress and any additional stress caused by the use of respiratory protective devices should be minimized. This can be done by using low-weight devices with low breathing resistance. Supplied air respirators and hoods and suits with an adequate supply of cool breathing air are recommended.

3.4.8   Use of dust canister and cartridge respirators

Before entering a contaminated atmosphere, steps should be taken to ensure that it is not deficient in oxygen and that the contaminant level is appropriate to the protection. The filters of dust respirators will, after a period of time, gradually become choked by trapped particles during use and breathing may become progressively more difficult. They should be replaced outside the contaminated atmosphere.

When using canister or cartridge respirators, it is essential to select the correct canister or cartridge for the given hazard. The instructions with the canister should always be read before use and if possible the user should ascertain the useful remaining life of the canister and ensure that he does not remain in the contaminated atmosphere for longer than that time. If the remaining life is not known, the canister should be replaced.

Some types of canisters have a seal over the air entrance that must be removed to permit breathing. It has to be remembered that after the seal is broken the contents can deteriorate and become ineffective.

The replacement of canisters should be the responsibility of a competent person with knowledge of their shelf life and of the use to which they are being put. Cartridges have a very limited life and should be replaced regularly according to the circumstances of their use. If the mask leaks or the canister is exhausted, the user will usually know by odor, taste, or irritation of eye, nose, or throat, and should immediately return to fresh air.

If the canister is used up, it should not be left attached, but removed, and a new one should be selected and fastened in place. When a respirator is worn in a gas or vapor that has little or no warning properties, like carbon monoxide, it is recommended that a fresh canister be used each time a man enters the toxic atmosphere.

Note: Cartridge and canister type respirators should not be used in confined spaces.

3.4.9   Use of air line breathing apparatus

- Air lines and air hoses: Before being put into use, air lines and air hoses should be examined externally for defects and tested for freedom from blockage. The intake end of fresh-air hoses should be positively fixed in a position from which clean, fresh air can be drawn. Precautions against interference with them and against contamination of the air supply by vehicles and mobile equipment may also be necessary. Precautions should be taken to prevent the air line or hose fouling on projections or being damaged.
- Compressed air supply: There are three methods of providing air for breathing purposes:
  1. Separate breathing air service: The addition of an air service separate from the normal air service is the best method of supplying air for personal protection and should always be considered for installation in new works or where major alterations justify it.
  2. General works air service: The breathing air service may be taken from the general works air supply but only after special precautions against any contamination have been taken.
  3. Portable air-supplying units: Where breathing air is required infrequently, or in an emergency, or in remote places, a portable air-supply unit should be used.

3.4.10   Requirements for all compressed-air systems for air line breathing apparatus

- Air purity: Air being supplied to the user should not contain impurities in excess of the following limits:

Carbon monoxide	5 parts per million (5.5 mg/m^3)
Carbon dioxide	500 parts per million (900 mg/m^3)
Oil mist	0.5 mg/m^3

- Odor and cleanliness: The air must be free from all odor and contamination by dust, dirt, or metallic particles and should not contain any other toxic or irritating ingredients.
  Note: Odor and cleanliness of compressed air is difficult to check accurately without special equipment. A rough check may be made by smelling the delivered air and by noting any discoloration or wetness when the air is passed gently through a wad of tissue or filter paper. An absorption filter may be necessary to remove odor. There should be no free water in the air supply.
- Compressors: Compressors, particularly the exhaust valves, should be well maintained and should not be allowed to overheat as a dangerous amount of carbon monoxide or other toxic substances may be produced by the decomposition of lubricating oils. Figure 3.3 shows a compressed air line BA continuous-flow type.
- Air supply: The capacity of any air service for personal protection should be calculated at a minimum requirement of 120 LPM. The pressure of air admitted to the kink-resistant tubing connected to personal-protective apparatus should be within the safe working pressure for the tubing and should never be at a pressure less than 345 m bar (5 psi).
- Air temperature and humidity: Air supplied to the mask, hood, or other device should normally be at a comfortable breathing temperature within the range of 15 to 25°C. The user's comfort is influenced by the humidity of the air breathed, and it is recommended that 85% relative humidity should not be exceeded.
- Air intake: The air intake provided for any breathing air service should be sited and constructed so as to avoid the entry of contaminated air into the system and ensure a sufficient supply of air suitable for respiration. The use of filters on any intake should be of secondary importance to the foregoing requirements.
- Recharging of air cylinders: When an air compressor is used to charge air cylinders, the manufacturer's instructions should be carefully followed. These will usually provide for the compressor to be run for a few minutes with drain valves open before cylinder charging commences and for opening the drain valve on completion of charging. After use, compressors should not be left under pressure for long periods.
- Air supply in an emergency: Every system of air supply employed should incorporate a receiver of sufficient capacity to enable people to escape from an irrespirable atmosphere in the event of a failure of the equipment supplying the air.
- Warning device: Arrangement should be made to warn the user whenever the air pressure falls to the minimum safe working level.

3.5 Use of self-contained breathing apparatus

Air and oxygen supplied from cylinders should be in accordance with standards and should preferably have a dew point not exceeding −50°C at atmospheric pressure. With every type of self-contained BA it is essential that the manufacturer's instructions are followed and that the apparatus is put on in air suitable for respiration. The apparatus should be checked immediately before entry into a contaminated atmosphere.

Once a cylinder has been used, for however short a period, the cylinder should be recharged as soon as possible. The valves of empty cylinders should be kept closed until the cylinders are recharged. Keeping in mind the limited duration of the apparatus, care should be taken to allow sufficient time to reach air suitable for respiration. The mouthpiece and nose clip, or the facepiece, should not be removed until the user is certain that he can do so without danger.

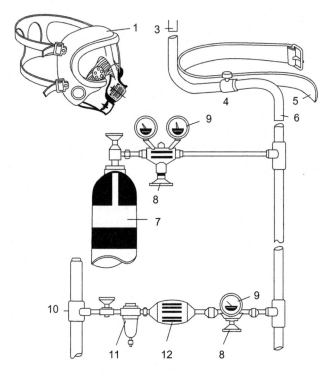

Figure 3.3 Compressed air line breathing apparatus continuous flow type. (1) Facepiece; (2) Equipment connector; (3) Breathing hose; (4) Coupling and continuous flow valve; (5) Belt or body harness; (6) Compressed air supply tube; (7) Compressed air cylinder; (8) Pressure reducer with warning device; (9) Pressure gage; (10) Compressed air line; (11) Separator; (12) Filter.

3.6   Care and maintenance of respiratory protective equipment

The program for the care and maintenance of equipment should be appropriate to the type of plant, working conditions, and hazards involved and should ensure that the equipment is properly maintained to retain the original performance standards. Arrangements should include those for:

- Inspection for defects
- Cleaning and decontaminating
- Repair
- Storage
- Issue

Where many respiratory protective devices are in use, a central station for care and maintenance under a suitably instructed supervisor is desirable.

3.6.1   Cleaning and decontamination

Equipment that is used regularly should be collected, cleaned, and decontaminated as frequently as necessary to ensure that proper protection is provided for the user. It should be cleaned as soon as possible after each use as moisture allowed to dry on the valves will interfere with their correct functioning.

The facepiece and breathing tube should be removed from the rest of the apparatus and cleaned by washing with soap and warm water and then thoroughly rinsed. In addition, wiping out the facepiece with a dilute solution of rinsing disinfectant may make the equipment more acceptable to the user. The equipment should then be dried out by direct sunlight. Proper cleaning and decontamination requires care and the manufacturer's instructions should be followed; in particular, temperatures above 85°C should never be used for decontamination. Many cleaning agents commonly used are irritating to the skin and if not completely removed from the facepiece by thorough rinsing, may cause skin rashes. Each employee should be trained on the cleaning procedures and have confidence that he will always be issued clean and uncontaminated equipment.

Respiratory protective devices may be contaminated by toxic materials such as organic phosphates, pesticides, and radionucleic. If the contamination is light, normal cleaning should provide satisfactory decontamination; if heavy, a separate decontamination process may be required before cleaning.

3.6.2   Servicing

Servicing should be carried out only by experience persons using parts designed for the particular respiratory protective devices. No attempt should be made to replace components or to make adjustments or repairs beyond the manufacturer's recommendations. Valves and regulators should be passed to a competent person for adjustment or repair.

Dust respirator filters should be renewed as soon as increased breathing resistance becomes evident. Gas respirator canisters and cartridges should be renewed no later than the expiration date of the maximum service life (assuming exposure to maximum concentration), as stated by the manufacturer.

Immediately after use, used and partly used cylinders or other containers of compressed air or oxygen on self-contained BA should be replaced by fully charged cylinders and the absorbents of closed-circuit apparatus renewed. It is dangerous to use organic-based oils and greases for lubricating cylinder valves, gauges, reducing valves, or other such fittings on oxygen equipment.

3.6.3   Storage

After inspection, cleaning, and any necessary repair, respiratory protective equipment should be stored in suitable holder boxes to protect against dirt, oil, sunlight, extreme heat and cold, excessive moisture, and harmful chemicals. The equipment should be stored so that it is not subjected to distortion.

Respirators placed ready for emergency use should be stored in special clearly marked cabinets and may also be enclosed in plastic bags to protect them from corrosive atmospheres.

To prevent tampering with canisters it may be desirable to store them in sealed containers with the date of the last use. Care should be taken to replace stored canisters before their shelf life is exceeded. The ends of air lines should be sealed to keep the lines internally clean.

Each respiratory protective device should be given a distinguishing number, and a record of cleaning, inspection, and maintenance should be kept. A record should also be kept of each canister and cartridge showing when it was opened and the duration and conditions of use.

3.7   Special problems

3.7.1   Beards

It is unlikely that a facepiece fit to the specified standards will be achieved by men with beards or with side-whiskers that interfere with the faceseal.

3.7.2   Corrective glasses and protective eyewear

A satisfactory fit of a full facepiece cannot be expected when glasses are worn unless the glasses are specially made or adapted. As a solution, glasses without side arms or with short side arms may be taped to the user's head. If it is necessary to wear eye protection with half-masks, care should be taken in selection that no mutual interference results.

3.7.3   Communication

Normal facepieces distort the voice to some extent but the exhalation valve usually provides some speech transmission over short distances in relatively quiet conditions. However, speech can cause facepiece or component leakage and should be limited, especially when wearing a half-mask.

Mechanical speech transmission devices are available as an integral part of some respirators. These consist of a resonant cavity and diaphragm that amplifies the sound in the frequency range most important to speech intelligibility. The diaphragm acts as a barrier to the ambient atmosphere and should be carefully handled and protected by a cover to prevent puncture.

Various methods of transmitting speech electronically from the facepiece are available and usually make use of a microphone connected to a telephone or radio transmitter. Usually the microphone is mounted in the facepiece, with the amplifier, power pack, and loudspeaker or transmitter attached to the outside of the facepiece, carried on the body, or remotely located. Facepieces with electric or electronic speech transmission devices having an integral power supply or one attached to the body should be certified as intrinsically safe or flameproof if they are to be used in flammable or explosive atmospheres. Speech transmission means may be incorporated in full facepieces for use with self-contained and air line BA.

3.8 Inspection, care, and maintenance of respiratory equipment

Strict attention to the care and maintenance of all types of respiratory equipment used is of vital importance for the safety of users. More damage is probably caused by incorrect maintenance and during storage than incurred during actual work operations. It is also essential that respiratory equipment be thoroughly examined for damage and defect before and after every occasion on which it is used and all defects to be rectified before the equipment is used again. Equipment to be checked include:

- Air-purifying respirators
- Atmosphere air, supplied respirators
- Air line respirator
- Self-contained BA

The recommendations made in the following sections concerning the handling of the various components should be most carefully observed and the manufacturer's instructions with regard to the assembly, adjustment, replacement of components, and maintenance should be obtained and followed.

3.8.1 Respiratory equipment

Supervisors should be responsible for daily inspections, particularly of functional parts such as exhalation valves and filter elements. They should make sure the edges of the valves are not curled and that valve seats are smooth and clean. The diaphragm of inhalation and exhalation valves should be replaced whenever they are deteriorated.

In addition to daily inspections, respirators should be inspected weekly. During the weekly inspection, rubber parts should be stretched slightly for detection of fine cracks. The rubber should be worked occasionally to prevent setting. One of the causes of cracking, and the stretch of headband to be checked is that the user has not stretched it in an attempt to secure a snug fit.

In an effort to reduce resistance to breathing, employees will punch holes in the filter, the rubber facepiece, or other parts, which is a dangerous practice and must be stopped.

In cleaning respirators, dirt and dust should first be blown from them using compressed air at not more than 1.5 bar pressure through a fixed nozzle directed toward an exhaust hood. Dust filters should not be cleaned by brushing.

Filters, screens, headbands, and cotton facelets should be removed if respirators are coated with paint or other foreign matter. They should be soaked for 3 hours in a cleaning solution of 1 kg of commercial alkaline base cleaner and 30 L of water. Fresh paint can be wiped off with a clean rag moistened with alcohol.

Respirators having no visible accumulation of foreign matter should be scrubbed in warm soapy water, rinsed, disinfected, then rinsed again and dried.

Dirty or oily elastic headbands should be washed in warm soapy water and rinsed. Cleaning of the filter depend on the design. Respirators that have replaceable filters can be cleaned and reused. Cleaning methods vary with filter composition and design. Only cleaning methods that are recommended by the manufacturer should be used.

Employees should be instructed to wipe off oil, grease, and other harmful substances from headbands and other parts of the respirators as soon as they collect. They should be instructed not to use solvents to clean plastic or rubber parts.

Most face and mouthpieces for respiratory protective devices are made of rubber or rubber-like compounds. Usually hand brushing or agitation in soapy water is sufficient to clean them. Aqueous solution or any other disinfectant in solution in warm water can be used to disinfect the parts. The parts should then be rinsed in clean water and dried quickly. Hot water, steam, solvents, and ultraviolet should not be used to clean and disinfect rubber parts because they have a deteriorating effect.

Before being stored, a respirator should be carefully wiped with a damp cloth and dried. It should be stored without sharp folds or greases. It should never be hung by the elastic headband or put down in a position which will stretch the facepiece.

Since heat, air, light, and oil cause rubber to deteriorate, respirators should be stored in a cool, dry place and protected from light and air as much as possible. Wood, filter, or metal cases are provided with many respirators. Respirators should be sealed in clean plastic bags.

3.8.2   Canister and cartridges

Although a year is usually given as the maximum effect life of a canister, it is possible for an unused canister to deteriorate in less than a year to the point where it becomes unusable. The canister should be replaced, therefore, not more than 1 year after the date the seal is initially removed. Canisters stored with seals intact should be replaced on or before the recommended use and the date stamped on each canister.

- Storage: Safe storage life for nonwindow indicator universal masks canisters stored in a dry place is 5 years, since even sealed canisters take up some moisture during storage. Tests show that a sealed canister loses much of its effectiveness against carbon monoxide after it has gained 45 g in weight. It is recommended, therefore, that sealed canisters (except the window-indicator types) be weighed as soon as received from the manufacturer and the weight should be marked indelibly on each canister. Stored canisters should be reweighed from time to time, when the weight increases to 45 g, the canister should be discarded. A card should be set up for each mask to indicate the date of latest inspection and replacement of the canister and the amount of use the canisters has had. It is wise to replace canisters after each emergency use.
- Cartridge respirator: Cartridge respirators are for use only in nonemergency situations, i.e., for atmospheres that are harmful only after prolong exposures. Filters should be replaced whenever breathing becomes difficult due to plugging of filters by retained particulates.

3.8.3   Atmosphere air supplied respirator

- Air line respirators: The air line and breather tube should be blown out with air before a mask is put on to eliminate dust and fumes that may have accumulated within the mask. The couplings in the hose line should be tested for tightness. The body harness needed to pull the hose lines require inspection prior to each use. The parts of the harness should withstand a pull of at least 120 kg. Parts that have to be used should be examined and checked for signs of wear and deterioration. The limitations of air line respirators of continuous-flow or demand-flow types should be understood. A trap and filter must be installed in the compressor line ahead of the mask to separate oil, water, scale, or other extraneous matter from the air stream. A pressure regulator with an attached gauge is required if the pressure in the compressor line exceeds 1½ bar. There should be a pressure-relief valve set at a predetermined setting that will operate if the regulator fails. The air supply must be free of carbon monoxide or other gaseous contaminations. For safety of user the components of all types of hose-line respirators should be inspected before each use. Low-pressure blowers that do not use internal lubricants are preferred. Components of air line respirators using compressed supply air cylinders should also be checked and inspected frequently and before each use. Air must be free of contaminants.

3.8.4 Self-contained breathing apparatus

Self-contained BA requires rigid inspection and maintenance because it is usually used under the most adverse circumstances. Periodic inspections should be made and records kept. All connector valves and hoses should be inspected to assure proper function when needed, and manufacturer's instructions should be followed. A preventive maintenance program for BA is the only assurance that the device will operate properly when needed. It is important to keep in mind that more damage is probably caused by incorrect maintenance and during storage than is incurred during actual work operations.

The components covered are cylinders, valves, demand regulators, pressure-reducing valves, manifolds, pressure gauges, flexible hoses, and rubber, fabric, or plastic components. The recommendations made in the following sections concerning the care and maintenance of these components should be carefully observed and the manufacturer's instructions with regard to assembly, adjustment, replacement of components, and maintenance should be obtained and followed. Mistreatment and careless manipulation with improper tools may also result in further maintenance expense.

3.8.5 Cylinders

- Handling: Cylinders should be handled with care and should not be dropped or roughly treated; when being transported they should be firmly secured so that they cannot move about.
- Storage: The condition of the inside of the cylinder can be maintained by keeping it dry at all times. The cylinder should be filled with dry air, and never completely discharged. Cylinders should be stored, preferably in the vertical position, in a cool, dry place adequately protected from the weather and away from excessive heat and direct exposure to the sun. Once a cylinder has been put into service, it should never be left completely discharged. A slight positive pressure should always register on the gauge.
- Maintenance: Protection against corrosion is important. The paintwork, metal spray undercoating (where applied), and fittings should be kept in good condition. Scratching of cylinders should be avoided. Protection by electro-plating methods is not recommended. Heat

or chemical strippers should not be used to remove old paint from any type of cylinder, and cylinders should not be modified under any circumstances, since this may result in serious weakening of the cylinder and lead to accident. The threads in the cylinder neck should not be altered in any way. Bushes or adapters should not be used. If the cylinder is not required for a long period (e.g., 6 months) it is recommended that it be made ready by a responsible employee for discharging, removal of the valve, extraction of any oil or water, drying out, and refitting of the valve. The cylinder should then be recharged to a slight positive pressure. If the cylinder is not to be recharged immediately, it should be left with the valve closed. A cylinder that has failed on inspection should be left with a responsible person who is to destroy it.

- Recharging: Recharging should be undertaken only with proper equipment that ensures that the compressed air is free from moisture, oil, and other impurities, and is fit for breathing purposes. Before recharging a cylinder, it is the responsibility of the supervisor to ensure that the cylinders are retested hydraulically unless this has been carried out within the prescribed retesting period. Cylinders should be visually examined and hydraulically tested. A certificate should be obtained after each test. It is essential that cylinders be charged carefully and slowly to prevent overcharging, and that the charging pressure be such that, after cooling to ambient temperature, the rated pressure for the cylinder is not exceeded. Before recharging, the valve should be cracked open to blow clear any dust or moisture in the valve passages. The rated working pressure at 15°C, in bars, should be stamped on the cylinder. It should be noted that, if a cylinder is subjected to heat from any source, the pressure inside it will increase.

- Identification of cylinders: It is essential that each storage cylinder be painted with identification colors for "air," i.e., grey body with black and white quarters at the valve end, and the words "breathing air" clearly stenciled or painted on the cylinder in a contrasting color.

- Cylinder valves: Cylinder valves should be lubricated only with lubricants recommended by the manufacturer of the apparatus. Other lubricants should not be applied to any of the valve parts or to the connecting fittings, since this may result in an explosion, or may contaminate the air, making it unsafe for breathing.

- Faulty valves: A cylinder with a damaged valve should never be used. If a valve is leaking, this should be corrected. Valve leaks may be broken down into two categories, both of which may be identified by an underwater bubble test.

  Valve seat leaks: If the valve of a fully charged cylinder held underwater is seen to emit gas bubbles, this is an indication of seat leakage. If tightening the handwheel fails to stop the bubble flow, the cylinder should be discharged and sent to a responsible person for the valve to be repaired.

  Gland leaks: If the test shows the valve seat is gas tight, the outlet should be carefully dried and then blanked off by attaching the regulator that would normally be fitted to it, or with a suitable high-pressure plug. The valve should again be immersed in water and the handwheel should be opened. Bubbles forming at the point where the valve spindle enters the main valve are indicative of a gland leak, in which case the cylinder should be discharged and sent to a responsible person for repair.

- Repair: No attempt should be made to remove a valve. Where bent, broken, or damaged valve spindles are found or the valve is excessively stiff in operation, the valve cylinders should be returned to a competent person for the valve to be serviced or replaced.

- Care after use: Immediately after use, the valve should be closed finger tight. It is recommended practice to close the valve while there is still a slight positive pressure in the cylinder.

- Maintenance: Maintenance and fitting of cylinder valves should be carried out only by authorized people. In the interests of safety, it is essential that the valves be inserted into the cylinder necks with the correct torque. After a valve has been fitted, the cylinder should be air tested to its full working pressure so that leaks and correct valve functioning can be checked.

3.9   Demand regulators, pressure-reducing valves, and manifolds

3.9.1   Examination before use

The following procedures should be carried out before use:

Read the manufacturer's manual.

Check the seal for defects before fitting the demand regulator or reducing valve to the cylinder or manifold and renew if it is damaged or badly worn.

Check that the filter, if fitted, is clean and in good condition. If found to be in poor condition, it should be replaced.

Check the mechanical air reserve valve, if fitted, for freedom of operation and ensure that it is returned to the "normal" or "main supply."

Before fitting the demand regulator to the cylinder, position the cylinder so that the valve is at the bottom end, and crack the valve to blow out any dirt.

Carry out a high-pressure test as follows:

Open the cylinder valve and check that there are no audible leaks.

Check the cylinder pressure. It is recommended that the cylinder be fully charged. At no time should any emergency work in confined spaces be carried out without an adequate air supply. The cylinder pressure should again be checked immediately. Caution should be exercised when opening the valve.

Close the cylinder valve and check that the pressure in the high-pressure system does not fall more than 10 bar in 1 min.

Locate any leaks by submerging the set in water.

Carry out a low-pressure test as follows. Breathe from the apparatus for 1 min to 3 min and check by taking several deep sharp breaths that there is no restriction to breathing. Close the cylinder valve and continue breathing until all air in the tube is exhausted. If it is then possible to draw in any air, there is a low-pressure leak that should be found and rectified.

3.9.2   Cleaning

Cleaning should be carried out in accordance with the manufacturer's instructions. Where the apparatus has been used under very dirty or oily conditions, it may be advisable to remove the regulator diaphragm to facilitate cleaning. Warm, soapy water should be used to remove any oil. Do not use solvents or detergents. Particular attention should be paid to the sealing surfaces of any associated nonreturn valves to make sure they are clean, dry, and free from deposits.

3.9.3   Maintenance

Nonmetallic components such as nonreturn valves, regulator valves, diaphragms, etc., are liable to deterioration, and this can interfere with the operation of the apparatus. Great care should be taken to observe the manufacturer's instructions. Demand regulators should be serviced regularly by an authorized person at intervals of not more than 1 year.

3.10   Pressure gauges

3.10.1   Examination before use

The following procedures should be carried out before use:

Check for signs of visible damage, e.g., broken glass, bent gauge needle, reading off zero, dents in casing ingress of water, etc.

Connect the gauge to the breathing regulator and source of high-pressure air. Open the cylinder valve slowly and check that the gauge needle moves smooth indicate the cylinder pressure; this should remain steady.

Close cylinder and watch the pressure gauge needle. If the indicated pressure drops and the high-pressure side of the apparatus has been tested as outlined in 10.6.1(f) (3) and (4), check for leaks in the gauge or gauge tubing.

If the test under (c) is satisfactory, vent the air from the downstream mouthpiece side of the closed cylinder valve and check that the gauge reading returns smoothly to zero.

Any pressure gauge that is suspect after carrying out the above should be examined by a qualified person before use. The gauge should be tested at regular intervals for accuracy and water tightness at intervals of not more than 1 year.

3.11   Flexible hoses

It should be ensured that hoses are suitable for the purpose for which they are to be used. Hoses are susceptible to attack by direct sunlight, oil contamination, high temperature, humidity, and seawater. Any prolonged tension on a hose increases its liability to cracking, crazing, and perishing. Hoses should be carefully examined at frequent intervals and before use for signs of crazing or cracking; if this appears on the surface, bending the hose will indicate the depth of cracking. If the crack penetrates the reinforcing ply, the hose should be replaced.

3.11.1   Storage

When not being used, hoses should be stored in cool, dry conditions, and in a circulating atmosphere, if possible. It is recommended that hoses should be loose in storage. Freshwater rinsing and drying is recommended after use. If the hoses are to be stored for a period exceeding 1 month, it is recommended that they be blown through with clean, dry air before storage and again before reuse.

3.11.2   End fittings

Regular attention to metal end fittings and their attachment to the hose is advisable. Screw threads and sealing surfaces should be kept clean and any sealing washers examined and replaced, if faulty. Care should be taken to prevent water entering the interior of the hose. The hose should be examined and pressure tested at intervals of not more than 2 years.

3.12   Rubber, fabric, and plastic components

Rubber, fabric, and plastic components of the breathing set and harness may be susceptible to attack by direct sunlight, high temperature, humidity, and oil contamination.

3.12.1   Examination

The external signs of deterioration in rubber, fabric, and plastic are crazing, cracking, stickiness, lack of elasticity, changes in color, or signs of abrasion. The appearance of one or more of these symptoms means that the end of the useful life of the component is near, and it should be replaced. Particular attention should be paid to components having a thin cross-section.

3.12.2   Storage

Before storage, all rubber, fabric, and plastic components should be rinsed in clean fresh water after use, especially where dirty conditions (oil or sewage contamination) have been encountered. Particular care should be taken at such points as buckles, eyelets, etc., where contaminants will tend to lodge. Rubber compounds should be lightly dust treated as recommended by the manufacturer.

Storage conditions should be cool, dry, shaded, and with a circulating atmosphere. Any prolonged tension on a rubber component increases its liability to cracking, crazing, and perishing; it is therefore advisable to avoid creasing as far as possible.

All webbing should be carefully inspected at periodic intervals for wear and tear, particular attention being paid to the stitching, fastening of buckles, etc., and to the operation of quick release devices.

3.13   Underwater breathing apparatus

The corrosive action of seawater and water-borne contaminants should never be underestimated and if precautions are not taken to clean the apparatus properly after use, serious damages may be caused to all parts of the apparatus while it is stowed away. It is worth remembering that even, when diving in apparently fresh water, there may be corrosive substances in solution such as chemical and petroleum wastes that are not noticeable at the time, but which will start corrosive action if left in contact with the apparatus.

3.13.1   Cylinders

- Care of cylinders: Accessories fitted to the cylinder, even if they are plated or of stainless steel, should be insulated from the cylinder by suitable means, either a plastic or nylon coating or a rubber sleeve. After use of underwater BA, particularly in seawater, the cylinder should be removed from its harness and boot and then washed carefully in clean, fresh water to remove all traces of saltwater and dirt, especially from any cracks. The cylinder and valve should be thoroughly dried. To ensure the good service of the cylinder valve, care should be taken when opening and closing it to ensure that excessive forces is not applied once the stop has been reached, as this causes damage to the internal components of the valve and could result in a leak through the valve. Before storage, or when the cylinder has been completely discharged and seawater may have entered the cylinder, the cylinder valve should be removed and the cylinder washed internally and externally in clean fresh water and thoroughly dried. This operation should normally be undertaken by an authorized person. The cylinder should not be stored with the valve downwards. Cylinders should be retested periodically in accordance with manufacturer's recommendations.

- Care after diving: The demand regulator (with the high-pressure inlet blanked off) reducing valve gauge and associated assemblies should be thoroughly rinsed in clean fresh water, and then allowed to dry naturally. Artificial heat or sunlight should not be used to accelerate the drying process.

3.14   Compressed air for human respiration

- Preparation of Compressed Air Cylinders: Cylinders should be internally and externally clean and free of scale or other foreign matters.
- Compression of Atmospheric Air: Atmospheric air may be compressed by means of suitable compressors to attain the air purity and pressure desired. Precautions should be taken to ensure that only uncontaminated air is admitted into the compressor intake. Attention should be paid to the location of the compressor intake and to the provision of suitable intake screening or filtration. Where compressors are driven by an internal combustion engine, every care should be taken, by extending the exhaust of the engine or the inlet of the compressor, to avoid the compressor drawing in the exhaust gases of the engine. The compressor manufacturer should be consulted concerning the maximum length and the minimum cross-sectional area' of such an extension to avoid reducing the efficiency of the engine or compressor. When compressors are being run in the vicinity of other machinery, adequate precautions should be taken to avoid intake of fumes from these machines. The maintenance and operation of compressors should be carried out in accordance with the manufacturer's instructions, particular attention being paid to the condition of piston rings, driers, filters, and accessories. No lubricant other than that recommended by the compressor manufacturer should be used. The air discharged from the compressor should be subjected to the processes necessary to achieve the degree of purity. At regular intervals, not exceeding 6 months and after major overhaul, a sample of the compressed air delivered by the compressor should carefully be tested by laboratory. Figures 3.4, 3.5, and 3.6 show different BA systems.

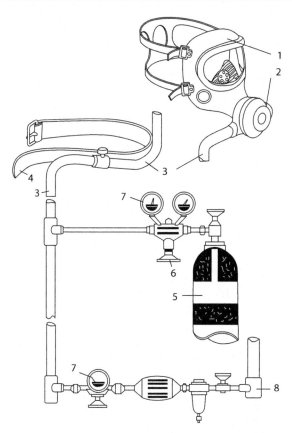

Figure 3.4 Demand type breathing apparatus system. (1) Facepiece; (2) Demand valve (Lung governed); (3) Medium pressure connecting tube; (4) Belt or body harness; (5) Compressed air cylinder; (6) Pressure reducer; (7) Pressure gage; (8) Compressed air line; (9) Filter.

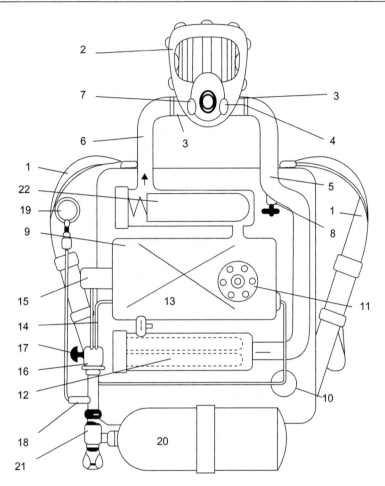

Figure 3.5 Self-contained closed circuit oxygen breathing apparatus compressed oxygen type. (1)Body harness; (2) Facepiece; (3) Equipment connector; (4) Exhalation valve; (5) Exhalation hose; (6) Inhalation hose; (7) Inhalation valve; (8) Saliva trap; (9) Breathing bag; (10) Warning device; (11) Relief valve; (12) Regeneration cartridge; (13) Flushing device; (14) Oxygen supply tube; (15) Demand valve (Lung governed); (16) Pressure reducer; (17) Supplementary oxygen supply valve; (18) Pressure gage tube; (19) Pressure gage; (20) Oxygen cylinder; (21) Cylinder valve; (22) Cooler.

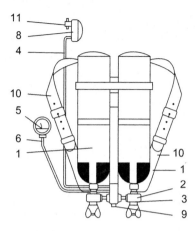

Figure 3.6. Self-contained open-circuit compressed air breathing apparatus demand type. (1) Compressed air cylinder; (2) Cylinder valve; (3) Pressure reducer; (4) Medium pressure connecting tube; (5) Pressure gage; (6) Pressure gage tube; (7) Facepiece; (8) Demand valve (Lung governed); (9) Warning device; (10) Body harness; (11) Equipment connector; (12) Breathing hose.

Personnel safety and protective equipment

Personal Protective Equipment (PPE) for firefighters refers to protective helmets, masks, boots, or other garments designed to protect firefighters or rescue workers from injury. Firefighter PPE is technically advanced to protect against the demands of firefighting and rescue. PPE addresses hazards from physical, electrical, heat, chemicals, biohazards, and airborne particulate matter. PPE is also used by rescue and emergency teams for search and rescue.

Firefighter boots or fire boots are designed to meet the tough demands of a firefighter providing comfort and high levels of protection. Firefighter boots are generally waterproof and provide breathable qualities while ensuring fire fighter safety in the harshest of environments.

Firefighter gloves need to offer firefighters a high level of heat resistance and protection against other risks including falling debris and potentially harmful chemical risks. Firefighter gloves also need to offer flexibility, dexterity, and overall comfort to the fire services.

Firefighter helmets are made from tough fiberglass are heatproof and worn when firefighting or carrying out rescue work. A fire helmet protects the firefighter's head from falling objects or banging their head on low beams. At the front of the firefighter helmet there is a visor that moves down to protect the firefighter's face from the heat and sparks of the fire.

Fire-resistant fabric also known as fire-retardant fabric is used in protective clothing for firefighters and primarily designed to protect fire fighters from flames, heat, and heat stress when fighting fires. Fire fabric has to meet tough international safety standards.

PPE storage products include boxes, cabinets, and racks that all help protect the PPE equipment when not in use. Employers are legally required to provide suitable storage to protect PPE from damage, contamination or loss, covered by the Personal Protective Equipment Regulations. PPE storage is therefore essential for any company where PPE is used.

4.1 Head protection

In general, a safety helmet must be worn where a person may:

Be struck on the head by a falling object
Strike his/her head against a fixed object
Inadvertently come into contact with electrical hazards

Personnel Protection and Safety Equipment for the Oil and Gas Industries.
DOI: http://dx.doi.org/10.1016/B978-0-12-802814-8.00004-3
© 2015 Elsevier Inc. All rights reserved.

4.1.1 Accessories

A wide range of accessories can be fitted to helmets to make them more suitable for variable working conditions. Examples include:

- Retaining strap worn under the chin or at the nape of the neck
- Bracket and cable clip for the attachment of a lamp
- Eye shield, face shield, or welding shield
- Wide brim for additional shade in hot climates
- Neck flaps for protection against weather, molten metal splash, hot substances, etc.
- Lining for cold conditions
- Ear muffs

Care should be taken to ensure that accessories and their attachment systems do not reduce the safety characteristics of the helmet nor adversely affect the balance or comfort of the helmet. Particular care should be given to the electrical resistance.

4.1.2 Selection

The following should be considered:

- Nature and location of the work
- Extent of adjustment for comfort
- Accessories must be compatible with the make of helmet used
- Sweat bands
- White helmets will provide better heat reflection and are easily seen in poor lighting conditions

4.1.3 Unsafe practices

The following practices are considered detrimental to the safe working life and performance of the helmet and must be avoided:

- Storage or placement of helmets near any window, particularly the rear window of motor vehicles, through which excessive heat can be generated. Helmets placed on the rear window ledge of motor vehicles may also become dangerous missiles in the event of an accident or when sudden braking occurs.
- Failure to follow manufacturer's cleaning instructions. The helmet may be damaged and rendered ineffective by chemicals such as petroleum and petroleum products, cleaning agents, paints, and adhesives, without the damage being visible to the user.
- Alteration, distortion, or damage to the harness or to the shell such as splits and cracks.
- Use of safety helmets for any other purpose than that for which they are designed, e.g., as seats, liquid receptacles, or wheel chocks.

4.1.4 Cleaning

It is recommended that safety helmets be cleaned regularly. In general, normal washing methods using warm water and soap are adequate. The use of solvents, very hot water, or harsh abrasives is not advisable.

4.1.5   Inspection and maintenance

All safety helmet components and accessories should be visually inspected prior to use by the user for signs of dents, cracks, penetration, or other damage due to impact, rough treatment, or unauthorized alterations that could reduce the degree of safety provided. Helmets showing damage or deterioration to the shell should be immediately withdrawn from service and discarded (completely destroyed). Helmets with sound shells but with damaged or defective harness components should be withdrawn from service and the complete harness and cradle replaced.

4.1.6   Working life

Excessive discoloration of the shell color or weathering of the surface may indicate a loss of strength. Helmets that have been in service for longer than 3 years should be thoroughly inspected and replaced as necessary. Plastic components of harnesses may deteriorate more rapidly under aggressive service conditions and in these cases harnesses should be replaced at intervals not longer than 2 years.

4.1.7   Helmet types

•   Type 1: These helmets should have a full brim.
•   Type 2: These helmets have no brim but may include a peak.

4.1.8   Helmet classes

•   Class A: These helmets are intended to reduce the force of impact of falling objects and to reduce the danger of contact with exposed low-voltage conductors. Representative sample shells are tested at 2200 volts (phase to ground).
    Note: This voltage is not intended to be an indication of the voltage at which the headgear protects the user.
•   Class B: These helmets are intended to reduce the force of impact of falling objects and to reduce the danger of contact with exposed high-voltage conductors. Representative sample shells are proof-tested at 20000 volts (phase to ground).
    Note: This voltage is not intended to be an indication of the voltage at which the headgear protects the user.
•   Class C: These helmets are intended to reduce the force of impact of falling objects. This class offers no electrical protection. All materials that come in contact with the user's head should be those generally known to be nonirritating to normal skin.

4.1.9   Construction

The construction of the helmet should be in the form of a hard shell having a smooth outer surface and fitted with a harness. The outer surface should be smoothly finished. All edges should be smooth and rounded. The shell may be shaped to form a brim and/or peak. If the shell is pierced with holes for any purpose other than for the attachment of the means of energy absorption, no internal chord of any such hole should exceed 4 mm, and the total area of such holes on either side of the helmet should not exceed 160 mm², making a total on both sides not exceeding 320 mm².

4.1.10   Physical requirements

Each helmet should consist of a shell and a means of absorbing energy within the shell. The harness should be securely attached to the shell. Provision should be made for ventilation between the headband and the shell. The shell should be generally dome shaped. There should be no holes in the shells of Classes A and B helmets that would cause the helmet to fail the electrical insulation test. Identification markers used on shells for Class B helmets should be affixed without making holes through the shell and without the use of any metal parts or metallic labels. The area under the peak or the front of the brim may be covered with a nonconducting antiglare material.

4.2   Headband, sweatband, crown straps, and protective padding

The headband, sweatband, crown straps, and protective padding should be made of any suitable materials that are comfortable.

Headbands should be adjustable in at least 1/8 hat size increments. The approximate size range that can be accommodated should be marked on the helmet in a permanently legible manner. When the headband is adjusted to the maximum designated size, there should be sufficient clearance between the shell and the headband to provide ventilation. Headbands should be removable and replaceable.

Sweatbands may be of the removable-replaceable type or may be integral with the headband. The sweatband should cover at least the forehead potion of the headband.

Crown straps, when assembled, should form a cradle for supporting the helmet on the user's head so that the distance between the top of the head and the underside of the shell cannot be adjusted to less than the manufacturer's requirements for that particular helmet. Protective padding may be used in conjunction with or in place of crown straps.

Mass: The mass of each helmet, complete with harness but exclusive of accessories, should not exceed 0.44 kg for Classes A, B, and C helmets.

4.2.1   Accessories

- Chin strap and nape strap: The chin strap and nape strap should be made of suitable material no less than 12.7 mm in width.
- Winter liners: The winter liner should be made of suitable materials. Colored materials should be colorfast. The outer surface may be water-resistant. There should be no metal parts in winter liners intended for use with Class B helmets.
- Lamp brackets: Headwear equipped with a lamp bracket should have a low-crown clearance for work in low-ceiling areas and should be made of lightweight, tough polycarbonate plastic material.
- Instructions: Each helmet should be accompanied by instructions explaining the proper method of adjusting the harness.

- Marking: Each helmet conforming to the requirements of this standard should bear identification on the inside of the shell stating the name of the manufacturer, the standard designation, and the class of the helmet.
- Labeling: A label should be attached to each helmet bearing the following information:

  The following words "for adequate protection this helmet must fit or be adjusted to the size of the user's head."

  This helmet is made to absorb the energy of a blow by partial destruction or damage to the shell and the harness or protective padding, and even though such damage may not be readily apparent, any helmet subjected to severe impact should be replaced.

  The attention of users is also drawn to the danger of modifying or removing any of the original parts of the helmet other than those recommended by the helmet manufacturer, and helmets should not be adapted for the purpose of fitting attachments in any way not recommended by the helmet manufacturer.

  Do not apply paint or solvents or adhesives or self-adhesive labels except in accordance with instructions from the helmet manufacturer.

4.2.2 Performance

Helmets should be certified by the manufacturer for the following tests in accordance with BS 5240 or ANSI Z89-1. No helmet that has been subjected to the testing should be offered for sale:

Shock absorption
Resistance to penetration
Electrical insulation
Resistance to flame
Water absorption

4.3 Recommendations for the material and construction of helmets

The materials used in the manufacture of helmets should be of durable quality, i.e., their characteristics should not undergo significant alteration under the influence of aging or of the circumstances of use to which the helmet is normally subjected, e.g., exposure to sun, rain, cold, dust, vibrations, contact with the skin, effects of sweat, or of products applied to the skin or hair. For those parts of the harness coming into contact with the skin, materials that are known to cause irritation should not be used.

For a material not in general use, advice as to its suitability should be sought before use. Any devices fitted to the helmet should be so designed that they are unlikely to cause any injury to the user in the event of an accident. In particular, there should be no metallic or other rigid projections on the inside of the helmet.

No part of the helmet should have sharp protruding edges. Where stitching is used to secure the harness to the shell, it should be protected against abrasion. No part of the shock-absorbing device should be capable of being easily modified by the user. A chin strap should have sufficient strength to maintain the helmet on the user's head in circumstance where helmet retention would otherwise be unreliable. If other protective equipment is designed to be used with a particular industrial helmet, that helmet should still comply with this standard when worn in conjunction with the designed equipment.

4.4   Method for measuring wearing height, vertical distance, horizontal clearance, and precautions concerning helmet use, maintenance, and testing

4.4.1   Headforms

Headforms for these measurements are in accordance with BS 6489 and are of sizes B, D, F, J, L, and N.

4.4.2   Procedure

Mount the helmet on a headform of appropriate size, leveled, and in the normal wearing position. If the size of the harness is adjustable to such an extent that the helmet can fit more than one of the sizes of headform, carry out these measurements twice, once at each extreme of the range of appropriate sizes of headform. Measure the wearing height, the vertical distance, and the horizontal clearance.

4.4.3   Painting

Caution should be exercised if shells are to be painted, since some paints and thinners may attack and damage the shell and reduce protection. The manufacturer should be consulted with regard to paints or cleaning materials.

4.4.4   Periodic inspection

All components, shells, suspensions, headbands, sweatbands, and accessories, if any, should be visually inspected daily for signs of dents, cracks, penetration, and any damage due to impact, rough treatment, or wear that might reduce the degree of safety originally provided. Any industrial helmet that requires replacement or the replacement of any worn, damaged, or defective part should be removed from service until the condition of wear or damage has been corrected.

   Note: All items constructed of polymeric materials are susceptible to damage from ultraviolet light and chemical degradation, and safety helmets are no exception. Periodic examinations should be made of all safety helmets and in particular those worn or stored in areas exposed to sunlight for long periods. Ultraviolet degradation will first manifest itself in a loss of surface gloss, called chalking. On further degradation the surface will craze or flake away, or both. At the first appearance of either or both of the latter two phenomena the shell should be replaced immediately for maximum safety.

4.4.5 Cleaning

Shells should be scrubbed with a mild detergent and rinsed in clear water approximately 60°C. After rinsing, the shell should be carefully inspected for any signs of damage. Removal of tars, paints, oils, and other materials may require the use of a solvent. Since many solvents may attack and damage the shell, the manufacturer should be consulted with regard to an acceptable solvent.

4.4.6 Precautions

Because helmets can be damaged, they should not be abused. They should be kept free from abrasions, scrapes, and nicks and should not be dropped, thrown, or used as supports. This applies especially to helmets that are intended to afford protection against electrical hazards.

Industrial protective helmets should not be stored or carried on the rear-window shelf of an automobile, since sunlight and extreme heat may cause degradation that will adversely affect the degree of protection they provide. Also, in the case of an emergency stop or accident, the helmet might become a hazardous missile.

The addition of accessories to the helmet may adversely affect the original degree of protection, then precautions or limitations are indicated by the manufacturer, they should be transmitted to the user and care taken to see that such precautions and limitations are strictly observed.

4.5 Impact system calibration procedures

4.5.1 Medium calibration

This calibration step should be carried out with a guided-fall system and an accelerometer mounted on the 3.64-kg falling mass. The accelerometer should have the following characteristics:

Minimum Range	0–125	g's
Maximum Resolution	1	g
Minimum Frequency Response (± 0.5 dB)	0.1–2000	Hz
Minimum Resonant Frequency	20	kHz
Linearity	1%	full scale
Repeatability and Stability	0.5%	full scale

The accelerometer should be mounted, according to the manufacturer's instructions, on the falling mass within 5° of true vertical. A suitable amplifier and peak meter (or equivalent devices) are required; a storage oscilloscope is recommended but not required. Mount a calibrating medium over the load cell. Drop the mass from at least 915 mm to strike the medium. The centers of the load cell, medium, mass, and accelerometer must be co-linear. A means to verify the velocity at impact should be used.

The values shown on two peak meters should read such that the acceleration value a, in g's, times the weight of the falling mass m equals the force value F within 2.5% (F = ma). This accuracy must be repeatable through at least five impacts.

4.5.2 System calibration only

A calibrating medium that has been tested in accordance with A.3.1 may be used without the accelerometer or guided mass. The force value obtained when the medium was tested according to A.3.1 should be recorded and this information provided with the calibrating medium. The calibrating medium is mounted over the load cell with the centers of both aligned. The mass is then dropped directly on the center of the medium. The force value obtained should be within 2.5% of that achieved during testing. The calibrating medium should be retested at least three times a year and more often if a significant change in force becomes apparent.

4.5.3 Static calibration

A rough determination of the calibration of the system may be obtained as follows:

1. Apply a known weight of at least 45 kg to it
2. Zero the peak meter and amplifier
3. If the amplifier has an adjustable time constant, move it to the longest setting available
4. Slowly add the weight, being careful not to impart acceleration to it

The peak meter should indicate the weight. This method should be used before each series of tests. An error in weight can indicate the need for a more sophisticated calibration check.

4.6 Application of safety hats and caps

Hats have a full brim and are designed primarily for use in industrial situations requiring additional protection around the back and sides (1) from falling objects and (2) from the weather (rain) (see Figure 4.1). Caps have a peak without a full brim and are designed primarily for use in tight or confined areas. Cap configuration minimizes the possibility of accidentally dislodging the cap from the head.

The cap provides head protection from impact hazards in industrial plants and, at the same time, provides capability for wearing hearing protection devices, and faceshields or welding helmets. They are widely used in construction industry, government, utilities, and manufacturing plants (see Figure 4.2).

Figure 4.1 Safety hat full brim.

Figure 4.2 A safety cap.

Figure 4.3 The hat suitable for linemen and utility crews.

Hats and caps provide head protection for personnel likely to encounter electrical contact as well as impact hazards, such as linemen, electric utility, maintenance crews, and electricians. The hat model is especially suitable for linemen and utility crews because of the more complete protection given to the neck and back (see Figure 4.3).

Aluminum hats and caps are used for head protection in various industries, especially for workers exposed to hot weather conditions, such as those in the petroleum, forestry, and construction industries (see Figure 4.4). Winter liners can be worn alone or under protective hats and caps to provide warmth in cold weather (see Figure 4.5).

Headwear with a lamp bracket for work in low ceiling areas (see Figure 4.6)

Figure 4.4 Aluminum Hats.

Figure 4.5 A Winter Liner.   **Figure 4.6** Headwear with a lamp bracket.

4.7 Eye protection

Eye-protective devices must be considered as optical instrument and they should be comfortable and carefully selected, fitted, and used. To give the widest possible field of vision, goggles should be fitted as close to the eyes as possible, without bringing the eye lashes in contact with the lenses. This section specifies material, design, and performance requirements of personal eye protection for industrial use and covers the following:

1. Eye protection for impact, dust, gas, liquid splashes, and combination of these that cover:
   Glasses type with and without side shield of plastic or tempered glass lenses
   Goggle type

Prescription lenses (glasses)
Sun-glare eye protection
2. Eye protection and backing lenses for welding and similar operations are included in Section 4 (face protection) of this standard.

4.7.1 Materials

- Corrosion resistance: Samples of all metal parts used in the eye-protector should show no sign of corrosion when viewed by the unaided eye of a trained observer and should be in a serviceable condition.
- Ignitability: When tested, no part of the eye-protector apart from headbands and textile edging should ignite or continue to burn after removal of the rod.
- Cleaning: When cleaned by the method recommended by the manufacturer, the eye-protector should show no visible deterioration.
- Skin irritation: All materials that come into contact with the user should be of a kind that is not known to cause skin irritation.
- Plastic material: Plastic material should have strength and elasticity suitable for the use and should not be flammable such as cellulose.

4.7.2 Design and manufacture

Eye-protectors should be free from patent defects. Eye-protectors should have no sharp edges and should be free from projections or other features likely to cause discomfort in wear. Headbands or harnesses, where provided, should have a width of no less than 9.5 mm. Adjustable parts or components incorporated in eye-protectors should be easily adjustable and replaceable. Where provided, ventilation features should be designed to prevent the direct access of any particle to the eye from any angle forward from the frontal plane of the eye-protector. Where eye-protectors have rims secured by a screw or screws, these should be penned, coated with adhesive, or otherwise treated or designed to ensure that they should not become loosened in use.

4.8 Lenses

The lens appearance should have smooth surfaces and have no visible flaws, striae, bubbles, waves, and other foreign objects in or on to it. Lenses both serve to afford the vision required for work and to protect the eyes during the performance of specific activities. There are limits to which both sets of requirements can be met at one and the same time. Since the use of eye-protectors always involves a certain degree of inconvenience of restriction in movement, in order to guarantee reliable protection, it is imperative that the properties of lenses undergo no substantial alteration during use.

Lenses should be made of plastic materials, of toughened glass or laminated glass, or any combination of these materials, or untreated glass.

4.8.1 Optical properties

- Conditioning: Lenses should be conditioned in accordance with BS 2092.
- Light transmittance: Lenses should transmit no less than 80% of the light energy within the visible spectrum unless they are in the impact resisting group and are double-layered, in which case the transmission should not be less than 70%. These limits should not apply to lenses claimed to be tinted.
  Note: Tinted lenses include those with metal coatings applied.
- Quality: Lenses should be free to within 3 mm of their edges from inherent faults that can be observed by the user when the eye-protector is worn. Inspection for faults should be done by the user with his eyes focused at a variety of focal distances likely to be encountered at work, i.e., the user should not attempt to focus on the lens itself. Where mold or crease lines are a design feature of the lens they should not occur within the minimum dimensions.

4.8.2 Construction and dimensions

The eye protection should conform to the following general requirements:

Eye-protector should not be excessively uncomfortable to the user.
Lenses of eye-protector should not easily come off from the frame nor reform their curve.
Each part of the eye-protector should be easily replaced.
- Eye-protector similar to glasses: Eye-protector of this type should consist of two lenses, frame, and two bows.
- Eye-protector with side-shield: Eye-protector of this type should be the one similar to the usual glasses with side-shield attached in a way that does not excessively obstruct the user's view.

4.8.3 Dimensions of lenses

The minimum dimensions of lenses should be as follows:

- For circular lenses: 48 mm diameter with a minimum aperture size of 40 mm diameter.
- For shaped lenses: 42 mm horizontal datum length × 35 mm mid. datum vertical depth, using the system of measurement described in BS 3199.
- For one piece rectangular lenses: 105 mm × 50 mm.
- For one piece shaped lenses: such that two circles 48 mm in diameter can be spaced symmetrically about the vertical center line of the eye-protector with the centers being 66 mm apart measured in the horizontal front plane of the eye-protector as worn. Refractive, astigmatic, and prismatic power for focal lenses, when measured by telescope in accordance with BS 2092. Eye-protectors should comply with the tolerances given in Table 4.1.

Spherical and astigmatic powers should be within the specified limits at all points on the lens lying within 25 mm of the test point. Individual lenses for glasses or goggles having separate eyepieces, should comply with Table 4.2. For impact-resistant eye-protectors the combined prismatic imbalance in the vertical direction should not exceed 0.30 for refractive, astigmatic, and prismatic powers for prescription lenses. Prescription lenses should comply with relevant standard.

Table 4.1 **Tolerances for eye-protectors**

Type of protector	Spherical effect D	Astigmatism D	Prismatic Effect
Impact All other eye-protectors	±0.12 ±0.06	±0.12 ±0.06	0.25 0.15

Note: The unit of power is the dioptre (symbol). (see BS 3521).
The unit of prism power is the prism diopter.

Table 4.2 **Classification and uses of sunglasses**

Classification	Use
Cosmetic Spectacles	Lightly tinted spectacles not intended to give significant protection against sun glare and worn largely for their fashion properties.
General purpose	Sunglasses intended to reduce sun glare in bright circumstances including the driving of motor vehicles in daylight.
Special purpose	Sunglasses intended to reduce sun glare in abnormal environmental conditions, e.g. near large expanses of water or in snow and mountain altitudes, or for persons who may be abnormally sensitive to glare as a result of medical treatment or otherwise. Non-photochromic filters having a shade number of 4.1 are not considered suitable for use by persons when driving motor vehicles.
Refraction Class 1	Equivalent to prescription lens quality and recommended for continuous daytime wearing.
Refraction Class 2	Suitable for intermittent wearing.
Break resistant sunglasses	Suitable for conditions where mechanical abuse is possible but will not be severe, e.g. driving, cycling, walking, camping or boating.

4.8.4   Performance

Eye-protectors should be subjected to tests. Replacement lenses should be subject to the relevant tests when mounted in an appropriate housing. Prior to testing eye-protectors should be conditioned as described in BS 2092.

4.8.5   Robustness of construction

When tested as described in BS 2092, eye-protectors should not shown any of the following defects:

Lens fracture
Lens deformation
Lens housing and/or frame failure
Lateral protection failure

4.8.6   Protection against impact

- Type of eye-protector: Grade 1 impact eye-protectors should be goggles or face shields only.
  Note: Glasses are specifically excluded from grade 1.
- Impact eye-protectors: When tested as described in BS 2092 using a velocity of impact of
  45 m/s for grade 2 and a velocity of impact of 120 m/s for grade 1, impact eye-protectors
  should not show any of the following:
  Lens fracture
  Lens deformation
  Lens housing and/or frame failure
- Lateral protection of impact eye-protectors: The lateral protection of impact eye-protectors
  should comply with the requirements for either grade 1 or 2 as in 6.6.4.2 or for robustness
  of construction as in 6.6.3. If the lateral protection of any eye-protector has a lesser impact
  resistance than that of its lenses, the eye-protector should be marked accordingly. When
  the lateral protection is tested in accordance with BS 2092 it should be considered to have
  failed to meet the particular impact grade or general robustness claimed with respect to the
  associated lens if it shown any of the defects listed in 6.6.3.(d).

4.8.7   Protection against molten metals and hot solids

- Type of eye-protector: Molten metals eye-protectors should be nonmetallic or should be
  treated to prevent molten metals adhering to the lenses or other parts of the eye-protector
  when tested as described in BS 2092. They should include goggles and face screens.
- Ocular area (face screens): Face screens should cover the ocular area defined in BS 2092.
  When assessed by the method described in 6.6.4.1 should apply only to that part of the face
  screen that provides protection to this ocular area.
- Hot-solids penetration: Complete penetration of the lenses and housings of goggles or brow
  guards and helmet mountings of face-screens should not occur within 7s when tested as
  described in BS 2092. Complete penetration of face-screens should not occur within 5s.

4.8.8   Protection against liquids

- Liquid droplets: When tested as described in BS 2092 eye-protectors for protection against
  liquid droplets should be deemed to comply with this standard if there is no coloration of
  the paper representing the ocular areas.
- Liquid splashes: When tested as described in BS 2092 eye-protectors for protection against
  liquid splashes should be considered to comply with this standard if they cover the ocular
  area as described.

4.8.9   Protection against dusts

When tested as described in BS 2092 eye-protectors for protection against dusts
should be deemed to comply with standard if the reflectance of the white test paper is
no less than 80% of that before the test.

4.8.10   Protection against gases

When tested as described in BS 2092 eye-protectors for protection against gases
should be deemed to comply with this standard if no staining appears on the area
enclosed by the eye-protector beyond the permitted limits.

Note: Eye-protectors for protection against liquid droplets, dusts, and gases are tested for resistance to entry. Face screens for molten metal and liquid splashes are assessed for coverage on a head-form that does not attempt to cover all head sizes. Great care should be taken to ensure a proper fit or adequate coverage on the individual user.

4.9 Marking

Eye-protectors complying with this standard should be clearly and permanently marked. The marking should not be placed in such position that it might be confused with other information. Adhesive labels, if used, should not be easily removed. The manufacturer's name or its abbreviation should be marked on the lens surface in a permanent way but not affecting the user's view. The following should be marked:

> Manufacturer's name
> Date and the standard used

4.10 Sun-glare eye protection

The main purpose of sun-glare filters is to protect the eyes from excessive solar radiation so as to reduce eye strain and increase visual perception in order to ensure fatigue-free vision, especially for prolonged use. The choice of filter depends on the ambient light level and the individual's sensitivity to glare. Table 4.2. shows the classifications and uses of sunglasses.

4.10.1 Transmittance

- General requirements
  Shade numbers and transmittance values: Shade numbers and transmittance values of filters should be as given in BS 2724. Transmittance values should be determined in accordance with BS 2724
  Spectral transmittance: The mean spectral transmittance for the wavelength range 380 nm to 500 nm should not exceed $1.2\,\hat{o}v$ when determined in accordance with BS 2724.
  Note: Filters should have a mean spectral transmittance over this spectral range less than the spectral transmittance for the wavelength range 450 nm to 650 nm should be less than $0.2\,\tau$ v when determined in accordance with BS 2724.
  Uniformity of luminous transmittance: Apart from a marginal zone 5 mm wide, the difference in luminous transmittance when determined in accordance with BS 2724 between any two points on the filter should not be greater than 10% of the higher value.
  Note: For gradient filters this requirement applies in a section perpendicular to the gradient. For mounted filters the difference between the luminous transmittance of the filters at the visual center for the right and left eye should not exceed 20% of the higher value.
  Recognition of signal lights and colors: Each tinted filter should have a relative visual attenuation coefficient no less than 0.8 for each of the four signal colors specified when determined in accordance with BS 2724.

4.10.2   Additional requirements for special filters

- Photochromic filters: When tested in accordance with BS 2724 photochromic filters should be classified according to their luminous transmittance in the clear state, $\tau0$, and in the darkened state, $\tau1$, and their spectral transmittance values should be as given in 6.9.2, where $\tau1$ enable the filters to comply with c and d in clear and dark states. The ratio of luminous transmittances $\tau0$ /$\tau1$ should be greater than 1.25. When a representative sample of photochromic filter is tested in accordance with BS 2724, the relative change in luminous transmittance $\tau1$ /$\tau2$ should not exceed 5% for the determined value for the clear state and 20% for the determined value for the darkened state.
- Polarizing filters: When tested in accordance with BS 2724 sunglasses fitted with polarizing filters should not show a deviation from the vertical of greater than ±5° for the plane of polarization of the filters in the frame. The misalignment between the plane of polarization of the left and right filters should not be greater than 6°. The ratio of values of luminous transmittance determined with light polarized parallel and perpendicular to the plane or polarization of the filter should be greater than 20:1.
- Gradient filters: Shade numbers for gradient filters should be determined by the highest and lowest values of transmittance within a distance 15 mm above and below the center of the filter for nonmounted filters or the visual point for mounted filters.
- Infra-Red filters: If a filter is claimed to attenuate infra-red radiation associated with daylight, the mean infra-red transmittance should not exceed τv when determined in accordance with BS 2724.

4.10.3   Refraction properties

- Unmounted filters: When tested in accordance with BS 2724, the values for refractive, astigmatic and prismatic powers for unmounted filters should be as given in BS 2724. When examined in accordance with BS 2724, Class 1 unmounted filters should show no local distortion effect to within 2 mm of the edge of the filter.
- Mounted filters: When tested in accordance with BS 2724, values for mounted filters should be as given in BS 2724 Tables 2 and 3 for differences between prismatic power of each pair of mounted filters. When examined in accordance with BS 2724, Class 1 mounted filters should show no local distortion effect within the full clear aperture of the filters.

4.10.4   Quality of filter material and surface

- Freedom from visible defects: When filters are examined in accordance with BS 2724, they should be free, within an area 15 mm in radius about the visual point, from defects that affects their suitability for use, e.g., bubbles, striae, inclusions, scratches, digs, mold marks, and distortion due to surface irregularity.
- Light diffusion: When tested in accordance with BS 2724, the reduced luminance coefficient of unused filters should not exceed 0.5 cd/m²lx.

4.10.5   Stability

- Thermal stability: After treatment in accordance with BS 2724, filters should not experience any change in their properties. The relative change in luminous transmittance should be less than 5% for shade numbers 1.1. to 3.1 and less than 10% for shade number 4.1.

- Radiation stability: After exposure to radiation in accordance with BS 2724, the filter should comply with standards. The relative change in luminous transmittance should be less than 5% for shade numbers 1.1 to 3.1 and less than 10% for shade number 4.1.
- Flammability: When tested in accordance with BS 2724, plastics filters should neither ignite nor continue to glow when removed from the oven.

4.10.6   Frames

- Design and manufacture: Frames should be free from obvious defects and should be smoothly finished with no sharp edges or projections likely to cause discomfort or injury to the user.
- Materials: All materials that come into contact with the user should not be known to cause skin discoloration or irritation.
- Flammability: When tested in accordance with BS 2724, frames should neither ignite nor continue to glow when removed from the oven.
- Security of filters: Filters should be firmly and securely fitted to the frame. If the frame is fitted around the filters there should be no gaps greater than 0.1 mm between the filter edge and the frame, unless they are a design feature.

4.10.7   Mechanical strength of break-resistant, robust, and impact-resistant sunglasses

- Break-resistant sunglasses: When tested in accordance with either BS 2724 (unmounted filters) or (mounted filters), a filter should not break through its thickness into two or more pieces, and more than 30 mg of filter material should not become detached from the side remote from that receiving the load or from the concave side, as appropriate.
- Robust sunglasses: When tested in accordance with BS 2724, sunglasses should not suffer filter displacement, the filter should not break through its thickness into two or more pieces, and more than 30 mg of filter material should not become detached from the side remote from that struck by the ball.
- Impact-resistant sunglasses: When tested in accordance with BS 2724 sunglasses should not suffer filter displacement, the filter should not break through its thickness into two or more pieces, and more than 30 mg of filter material should not become detached from the side remote from that struck by the ball. The side-pieces of the sunglasses should not permit penetration of the ball through the material of the side-piece and they should not fracture through their thickness or produce jagged projections of side-piece material that could reasonably be expected to damage the eye.
- Frames: When tested in accordance with BS 2724, the frame and bridge of break-resistant, robust, and impact resistance sunglasses should not show any breaks, tears, hairline cracks, sharp edges,: or points.

4.10.8   Information and labeling

- Information: The manufacturer or supplier of sunglasses should provide the following information to be used on the packaging, as leaflets, by means of labeling or on a display card so that the information is available:
  Manufacturer's or supplier's identification mark
  Classification for recommended use and shade number(s) or nominal luminous transmittance(s)

Notes:

1. Additional information on transmittance values is desirable but not mandatory.
2. For gradient or photochromic filters, classification for recommended use should be determined by the lesser value of luminous transmittance.
   Classification of refractive quality
   Classification of mechanical strength
   Warning stating that sunglasses are not intended to be used to view the sun directly
   Where appropriate, warnings related to use, e.g., if the lenses are too dark for driving or if the sunglasses are unsuitable for use in solaria
   • Labeling: Sunglasses should be marked or labeled to show the following:
   Number and year of official standard
   Name, trademark or other identification of the manufacturer or supplier
   Classification for recommended use
   Classification of refractive quality
   Classification of mechanical strength

4.10.9 Performance and quality assurance

Manufacturers of sun-glare eye-protectors should certify in writing that sun-glare eye-protectors are tested in accordance with relevant official of this standard or BS 2724 for the following:

- Test for determination of transmittance for all sun-glare filters, the axis of polarization of polarized filters, and fatigue of photochromic filters standard; BS 2724
- Test of refractive, astigmatic and prismatic power; BS 2724
- Test for quality of material and surface; BS 2724
- Test for filter material stability and mechanical strength; BS 2724

4.11   Prescription safety lens glasses

Employees using prismatic, astigmatic, and refracture prescription lenses working in an area and performing any types of work that require eye protection such as chemical handling, chipping, welding, grinding, laboratory, machining, spot welding, furnace operation, and in a risk of harmful effect of ultraviolet, infra-red, and laser beams should be protected by either safety prescription goggles or safety swing-up type lenses or cover goggles to be worn over ordinary prescription lens glasses.

4.11.1 Optical tests

Employees who should wear corrective lens glasses should be tested and prescribed by ophthalmologist. Glasses should be fitted with prescribed lenses in accordance with specifications covered in ISO 4855. The supplier should certify in writing that the safety glasses are tested as prescribed and meets all requirements for impact protection.

4.11.2 Classification

Prescription lens glasses should be classified into two types according to the shape of frame mounting:

1. Conventional glasses type
2. Glasses type with side shields

4.11.3 Material

Material, exclusive of tempered glass lenses, should meet the following requirements:

They should have suitable strength and elasticity for intended use
Material of parts to contact the skin should be nonirritating and capable of being disinfected
Metal parts should be made of corrosion-resistant material or treated as corrosion-resistant
Plastic material should not be fast burning

4.11.4 Construction

The general construction of the glasses should satisfy each of the following requirements:

It should be simple of handling and not to break easily.
It should not give remarkable discomfort to the user.
It should be free from sharp edges or projections likely to cause cuts or scratches to the user.
Every part of glasses should be easily removed and replaced.

Conventional glasses type should be composed of two lenses a frame and temples.
Glasses with side shields should be of conventional Glasses type fixed firmly with side shields that should obstruct visual field as little as possible.

4.11.5 Quality

• Impact resistance: The glasses if/when subjected to testing should have neither the lens edge chipped nor the lens displaced from the frame by an impact.

4.11.6 Lenses

Lens should be free from any visible flaws, striae, bubbles, waves, and foreign bodies, and both surfaces of it should be well polished. The lenses should be checked by ophthalmologist to make sure they are made as prescribed before issuing to the employees. Lenses if/when subjected to the tests specified in accordance with BS 2738 should not be fractured. If lenses supplied as pair, the two lenses should be reasonably matched in shape, size, and form.

4.11.7 Cover lens

Ordinary prescription lens glasses can be fixed with a swing-type cover safety lenses or employees using prescription lenses. Glasses can be protected by all plastic soft sided cover goggle with shielded vents or appropriate face shields.

4.11.8 Description of faults sometimes found in glass and plastics used for eye-protectors

- Thermoplastics: The following faults may be found in transparent thermoplastics:

  Bubble: This can be totally within the sheet material or complete and on the surface. If broken and on the surface the bubble appears pit-like.

  Dust: Fine inclusions in the surface coating are sometimes referred to as haze.

  Gel: Nonhomogeneous areas in the sheet, normally undissolved polymer. In surface coating, usually undissolved lacquer, etc.

  Inclusion: This is a foreign particle usually in the sheet; commonly 'black species' are small particles of degraded polymer.

  Line: A mark usually in the direction of extrusion with the surface unbroken. Also known as extrusion lines or hairlines.

  Organ peel: The surface has an appearance as indicated by the description; a term associated with the poor surface finish of a molding tool.

  Ridge: This is an undulation normally at right angles to the direction of the sheet. It can be associated with local thickness variations.

  Run: This is a coating run on a coated sheet.

  Scratch: This is a surface mark of an open nature from mechanical damage if at the sheet production stage, normally running in the direction of extrusion.

  Surface gel: There is no common description. This is a small area of molten polymer on the surface of sheet.

  Watermark: This is a less common fault with extruded sheet. Normally it is a repeated pattern at right angles to the direction of extrusion, invariably in edge areas of original extrusion widths.
- Glass: The following faults may be found in glass:

  Bubble: This is a large gaseous inclusion generally over 0.25 mm in size.

  Check: This is a small fracture penetrating below the surface generally less than 1 mm in length.

  Chip: This is evidence of small fragments broken from a surface.

  Draw line: Straight lines that occur in drawn sheet glass.

  Fled: This is a fracture like a check but penetrating generally more than 1 mm.

  Inclusion: This is a nonglassy particle within glass. This may be further identified using the terms: bubble, seed, or stone.

  Organ peel: A term associated with imperfectly polished glass giving the surface an appearance as indicated by the description.

  Seed: A small gaseous inclusion generally less than 0.25 mm in size.

  tone: An opaque solid inclusion.

Note: Stones result from unmelted glass raw material or fragments of refractory material used in the construction' of the melting tank. These gradually dissolve into the glassy mix, and on occasion they may be completely dissolved but leave a transparent sac within the glassy mix. Such an inclusion is known as a knot.

  Vein: Veins are glass inhomogeneities within glass, usually having directional characteristics associated with the method of forming a particular glass product.
- General terms used for faults in optical materials: The following general terms are used for faults in optical materials:

  Dig: This is a small defect in the polished surface due either to an inclusion breaking surface or to some abusive damage.

  Scratch: This is the rupture of the polished surface through abrasion by one or more hard particles.

  Wave: This is the local geometrical distortion of the lens surface deviating rays of transmitted light. Table 4.3 shows kinds and types of eye-protectors.

Figure 4.8 shows different types of safety glasses.

Table 4.3 **Kind and types of eye-protectors**

Kind	Type			Symbol for Figure 4.7
Shield Eyeglasses	Spectacles Type	Without side Shield	Common eyeglasses type Single swing up type Double swing up type Safety helmet mount type	A-1 A-2 A-3 A-4
		With side shield	Common eyeglasses type Single swing up type Double swing up type Safety helmet mount type	B-1 B-2 B-3 B-4
	Front type		Fixed type Swing up type	C-1 C-2
	Goggle type		Box type Cup type	D-1 D-2

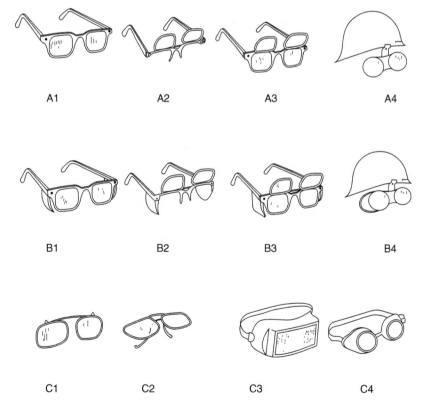

A1   A2   A3   A4

B1   B2   B3   B4

C1   C2   C3   C4

Figure 4.7 Kind and types of eye-protectors.

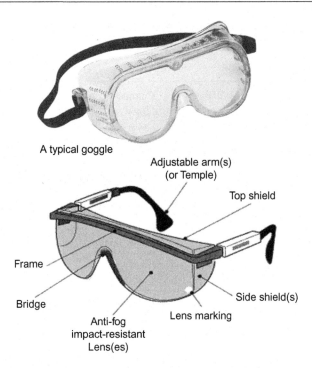

Figure 4.8 Different types of safety glasses.

4.12   Requirements for eye, face, and neck shield protection

This section specifies requirements for equipment to protect an operator above the shoulders against harmful splashes, flying particles, and radiations when engaged in welding, cutting and similar operations. The equipment is designed to use protective filters with or without filter covers. Eyes and face can be protected from injuries caused by the above named factors if any appropriate equipment is made available and worn by the employees. Face and neck protections against harmful liquid splashes and other hazards such as sand blasting will be covered in body protection.

4.12.1   Classification of protection requirements

For the purposes of this standard, operations in welding should be grouped into the following classes in ascending order of protection requirements:

- Class 1: Covering work, other than actual welding, in the vicinity of welding operations, where some protection from harmful radiation is required, but where good general vision is also necessary, e.g., work of supervisory staff and erectors.

   For Class 1 operations, protection is provided by glasses, goggles, face shields, hand shields, helmets, or fixed shields.

- Class 2: Covering gas welding and cutting, which involve direct exposure to radiation of heat and light, sparks and particles of metal and where moderate reduction of transmitted, ultraviolet and visible radiation is necessary. For Class 2 operations, protection is provided by goggles, face shields, hand shields, helmets, or fixed shields.
- Class 3: Covering electric-arc welding, cutting and similar processes involving direct exposure to high-intensity radiation, sparks and particles of metal, together with the risk of electric arcing from tools. In this work, a large reduction in the ultraviolet infra-red and visible radiation is necessary.

  For Class 3 operations, protection is provided by face shields, hand shields, helmets, or fixed shields. Neck shields may also be necessary.
- Class 4: Covering gas-shielded arc welding and cutting involving exposure to large amounts of ultraviolet infra-red and visible radiation both from direct radiation and by reflection together with particles of metal ejected from the arc region.

  For Class 4 operations, protection is provided by helmets as for Class 3 but with provision for an auxiliary heat-absorbing filter. Neck shields are also sometimes necessary.

4.12.2   Optical quality

Filters, cover lenses, and backing lenses should be free to within 5 mm of their edges from inherent faults that can be observed by the user when the equipment is worn. Inspection for faults should be made by the user with his eyes focused at a variety of focal distances likely to be encountered at work. There should be not attempt to focus on the filter cover lens or backing lens itself.

4.12.3   Protection against radiation

- Filters and backing lenses: Each filter and backing lenses incorporated or intended for use with the equipment should comply with BS 679.
- Replacement: Except for one-piece goggles, filters and filter covers should be capable of replacement without the use of special tools.

4.13   Design and manufacture

The field of view should not be obstructed except by the boundaries of the filter holders, if any. All parts of the equipment should be free from sharp edges or projections that could cause harm or discomfort to the user. Where equipment is designed for replaceable filters, the housing provided should take filters of one of the sizes specified in relevant standards. Filter housings and, in the case of glasses, frames, and lateral protection should be provided at least the same protection against radiation as that given by the filters.
Note: A consequence of this requirement is that users should exercise care when replacing filters with those of a darker shade than those originally fitted or supplied.

In the case of equipment for Class 4 duties, provision should be made for accommodating a filter cover and two filters, the latter separated by a 1 mm thick spacer. The design of the equipment should not allow the direct entry of any particle or stray radiation to the inside from any angle forward of the plane of the rear edge of the equipment. All metal fittings that are likely to be exposed to radiation during use and that come into contact with the operator should be insulated to reduce thermal conductivity.

4.14   Robustness

All equipment should be designed and constructed to withstand a test for general strength of construction consisting of the impact of a 6.35 mm steel ball traveling at a velocity of 12.2 m/s. The test should be applied to the equipment fitted with filter(s) and backing lens(es) as normally offered for sale. When so tested, the equipment should not show any failure or deformation.

4.14.1   Filter/Backing lens failure

A filter or backing lens should be considered not to comply with this standard if the ocular nearest to the eye of the test headform cracks through its entire thickness into two or more separate pieces or if more than 30 mg of material becomes detached from the surface of the filter or backing lens nearest to the eye of the test headform.

4.14.2   Filter/Backing lens deformation

A filter or backing lens should be considered to have deformed when a mark appears on the white paper appropriate to the striking face of the ball.

4.14.3   Filter housing or frame failure

A filter housing or frame should be considered not to comply with this standard if it is fractured, if its parts separate or if it allows the filter or backing lens nearest to the eye of the test headform to be knocked from it housing or frame.

4.14.4   Glasses frame failure

A frame should be deemed not to comply with this standard if it fractures or its parts separate when the steel ball is projected three times on the hinge pin of the earbows and three times on the junctions of the bridge and lens holders.

4.15   Faceshields and helmets

Where movable filters are employed, e.g., filters hinged at the top, the design should be such that on failure of the device, the user should be protected against radiation, i.e., it should be "fail safe."

4.15.1   Variable-shade observation lens

Where a variable-shade observation lens is employed in addition to the welding filter.

Variable-shade observation lenses enable the welder to commence the welding operation with the helmet or hand shield in position but with a clearer view than would be obtained through a high-shade filter but should not be used to view welding operations.

Table **4.4** **Shade numbers and permitted response times**

Shade no. in "DARK" state	Maximum response time Ms
3 to 10	100
11 to 12	10
13 to 14	1
15 to 16	0.1

The shade number of the observation lens in its darkened state should be stated on the equipment, together with a warning to the effect that welding filters of a higher shade number than this should not be used in the equipment. The variable-shade observation lens should revert to its darkened state in the event of failure, i.e., the device should be "fail safe" in the event of loss of power or obstruction of the sensing device.

The response or activation time, i.e., the time between that taken to reach the 50% level of the maximum arc intensity and the time at which the variable-shade observation lens attains a density within a visual density of 0.5 of that in the dark state, from light to darkened state of the variable-shade observation lens should not exceed the values given in Table 4.4.

The transmission properties of the variable-shade observation lens should be as follows:

The nonvisual densities in light or darkened state should be greater than or equal to the minimum values specified in BS 679: 1977 for the darkened state shade of the device.

The visual density in the darkened state should be greater than or equal to the minimum values given in Table 4.4 for that shade number. The visual density in the light state should not be less than that of a shade number 3 filter.

4.16   Size of filter holder and filters

A filter holder should hold securely filters and filter covers of the appropriate size. Filters should have minimum dimensions of 105 mm × 50 mm, nominal.
Note: The preferred sizes of filters are as follows:

108 mm × 51 mm, nominal
108 mm × 82 mm, nominal

The actual cutting size should not differ from the nominal size by more than 1 mm.
Note: Care should be taken to minimize confusion between sizes.

- Hand grips for hand shields: Hand shields should be provided with a hand grip that should either be fixed inside the shield or be provided with other means of protection for the hand.
- Electrical insulation: The value of electrical insulation between any metal part of the exterior and any part of the interior surface should not be less than 500 000 when tested as described in 7.12.(3) using a testing voltage of 500 V d.c. With the exception of the harness

attachment to the shield, metal components used in the construction of the head harness should be exempt from this requirement.

- Neck Shields: Neck shields should comply test as described in Appendices A to E of BS 679.
- Fixed Shields: Fixed shields are for special needs and should be exempt from the requirements of 7.4.3 and disinfection. Table 4.5 shows visual density for various shade numbers.

Note: Filters, cover lenses, and backing lenses for use during welding and similar operation should comply with BS 679-1989 or ISO 4850, 4851, and 4852.

- Marking: All equipment should be clearly and permanently marked with the following:
  Manufacturer's name, trademark, or other identification mark
  Highest class number for which the article is suitable
  Number of the standard
    - Filters should be permanently marked with the following:
  Manufacturer's name and trademark

Table 4.5 Visual density

Shade no.	Visual density	
	Min.	Max.
1.2A	0.00	0.13
1.2	0.00	0.13
1.4	0.13	0.24
1.7	0.24	0.36
2.0	0.36	0.54
2.5	0.54	0.75
3	0.75	1.07
4	1.07	1.49
5	1.49	1.92
6	1.92	2.36
7	2.36	2.80
8	2.80	3.21
9	3.21	3.64
10	3.64	4.07
11	4.07	4.49
12	4.49	4.92
13	4.92	5.36
14	5.36	5.80
15	5.80	6.21
16	6.21	6.64

Note: Filters, cover lenses and backing lenses for use during welding and similar operation should comply with BS 679-1989 or ISO 4850, 4851 and 4852.

Figure denoting shade number or if variable-shade filter, the shade number in the light and dark states. If dual-shade filter is used the shade number of each zone. If variable shade, the following warning:

Do not use a variable-shade observation lens for viewing welding operations.

• Disinfection: Each piece of equipment should be disinfected by the user by immersion in a 1% (v/v) solution of dodicyl (deaminoethyl) glycine hydrochlorident or similar in tap water for 10 minutes.

• Tests: The manufacturer should certify in writing that the following tests have been carried out:

1. Resistance to corrosion test for metal parts (BS 1542)
2. Ignitability (heated rod) test (BS 1542)
3. Electrical insulation test for helmet and hand shield (BS 1542)
4. Filter test Appendices A to E (BS 679) or equivalent (BS 1542)

4.17 Hand protection

This section specifies the minimum requirements for material manufacturing details and performance requirements for gloves that afford protection to the hands and as appropriate to the wrists of the user when carrying manual operations that are common in oil and gas industries work places. For the purposes of this standard, industrial gloves are divided into the types given in Table 4.6. Figure 4.9 shows additional design features for leather and fabric gloves. Table 4.7 shows different classification of hazards.

Table 4.6 Glove types

Type number	Description
1	Flesh split leather inseam gloves, gauntlets, mitts and one finger mitts
2	Grain leather inseam gloves, gauntlets, mitts and one finger mitts
3	Fabric gloves with leather palms
4	Inseam gloves and gauntlets made wholly from fabric
5	Leather outseam armoured gloves and gauntlets
6	Lightweight PVC supported gloves with a rough finish
7	Lightweight PVC supported gloves with a smooth finish
8	Standard weight PVC supported gloves with a smooth finish
9	PVC gloves with a granular finish
10	Flock lined unsupported PVC gauntlets
11	Unflocked, matt finish, unsupported PVC gauntlets
12	Unlined rubber gloves or gauntlets
13	Flock lined rubber gloves or gauntlets
14	Fabric lined rubber gloves or gauntlets
15	Rubber gloves or gauntlets, fabric or flock lined or unlined with additional rubber reinforcement over the whole or part of the hand

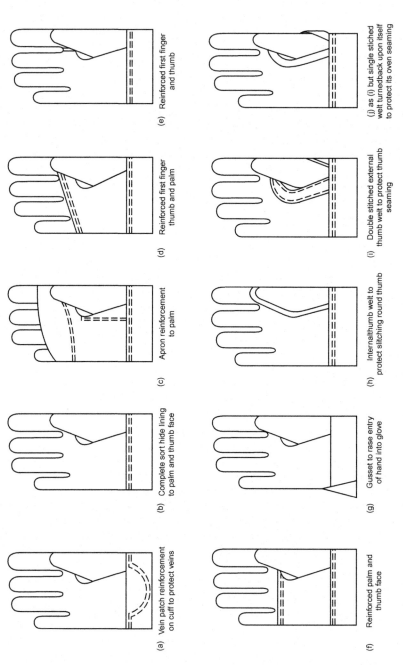

Figure 4.9 Additional design features for leather and fabric gloves.

(a) Vein patch reinforcement on cuff to protect veins

(b) Complete sort hide lining to palm and thumb face

(c) Apron reinforcement to palm

(d) Reinforced first finger thumb and palm

(e) Reinforced first finger and thumb

(f) Reinforced palm and thumb face

(g) Gusset to rase entry of hand into glove

(h) Internal thumb welt to protect stitching round thumb

(i) Double stitched external thumb welt to protect thumb seaming

(j) as (i) but single stiched welt turnedback upon itself to protect its oven seaming

Table 4.7 Classification of hazards

Hazard	Typical operations	Suitable gloves	Glove type no.
Heat but no serious abrasion	Furathue work, drop stamping, casting, Forging, handling hot objects	Heat resistant leather wrist gloves	1,2
		Heat resistant leather inseam mitts	1,2
		Heat resistant leather gauntlets	1,2
		Heat resistant leather gauntlets with canvas cuffs	
		Felt mitts, palms faced with canvas or heat resistant leather	1,2
		Loop-pile gloves or gauntlets	3,4
			4
Heat and abrasion	Riveling, hot chipping	Heat and abrasion resistant leather gauntlets and mitts	1,2
		Fabric mitts with palins faked with convas or heat	3,4
Heat when a far degree of sensitivity is required and splashes or splatter of molten metal may occur	Welding, casting, galvanizing	Heat resistant leather inseam mitts	1,2
		Heat resistant leather gauntlets	1,2
		Heat resistant leather inseam gauntlets with cuffs	1,2
Sharp edge materials and objects	Swarf, guilloining metal, blanking, handling metal sheets, handling undressed casungs	Leather inseam mitts and gauntlets	1,2
		Fabric gloves with palms faced with canvas or leather	3,4
		Supported PVC gluves with granular finish	9
		Reinforced rubber gloves heavy-weight	15
Sharp materials or objects in an alkaline degreasing bath		Supported PVC gloves with granular finish	9
		Reinforced rubber gloves heavy-weight	15
Glass or limber with splintered edges	Glass handling, hmber handling, huilding demulition	Leather gloves and mitts	1,2,3
		Loop pile gloves	4
		Supported PVC gloves with granular finish	9
		Reinforced rubber gloves heavy-weight	15
Very heavy abrasion	Shot blasung	Reinforced natural rubber heavy-weight	15

(Continued)

Table 4.7 (Continued)

Hazard	Typical operations	Suitable gloves	Glove type no.
Heavy abrasion	Handling dressed casting or forgings, bricks, concrete cement, steel block, heavy duty packaging	Abrasion resistant leather inseam mitts	1.2
		Abrasion resistant grain hide palm split leather back inseam gloves	2
		Supported loop pole gloves	4
		Abrasion resistant leather stapled double palm gloves	5
		PVC gloves, granular surface heavy-weight	9
		Reinforced natural rubber gloves	15
Light abrasion	Handling of Package for General Handling	Leather wrist gloves and mitts	1
		Fabric gloves	4
		Fabric gloves with leather palms	3
		Loop pile gloves	4
		PVC gloves	6,7,8,9,10,11
		Rubber gloves	12,13,14,15
Solvents*	Degreasing, printing, chemical manufacturing, paint spraying	Supported PVC gloves rough, sinchith, lightweight (excluding open back and knitted wrist styles)	0.7
		PVC lined gloves smooth and granular finish (excluding open back and knitted wrist styles)	7,8,9,10
		Natural and synthetic rubber gloves and gauntlets	12,13,14,15
Chemicals*	Acids, alkalis, dyes and general chemical hazards not involving	Standard weight PVC gloves	8
		PVC gloves with a granular finish (excluding open back and knitted wrists styles)	9
		Tubber gloves	12,13,14,15
Fats. Oils*	Chemical hazards involving contact with oils	Standard weight supported PVC gloves and gauntlets	8
		Granular finish PVC gloves and gauntlets (excluding open back and knitted wrist styles)	9
		Natural and synthetic rubber gloves and gauntlets	12,13,14

*It is important that the Purchaser or user should seek advice from manufacturer before making a final selection of the type of glove to meet his particular needs. Attention is drawn to the method of tests and the type of glove under consideration.

4.18   Materials

Materials that are known to be likely to cause skin irritation or any adverse effect on health of the user should not be considered as materials of construction.

4.18.1   Leather

The leather should not contain cuts, holes or grain damage. Grain leather should not crack when subjected to a double fold grain-side outward. The second fold should be at a right angle to the first. The leather should be full chrome tanned, semichrome tanned or combination tanned. The tear strength of the leather should not be less than 108 N (11.0 kgf) when tested. The grease content of the leather when a sample is dried should not be less than 5% and no greater than 25%.

> The chromic oxide, zirconium oxide, and aluminum oxide content of the leather, calculated on a fat-free basis, should be in accordance with the following when a sample of the leather is dried and subsequently tested:
> Full-chrome leathers should contain no less than 3.5% chromic oxide (Cr_2O_3) and semi-chrome leathers no less than 2% chromic oxide.
> Combination tanned leathers not containing an organic tanning agent should contain no less than 3.5% of metal oxides-derived from mineral-tanning agents, i.e., chromic oxide, zirconium oxide, and aluminum oxide. Where mineral-tanning agents are used in conjunction with an organic tanning agent the leather should contain no less than 2% metal oxide in total.
> The pH value of an aqueous extract of the ground leather should not be lower than 3.2.
> When two specimens of the leather are tested at temperature of 90°C and a test time of 1 min., neither specimen should shrink by more than 10% of its original area.
> No chromate should be detected when a sample of the aqueous extract of the leather is tested.

4.18.2   Woven fabrics

The breaking strength of woven fabric should be in accordance with Table 4.8.

4.18.3   Loop-pile fabric (terry fabric)

When tested the cloth should have a minimum mass per unit area of 670 g/m² for medium weight reversible loop-pile fabric gloves and mitts and 930 g/m² for heavy weight reversible loop-pile fabric gloves, and mitts, using the largest sample possible

Table 4.8 **Breaking strength of woven fabrics**

Direction	Breaking strength	
	Unraised	Raised
	N	N
Warp Weft	1100 580	980 350

and adapting the procedure and the calculation of results accordingly (as described in methods of BS 2471). The fabric should be constructed with 5.5 to 7 courses per centimeter and 3 to 4 wales per centimeter. The abrasion resistance of the fabric should be such that the loss in mass is not greater than 0.18 g.

4.18.4   Sewing threads

The sewing thread for leather and fabric gloves (Types 1 to 5) should be a polyester and cotton-core spun thread or should be a cotton or linen thread of equivalent tensile strength.

4.18.5   Stitches

All stitching should be lock stitch or double thread chain stitch, and there should be 27 to 35 stitches per 10 cm for fabric gloves and 23 to 31 stitches per 10 cm for all-leather gloves.

4.19   Construction and design

4.19.1   Type 1

Flesh-split leather inseam gloves and gauntlets, mitts and one-finger mitts.

- Style: Type 1 gloves and gauntlets should be of the clute or gunn pattern and be inseam apart from the back, which may be either inseam or outseam.
- Materials: The leather should be flesh splits complying with 8.2.1. The thickness for gloves should not be less than 1.2 mm and no more than 1.6 mm. For one-finger mitts and mitts, the thickness should not be less than 1.4 mm. The leather of gloves, gauntlets, and mitts marked "heat-resistant" should comply with BS 1651. The sewing thread should comply with 8.2.4.
- Design: For clute-pattern gloves and one-finger mitts the palm and the leather covering the back of the first finger and the front of all fingers should be cut from one piece of leather. The cuff should be cut from no more than two-pieces of leather and should be joined to the glove by a double row of stitching.

Notes:

1. Provision for a wider opening to the gloves, if required, may, for example, be by a triangular leather gusset fitted at the side seam.
2. In the case of gunn pattern gloves and inseam mitts, the palm, fronts of the first and fourth fingers and thumb should be cut from one piece of leather. The fronts of the second and third fingers should be cut from one or two-pieces of leather. The back should be cut from one piece of leather.
3. Where a wing thumb is used, a seam wholly or partly along the join between the first finger and palm is permissible.

4.19.2   Type 2 grain leather inseam gloves, gauntlets, mitts, and one-finger mitts

- Style: Type 2 gloves and gauntlets, inseam mitts and one- finger mitts, should be of the clute or gunn pattern, and inseam apart from the back, which may be either inseam or outseam.

- Materials: The leather should comply with 8.2.1. In the clute pattern the palm, the fronts of the fingers and the whole of the thumb should be of grain leather no less than 1.2 mm and no more than 1.7 mm thick. The backs of the fingers should be either of grain leather or of flesh-split leather.
- In the gunn pattern the palm and the fronts of the fingers should be of grain leather no less than 1.2 mm and no more than 1.7 mm thick. The back should be of the flesh split leather, cut from one piece and should cover at least three fingers of the backs up to the cuff. The sewing thread should comply with 8.2.4
- Design: The cuff should be made from no more than two-pieces and should be joined to glove by a double row of stitching. Where a vein patch, comprising a semi-circular piece as shown in Figure 9 of grain leather, is fitted it should be sewn midway in the palm of the glove at the cuff seam and sewn on to the cuff. The patch should not be less than 75 mm in diameter.

Note: Provision for a wider opening to the gloves, if required, may, for example, be by a triangular leather gusset fitted at the side seam.

4.19.3   Type 3 fabric gloves with leather palms

- Style: Type 3 gloves should be of the gunn or clute pattern with a fabric back and cuff and be inseam sewn.
- Materials: The leather should be flesh splits no less than 1.0 mm thick or grain leather no less than 1.2 mm thick and should comply with 8.2.1. Woven fabric should comply with 8.2.2. Loop-pile fabric, except for knitted wristing fabric, should comply with 8.2.3. The sewing thread should comply with 8.2.4.
- Design: clute pattern gloves should be one of the following three designs:
  Leather palm and thumb with all four finger backs in a textile fabric.
  Leather palm, thumb, and part wrap around first finger. The second, third, and fourth fingers to be from a textile fabric and the leather from the palm should cover the thumb side of the first finger without seam with the remainder of the first finger to be made from fabric.
  Leather palm, thumb, and first finger with no seam between the first finger and palm. The second, third, and fourth fingers should be made from a textile fabric. If a woven cuff is used the length after sewing should be no less than 50 mm. The edges should be hemmed or over-locked.

In the gunn pattern, the back should be of textile fabric in one piece and should cover at least three fingers of the backs up to the cuff.

4.19.4   Type 4 fabric gloves

- Style: Type 4 gloves should be inseam gloves and gauntlets made wholly from fabric.
- Materials: Type 4 gloves should be made from woven or loop-pile fabric complying with 8.2.2 or 8.2.3, respectively. The sewing thread should comply with 8.2.4.
- Design: The glove should be closed inseam and the cuff edge if of woven fabric should be hemmed or over-locked. The cuff should be fabricated by one of the following methods:
  In the same material as the glove
  In a canvas material of mass per unit area no less than 320 g/m²
  By use of a fabric to produce a double-rib cuff of minimum length 50 mm. The wristing should be of mass per unit area no less than 230 g/m²

4.19.5   *Type 5 leather outseam-armored gloves and gauntlets*

- Style: Type 5 gloves or gauntlets should be of the clute or Montpelier pattern. The seams of the fingers and thumb should be wire-stitched. In the clute pattern the seams on the back of the glove should be sewn with thread.
  Notes: The fingers and thumb may additionally be sewn with thread. The seams may additionally be wire stitched.
- Materials: The leather should be flesh splits complying with 8.2.1. The thickness for the palm/finger should not be less than 1.4 mm and no more than 2.0 mm. The sewing thread should comply with 8.2.4.
- Design: The glove palm, fingers, and working surface of thumb should be reinforced by galvanized steel staples. The closed staples should not be less than 2.5 mm wide, no less than 0.5 mm thick and no less than 8 mm long.

The stapling should be applied diagonally or horizontally to the palm, finger and the front face of the thumb, and should consist of rows of staples running from the tips of the fingers to the seam joining the cuff. there should be 130 ± 15 staples on each glove arranged so as to give maximum protection to the user. All staples should be firmly closed. The palm should be lined to ensure that the staples do not come into contact with the hand.

Note: For this purpose the lining should be of leather or heavy-weight fabric of appropriate thickness.

4.19.6   *Sizing and dimensions*

- Sizing: Leather and fabric-fingered gloves should fit on a glove iron, the minimum palm circumference of which is 254 mm for men's gloves.
- Dimensions: The minimum outside dimensions of leather and fabric gloves should be as given in Tables 4.9, 4.10, and 4.11 and as shown in Figure 4.10. Gauntlet cuffs should not be less than 100 mm nor greater than 250 mm in length.
  Note: Dimension C, together, if appropriate, with dimensions G and H, apply to the cuff, particularly when the latter is a discrete extension to the glove or mitt.

Table 4.9 **Minimum outside dimensions of glove types 1, 2, 3 and 4**

Position where measured	Fig. 10 Reference	Men's size Mm
From tip of second finger position to top of cuff	A	225
From tip of second finger position to bottom of cuff	B	205
Cuff length	C*	50
From tip of forefinger position to crotch of thumb	D	125
From tip of thumb to crotch of thumb	E	75
Across palm at crotch of thumb	F	125
Across bottom of cuff	G	125
Across cuff opening	H*,†	125

*Applicable only to wrist gloves.
†Not applicable to knitted wrist cuffs.

Table 4.10 **Minimum outside dimensions of type 5 gloves, clute pattern**

Position where measured	Fig. 10 Reference	Men's size Mm
From tip of second finger to top of glove	A	225
From tip of forefinger to crotch of thumb	D	135
From tip of thumb to crotch of thumb	E	75
Across palm at crotch of thumb	F	135
Across cuff opening	H*	135

*Applicable only to wrist gloves.

Table 4.11 **Minimum outside dimensions of type 5 gloves, montpelier pattern**

Position where measured	Fig. 10 Reference	Men's size Mm
From tip of second finger to top of glove	A	225
From tip of forefinger to crotch of thumb	D	125
From tip of thumb to crotch of thumb	E	75
Across palm at crotch of thumb	F	135
Across opening at top of glove	H*	135

*Applicable only to wrist glove.

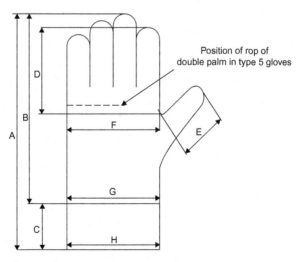

Position of rop of double palm in type 5 gloves

Figure 4.10 Dimensions of leather and fabric gloves. **Note:** Dimension C, together, if appropriate, with dimensions G and H, apply to the cuff, particularly when the latter is a discrete extension to the glove or mitt.

4.20   PVC gloves (Types 6 to 11 of Table 4.5)

The gloves should be manufactured by a dipping process from a PVC plastisol or organosol complying with relevant standards.

4.20.1   Materials

The PVC coating should be virgin homogeneous PVC plasticized compound.
Note: Substances that are known to the likely to cause skin irritation or any adverse effect on the health of the user should not be incorporated within the materials from which the gloves are made.

4.20.2   Construction and design

- Appearance: Gloves for protection against liquids should be free from patched areas, embedded foreign matter, perforations, porosity, blisters, and exposed fiber on the external surface of the glove.
- Fingers: The fingers should be entirely separate and should not be interconnected in any way by PVC.

4.20.3   Supported PVC gloves

- Classification: Supported PVC gloves should be classified in accordance with Table 4.12.
- Design: Should be as follows:
  Size: The sizes of supported PVC gloves, the minimum length of the gloves and, where appropriate, the maximum width and length of wrist should be as given in Table 4.13.
  Gauntlet gloves: All sizes of gauntlet gloves should have a minimum length of 260 mm.
  Liner: There should be no seams on the effective wording surface of the liner.
  Thumb: There should be no seam triple union in the crotch area of the thumb.

Table 4.12   Classification of supported PVC gloves

Type	Weight/description	Style patterns	Average palm thickness[*] mm
6	Light weight rough finish.	Knitted wrist, gauntlet, or palm coated open back.	Not less than 0.75, not greater than 1.2.
7	Light weight smooth finish.	Knitted wrist, gauntlet, or palm coated open back.	Not less than 0.75, Not greater than 1.2.
8	Standard weight smooth finish.	Knitted wrist, gauntlet, or fully coated.	Not less than 1.2, Not greater than 1.7.
9	Granular finish.	Knitted wrist, gauntlet, or palm coated open back.	Not less than 1.1.

[*]Average palm thickness measured by the method described in of BS 1651.

Table 4.13 **Size designation of supported PVC gloves**

Sizes	Minimum length mm	Maximum width of wristing* if used mm	Minimum length of wristing if used mm
6, 6½ and 7	215	85	50
7½, 8, 8½ and 9	240	90	55
9½, 10 and 10½	255	95	55

*Measured at base of wristing.

Table 4.14 **Classification of unsupported PVC gloves**

Type	Description	Style pattern	Average palm thickness* mm
10	Flock lined	Gauntlet	Not less than 0.70
11	Unflocked matt finish	Gauntlet	Not less than 0.076

*Measured by the method described in BS 1651.

Seams: All seams should be covered by the PVC coating except the back wrist/hand and part of the side seam opposite the thumb of a palm coated open-back glove.

Coating: The coating should show no signs of penetration of PVC into the interior of the glove resulting in nodules that could give rise to abrasion of the skin of the user.

Average palm thickness: The average palm thickness of the glove when measured should be in accordance with Table 4.12.

4.20.4   Unsupported PVC gloves

- Classification: Unsupported PVC gloves should be classified in accordance with Table 4.14.
- Sizes: The minimum length of all sizes of unsupported PVC gloves should be 230 mm.
- Average palm thickness: The average palm thickness of the glove, when measured, should be in accordance with Table 4.14.

4.21   Rubber gloves (Types 12 to 15 of Tables 4.5 and 4.17)

4.21.1   Material

The glove should be made from natural or synthetic vulcanized elastomers of natural rubber (NR), chloroprene (CR), or butadiene/acrylonitrile (NBR) or mixtures of these materials.

Note: Substances that are known to be likely to cause skin irritation or any adverse effect on the health of the user should not be incorporated within the materials from which the gloves are made. When the glove is produced from a mixture of polymers the requirements of BS 1651 should be applied for the principal polymer used. Sewing threads for liners should be of cotton 50/2 cord or threads of equivalent strength.

4.21.2   Construction and design

- Appearance: Gloves for protection against liquids should be free from patched areas, embedded foreign matter, perforations, porosity, blisters, and exposed fiber on the external surface of the glove.
- Classification: Gloves should be classified for type and thickness in accordance with Tables 4.15 and 4.16. The minimum length and the wall thickness of each type should be as given in those tables. Thickness should be measured as described in method A1 of BS 903: Part A38 using a gauge with a circular foot of 5 ± 0.1 mm diameter.
- Design: The sizes of rubber gloves and the palm circumference of the glove that should be measured around the inside of the glove starting and finishing at the thumb crotch, should be as given in Table 4.16.

Table 4.15 Thickness classification of rubber gloves

Type	Weight	Symbol	Single wall thickness* mm
12 and 13	Ultra-lightweight	U	Not greater than 0.5
12 and 13	Lightweight	L	Greater than 0.5, Not greater than 0.9
12 and 13	Medium-weight	M	Greater than 0.9, Not greater than 1.3
12 and 13	Heavy-weight	H	Greater than 1.3
14	Lightweight	L	Greater than 0.5, Not greater than 1.0
14	Medium-weight	M	Greater than 1.0, Not greater than 1.5
14	Heavy-weight	H	Greater than 1.5

*For type 14 gloves the specified thickness is that of the elastomer and fabric combined.

Table 4.16 Size designation of rubber gloves

Size nomenclature	Numerical size designation* in.	Internal palm circumference (Tolerance embed[+10] mm) −6
Small (S)	6½	165
Medium (M)	7½	191
Large (L)	8½	216
Extra-large (XL)	9½	241
Extra-extra-large (XXL)	10½	267

*The figures given in this column are the same as the numerical values of internal palm circumference expressed in inches.

4.22 Performance requirement

Gloves should be tested in accordance with Appendices A to K of BS 1651 covering the following:

Method for detection of soluble chromate
Method of test for abrasion resistance of loop-pile and PVC gloves
Additional design feature of leather and fabric gloves
Method of test for abrasion resistance of leather gloves
Calculation of percentage area loss after testing in accordance with BS 3144
Preparation of test specimens and condition
Thermal contact test
Method of test for the determination of thickness of PVC gloves
Method of test for degree of gelatin of PVC gloves
Method of test for flex cracking of PVC gloves
Method of test for air permeability of PVC and rubber gloves

Table 4.17 shows type designation of rubber gloves.

• Marking: Gloves or their immediate packaging should be clearly marked with the following:
Gloves type number
If appropriate, the code letters defining the type of elastomer used or its full name
If appropriate, for rubber gloves, the words "ultra-lightweight," "lightweight," "medium-weight" or "heavyweight" or the symbol letters given in Table 415
If appropriate, for leather and fabric gloves, the words "heat-resistant"
If appropriate, for leather and fabric gloves, the words "abrasion-resistant"
If appropriate, for PVC and rubber gloves, the words "pressure tested"

Table 4.17 Type designation of rubber gloves

Type	Description	Style	Minimum length mm
12	Unlined	Wrist	265
13		Gauntlet	305
	Flock lined	Wrist	265
14		Gauntlet	305
	Fabric lined	Wrist	265
15		Gauntlet	305
	Unlined, flock	Wrist	265
	Lined or fabric	Gauntlet	305
	Lined with additional		
	Rubber Reinforcement over		
	The whole or part of the hand		

4.23   Rubber gloves for electrical purposes

4.23.1   Rated potential

The rated potential (a-c(r-m-s) or d.c.) between any conductors and earth in a system does not exceed the following:

650 V
1000 V
3300 V
4000 V

* Composition: Gloves should be made from good quality raw natural rubber or raw synthetic rubber or from a mixture of these, in conjunction with suitable compounding ingredients.

4.24   Construction

Gloves should be made by a one-piece process or should be built from sheets. Gloves should be free from patched areas, embedded foreign matter, blisters (other than shallow broken blisters), and other physical defects that may arise from any lack of physical homogeneity or continuity in the glove material, when inspected in a well-lit area by the naked eye (aided by glasses if necessary to ensure normal vision) of a designated person. **Note:** Minor surface irregularities that do not cause hazard or significant degradation in quality or life may be disregarded.

* Length: The minimum internal length from the tip of the second finger to the edge of the cuff, denoted as dimension in Figure 4.11, should be 265 mm for the wrist type and 355 mm for the gauntlet type.

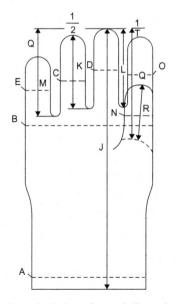

Figure 4.11 Outline of typical standard glove (internal dimensions).

4.25   Typical dimensions

Two types of former may be used in the manufacture of rubber gloves, namely, a flat type and a shaped type. Gloves made on the shaped type of former are generally more comfortable. Table 4.18 gives typical values of the principal internal dimensions of well-proportioned gloves. The external dimensions will depend on the thickness of the rubber used.

Note: If required, a reinforced extension for suspension purposes may be incorporated on the back of the wrist of wrist-type rubber gloves but an extension should be the subject of agreement between the buyer and the manufacturer.

4.26   Color codes

If gloves are color-coded to indicate the rated potential, the colors used should be in accordance with Table 4.19.

Table 4.18 **Typical internal dimensions**

Detail	Size			
	8 mm	**9 mm**	**10 mm**	**11 mm**
Circumferences				
A	218	236	254	271
B	218	236	254	271
C*	58	62	67	72
D*	60	65	70	75
E*	57	60	65	69
N*	72	78	84	90
Lengths				
J (Minimum)† Wrist	265	265	265	265
Gauntlet	355	355	355	355
K	67	70	74	78
L	75	80	84	88
M	57	60	63	67
P	110	116	122	128
Q	28	31	31	33
R	57	59	62	65
T	8.5	9	9.5	10

*Circumference is measured half-way between crotch and tip.
†The values for dimension J are the minimum requirements.

Table 4.19 **Color codes**

Rated potential V (r.m.s.)	Color
650	White
1000	Red
3300	Green
4000	Blue

4.27 Performance

Rubber gloves for electrical purposes should be tested by manufacturer and subsequently by user in accordance with BS 697 for the following tests:

Measurement of thickness
Electrical resistance
- Instructions: Instructions should accompany each pair of gloves and should include the following information:
Recommendations for storage and cleaning (including maximum washing and drying temperatures)
Appropriate details of inspection and re-testing procedures
- Marking: Each glove should be marked with the following:
Number and date of relevant standard
Name, trademark or other means of identification of the manufacturer
Month and year of manufacture
Rated potential followed by the word "working" in brackets
Size

The marking should be durable and should not impair the properties and characteristics of the glove.

4.28 Guidance concerning the maintenance, storage, inspection, re-testing, and use of rubber gloves after purchase

- Storage: Gloves should be stored unfolded in a container in a dry, dark place where the temperature is between 10°C and 21°C. Gloves that have been issued for service but that are not in use should be kept in containers used solely for that purpose or in a place where they will not be subject to mechanical or chemical damage.
- Issue: Gloves intended for linesmen and other outdoor workers should be issued in a protective container free from grease and oil, and suitable for the class of work for which they will be used. Canvas or leather bags that can be attached to the linesmen's belts are suitable for overhead line work. Fiber boxes are appropriate when gloves need to be kept in toolboxes. Gloves issued for emergency use only should be kept in waterproof containers.
- Examination before use: Each glove should be examined inside and outside before each occasion of use. If either of a pair of gloves is thought to be unsafe, the pair should be re-tested.

4.28.1 Precautions in use

Care should be taken to avoid mechanical damage caused by abrasion or sharp edges. Gloves should not be exposed unnecessarily to heat or light or allowed to come into contact with solvents, oils, or other chemical agents. If other protective gloves are used at the same time as rubber gloves for electrical purposes, they should be worn over the rubber gloves. If the outer protective gloves become damp, oily, or greasy, they should be removed. They should also be removed from the rubber gloves when the latter are not in use.

When rubber gloves become soiled they should be cleaned by washing with soap and water at a temperature not exceeding that recommended by the glove manufacturer, thoroughly dried and dusted with talc. If insulating compounds such as tar and paint continue to stick to the glove, the affected parts should be wiped immediately with a suitable solvent, avoiding excessive use, and then immediately cleaned as described in the preceding sentence. In case of difficulty, advice should be sought from the manufacturer. Gloves that become wet in use or by washing should be dried thoroughly, but not in a manner that will cause the temperature of the gloves to exceed 65°C.

4.28.2 Inspection and re-testing of gloves

Gloves that are used frequently should be re-tested at intervals of no more than 6 months. Gloves that are used only occasionally should be re-tested after use and, in any event, at intervals of no more than 12 months. Gloves held in store should be re-tested at intervals not exceeding 12 months.

Surface defects may develop with use, resulting from the breaking of blisters in the rubber, or from foreign matter breaking through the surface. Gloves showing any such defects on return to store should be destroyed or rendered unusable. Each finger of each glove should be stretched by hand to ascertain that its mechanical strength is adequate. Those that appear to be in good condition should be re-tested by being given a single electrical test in accordance with the appropriate test potential specified in standards (i.e., according to the rated potential), and in the manner described in BS 697. In the re-test, no glove should break down or show a leakage in excess of the specified maximum value shown in Table 1 or 3, BS 697, as appropriate. Only those gloves that pass the re-test should be accepted as satisfactory. Other gloves should be rejected and destroyed or rendered unusable for electrical purposes.

- Salvage: When only one glove of a pair is rejected, the other, where possible, may be mated with a similar glove of same size and make; the resulting pair, after re-testing, may be placed in serviceable stock. No glove should be turned inside out for mating.

4.29 Protective leather gloves for welders

4.29.1 Types

The types of gloves are shown in Table 4.20, according to the materials, shapes, and purposes.

Table 4.20 **The types of gloves**

Type		Material		Shape	Purpose
Class 1	No. 1	Palm and back	Cow-leather	2-Finger	For chiefly arc
	No. 2	Cuff	Back split	3-Finger	welding
	No. 3		Cow-leather	5-Finger	
Class 2	No. 1	Palm and back	Back split	2-Finger	For chiefly gas
	No. 2		Cow-leather	3-Finger	welding, fusion
	No. 3		Back split	5-Finger	cutting
		Cuff	Cow-leather		

Table 4.21 **The parts of glove where such additional pieces are to be used**

Type		Applicable part		
		Welt leather	**Reinforcing leather**	**Inset leather**
Class 1 and Class 2	No. 1	Seam of palm and back	Boundary of crotch of thumb	–
	No. 2	Seam of palm and back	Boundary of crotch of thumb	Sides between Forefinger and middle finger
	No. 3	Seams at crotches of middle and ring finger and thumb	–	–

Note: The inset leather may be omitted according to the construction for manufacturing convenience.

4.29.2 Construction, dimensions, and thickness of leather

- Construction of gloves: The construction of gloves should be in 2-finger, 3-finger and 5- finger types, for both Classes 1 and 2, with the seam between the palm and back stitched together with a welt leather inserted. For the welt, chrome-tanned cow leather or back split of cow leather should be used. For inset and reinforcing leather, the same leather as for the palm and back should be used, and the width of reinforcing leather should not be less than 15 mm. The parts of the glove where such additional pieces are to be used should conform to Table 4.21. Dimensions: The minimum standard dimensions of the gloves should conform to Table 4.22. In the Table, A is the outside length from the tip of middle finger to the bottom of cuff, B is the outside length from the tip of middle finger to the top of cuff, C is the length of cuff, and D is the width across palm at the crotches of forefinger and little finger.

4.29.3 Thickness

The thickness of leather should conform to Table 4.23.

Table 4.22 **The minimum standard dimensions of the gloves**

Type		Length			Unit: mm
					Width
		A	**B**	**C**	**D**
Class 1 and Class 2	No. 1	350	200	150	130
	No. 2	350	200	150	130
	No. 3	350	200	150	130

Table 4.23 **The thickness of leather**

Applicable part	Kind of leather	Unit: mm
		Thickness
Palm and back	Cow leather	1.5 min.
	Back split cow leather	1.5 min.
Cuff	Back split cow leather	1.0 min.

Table 4.24 **The main materials of the gloves**

Item		Cow leather (Chrome tanned)	Back split cow leather (Chrome tanned)
Tensile strength	(kgf/mm²) {MPa}	2.0 {19.16} min.	1.0 {9.81} min.
Elongation	(%)	40 min.	30 min.
Tear strength	(kgf/mm) {N/mm}	5.0 {49.03} min.	3.0 {29.42} min.
Grease content	(%)	6.0 min.	2.0 min.
Chrome content	(as Cr_2O_3) (%)	2.5 min.	2.5 min.

4.29.4 Material

- Leather: The main materials of the gloves should be chrome-tanned cow leather and back-split thereof, which should satisfy the values in Table 4.24. For the palm and back of the gloves, in particular, leather of nearly uniform thickness, being free from uneven splitting, and being flexible and strong should be used, and the leather for the cuff should be of moderate elasticity.
- Sewing thread: The sewing thread used for sewing the gloves should be the spinned thread of synthetic fiber such as nylon, polyester, vinylon, etc., of 20-count or the equivalent, being free from irregularity in twist, flaw, etc., and its tensile strength should be no less than 22.56 N (2.3 kgf).

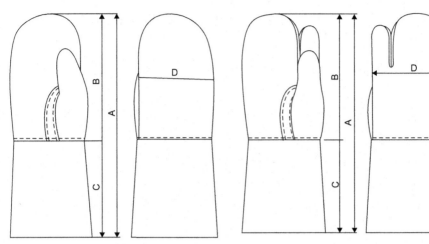

Figure 4.12 Class 1, No. 1 and Class 2, No. 2 (2-Finger Type).

Figure 4.13 Class 1, No. 2 and Class 2, No. 2 (3-Finger Type).

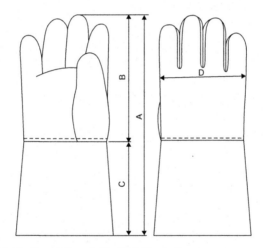

Figure 4.14 Class 1, No.3 and Class 2, No.3 (5-Finger Type).

4.30   Ear protection

This section specifies ear protectors that are used to protect employees from the harmful effects of noise in working environment with high-noise level. This standard is divided in three sub-sections as follows:

1. Industrial hearing protection: Ear Muffs
2. Industrial hearing protection: Ear Plugs
3. Sonic Ear Valve

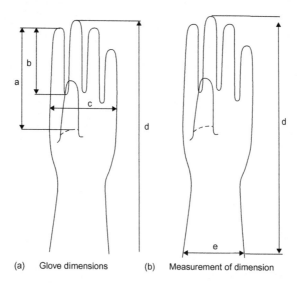

(a) Glove dimensions   (b) Measurement of dimension

Figure 4.15 Dimensions of leather and fabric gloves.

4.30.1 Materials

Ear-muff materials that can be cleaned should not be done so with substances that are known to be harmful to health. Materials of the cushions that may come into contact with the skin should be nonstaining, soft, pliable, and not known to be likely to cause skin irritation or any adverse effect on health. All materials should be visibly unimpaired after cleaning by the method specified by the manufacturer. Ear muffs should have an ignitability (heated rod) index of P when tested in accordance with BS 6344. The metal part should be processed with a suitable rust prevention and should be capable of being disinfected.

4.30.2 Construction

All parts should be designed and manufactured such that they are not liable to cause physical damage to the user when used as intended. All edges that are in contact with plastic cushions should be radiused, finished smooth, and be free from sharp edges that could cause damage to the cushions. Where the cushions are not intended to last the lifetime of the muff, the cushions should be replaceable without the use of a special tool(s). In the case of ear muffs that should be suitable for wearing other than over the head, a head strap should be provided.

4.30.3 Fit and size

- Adjustability and size: **Note:** The requirements of this section give rise to and accord with the classification of ear muffs given in Table 4.26. Unless user information is provided, the adjustability of the ear muffs should comply with 7.2.3 (3) 7.2.3 (4) as appropriate.

Table 4.25 Head dimensions (Over-the-head ear muffs)

Head height mm	Head width mm		
	130	150	160
150	×	t	×
130	t	t	t
140	×	t	×

t Indicates ear muffs to fit this size.
× Indicates no requirement for this size to be fitted.

Table 4.26 Head dimensions (Behind-the-head ear muffs)

Head height mm	Head width mm		
	130	150	160
90	×	t	×
105	t	t	t
115	×	t	×

t Indicates ear muffs to fit this size.
× Indicates no requirement for this size to be fitted.

- Over-the-head ear muffs: For each of the combinations of head dimensions for over-the-head ear muffs shown in Table 4.25, the range of adjustment of the head band and of the width between the cushions should enable the ear muffs to be fitted to the apparatus (see BS 6344).
- Behind-the-head ear muffs: For each of the combinations of head dimensions for behind-the-head ear muffs shown in Table 4.26, the range of adjustment of the headband and of the width between the cushions should enable the ear muffs to be fitted to the apparatus (see BS 6344).

Note: The dimensions quoted in Tables 4.25 and 4.26 have been chosen to cover the appropriate combinations of the head width with the head height or with the head depth of the 5th, 50th, and 95th percentiles of the industrial population.

- Universal ear muffs: The range of adjustment of the headband and of the width between the cushions should enable the ear muffs to be fitted to the apparatus (see BS 6344).
- Head straps: Head straps, where provided, should be continuously adjustable, should accommodate the range of head sizes in Tables 4.24 and 4.25 and should not disengage under normal head movements.
- Cup rotation: Each cup should be capable of movement through $\pm 10°$ about two orthogonal axes (one of which is horizontal) set in the plane defined against which a cup rests with the headband set appropriate to the width and height or depth. The contact between the cushions and test-mounting plates should be continuous throughout this range.

Table 4.27 Ear muff settings for the measurement of cup rotation, headband force and cushion pressure

Classification	Type	Corresponding setting of apparatus and ear muffs (See BS 6344: part 1)	
		Head width mm	Head height or head depth mm
Ear muffs which do fit the head dimensions that cover those given in Tables 4.24 and 4.25 when tested in accordance with BS 6344	Over-the-head ear muffs	150	130
	Behind-the-head ear muffs	150	Mid-point of range of cup adjustment
Ear muffs which do not fit the head dimensions that cover those given in tables 4.24 and 4.25 when tested in accordance with L-BS 6344	Over-the-head ear muffs	Mid-point of range of head dimensions stated by the manufacturer in accordance with 9.5.1(f)	
	Behind-the-head ear muffs		

• Size of openings in cups: The maximum length (measured in the plane of, or in a plane parallel, to the face of the cup) of the opening in the cups of the ear muffs, with the cushions fitted, should be no less than 50 mm and the maximum width of the opening, measured along an axis that is in the plane of, or in a plane parallel to, the face of the cup and orthogonal to the line defining the length, should not be less than 35 mm.

4.30.4 Comfort

Ear muffs with a total mass greater than 400 g should be marked with their total mass. The headband force should not be greater than 16 N when measured with the headband set in accordance with 9.2.4.4 if appropriate to the width and height or depth as specified in Table 4.27. The cushion pressure should not be greater than 4000 Pa when measured with the headband set. In the case where ear muffs claim to be universal the tests should be undertaken according to the requirements for both over-the-head and behind-the-head ear muffs.

4.31 Performance tests

The manufacturer should certify in writing that the following tests have been carried out:

1. Sampling, conditioning and order of testing
2. Ignitability (heated rod) test
3. Method of measurement of rotation of cups
4. Method of test for head band test
5. Cushion pressure test (see Note 1)

6. Drop test (see Note 2)
7. Vibration test
8. Headband durability test
9. Objective method for measurement of insertion.
10. Loss of ear muffs
11. Cushion leakage test
12. Test for adjustability

Notes:

1. Any cushion deformation that is present on first unpacking the ear muffs should disappear within 30 minutes of ear muffs being used.
2. Attenuation. Values of the attenuation of the ear muffs as measured and presented by the procedures described in BS 5108 should be provided for the type of ear muffs being assessed against this standard. In the case of universal ear muffs, attenuation values should be provided for both modes of use.

4.32   Test result

For the following tests:

Resistance to damage when dropped
Resistance to damage when vibrated
Headband durability

The ear muffs should not crack, rupture, or otherwise suffer damage likely to affect performance.

• Headband force: The headband force should not change by more than ±25% from that measured after the performance of the ear muffs has been assessed.
• Change in insertion loss: The change in insertion loss measured after carrying out the performance tests on each cup of the designated sample should not be greater than 4 dB at more than one test frequency.
• Resistance to leakage: In the case of liquid or gas-filled cushions, the cushions fitted to ear muffs should not leak when tested.

4.33   Information

4.33.1   User information

The following information for the user should be supplied with the ear muffs by manufacturer:

Method of fitting/adjustment
Method of cleaning that should not require the use of cleaning agents that are known to be harmful to the health of the user
Method of maintenance

Address from which further background information can be obtained

Statement "Hearing protector cushions may deteriorate with use and will then need replacement

Head dimensions the ear muffs are designed to fit if the dimensional range does not cover those given in Tables 4.24 and 4.25, when tested

Tabulated mean attenuation at each frequency and tabulated standard deviation of attenuation at each frequency both measured by the procedures given in BS 5108

4.33.2 Marking

Each pair of ear muffs should be durably marked with the following information:

Name of manufacturer's or trademark

Referenced standard followed for manufacturing

Year and month of manufacture or their abbreviation

Sound insulation performance

Requirement given in 9241 for mass of ear muffs

If particular orientation of the ear cup is intended by the manufacturer, the ear muff (cups) should be labeled by top and/or front

4.34 Ear plugs

4.34.1 Materials of construction

Materials used in parts of ear plugs coming into contact with the user's skin should comply with the following requirements:

Materials should be nonstaining

Materials should not be known to be likely to cause skin irritation, skin disorders, or any other adverse effects to health within the lifetime of the use of the ear plugs

When subjected to contact with sweat, ear wax, or with other materials likely to be found in the ear canal, the materials should not be known to undergo changes within the lifetime of the use of the ear plugs that would result in:

1. Significant alteration to those properties of the ear plugs that are required to be assessed when the ear plugs are examined for compliance with relevant standards
2. Such changes as would be expected to cause significant alteration to the attenuation characteristics

4.34.2 Design

All parts of ear plugs should be designed and manufactured such that they are not liable to cause physical damage to the user when fitted and used according to the manufacturer's instructions. Any part of the ear plug that is likely to protrude outside the ear canal when fitted in accordance with the manufacturer's instructions should be of such a construction that mechanical contact with the ear plug is unlikely to cause any injury to the ear. When inserted in accordance with the manufacturer's instructions, ear plugs should be capable of being readily and completely removed from the ear canal by the user without the use of special instruments. It should be well intimate with ear and not easily slip off.

Table 4.28 Nominal size designations of ear plugs

Normal size designation	5	6	7	8	9	10	11	12	13	14
Diameter of circular holes in gage (in mm)	5± 0.1	6 ± 0.1	7 ± 0.1	8 ± 0.1	9 ± 0.1	10 ± 0.1	11 ± 0.1	12 ± 0.1	13 ± 0.1	14 ± 0.1

The dimensions of the ear plug should be such as to enable the ear plug to be assigned at least one of the nominal size designations given in Table 4.28 and should be free from significant discomfort feeling in use. The total length including nipple, tongue, etc., (except for the connecting element between two plugs) of ear plugs to be worn in the ear canal should not exceed 35 mm.

4.34.3　Storage

Where ear plugs are marked "re-usable," they should be supplied in suitable packaging to ensure hygienic storage between use.

4.34.4　Cleaning

If ear plugs are marked "re-usable," there should not be:

Any significant alteration to those initial properties of the ear plugs that are required
Any changes that would be expected to cause any significant alteration to the attenuation characteristics when the ear plugs are cleaned

4.34.5　Ignitability

The ear plugs should have an ignitability (heated rod) index of "P" when tested as described in Appendices A and B, of BS 6344: P.2.

4.34.6　Assumed protection

The algebraic mean value of assumed protection of the ear plugs at the test frequencies of 500 Hz, 1 kHz, and 2 kHz hall should be greater than or equal to 12 dB.

4.34.7　Information

- User information: The following information for the user should be supplied with the ear plugs:
  Tabulated mean attenuation at each frequency and tabulated standard deviation of attenuation at each frequency guidance on ensuring correct fit of the ear plugs
  If the ear plugs are marked "re-usable," the method of cleaning that should not require the use of cleaning gents that are known to be harmful to the health of the user
  If the ear plugs are marked "re-usable," the number of times the ear plug may be cleaned.

4.35 Sonic ear valve

Sonic ear valves are insert-type ear protectors that attenuate high-level noises while allowing low level environmental and air to pass through. They are specially effective against impulsive or repetitive-impulsive noises like those generated by drop forces, jackhammer, punch press, piston engines, riveting stamping, and chipping operating (see Figure 4.16).

4.35.1 Design and construction

Material should be of nontoxic and nonallergic

Sizes to be of small-medium-large and marked with letter S-M-L respectively

Sonic ear valve should provide at least as much protection as commonly used by ear muffs

Air should easily pass through sonic ear valve for normal circulation

Sonic ear valve should cause diffraction and attenuation of continuous high-frequency noises to impede the passage of the hazardous components of sound energy, so the user can communicate with others while wearing

Precision metal mechanism inside silicone rubber unit permits normal conversation while providing protection against the loud harmful noises of guns, factories, airplanes, tractors, construction equipment, rock music, traffic, automobiles, and motorcycles plus other high-frequency, impulse-impact type noises

• Filters Noise Instead of Plugging: Sonic ear valves leave ear canals open to healthful, comfortable air circulation and pressure equilization. Unlike conventional ear plugs the Sonic ear valve does not cause that "plugged-up" sensation.

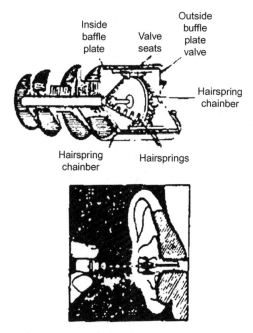

Figure 4.16 Sonic ear valves.

Figure 4.17 Ear muffs.

Figure 4.18 Cap-mounted hearing protection.

4.35.2　Marking

- Ear plug: The ear plugs or immediate packaging or dispensers should carry the following information:
  Name, trademarkm, or other identifying mark of the manufacturer
  Number of standard marked by manufacturer
  Model designation
  Whether the ear plugs are disposable or re-usable
  Onstructions for fitting and use, which should indicate the need for proper fitting
  Nominal size designations of the ear plugs
  Attenuation chart
- Sonic ear valves: The ear valve should be supplied in package of 12 and each pair with hand carrying case and chain. The carrying case hall carry the following information:
  Name, trademark, or other identifying mark of the manufacturer
  Number of standard followed
  Instruction for fitting and use
  Size
  Attenuation chart

4.35.3　Ear muffs

Ear muffs are designed to reduce the effects of excessive noise found typically in plants, factories, airports, and areas where air compressors, jackhammers, and turbines are used (see Figures 4.17 and 4.18).

Ear plugs, made of a soft elastomeric material, are manufactured in five sizes; the size is marked on each plug. These small, featherweight plugs effectively seal the outer quarter of the ear canal. A molded tab permits easy insertion and removal. Plugs clean easily with mild soap and warm water (see Figure 4.19 (a–d)).

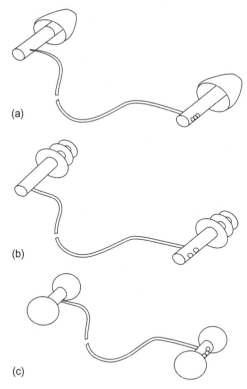

(a)

(b)

(c)

Figure 4.19 Ear plugs.

4.36 Foot protection

This section of standard specifies the minimum requirements for protection of occupational footwear designed primarily to protect the user's foot against injuries. Protective footwear introduced in this standard include:

Safety footwear (general requirement)
Conductive safety footwear
Electrical hazard footwear
Rubber safety footwear
Puncture resistance

Protective footwear is intended to provide protection for the toes against external forces by the use of a protective toe box incorporated in the footwear that is capable of complying with the requirements of this standard (Figure 4.20). The footwear should be felt comfortable and suitable for work by the user. The footwear should be firmly made of the materials to have well balanced shape, and the upper leather, outsole, etc., should be carefully finished. The inner side of toe cap should be lined with cloth, leather, rubber, plastic, etc., and the inner side of the rear end portion should

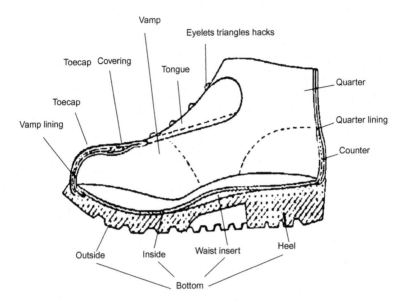

Figure 4.20 Safety footwear components.

be reinforced. The tongue should be a sheath tongue, as possible. The toe box should be incorporated into the footwear during construction and should be an integral part of the footwear.

4.36.1 Material and workmanship

- Footwear material: Protective footwear should be constructed of materials suitable for the exposure it is intended to receive and should provide protection, comfort, and wearability.
- Construction of each part and dimensions
- Upper leather: The upper leather should be uniform in thickness and free from such defects as flaws, and the thickness should be 1.5 mm or more.
- Toe cap: All the surfaces of toe cap should be finished smooth, the edges and corners should be rounded, and all the surfaces of steel toe cap should be treated deterrent to rusting.
- Dimensions of toe cap: The dimensions of toe cap should be as follows (see Figure 4.20):
  Horizontal distance between the upper center of arch and the tip of toe "a" should be 40 to 60 mm.
  Hight at the highest point of toe cap "b" should be at least 33 mm.
  Flanged bottom edge should be bent almost horizontal and the width of such horizontal bottom edge "c" should be at least 3 mm.
  Such parts as sole, heel, leg, and upper that are stuck or stitched together should be complete so that such defects as water leakage, rubber separation, cloth separation, and rubber float-on do not occur.
  Such defects as scar, crack, froth, air form, mixing of foreign particle, which are detrimental to service, should not be found.

Thickness of outsole, leg, and upper: The thickness of the outsole, leg, and upper should be measured and the result should comply with Table 4.29. The outsole

Table **4.29** **The thickness of the outsole, leg, and upper**

Part	Outsole		Leg	Upper
Classification	Non thread part	Main part of thread (including projection)		
Boots	2.8 min	8.0 min	1.2 min	–
Shoes	2.5 min	7.0 min	–	1.2 min

(including heel) should have a pattern effective to prevent slip, and the thickness, when the thinnest part is tested should be at least 3.5 mm. The outsole (including heel) should be of homogeneous synthetic rubber appropriate for the purpose of use. Figure 4.21 shows toe cap and safety shoes components.

4.36.2 Types

The different types of safety footwear are shown in Figure 4.22.

4.36.3 Sizes of safety boots

The metric sizes of safety footwear should be from 36 to 45 and the minimum toe-cap sizes should be as follows:

Size	Toe-cap size
Up to 38	6
39–40	7
41–42	8
43–44	9
Over 44	10

4.37 Conductive safety footwear

Conductive footwear intended to provide protection to user from static electricity accumulated on the body of the user. Such shoes are designed primarily to dissipate the static charged. Accumulation of static electricity in the body cause spark when the body comes in contact with materials.

4.37.1 Classification

Conductive protective-toe footwear should be of two types, designated Types 1 and 2. The requirements for Type 1 footwear are applicable to conductive shoes intended to protect the user and the environment where the accumulation of static electricity on the body is a hazard. Type 1 shoes are designed to dissipate static electricity to the

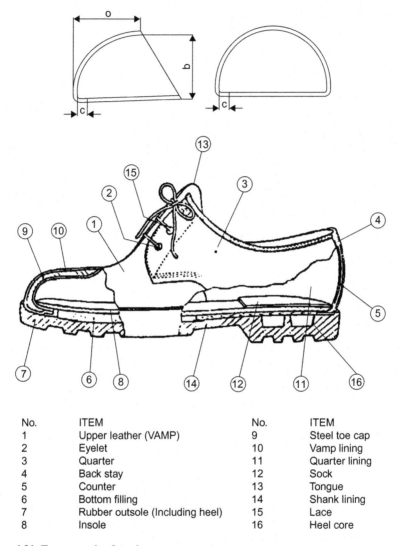

No.	ITEM	No.	ITEM
1	Upper leather (VAMP)	9	Steel toe cap
2	Eyelet	10	Vamp lining
3	Quarter	11	Quarter lining
4	Back stay	12	Sock
5	Counter	13	Tongue
6	Bottom filling	14	Shank lining
7	Rubber outsole (Including heel)	15	Lace
8	Insole	16	Heel core

Figure 4.21 Toe cap and safety shoes components.

ground and prevent the ignition of sensitive explosive mixtures. These shoes should not be used by personnel working near open electrical circuits.

The requirements for Type 2 footwear are applicable to conductive shoes intended for use by linemen or personnel operating on high-voltage lines where the potential of the person and the energized parts must be equalized. Type 2 conductive shoes are designed to protect personnel working on Faraday-type shielded aerial lift equipment, or similar types of equipment on high-voltage lines, or where induced voltage is a problem.

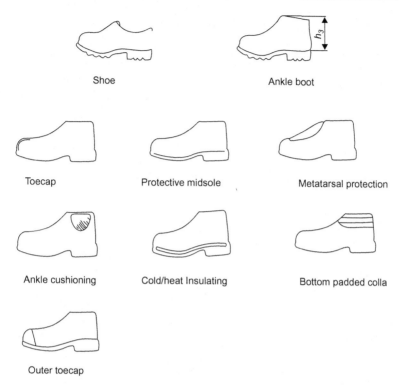

Figure 4.22 Protective features.

Conductive protective-toe footwear, Types 1 and 2, should meet the requirements given in standards. Types 1 and 2 footwear should be of any construction that facilitates a stable conductive path. All exposed external metal parts should be nonferrous.

4.37.2 Material and workmanship

- Toe box: The toe box should be incorporated into the footwear during construction and should be an integral part of the footwear. The toe box should comply with ANSI Z 41.
- Uppers, linings, and outsoles: The uppers, linings, and outsoles should be of materials that facilitate the performance requirements stated in this Section.
- Heels: The heels should be nonmetallic half heels or full heels composed of conductive rubber or any combination of materials and construction that facilitate conductance and transfer of electricity to ground. The tread surface should be smooth. Washer-type heels should not be used. The area of the conductive surface in contact with the ground should be 256 mm² or greater. The heel should be attached to assure permanent conductivity. The nail heads should be below the tread surface, they should be covered by the rubber (blind nailing), and they should not be visible.
- Strap connector: When required, Type 2 footwear should contain a conductive strap connector that fits around the calf of the user. The strap connector should be electrically connected to the back part of the shoe and should facilitate a path for electricity through the heel and sole.

4.38   Conductance (electrical resistance inverse)

4.38.1   Type 1: Footwear

The electrical resistance of new or unworn Type-1 shoes should range between 0 and 500,000 ohms when measured. Reference should be made to BS 5145.

4.38.2   Type 2: Footwear

The electrical resistance of Type 2 shoes for each conductive component and sock lining should not exceed 10000 ohms when measured.

4.38.3   Performance

Conductive safety footwear should be tested in accordance with JIS-T 8103, ANSI Z 41, or BS 5451. For antistatic purposes (low voltage) the discharge path through a product should normally have an electrical resistance of less than $10^8 \Omega$ at anytime throughout its useful life. A value of $5 \times 10^4 \Omega$ is specified as the lowest limit of resistance of a product when new. During service the electrical resistance of footwear (boots, shoes, and overshoes) made from conductive or antistatic materials may change significantly due to flexing and contamination, and it is therefore necessary to ensure that the product is capable of fulfilling its designed function of dissipating electrostatic charges and also of giving any protection during the whole of its life. The user is required to carry out the test for electrical resistance at regular and frequent intervals.

4.39   Electrical hazard footwear

The electrical hazard footwear described in this section should also comply with the requirements given in ANSI Z 41. The construction should provide an assembly that assures prolonged insulation against electricity when tested. No metal parts should be present in the sole or heel of the footwear. A protective-toe box should be incorporated into the shoe during construction and should be an integral part of the shoe or boot.

4.40   Material and workmanship

- Uppers and insole: The uppers and insoles should be of any suitable material.
- Outsole and heel: The outsole and heel should be of a suitable material that should meet the requirements of 10.2.6.5
- Electrical properties: Each shoe should be capable of withstanding the application of 14000V (root mean square [r.m.s.] value) at 50Hz for 1 minute showing no leakage current in excess of 5.0mA when tested.
- Apparatus: A 0.5KVA (500-VA) transformer or larger should be used and the impedance value of the measuring system should not exceed 280000 ohms.

4.40.1   Procedures

Blotting paper as specified in Blotting Paper (Laboratory) should be cut to cover 65% or more of the insole, but not touch the upper when the paper is inserted in the footwear. Immerse the cut paper in a 1% solution of sodium chloride for 15 to 30 sec or until completely saturated. Insert the wet paper on the insole avoiding the upper and lining of the upper and test for current leakage after 5 min.

The shoe should be mounted on a metal base (larger in width and length than the test-shoe outsole) electrode, and a 2.265 kg metal foil electrode should be placed inside the shoe so that it is in contact with at least 65% of the surface of the insole including the wet blotting paper. Voltage is applied in accordance with 10.4.3.

4.40.2   Rubber safety boots

Rubber boots are of three types as illustrated in Figure 4.14 and 4.15:

Ankle
Knee and thigh types

4.40.3   Main material and parts

*   Rubber: The rubber should be manufactured into uniform composition so that it fits for the purpose of use.
*   Cloth: The cloth such as knit and flannel used as the insole cloth padding and lining cloth of lining product should have uniform density suitable composition for the purpose of use and tearing strength of 0.39 MPa (3.9 bars) or more when it is tested.
*   Toe cap: The toe cap should be of material having suitable strength for the purpose of use.

4.40.4   Construction

Rubber safety boots should be designed in such a manner as to protect a user's toes from compression and impact by fitting of toe caps on the toe box in the production process and should comply with the following:

It should be comfortable and be easy to work for user.
It should be made of material, and into robust structure and well-balanced shape.
Such parts as sole, heel, leg, and uppers that are stuck, vulcanized, or stitched together should be complete so that such defect as water leakage, rubber separation, cloth separation, and rubber float-on do not occur.
The inside of toe cap should be lined with cloth, rubber, or plastic and specially the inside of rear should be reinforced.
Such defects as scar, cracks, froth, air form, mixing of foreign particles, which are detrimental to service, should not be found.
The outsole should be shaped in such a manner as to prevent a user from slipping.
The rubber boot used in oil production facilities should be made of oil-resistant materials.

4.40.5   Puncture resistance footwear

The purpose of this requirement is to reduce the hazard of puncture wounds caused by sharp objects that could penetrate the sole of footwear. Puncture-resistant footwear (protective midsole) should not be removed from the bottom and should comply with ANSI Z 41.

4.40.6   Protective device requirements

The protective device should cover the maximum of the insole allowed by the construction of the footwear and should at least extend from the toe to overlaps the breast of the heel. The footwear should withstand an average of force on each device of no less than 150 kg to penetration when tested. Manufacturer should certify in writing that the type of foot protection supplied have been tested in accordance with this standard.

- Marking: Footwear should be indelibly and legibly marked with the following:
  Size
  Manufacturer identification
  Number of standard
  Type of protective footwear (conductive-electrical hazard puncture resistance)

For electrically conducting footwear, each piece should have a red label bearing the words "electrically conducting", bonded or otherwise securely fixed, and in a suitable position to the outside of the footwear "tested regularly" should appear on each piece on or near the label. Figure 4.23 shows typical industrial vulcanized rubber boots (ankle, knee, and thigh types).

For antistatic footwear each article should have a yellow back strip together with a yellow label bearing the word "antistatic" bonded or otherwise securely fixed in a suitable position to the outside of the footwear. The words "tested regularly" should appear on each article on or near the label.

- Labeling: With each pair of conductive or electrical hazard protective footwear an information label should be supplied. This label should state: "Flexing, contamination, damage and wear can cause changes in electrical resistance (tested regularly)."

4.41   Body protection

In this section personnel body protection covering the following titles is discussed:

- Specifications for general industrial workwear
- Protective apron for wetwork
- High-visibility garments and accessories

4.41.1   Specifications for general industrial workwear

This sub-section specifics materials, standards of manufacture, sizing, and marking requirements for workwear used by general industrial sectors and also gives requirements for a lightweight two-piece working rig.

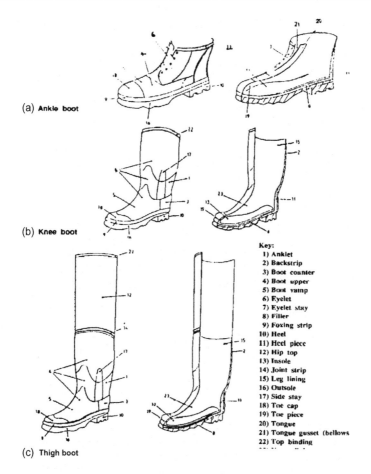

(a) Ankle boot

(b) Knee boot

(c) Thigh boot

Key:
1) Anklet
2) Backstrip
3) Boot counter
4) Boot upper
5) Boot vamp
6) Eyelet
7) Eyelet stay
8) Filler
9) Foxing strip
10) Heel
11) Heel piece
12) Hip top
13) Insole
14) Joint strip
15) Leg lining
16) Outsole
17) Side stay
18) Toe cap
19) Toe piece
20) Tongue
21) Tongue gusset (bellows
22) Top binding

Figure 4.23 Typical industrial vulcanized rubber boots: ankle, knee and thigh types.

- Manufacture
  Seams: All visible seams (i.e., those that are visible on the surface or inside the garment) should be one of the following:
  Seam type overedged, sewn in one or more operation (see BS 3870)
  Bound seam e.g., seam types (see BS 3870)
  Lapped seam with two or more rows of sewing e.g., seam type (see BS 3870)

  There should be no raw edges, but a single row of stitching where the edges are selvedged. The hidden seams (e.g., inside collar) seam type shown in (BS 3870) should be used. The minimum seam allowance should be 8 mm for all fabrics.

- Stitching: For garments with the exception of the lightweight two-piece working rig, the stitching should be one of the following:
  Multithread chain stitch (see BS 3870)
  Lock stitch (see BS 3870)

  There should be no less than 3.2 and no more than 4 stitches per centimeter. For the lightweight two-piece working rig, for side sleeve head, sleeve, shoulder, yoke, inside leg and seat seams, side pockets and pocket bags stitching should be one of the following:

Multithread chain stitch (see BS 3870)
Combination stitch (see BS 3870)
Six thread safety stitches

- Sewing threads: Sewing threads should be spun polyester fiber or polyester cellulosic core spun or 100% polyester core spun.
- Facing: Where facing of coats and jackets are stitched on, the inner edges should be stitched to the foreparts using seam type 5.31.01 as described in BS 3870 Part 2. The stitching should be no more than 0.5 cm from the edge.
- Fronts: Where studs and buttons are used, the fronts of jackets and coats should be finished with the button stand of no less than 20 mm.
- Closures and fasteners: Closures and fasteners should be suitable for the care, maintenance process to which the garment is to be subjected, and should be in correct register. Buttons, buttonholes, and studs should be affixed through a minimum of two thickness of fabric or through one thickness and a fused patch. Buttonholes should not be less than 2 mm and no greater than 4 mm longer than the diameter of the button used. The slide fastener if used should be centrally located and open-ended. It should have a locking slider and should be made from polyamide.
- Pockets: All pockets should be fastened by bar tacking, back tacking, triangular tacking, or riveting.
- Hanger: All garments should be provided with hanger. The hanger strength should not be less than 30% of the breaking strength, in the warpwise direction of the fabric used for the manufacture of the garment.
- Garment presentation: Garments should be clean and free from loose threads and ends of thread.
- Seam strength: Seams should have the minimum strength of 185 N for load-bearing seams and 135 N for all other seams when measured using the standard method.
- Skirt hemigirth: For a coat of 100 cm length, the hemigirth should be equal to the user's chest measurement plus 37 cm. This allowance should be increased or decreased by 3 cm for each 8 cm change in length. For a jacket of 74 cm length the hemigirth should be equal to the user's chest measurement plus 20 cm. All measurement of garments should be carried out in accordance with standard method.
- Marking: Labels should be attached to each garment, sewn on the inside 2 cm below the collar, the label should be durable to the appropriate cleaning process. The label should give the following information:

Name, trademark, or other means of identification of supplier.
Indication of body measurement of the person. The garment is intended to fit-by giving the size designation. An indication should be given of the height of person the garment is intended to fit. Garments designed specially for short, regular, tall, or very tall people should be provided with indication to that effect by means of suffixes S, R, T, or XT, respectively.
Instructions for care and cleaning.

4.41.2 Particular requirements for lightweight two-piece working rig

The lightweight two-piece working rig (Figure 4.24) consists of a jacket and trousers. The jacket is fastened at the front by a centrally located slide fastener. The lightweight

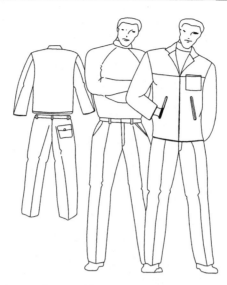

Figure 4.24 Lightweight two-piece working rig: example of seam and pocket details.

working rig should comply with the requirement of this section in addition to the requirements of standards.

- Jacket: The jacket should be single-breasted and hip length with front yoke, center slide fastener, two slanted pockets, and breast-patch pocket.
  The yoke should be 22 cm deep measured from the neck point.
  The front should have two slanted pockets, with the overall size of the pocket opening 18 cm. The bottom of the pocket opening should be 13 cm from the hem.
  The top of the pocket opening should be 30 cm from the hem. The pocket bag should be fully overlocked. The slide fastener should be centrally located and open-ended. It should have a locking slider and should be made of polyamide.
  A breast-patch pocket (or pockets) with a square-closing flap should be positioned on the jacket with the top of the pocket against the yoke. The pocket should be 14 cm deep and 13 cm wide overall. The pocket should be top stitched in matching color polyester thread and stress points should be bar tacked. Hems should be neatened.
  The top of the forepart above the slide fastener should be pressed back (lapel style).
  The collar should be one piece, with pointed ends bagged to the gorge (with no leaf-edge seam but bagged at the collar end). It should be pressed into a 3 cm stand with a 4 cm fall and should be 7 cm wide at the collar end. The inside of the top collar should be turned in and neatened.
  The top of the forepart above the slide fastener should be pressed back (lapel style).
  Edges on forepart, collar, patch pocket, and hems should be top-stitched in self-color polyester thread. The upturn at the bottom and sleeves should be neatened to 2 cm and machined through. The hanger should be sewn-in, with the stitching at the bottom of the collar, inside the center-back neck position.
- Trousers: Trousers should be made with 2 side pockets and 1 hip patch pocket with a button-down flap.
  Waistband: A one-piece outer waistband should be 4 cm wide. The inner waistband should be in a white polyamide with shirt gripper and should be 5 cm wide. The

waistband closure should be a 30 ligne (19 mm), four-hole button matching the color of the main fabric. The buttonhole should be machined through the outer waistband and bar tacked.

Belt loops: Six belt loops, each 1 cm wide, should be attached to each waistband.

Adjusting tabs: Adjusting tabs, of finished dimensions 9 cm long and 2 cm wide should be laid-on, 3 cm in advance of the side seams, turned back and sewn through the waistband, 15 mm away from the fold. Each tab should be fitted with nickel-plated-steel waist adjustment buckles.

Pockets: The top and bottom of the side pocket mouths and each top corner of the hip pocket should be bar tacked. The hip pocket should be 14 cm deep and 13 cm wide, positioned on the right hip, 9 cm from the top of the waistband. If required, a flap should be positioned 7 cm from the top of the waistband and should be 7 cm deep. The flap should be fastened by means of a 30 ligne (19 mm) matching-color button on a patch with a buttonhole in the flap.

Fly: All garments should be fitted with a fly slide fastener in polyamide, with locking slider. The finished width of the top fly should be 5 cm and the under fly 4 cm. A slide fastener tape should be inserted between the center front seam under the fly, top sewn, and overedged. The edge of the slide fastener tape at the top fly should be attached 3 cm from the fly edge by two rows of stitching, 5 to 6 cm apart. The under fly should be single thickness and arm locked on edge.

Inlays: The upturn of the bottom of the leg should be neatened to 4 cm and machined through.

- Marking: Labels should be cellulosic and should be sewn inside, the jacket label being sewn inside the back neck.

4.42 Protective apron for wetwork

4.42.1 Materials

Apron should be made from coated fabric, material composed of two or more layers at least one of which is a textile material (woven knitted, or nonwoven), and at least one of which is a substantially continuous polymeric film, bonded closely together by means of an added adhesive or by the adhesive properties of one or more of the component layers. The material may be specified either single-face coated or double-face coated fabrics with the following exceptions:

PVC-coated fabric should have a total mass per unit area no less than 140 g/m²
Silicon-coated fabrics should not be used
Eyelets and press-stud fasteners should be of nonferrous metal, nickel plated, or PVC

4.42.2 Dimensions

Aprons should not be less than 900 mm long and no less than 750 mm wide across the skirt. The bib top should not be less than 250 mm wide extending to the full width of the apron over a depth of no less than 200 mm and no more than 300 mm.

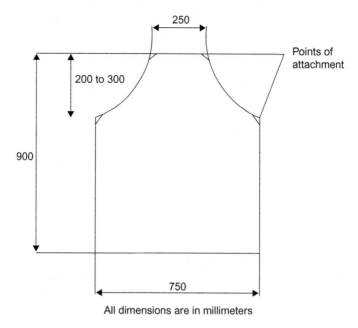

All dimensions are in millimeters

Figure 4.25 Minimum dimensions and style.

Note: The minimum dimensions of aprons and style details are given in Figure 4.25. Where straps are supplied with the apron they should have the following minimum length measured from the edge of the apron:

Waist straps	500 mm
Shoulder	750 mm
Neck loop (strap)	500 mm

4.42.3 Sewing

There should be no seams or joints in the main body of the apron. For the purpose of attaching straps or eyelets the corner of bib and waist should be reinforced with triangular pieces of self-material stitched or welded except where wieldable eyelets are used or where the fabric mass per unit area is greater than 270 g/m².

Sewing should not be less than four and no more than five stitches per 10 mm. The sewing threads should be appropriate for the fabric used. If provided, eyelets for tying-tapes should be set in at the upper corners of the bib and skirt and should be positioned clear of any stitching. If fitted, straps (either permanently fixed or with a quick release fastening) should be at least 10 mm in finished width and finished to prevent fraying.

4.42.4 Labeling

Aprons complying with this standard should be labeled with the following information:

Name, trademark, or other means of identification of the manufacturer
Standard reference for the fabric used
Length and the width of the apron
Cleaning and washing instructions

4.43 High-visibility garments and accessories

This type of body protection includes the retroreflective garments and accessories intended primarily to provide conspicuity of the user working on the roadway or other industrial premises.

4.43.1 Classification

Garment and accessories are categorized in three classes:

1. Class A: High-visibility aid providing the highest level of conspicuity
2. Class B: High-visibility aid, providing the intermediate level of conspicuity
3. Class C: High-visibility aid providing the lowest level of conspicuity
   - Performance of materials: Garments and accessories may be constructed from separate performance materials or from combined performance materials. Performance of materials should comply with BS 6629 (1985).

4.43.2 Construction of garments and accessories

For construction of garments and accessories reference should be made to BS 6629.

4.43.3 Labeling

Garments and accessories should be durably and legibly labeled with the following:

Name or identification of the manufacturer
Cleaning instructions
Class and level of conspicuity using one of the three following forms of words as appropriate:
1. Class A: High-Visibility Aid
   This garment provides the highest level of conspicuity.
2. Class B: High-Visibility Aid
   This garment provides the intermediate level of conspicuity.
3. Class C: High-Visibility Aid
   This accessory provides the lowest level of conspicuity.

4.43.4 Testing

Manufacturers should certify in writing that materials used have been tested in accordance with BS 6629 (1985).

4.44 Chemical protective clothing

Where chemical hazards exist in a work area, it is important to assess whether the risk of exposure to chemicals can be minimized or avoided. Relevant approaches are:

Use of alternative equipment or chemical involved in the event of an incident
Introduction of appropriate working practices and system of work to give early warning of possible exposure

After all reasonable efforts have been made to eliminate or minimize the hazard then consideration should be given to PPE (if required) for selection and use in work area. Care in selection of protective clothing will ensure proper use and correct protection. Suppliers or manufacturers should be consulted before buying chemical protective clothing.

4.44.1 Materials

The garments should be made from one or more of the coated fabrics specified in BS 3546. The primary function of protective clothing is to prevent, or reduce to an acceptable level, the exposure of the skin to a chemical hazard. The use of protective clothing offering a higher degree of protection that is necessary under routine conditions maybe advisable temporarily when the nature of a hazard is unclear, or when steps to reduce it cannot be taken immediately.

Clothing materials are classed broadly as air-permeable or air-impermeable. The two types of material have different applications and are considered separately in 12.2.2 and 12.2.7. Although general rules can be drawn up showing fabrics and materials that are likely to give suitable protection against different classes of chemicals, the adequacy of a material against a specific chemical can only be established by practical tests, as are referred to in 12.2.2. and 12.2.3.

- Air-Impermeable materials: Materials of a solid nature offering resistance to permeation by chemicals would, in general, be selected for the construction of protective gloves, boots and over garments that are likely to be exposed to organic solvents, liquid formulations of hazardous agents, toxic dusts, undiluted acids, and other corrosive or aggressive agents and formulated products.

4.44.2 Types of material of construction

- Coated textiles: Flexible nonabsorbent sheet materials with no pores to prevent penetration by liquids or gases. Relevant materials are made from a light, tightly woven textile base (commonly polyamide) with a suitable polymeric coating. The textile gives stability, strength, and durability to the composite is acceptable. If both faces of the textile are coated, the barrier is the more effective. The coating should be free from pinholes and there should be no exposed textile on the surface that could provide a path for liquid penetration by wicking.
- Polymer sheet: Unsupported plastic film (e.g., polyethylene) or rubber sheets can be used to construct aprons or similar garments, especially clothing designated as "disposable." The possibility of accidentally puncturing or tearing such films is greater than for a textile based

material and are less suitable for high-risk applications where significant mechanical stress on the garment can occur.

- Permeation by liquid: Even without any surface flaws or holes, coatings can absorb certain liquid chemicals, which are able to diffuse by permeation through the material. This process proceeds broadly in three stages:
  1. Initial absorption of the chemical by the polymer film or coating
  2. Solution of the chemical in the polymer film or coating
  3. Desorption from the opposite surface of the material into the internal environment of the garment

4.44.3 Classification of air-impermeable materials according to resistance to permeation

The rate of permeation of chemicals through air-impermeable materials is governed not only by the natures of the chemicals and polymers but also by the thickness of the protective coating or film and by the temperature. The measurements of breakthrough times and permeation rates under relatively artificial conditions in a laboratory should not be used as precise indicators of risk nor of the "safe" period of wear of a contaminated item nor of the effective life of a manufactured item in service. The principal application of such measurements lies with the process of selection of the most effective material from a group available for test. In a specific work situation, an acceptable time of wear can only be defined after appropriate consideration has been given to influencing factors within the workplace itself that affect the potential exposure to a chemical.

When there is risk of chemicals coming into contact with the skin air-impermeable forms of protective clothing are worn. This can be minimized by the actions indicated in Table 4.30.

4.44.4 Durability

Coatings and polymer films may be susceptible to attack by particular chemicals over a period (or repeated periods) of exposure, leading to degradation and eventual failure of the protective layer by brittle cracking. Resistance to permeation may be reduced by damage in use. Guidance from manufactures of garments and from chemical

Table 4.30 **Classification of breakthrough time in relation to type of application**

Breakthrough time	Action if contaminated	Application
up to 12 min.	Removes as soon as possible	Emergency use/disposable garments only
12 min. to 2 h	Wash off/clean immediately	Short term protection
2 h to 6 h	Wash off/clean at end of work period	Routine tasks
Over 6 h	Wash off/clean at end of work period	Long term continuous exposure

suppliers should be sought and, if necessary, tests should be performed after representative cleaning treatments and other processing to simulate the effects of continued use. It is essential to confirm that the barrier material will remain effective during its intended lifetime (see BS 903). In relation to various aspects of durability (reference should be made to BS 3424, BS 3546, and ASTM D2582-67).

4.44.5 Air-Permeable materials

Materials of a porous or semi-permeable nature (e.g., woven and spun-bonded fabrics, laminates incorporating a microporous film or coating) would normally be selected for the construction of overgarments for wear in circumstances where a compromise between protective efficiency and comfort is acceptable. This clothing would not be suitable for protection against hazardous undiluted liquid chemicals and formulations except perhaps in well-defined circumstances where contamination is limited to an occasional droplet or small drip. Their principal application would lie with the construction of overgarments for an acceptable but limited degree of protection against sprays, dusts, small drips or splashes of diluted chemicals, generally rated in a low to moderate category of chemical hazard.

- Textile fabrics: Air-permeable materials used in protective clothing act either by shedding liquids with minimum absorption and penetration, by delaying penetration sufficiently for the user to retreat to a safe place and remove the clothing, or, in the case of dustproof fabrics, by preventing penetration by solid particles.
- Fabrics are tightly woven or spun-bonded that allow air and moisture vapor to pass through them and promote user comfort. They give only limited protection against liquids and dusts, and cannot provide a satisfactory barrier to gases (some special absorbent materials containing activated charcoal are effective against many gases and vapor while the absorbent layer remains unsaturated).
- Semi-permeable materials: Semi-permeable or microporous materials, such as specially treated polytetrafluoroethylene films or polyurethane coated fabrics, allow air and water vapor to diffuse through them while presenting a barrier to the passage of liquids. They are penetrable by liquids of low surface tension and molecular size. Tests generally applicable to air-impermeable materials may apply also to semi-permeable materials.

4.45 Types and construction

4.45.1 Garments for localized protection

When there is a specific danger to part of the body only, local protection of that part may be adequate, such as:

Face protection, the face, eyes and respiratory tract may need protection as appropriate
Hand protection, to protect the hands, suitable gloves are required and the material and seams should be proof against the chemicals involved
Footwear, similar considerations should be given for the footwear
Aprons and bibs are appropriate wear if there is a clear danger of chemical attack to the front of the body only

Sleeves should be tubular in construction and should be designed to cover the fore-arm and/or upper arm so as to fit over other work wear (such as that part of a glove as may extend over the wrist). If the entries at each end of the sleeve are elasticated, the elasticated section should be covered by a wrap-over extension of the sleeve complete with adjustable grip. The upper closure should be sufficiently tight when closed so as to grip the underlying garment.

Notes:

1. The required elasticity may be obtained either by the use of a fabric that is inherently elastic or by the separate use of elastic materials, attached or inserted within the material of construction of the sleeve.
2. An example of a suitable design for a sleeve is given in Figure 4.26.

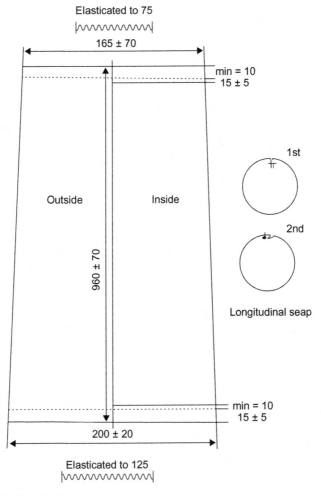

Figure 4.26 Design of a sleeve.

4.45.2 Jackets and coats

Special requirements: Jackets and coats should be capable of being closed to the neck.

Notes:

1. Jackets and coats may have an integral hood and should be designed to overlap the trousers by at least 20 cm.
2. If appropriate, a double cuff should be provided with the cuff being of sufficient length to allow the fitting of gauntlet type gloves. Both inner and outer cuffs should be elasticated or provided with other means of ensuring a close fit.
3. A double-thickness panel to act as reinforcement may be provided to the outside back of the suit to cover the area from shoulder to waist, so as to protect the suit against damage when breathing apparatus is worn.
4. Suits may be fitted with reinforced elbows and knees.
   - Collar: A collar, if provided, should be of self material and of minimum depth of 7 cm if a turnover collar and 4 cm if stand-up collar.
   - Pockets: Pockets weaken the resistance of the base fabric at the seams, present the risks of snagging, and may collect split or splashed chemicals. There should be no external pockets.
   - Hanger: A self-hanger should be provided. When the garment is suspended from its hanger, no permanent distortion nor damage to the garment should ensue.
   - Hood: A hood should be provided with the means of enabling it to be adjusted to the user's outline and to protect the throat. The distance from neck seam at one shoulder to neck seam at the other shoulder of the hood should not be less than 70 cm when measured over the crown of the head. The distance from edge to edge of the hood measured around the back of the head at eye level should not be less than 48 cm.

Notes:

1. A hood is intended to offer protection against chemicals to the head and neck.
2. For adjustment to the user's outline, a drawstring, or an elasticated closure may be incorporated into the garment.
3. In cases where elastic properties may be adversely affected by exposure to chemicals, elasticated sections should be protected by a suitable covering.
4. Where garments fitted with hoods are designed to be worn with industrial safety helmets, there is a need to increase these dimensions to values above the minimal in order to allow the weight of the garment to be carried on the shoulders rather than the head.
5. The hood should also be designed to accommodate the wearing of eye, ear, and respiratory protective equipment.

4.45.3 Trousers

Trousers should be supported, self-supported, or be of the bib and brace type. The trouser legs should be designed to permit a seal or overlap to be made with protective footwear so as to prevent the ingress of liquid.

Note: In a suitable design a double leg/ankle is provided with the leg being of sufficient length to allow the fitting of suitable footwear. Both inner and outer legs may be elasticated.

- Two-Piece suits: The overlap of the top garment over the trousers should be sufficient to ensure that protection of the body is maintained during the foreseeable range of articulation and movement of the user.

4.45.4   Complete-cover garments

Complete-cover garments can be worn with some form of visor and respirator or breathing apparatus to protect the eyes and face and to guard against inhaling chemicals. Where the danger to the skin is small, air-permeable garments together with specifically approved respiratory protection may be acceptable when dealing with chemicals in powder form. Otherwise an air-impermeable assembly should be adopted comprising one or two-piece plastics or elastomer coated coverall, gloves, boots and complete head protection. Hoods should be sufficiently large to accommodate goggles, etc., comfortably, and (if attached to a coat-like garment) to allow the weight of the garment to be taken by the user's shoulders, rather than by the head. For protection where particularly hazardous chemicals are not involved, and respiratory protection is not indicated, air-impermeable coveralls worn with gloves, goggles and boots are often adequate.

4.45.5   Air-supplied clothing

An enveloping garment (see Figure 4.27) inflated by an independent air supply presents a double barrier to the entry of chemicals. At any small holes or pores that may exist in the fabric, the excess pressure will tend to drive contaminants outward. However, the pumping action caused by the user's movements can still suck gases or particles into the suit through openings at neck, wrists, and ankles, or through pinholes. The protection afforded by air-fed suits is therefore increased by minimizing the apertures in them. Since the system does not eliminate the possibility of solvents and gases passing through the fabric by permeation, it still remains necessary to assess the resistance of the suit material to chemical permeation.

The air flow (which may be at a controlled temperature) will provide breathing air, and also maintains a tolerable temperature and humidity around the user. Any chemicals entering the suit, either by permeation or through holes, are likely to be inhaled. If the suit is to be worn for periods exceeding the known breakthrough time for permeation, the rate at which chemicals pass into the suit should be low enough, and the air flow high enough, to reduce the concentration of chemicals to well below the occupational exposure limit. Appropriate consideration should be given to the noise levels within air-supplied clothing.

4.45.6   Gas-tight suits

To isolate a user completely from his environment (e.g., from a toxic gas) an all-enveloping garment should have no pinholes and prevent passage of the gas by dissolution in the membrane (see Figure 4.27). If the breathing apparatus is isolated from the interior of the suit, the latter is effectively a sealed container. Any chemicals entering by permeation cannot be swept away and a greater concentration of contaminant will build up than in the case of the air-fed suit. Breathing apparatus (which may be either inside or outside the suit) is obviously necessary. Reinforcement of the part of the suit in contact with the apparatus may be desirable to give added support and to reduce the risk of abrasion to the suit. A gas-tight, air-supplied suit, in which the

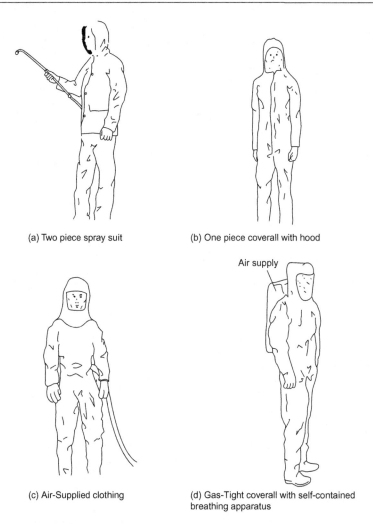

(a) Two piece spray suit   (b) One piece coverall with hood

(c) Air-Supplied clothing   (d) Gas-Tight coverall with self-contained breathing apparatus

Figure 4.27 Some examples of garments.

interior of the suit is purged and conditioned by an external air supply while breathing air is fed from a separate airline or self-contained breathing apparatus, provides maximum protection for both skin and lungs.

Note: Reinforcement should be provided to protect against damage caused by the breathing apparatus when this is worn externally.

4.45.7 Combinations of equipment

When protective clothing is worn together with other forms of PPE, such as respiratory protective equipment, eye protection, protective helmets, and/or hearing protection devices, whose primary function is other than protection of the skin as such, the following consideration apply:

Care should be taken that no new or additional hazard is introduced.
Whatever the primary function of each individual garment or item of PPE, the PPE as a whole should afford due protection to the skin.
As far as practicable, the PPE as a whole should fit the user and be comfortable in use.

4.45.8 Seams

Seams should be so constructed and sealed (by using a double overlap type of seam or other appropriate design) to prevent penetration of liquid through stitch holes or, by penetration or permeation, through other components of a seam. The performance of the seam in these respects should not be inferior to that of the material from which it is made. Seams should have a strength of no less than 150 N when tested as described in BS 2576. Where taped seams are used, the peel-attachment strength of the tape when tested as described in method 9B of BS 3424: Part 7: 1982 should not be less than 5 N/cm of tape width.

4.45.9 Closures

Button and eyelet holes should be adequate for the size of button to be used and should be through a minimum of two thicknesses of fabric. No button-hole edge should be less than 1.5 cm from the edge of a facing.
Note: Particular attention should be paid to the bight of button-holes. The bight of overlock and safety stitching should be adequate for its purpose.

The design and construction of the closure assembly should be such that there is no route for liquid to penetrate into the interior of the garment through the closure (see Figure 4.28). Any seams that form part of the closure should comply with 12.3.9. Since fastenings (zip fasteners etc.) are weak points, care is needed, particularly in high-performance garments, in the design (placement of fastenings, covering flaps, overlaps) to ensure adequate sealing. Attention is drawn to BS 3084: 1981

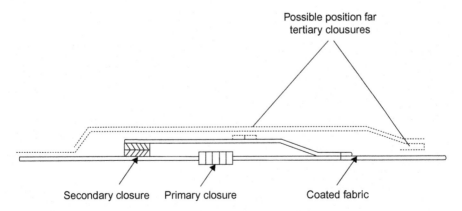

Figure 4.28 Cross section of components in a possible closure assembly.

(Codes C, D, E). All fasteners should be able to withstand the cleaning operations used on the garment.

Note: The provision of secondary or tertiary closures may some.

4.45.10 Openings

As far as possible, garment openings should be so placed as to allow easy putting on and removal, without undue strain on the material and without transfer of contamination to the user. In this respect, particular care should be taken in the removal of garments, such as blouses and ponchos, that are taken off over the top of the body because chemicals may thereby be readily transferred to the face and head. The primary garment closure may be supplemented by secondary or further closures to prevent the entry of liquids.

4.45.11 Apertures

Where separate garments combine to cover the body, good design is needed to avoid direct routes of entry for chemicals, particularly liquid jets, at the junctures, e.g., joins between respiratory protective mask and hood (or coverall); gloves and sleeves; jacket and trousers; trouser legs and boots. The direction from which the hazard is expected will dictate which of the components should be the outer (e.g., jacket outside trousers to guard against falling liquids). Elasticated cuffs and elasticated trouser bottoms are valuable in preventing liquids from running up sleeves and trouser legs but may allow liquids to come in contact with the skin. Further protection is given by interleaved double overlap junctures, especially if the two components can be held together or against the body by drawstrings, etc.

4.45.12 Protection against additional hazards

- Protection against hazards other than exposure to or contact with chemicals:
  Notes: The recommendations given in relevant standards must be taken into account when garments that are primarily designed to protect the skin against contact with chemicals are required to afford a degree of protection against other hazards. It should be kept in mind that some hazardous chemicals if spilled onto certain types of material might not be detected immediately and a hazard could be realized at some subsequent time.
- Explosion: Mixture of flammable gas or dust and air can be ignited by sparks. In such circumstances the build-up and effective ignition discharge of static electricity should be avoided by selecting nonspark clothing and footwear.
- Heat stress: The human body produces about 100W of heat energy at rest, increasing to perhaps 700W with vigorous exercise. This heat has to be dissipated to the surroundings by convection or other means. While sweating is the body's response to high temperatures, giving efficient evaporative cooling to regulate the body temperature, all enveloping clothing restricts heat dissipation by both convection and evaporation, and allows the heat stored in the body to increase. Heat stress results in discomfort (damp underclothing), lethargy, fatigue, loss of concentration, and eventual unconsciousness. It is possible for employees to be unaware of the danger until they are on the point of collapse. Although the hazard is worst

for clothing such as gas tight suits and other incompletely enveloping garments, where body moisture can not escape. Thus a PVC coverall worn on a warm day can cause distress to the user even when he is not working hard physically. For such clothing, particularly at high levels of physical exertion, work periods should be limited to a defined maximum time and should include enforced rests. If this is not possible suits with an outside supply of air for ventilation should be used. To overcome potential problems of heat stress, there may be advantages in designing garment interfaces that permit good air flow while still protecting against liquid drenching, e.g., trousers suspended by braces and overlapping jacket.

4.46   Use and maintenance

4.46.1   Limitations

Employees at all levels should be made familiar with the function and limitations of protective clothing, and they should be encouraged of wearing them.

4.46.2   Storage

Adequate space should be provided in a dry, well-ventilated room maintained at a moderate temperature for storage of garment. Protective clothing should be stored separately from personal clothing and chemicals and away from bright sunlight and from any equipment liable to produce ultraviolet radiation or ozone that might degrade it. Garments should be stored neatly, as far as possible free from creases or other types of distortion that could cause cracking.

Garments of different types and constructions should be kept apart from each other to avoid confusion. New garments should be kept separate from used ones. If possible, each employee should have his own clothing to make control easier, for hygiene, and to encourage personal responsibility. Wherever practical, a set practice should be laid down for the issue and storage of new and old garments, and records kept of the receipt, inspection and use of each one.

4.46.3   Inspection

Garments should be inspected on receipt, before and after use, and after repair. The garment should have no signs of damage or contamination (e.g., pinholes, abrasion, cuts, cracking discoloration, and lifting of seams or welds).

4.46.4   Use of protective clothing

Users should inspect the clothing for possible damage or soiling before putting it on.
Correct closure of all seals and fastenings should be checked.
Garments should be taken off in a set order that minimizes the chance of contaminating the user, with the help of an assistant if necessary.

Note: Since contamination on hands or clothing can be transferred to food, drink, tobacco, or cosmetics, and then ingested, contaminated clothing should not be worn in places where food and drink is consumed or where cosmetic application or smoking is permitted.

After using protective clothing users should practice exact personal hygiene and should not smoke, eat, drink nor use cosmetics nor use a toilet until they have washed at least face and hands and have moved to an area free of chemicals.

4.46.5 Cleaning

It should be considered preferable to dispose of contaminated garments rather than cleaning them. Information for safe handling of contaminated items of protective clothing should be provided whether disposal or cleaning is carried out at the work location or off the site.

- Cleaning facilities: Depending on the nature of the work and the chemicals involved, consideration should be given to the following recommendations:
  The cleaning station should be spacious, well-ventilated, and provided with running water and approved drainage.
  It should have a well-defined work-flow system to prevent cross-contamination.
  Consideration should be given to the provision of separate "clean" and "dirty" rooms, with intermediate areas where users can put on or remove their clothing, and pass through a shower on leaving the contaminated room.
  The cleaning station should be located as close as practicable to the working area:
  To minimize the time between contamination and cleaning
  To minimize risk of contaminant being transferred on clothing to a nominally clean area where unprotected people are working
- Cleaning operation: Garments should be cleaned according to the manufacturer's instructions. Any contaminated waste should be safely disposed of. Some possible sequences of cleaning operations are shown in the flow chart (Figure 4.29).

Note: Static soaking merely serves to redistribute the contaminant, and should be avoided. Solvents that may cause the garment material to swell or crack, or that may leach out certain components (such as plasticizers), should not be used for cleaning.

The staff responsible for cleaning should be well-trained and familiar with the properties of the chemicals and clothing in use. Where cleaning is performed by a separate organization, the cleaner should be informed of the recommended procedures and of any chemical hazard associated with the clothing, and should be required to certify that cleaning has been carried out according to the recommendations.

- Repair: The repair of damaged garments is not recommended, but if the damage is minor the repair should be followed in accordance with manufacturer's instructions. Repaired clothing should be inspected and tested before every use. Reference should be made to ISO 6530 for penetration test and BS 4724 Parts 1 and 2 for breakthrough time test.
- Disposal: Where a garment has been damaged so badly or highly contaminated it should be considered unserviceable and destroyed at once. Garments deteriorate slowly by wear, contamination, and cleaning, therefore an estimation of garments lifetime should be made in consultation with the manufacturer, and the garments should be destroyed well before the point of possible breakdown.

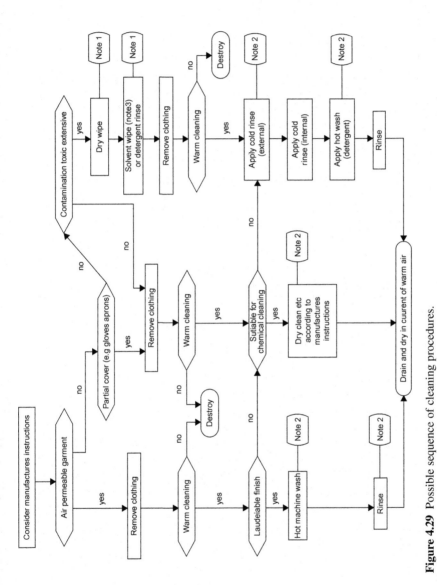

Figure 4.29 Possible sequence of cleaning procedures.

Notes:

1) Contaminated rags/cloths should be disposed off in a safe manner.
2) Care is needed with disposal of contaminated effluent and decontamination or disposal of cleaning equipment.
3) Use solvent that does not affect the garment material.

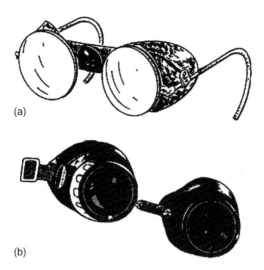

(a)

(b)

Figure 4.30 Types of eye protection for welding and cutting.

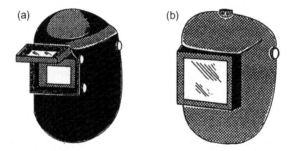

(a)   (b)

Figure 4.31 Arc welding helmet.

Figure 4.30 shows Types of eye protection for welding and cutting and Figure 4.31 shows arc welding helmet.

4.46.6   Records of use

Where appropriate, according to the type of hazard under consideration, records should be kept of the use of protective clothing under the following headings:

Garment type and specification
Dates of purchase and issue
Inspection
Names of previous users
Previous use (with appropriate details of chemical exposure, particularly when the garment has been used to protect against chemicals presenting a high degree of hazard)
Cleaning
Final disposal

These records should be in a form that is easy to update.

4.47   Sizing of jackets, coats, over-trousers, and one-piece suits

The finished garment measurements for jackets, coats, over trousers and one-piece suits, should be no less than those given in Tables 4.31, 4.32, and 4.33. Jackets and coats should have a minimum length of 86 cm. All sizes of over-trousers should have a circumference at the bottom of the trousers of at least 53 cm.

4.47.1   Instructions and marking

- Instructions: Instructions should accompany each garment or should be provided by the manufacturer in separate literature and should include the following information:
  1. The identity of the materials of construction
  2. If the clothing is marked to indicate the extent that the clothing offers protection against specific chemicals, either:

     If such chemicals are liquid, the results of tests (on the materials of construction) carried out as described in BS 4724: Part 1 or 2 or

     If such chemicals are not liquid, details of relevant characteristics of the clothing when exposed to such chemicals

Table 4.31 Sizing details for jackets and coats

Wearer's chest girth Cm	Size nomenclature	Chest measurement of fastened garment cm	Garment sleeve* length cm	Cuff circumference Cm
92	Small S	115	77	33
100	Medium M	123	81	33
108	Large L	131	85	33
116	Extra Large XL	139	89	33
124	Extra Extra Large XXL	147	93	33

*Sleeve lengths should be measured from center back to cuff.

Table 4.32 Sizing details for overtrousers including "bib and brace"

Wearer's chest girth Cm	Size nomenclature	Chest measurement of fastened garment cm	Garment sleeve* length cm	Cuff circumference Cm
80	Small S	106	70	104
92	Medium M	118	74	109
104	Large L	130	77	113
116	Extra Large XL	142	78	115
128	Extra Extra Large XXL	154	79	117

Table 4.33 Sizing details for one-piece suits

Sizing details for one-piece suits

Wearer's chest girth cm	Size nomenclature	Garment measurement back neck to crotch cm	Chest measurement of fastened garment cm	Garment sleeve length* cm	Garment inside leg cm	Cuff circumference cm	Ankle cuff circumference cm
92	Small S	91	115	77	67	33	53
100	Medium M	95	123	81	71	33	53
108	Large L	99	131	85	75	33	53
116	Extra Large XL	103	139	89	79	33	53
124	Extra Extra Large XXL	107	147	93	83	33	53

*Sleeve lengths should be measured from center back to cuff.

- Marking: Clothing should be labeled with the following information:
  Name, trademark, or other means of identification of the manufacturer
  Number and date of standard followed
  Month and year of manufacture
  Manufacturer's type number, identification number, or model number
  As appropriate, the waist, chest or head girth measurement of the user the garment is intended to fit and the size nomenclature, e.g., size medium to fit up to chest 100 cm.
  If appropriate an indication that boots of gas-tight suits are provided with protective mid-soles.
  If appropriate, a statement to indicate the extent that the clothing offers protection against specific chemicals and to explain which of the following sources of information should be referred to for details of relevant characteristics of the clothing when exposed to such chemicals:
  1. Manufacturer
  2. Manufacturer's instructions
  3. Appropriate test certificates

Notes: Consideration should be given to suitable additional marking:

Where the garment is designed to offer protection against gases or solids.
Where the garment is constructed from materials having special characteristics. Attention is drawn to BS 3424, BS 5438, and BS 6249.
In cold locations periodic test inspections as stated above should be made at least every month.

4.48   Face protection

Table 4.34 provides transmittances and tolerances in transmittance of various shades of filter lenses.

Arc welding creates artificial UV rays that will damage personnel cornea. It happens slowly but surely. It's called flash burn. It's an old school way of protecting the eyes during welding and cutting. They used in addition to the welding helmet. The user flips up the welding helmet lens so that he can perform other work without damaging the eyes (like cutting).

Image B shows another old school way of protecting the eyes, except these eye-cup type goggles are used in gas cutting and welding procedures. They should have a lens that filters out harmful rays from gas welding (which is infra-red). When arc welding or cutting a helmet that is fitted with the proper type of lens is needed. The welding helmets in Image C and D have flip-up windows that are actually lenses that filter out UV rays from welding and cutting.

- The first helmet (C) has a smaller lens. It is 2″ × 4.25″. This is a smaller opening.
- The second helmet (D) has bigger lens. It measures 4.5″ × 5.25″. A bigger lens is like having a bigger window to look out of and it's very handy because the user can see more. However, they usually do not flip up because they are so big, and that's why the small window is a popular choice, because being able to flip up provides the user with more air and to see work more clearly.

Table 4.34 Transmittances and tolerances in transmittance of various shades of filter lenses

Shade no.	Optical density			Luminous transmittance			Maximum infrared transmittance	Maximum spectral transmittance and violet for four wave lengths (Millimicrons)			
	Min.	Std.	Max.	Min.	Std.	Max.		313 / 405		334	365
							PERCENT				
1.5	0.17	0.214	0.26	67	61.1	55	25	0.2	0.8	25	65
1.7	0.26	0.300	0.36	55	50.1	43	20	0.2	0.7	30	50
2.0	0.36	0.429	0.54	43	37.3	29	15	0.2	0.5	14	35
2.5	0.54	0.643	0.75	29	22.8	18.0	12	0.2	0.3	5	15
3.0	0.75	0.857	1.07	18.0	13.9	8.50	9.0	0.2	0.2	0.5	6
4.0	1.07	1.286	1.50	8.50	5.18	3.16	5.0	0.2	0.2	0.5	1.0
5.0	1.50	1.714	1.93	3.16	1.93	1.18	2.5	0.2	0.2	0.2	0.5
6.0	1.93	2.143	2.36	1.18	0.72	0.44	1.5	0.1	0.1	0.1	0.5
7.0	2.36	2.571	2.79	0.44	0.27	0.164	1.3	0.1	0.1	0.1	0.5
8.0	2.79	3.000	3.21	0.164	0.100	0.161	1.0	0.1	0.1	0.1	0.5
9.0	3.21	3.429	3.64	0.061	0.037	0.023	0.8	0.1	0.1	0.1	0.5
10.0	3.64	3.857	4.07	0.023	0.0139	0.0085	0.6	0.1	0.1	0.1	0.5
11.0	4.07	4.286	4.50	0.0085	0.0052	0.0032	0.5	0.05	0.05	0.05	0.1
12.0	4.50	4.714	4.93	0.0032	0.0019	0.0012	0.5	0.05	0.05	0.05	0.1
13.0	4.93	5.143	5.36	0.0012	0.00072	0.00044	0.4	0.05	0.05	0.05	0.1
14.0	5.36	5.571	5.79	0.0004	0.00027	0.00016	0.3	0.05	0.05	0.05	0.1

Note: The values given apply to Class I filter glass. For Class II filter lenses, the transmittances and tolerances are the same, with the additional requirement that the transmittance of 589.3 millimicrons should not exceed 15 percent of the luminous transmittance.

Table 4.35 **Lens shade numbers for different welding projects**

Shade no.	Operation
Up to 4..	Light electric spot welding or for protection from stray light from nearby welding.
5..	Light gas cutting and welding.
6–7...	Gas cutting, medium gas welding, and arc welding up to 30 amperes.
8–9...	Heavy gas welding and arc welding and cutting, 30–75 amperes.
10–11..	Arc welding and cutting, 76–200 amperes.
12...	Arc welding and cutting, 201–400 amperes.
13–14..	Arc welding and cutting exceeding 400 amperes.

However, it's important that users also wear the flash-type goggles underneath in case there's another welder in the vicinity so the user isn't affected by their arc welding flashes. As mentioned above the flash goggles will also keep the eyes safe from bits of metal and slag. Good-quality helmets and some goggles have removable lenses that include both a filter lens and a clear protective lens. There is a protective lens in front of the filter lens that protects it from sparks and spatter, which will damage the filter lens. Table 4.35 shows lens shade numbers for different welding projects, and Table 4.36 gives more advanced lens shade chart for welding and cutting.

Two Reasons for A Filter Lens:

1. The user needs a filter lens to lower the intensity of the light so that it eliminates glare from welding and so he can see the weld puddle. The user can't weld if he can't see the puddle or the area he is welding.
2. Damage from UV rays. If the user doesn't have a lens that filters out UV radiation from arc welding hewill get flash burn or worse.

Welding Helmet Tip:

If the user is not sure about the lens shade for the welding job try this; fit the lens inside the helmet and then have him look at a fairly bright light bulb (that is turned on). Can he see the outline of the light bulb? If he can then he probably needs a darker lens. Therefore, then try it again with a darken lens. If The cannot see the outline of the light bulb then it is probably the right one.

Don't do this test in a home or office but in the area where the welding is to be done because the lighting is different. If the welding is done outside the user can try looking at a bright object in the area that is reflecting the sun.

Table 4.36 More advanced lens shade chart for welding and cutting

Lens shades for welding and cutting		
Welding or cutting operation	Electrode size metal thickness or welding current	Filter shade number
Torch soldering	–	2
Torch brazing	–	3 or 4
Oxygen cutting		
Light	Under 1 in., 25 mm	3 or 4
Medium	1 to 6 in., 25 to 150 mm	4 or 5
Heavy	Over 6 in., 150 mm	5 or 6
Gas welding		
Light	Under 1/8 in., 3 mm	4 or 5
Medium	1/8 to 1/2 in., 3 to 12 mm	5 or 6
Heavy	Over 1/2 in., 12 mm	6 or 8
Shielded metal-arc	Under 5/32 in., 4 mm	10
welding (stick)	5/32 to 1/4 in., 4 to 6.4 mm	12
electrodes	Over 1/4 in., 6.4 mm	14
Gas metal-arc welding (MIG)		
Non-ferrous base metal	All	11
Ferrous base metal	All	12
Gas tungsten arc	All	12
Welding (TIG)		
Atomic hydrogen welding	All	12
Carbon arc welding	All	12
Plasma arc welding	All	12
Carbon arc air gouging		
Light	–	12
Heavy	–	14
Plasma arc cutting		
Light	Under 300 Amp	9
Medium	300 to 400 Amp	12
Heavy	Over 400 Amp	14

4.49 Body-protection method for the determination of seam strength

A constant rate of extension tensile testing machine as described in BS 2576, except that the machine does not require means for indicating (or recording) extension.

4.49.1 Preparation of test specimens

Cut specimens from the garment from a double thickness of fabric that includes a seam, such that the seam lies midway between the ends and perpendicular to the major axis of the specimen when prepared and opened out as shown in Figure 4.32(a).

Cut the specimen 5 cm wide and of length/such as to provide a nominal gauge length of 100 mm. Cut one specimen from each main load bearing seam (up to a maximum of five) of the garment ensuring that the following seams are included:

For garments with sleeves, an armhole seam
For garments with trousers, the seat seam
For garments with two-piece back, the back seam
For bib and brace overalls, the waist seam.

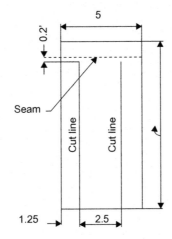

Note:
All dimensions are in centimeters.

(a) Double thickness specimen as cut from garment

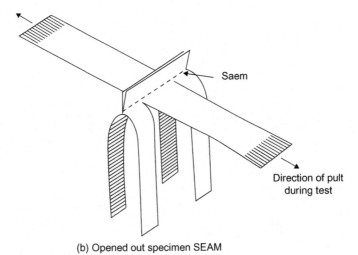

(b) Opened out specimen SEAM

Figure 4.32 Strength test Specimen.

Cut one specimen from each nonload bearing seam (up to a maximum of five) of the garment.

- Procedure: Carry out the test in accordance with 7.1 of BS 2576, additionally ensuring that the seam lies midway between the jaws and perpendicular to the direction of pull as shown in Figure 40b.
- Expression of Results: Record the seam strength, in Newtons, of each of the load bearing seams and calculate the mean value. Express the average strength of the load bearing seams to 0.1 N or to 1% of the sea strength, whichever be the greater, in Newtons.

4.50   Body-protection measurement and sizes

- Height: The length of the body measured in a straight line from the crown of the head to the soles of the feet. See Figure 4.33(a).
- Neck girth: The girth of the neck measured with the tape-measured passed 2 cm below the Adam's apple and at the level of the seventh cervical vertebra. See Figure 4.33(b)

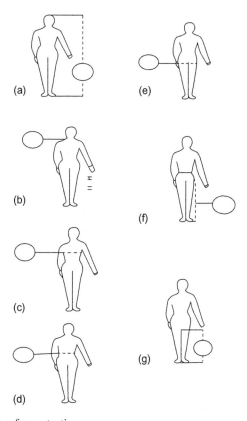

Figure 4.33 Body size for protection.

- Chest girth: The maximum horizontal girth measured during normal breathing with the subject standing erect and the tape measure passed over the shoulder blades (scapular), under the armpits (axillae), and across the chest. See Figure 4.33(c).
- Waist girth: The girth of the natural waist-line between the top of the hip bones (iliac crests) and the lower ribs, measured with the subject breathing normally and standing erect with the abdomen relaxed. See Figure 4.33(d).
- Hip girth: The horizontal girth measured round the buttocks at the level of maximum circumference. See Figure 4.33(e).
- Outside leg length: The distance from the waist to the ground, measured with the tape-measure following the contour of the hip. See Figure 4.33(f).
- Inside-leg length: The distance between the crotch and the soles of the feet, measured in a straight line with the subject erect, feet slightly apart, and the weight of the body equally distributed on both legs. See Figure 4.33(g).

Table 4.37 shows coveralls for men and youths and Table 4.38 shows coats and jackets for men and youth.

Table 4.37 Coveralls for men and youths

Size	Measurements of wearer		Measurements of garment when fastened						
	Inside leg max. cm	Chest Girth (Over Shirt) cm	Waist cm	Seat cm	Inside leg cm	Side Seam cm	Upper leg cm	Knee cm	Bottom of leg cm
72S	82	72	72	92	72	98	60	49	49
76S	82	76	76	96	72	98	62	49	49
80S	82	80	80	100	72	99	64	50	49
84S	82	84	84	104	72	99	66	51	49
88S	82	88	88	108	72	100	68	52	49
92S	82	92	92	112	72	100	70	53	49
96S	82	96	96	116	72	101	72	54	49
100S	82	100	100	120	72	101	74	55	49
104S	82	104	104	120	72	102	76	56	49
72R	90	72	72	92	78	105	60	49	49
76R	90	76	76	96	78	105	62	49	49
80R	90	80	80	100	78	106	64	50	49
84R	90	84	84	104	78	106	66	51	49
88R	90	88	88	108	78	107	68	52	49
92R	90	92	92	112	78	107	70	53	49
96R	90	96	96	116	78	108	72	54	49
100R	90	100	100	120	78	108	74	55	49
104R	90	104	104	120	78	109	76	56	49

(Continued)

Table 4.37 (Continued)

Size	Measurements of wearer		Measurements of garment when fastened						
	Inside leg max. cm	Chest Girth (Over Shirt) cm	Waist cm	Seat cm	Inside leg cm	Side Seam cm	Upper leg cm	Knee cm	Bottom of leg cm
108R	90	108	108	124	78	109	78	57	49
112R	90	112	112	128	78	110	80	58	49
116R	90	116	116	132	78	110	82	59	49
76T	98	76	76	96	84	112	62	49	49
80T	98	80	80	100	84	113	64	50	49
84T	98	84	84	104	84	113	66	51	49
88T	98	88	88	108	84	114	68	52	49
92T	98	92	92	112	84	114	70	53	49
96T	98	96	96	116	84	115	72	54	49
100T	98	100	100	120	84	115	74	55	49
104T	98	104	104	120	84	116	76	56	49

Table 4.38 Coats and jackets for men and youths

Size	Measurements of wearer		Measurements of garment when fastened		
	Height	Chest girth (Over shirt)	Chest	Sleeve length	Across back
	cm	cm	cm	cm	cm
92S	156 TO 170	92	104	76	42
100S		100	112	77	44
108S		108	120	78	46
116S		116	128	79	48
92R	170 TO 182	92	104	82	42
100R		100	112	83	44
108R		108	120	84	46
116R		116	128	85	48
124R		124	136	86	50
92T	182 TO 196	92	104	88	42
100T		100	112	89	44
108T		108	120	90	46
116T		116	128	91	48
124T		124	136	92	50

Table 4.39 **Bib and brace overalls for men and youths**

Size	Measurements of wearer			Measurements of garment when fastened							
	Inside Leg max. cm	Waist Girth (Over Shirt) max. cm	Waist Cm	Seat cm	Inside Leg cm	Side Seam cm	Upper Leg cm	Knee cm	Bottom of Leg cm	Crotch to Top of BIB cm	Width of Bib cm
76S	82	76	78	98	70	98	64	50	49	54	24
84S	82	84	86	106	70	99	68	52	49	57	25
92S	82	92	94	114	70	100	72	54	49	60	26
100S	82	100	102	122	70	101	76	56	49	63	27
108S	82	108	110	126	70	102	80	58	49	66	28
76R	90	76	78	98	76	105	64	50	49	57	24
84R	90	84	86	106	76	106	68	52	49	60	25
92R	90	92	94	114	76	107	72	54	49	63	26
100R	90	100	102	122	76	108	76	56	49	66	27
108R	90	108	110	126	76	109	80	58	49	69	28
116R	90	116	118	134	76	110	84	60	49	72	29
124R	90	124	126	142	76	111	88	62	49	75	30
76T	98	76	78	98	82	112	64	50	49	60	24
84T	98	84	86	106	82	113	68	52	49	63	25
92T	98	92	94	114	82	114	72	54	49	66	26
100T	98	100	102	122	82	115	76	56	49	69	27
108T	98	108	110	126	82	116	80	58	49	72	28

Table 4.40 Overalls trousers for men and youths

Size	Measurements of wearer		Measurements of garment when fastened						
	Inside leg max cm	Chest girth (Over shirt) cm	Waist cm	Seat cm	Inside leg cm	Side seam cm	Upper leg cm	Knee cm	Bottom of leg cm
72S	82	72	72	92	72	98	60	49	49
76S	82	76	76	96	72	98	62	49	49
80S	82	80	80	100	72	99	64	50	49
84S	82	84	84	104	72	99	66	51	49
88S	82	88	88	108	72	100	68	52	49
92S	82	92	92	112	72	100	70	53	49
96S	82	96	96	116	72	101	72	54	49
100S	82	100	100	120	72	101	74	55	49
104S	82	104	104	120	72	102	76	56	49
72R	90	72	72	92	78	105	60	49	49
76R	90	76	76	96	78	105	62	49	49
80R	90	80	80	100	78	106	64	50	49
84R	90	84	84	104	78	106	66	51	49
88R	90	88	88	108	78	107	68	52	49
92R	90	92	92	112	78	107	70	53	49
96R	90	96	96	116	78	108	72	54	49
100R	90	100	100	120	78	108	74	55	49
104R	90	104	104	120	78	109	76	56	49
108R	90	108	108	124	78	109	78	57	49
112R	90	112	112	128	78	110	80	58	49
116R	90	116	116	132	78	110	82	59	49
76T	98	76	76	96	84	112	62	49	49
80T	98	80	80	100	84	113	64	50	49
84T	98	84	84	104	84	113	66	51	49
88T	98	88	88	108	84	114	68	52	49
92T	98	92	92	112	84	114	70	53	49
96T	98	96	96	116	84	115	72	54	49
100T	98	100	100	120	84	115	74	55	49
104T	98	104	104	120	84	116	76	56	49

Table 4.41 Lightweight, two-piece working rig, jacket

Size	Measurements of wearer		Measurements of garment when fastened					
	Chest cm	Waist cm	Length of jacket cm	Sleeve length cm	Cuff cm	Chest* Cm	Waist* Cm	Back[†] cm
Small	90 to 96	80 to 84	69	78	28	Chest measurement of wearer plug 17	Waist measurement of wearer plug 17	44
Medium	100 to 104	90 to 94	69	79	28			45
Large	110 to 114	100 to 104	69	81	30			46
Extra large	120 to 124	110 to 114	71	83	32			47

*From edge to edge.
[†]Full width.

Table 4.42 **Lightweight, two-piece working rig, trousers**

Size*	Measurements of wearer		Measurements of garment when fastened						
	Inside leg cm	Chest girth (Over shirt) cm	Seat cm	Waist cm	Inside leg cm	Side seam cm	Upper leg cm	Knee cm	Bottom of leg cm
76R	90	76	93	76	76	102	62	54	50
80R	90	80	97	80	76	102	64	54	50
84R	90	84	101	84	76	102	68	54	50
88R	90	88	105	88	76	102	70	54	50
92R	90	92	107	92	76	104	72	58	54
96R	90	96	109	96	76	104	74	58	54
100R	90	100	113	100	76	104	74	58	54
108R	90	108	121	108	76	106	74	58	54
116R	90	116	127	116	76	108	78	62	56
124R	90	124	134	124	76	110	78	62	56
80T	cm	80	97	80	82	108	64	54	50
84T	98	84	101	84	82	108	68	54	50
88T	98	88	105	88	82	108	70	54	50
92T	98	92	107	92	82	110	70	54	54
96T	98	96	109	96	82	110	74	58	54
100T	98	100	113	100	82	110	74	58	54
108T	98	108	121	106	82	112	74	58	54
116T	98	116	127	116	82	114	78	62	56
124T	98	124	134	124	82	116	78	62	56

Table 4.43 **General purpose gowns (Garments made from polyester/cellulosic fabrics; not for use in anesthetizing areas)**

Size	Measurements of garment when fastened					
	Chest cm	Neck cm	Sleeve length* cm	Cuff cm	Length cm	Skirt hem girth cm
S	112	46	46	38	117	162
M	122	51	47	41	117	172
L	132	56	48	43	117	182

*The sleeve length given is measured from the neck point to the sleeve edge along the shoulder line.

Note: Tolerances are $^{+5}_{-0}$ cm for total length and $^{+3}_{-0}$ cm for all girth measurements.

4.51   Body protection: method of measuring garments

When taking measurements, place the garment flat on a smooth, flat surface of adequate size for measuring. Unless otherwise stated, garments should be measured when fastened. Figures 4.34 and 4.35 are provided to illustrate the locations at which measurements should be taken; they do not purport to indicate style.

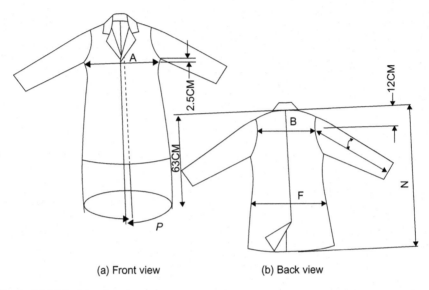

(a) Front view   (b) Back view

Figure 4.34 Man's coat.

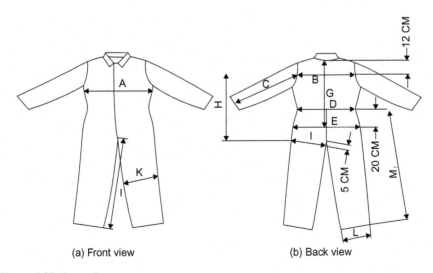

(a) Front view   (b) Back view

Figure 4.35 Coverall.

- Chest or Bust: Measure 2.5 cm below the underarm seam (location A, Figures 4.34 and 4.35 (a)), side seam to side seam, 30 cm down from the shoulder line. Multiply the value by two. Measure wrap-over overalls opened out flat.
- Across Back: Measure at a point 12 cm below the center back neck, straight across, between the armholes (location B, Figures 4.34(b) and 4.35 (b)).
- Waist: Measure the fastened waistband (location D, Figure 4.34(b)). If no waistband is present, measure the distance at the narrowest point between bust and hip locations. Multiply the value by two.
- Seat: Measure 20 cm below the waist-line (location E, Figure 4.35(b)). Multiply the value by two.
- Hips: Measure at a point 63 cm below center back neck (location F, Figure 4.34(b)). Multiply the value by two. Measure wrap-over overalls opened out flat.
- Back Neck to Crotch : Measure from center back neck collar seam to crotch (location G, Figure 4.35(b)).
- Crotch to Top of Bib : Measure from top of bib at center to crotch Figure 4.35(b)).
- Inside Leg: Measure from the crotch seam to the bottom of the leg hem.
- Upper Leg: Measure across the leg 5 cm below the crotch line. Multiply the value by two.
- Knee: Measure across the leg at the knee. Multiply the value by two (location K, Figure 4.35(a)). The position of the knee is taken as the distance up, from the bottom of the leg, of half the inside leg measurement plus 5 cm.
- G.13 Bottom of Leg : Measure across the leg along the edge of the hem (location L, Figure 4.35(b)). Multiply the valve by two.
- G.14 Side Seam: Measure from the top of the waistband to the bottom of the leg hem (location M, Figure 4.35(b)).
- G.15 Sleeve Length: Long sleeves Measure from the center back to the full length of the sleeve (location C +0.5 B, Figures 4.34(b) and 4.35(b)).
- G.16 Cuff: Measure across the sleeve along the edge of the hem. Multiply the value by two.

4.52 Body-protection guidance for the use of garments and accessories

Class A garments provide the user with a significantly higher level of conspicuity than Class B garments, and these in turn provide considerably higher conspicuity than Class C accessories. For this reason Class A garments should be selected wherever conditions allow their use. In particular, those users likely to be exposed to relatively high risk (e.g., pedal cyclists or motorcyclists) should wear Class A garments, and those exposed to particularly sever risk (e.g., traffic police officers or road workers) should consider wearing garments for which the performance is significantly better than the minimum specified for Class A.

The requirements for retroreflective performance in this standard are based on typical positions for head lamps and drivers of vehicles, and on typical positions for pedestrian and cyclist users, on public roads, railways and factory roadways. Articles complying with this standard are not necessarily suitable if the situation of lamps or observer or target differs significantly from these. Examples might be marine situations, or airfields, where specialist advice should be sought. Use of garments or accessories complying with this standard does not free the user from the normal duty to take all reasonable care.

4.53   Chemical-protection guidelines on selection and use of chemical-protective clothing

4.53.1   Assessment of chemical hazard

• Sources of data: Any chemical should be regarded as a potential health hazard to an extent dependent on the circumstances of use. For information on the chemical hazards presented by a specific chemical or formulation, the supplier should be consulted, in this respect.

Note: Suppliers have a legal duty to label the containers of dangerous substances with standard warning phrases appropriate to the hazard presented by the contents. Suppliers of dangerous substances also have a duty to ensure that their products are safe so far as is reasonably practicable and without risks to health when properly used, to advise users of their products of any hazards involved, and (were appropriate) to indicate suitable protective clothing.

• Assessment of nature of chemical hazard: Exposure of the skin to chemicals may cause harm both to the skin and to the body as a whole. The following considerations should be taken into account in appreciating the nature of the chemical hazard in the light of the information obtained from the sources.

Corrosive chemicals may destroy the skin and flesh by direct attack. Other chemicals such as petrol, paint solvents, and cleaning fluids can dissolve the skin's natural oils so leaving the skin dry and liable to from painful cracks. Such damage to the skin together with any existent cuts and grazes provide entry points for foreign substances and thus increase the risk of harm to the body.

Chemicals may pass through the skin and be carried by the bloodstream so as to cause injury to other parts of the body that may be remote from the initial point of contact.

Chemicals may gain access to the body via, for example, the eyes, respiratory or digestive tract.

The body's tolerance and rate of elimination of a foreign substance varies from person to person.

The harmful effects depend broadly on the amount of substance contacted or absorbed and hence are related to the concentration of the substance to which the body has been exposed or to the concentration in the environment, and to the duration of exposure.

The rate at which a chemical is taken up by the body may depend on whether it is swallowed, inhaled or absorbed through the skin.

Adverse effects on health may arise from a single exposure or repeated exposure to small amounts of a chemical and may be immediate, delayed or long term.

A mixture of chemicals may create a greater hazard than would the same chemicals separately.

4.53.2   Assessment of risk and danger

After steps have been taken to contain, minimize, or eliminate the chemical hazard and to reduce the risk, an assessment should be undertaken of how the chemical might inadvertently be released from within the plant or system in which it is contained and what the consequences of such release might be.

• Risk in relation to the physical form of chemicals: The type of risk associated with chemicals vs. with their physical forms, as indicated as follows:

Provided that they are free from volatile products or fumes or dust, solid materials in bulk may usually be contained without undue risk.

Liquids and free-running powders are mobile and can give intimate contact with the skin. Exposure may range from accidental splashes of laboratory reagents to deluge conditions. Although gases and vapors present relatively small amounts of matter in contact with the skin, they require more efficient barriers to exclude them. The danger associated with gases and vapors is high, especially if they are not detectable by the human senses.

The risk of release of airborne particles (fine dusts and liquid mists) and the consequent danger may be very high because particles in such physical forms are both pervasive and dense.

- Risk and danger in relation to storage and distribution: The danger associated with the inadvertent discharge of a chemical depends (for example) on the quantity, mode of transport and the manner of distribution of material present, the method of containment (flow pipes, glass bottles, etc.), the pressure and temperature at which it is held, and its proximity to working areas. The exposure to be considered may range from foreseeable incidents with moderate or high probability of occurrence (spillage while handling; contamination by sprays) to infrequent, but more serious, possibilities (for example, the fracture of a pipe in a chemical plant).
- Risk in relation to duration of exposure: The duration of exposure may increase when:
Contamination of the body is not apparent as soon as it occurs
An employee has to set emergency procedures in operation (e.g., switching off machinery) before leaving the hazard area
An employee is some distance from a place where the contaminant can be washed off

Note: The protection provided to employees and to rescue personnel should take into account the time needed to carry out necessary emergency actions.

4.53.3   Assessment of the need for protection

For this assessment of protection the following questions should be answered:

What is the chemical hazard, physical state, quantity, and mode of use of the chemical substance(s) involved?
Do these constitute a potential danger?
If so, can the hazard be removed, or the danger minimized by means other than the use of protective clothing.
How serious is the potential hazard?

If the answers to questions (a) to (d) indicate that, in addition to other precautions, the need for protective clothing should be considered, the following questions apply.

What form and extent of exposure is envisaged (e.g., spillage, liquid jets)?
What is the probability of exposure?
Will workers be immediately aware of exposure if it occurs?
What is the likely duration of exposure?
Is exposure likely to be restricted to specific parts of the body (i.e., eyes, ears, lungs, head, hands, and feet) or to specific areas of the skin?

4.53.4   Selection of protective clothing

It should be recognized that protective clothing is generally inherently prone to permeation and hence the form of protection provided should not be regarded as a complete barrier (to chemicals) under all circumstances.

Subsequent sections consider in detail the various factors that can influence the final selection and the test methods that are available to assess the relevant characteristics of protective clothing material. However, there are many factors involved in the selection and use of protective clothing that cannot be easily quantified, the relative importance of which may seem different to different people. Hence, the view of all interested parties should be considered to achieve the best balance. This should enable the features that determine the ability of the clothing to protect the user to be clearly defined and corresponding physical and chemical characteristics to be adequately checked using appropriate test procedures.

• Compatibility: Individual items of protective clothing should not be considered in isolation as they may have to be worn with other protective devices (respiratory protective equipment, goggles, etc.), special tools, or communications equipment. The user should not be isolated from other workers and should be able to acknowledge and respond to emergency procedures while wearing the clothing.
• Selection of material of construction: Suppliers of protective clothing should be able to advise on the general suitability and limitations of their garments, and on their practical value for protection against particular chemicals under defined conditions. In the selection of construction the following questions arise:
  What chemical resistance is required of the garment material, and for how long?
  What other requirements are there for the garment material (e.g., durability)?
  Is an air-permeable garment material acceptable?
  If there is a risk of significant chemical permeation through the garment material, will the concentration to which the skin is exposed be acceptably low throughout the period of work?

Discussion with chemical suppliers, safety experts, occupational hygienists, and garment suppliers will frequently be necessary to answer these questions and to make an initial choice of protective garments. Further consultations will normally be needed to ensure that the final choice meets the need to provide adequate protection under the circumstances that apply.

• Selection of design: Having made an initial choice of garments, the following questions then arise:
  Does the clothing give adequate protection against any other hazards (e.g., fire) that may be anticipated?
  Will the clothing chosen interfere unduly with the user's activity or subject him to stress and discomfort?
  Is the clothing compatible with the task in hand and with the use of any equipment or tools that are needed?
  Are workers sufficiently trained in the use of the clothing and any relevant safety procedures?
  Is there a risk of contamination being transferred to the user when putting on or removing the clothing?
  Are adequate cleaning procedures available?
  Is there a suitable system of maintenance?
  Is there an adequate management and supervision system?

4.53.5 Additional considerations

• Restrictive limitations: Where clothing that is clearly adequate for the danger cannot be obtained, it is not unusual for work to be permitted for restricted periods in the best available clothing. Such decisions require careful consideration of the relevant risks by the responsible people. Special safeguards, such as setting up showers adjacent to the workplace, may be needed.

- Possible disadvantages: It is possible for protective clothing to create a hazard, for instance, by limiting the user's movements or vision, or by preventing him from sensing spilt chemicals. All protective clothing causes some stress in the user, whether by discomfort, built-up of heat, or restriction of movement, and this should be kept in mind in the selection procedure. The presence of hazards other than chemical action on the body (e.g., high temperatures) may restrict the choice of clothing further.

4.53.6   Chemical protection: leak test for gas-tight suits

- **Procedure**: Lay out the suit, including gloves and boots, and face mask if appropriate, on a suitable flat and clean surface away from any sources of heat and/or drought. Remove any creases or folds in the suit as far as practicable. Make an inflation connection and carefully blank off the valves etc., with appropriate components as recommended for test purposes by the manufacturer. Inflate the suit carefully to a maximum pressure of 180 mm H_2O, and then allow it to settle for a period of at least 10 min to allow any creased areas to unfold, the suit to stretch, the temperature to stabilize and the pressure throughout the suit to reach equilibrium. Adjust the pressure in the suit to 170 mm H_2O. Allow a further period of 6 min to elapse and note any loss of pressure.

Note: Pay careful attention to the cleanliness and the refitting of valves that have been obstructed or removed in order to carry out the test, to ensure that they function satisfactorily after the test.

4.53.7   Chemical protection: examples of protection against a single hazard under differing degrees of danger

The selection of protection might be considered to apply to either concentrated hydrochloric acid or concentrated sulphuric acid. In either case the liquid poses a hazard to the exposed skin and eyes and the vapor mainly to the lungs and eyes. Dilute acid would not pose such a serious hazard to the lungs, unless present as a mist, but could still present a serious corrosion hazard to the skin and eyes. Thus respiratory protection is specified where the risk of vapor inhalation is high.

The protection concept in this case illustrates a possible approach to the problem of providing adequate protection for varying degrees of danger. More protection than suggested could be provided particularly where the activity has been over-simplified. Thus examples (n) and (o) assume a risk of serious all-over splashing and consequent inhalation of fumes. This implies large storage tanks and wide bore pipes. The level of protection would be likely to be considered excessive if, for example, the storage tank held 1 liter and the pipes were only 3 mm bore.

Sulphuric acid is more damaging to the exposed skin than hydrochloric acid, whereas the latter poses more of a fume problem. Thus in each example, particularly from (e) onwards, it would be a matter of informed judgment whether there is justification for increasing skin protection for the one and lung protection for the other using the suggestions made in the table as a starting point. Danger has been graded on an arbitrary scale from 1 to 10. The higher the number, the greater the chance of injury if no precautions are taken.

Personnel protection against radioactive

The use of sealed radioactive sources has become widespread. Safety is the prime consideration in establishing a standard for the use of sealed radioactive sources. However, as the application of sources becomes more diversified, a document is needed to specify the characteristics of a source and the essential performance and safety testing methods for a particular application and to maintain the record of safe use.

The objective of this chapter is to provide guidance for the protection of people from the undue risks and harmful effects of ionization radiation. The chapter covers the following topics:

- Basic concept of radiation specifications
- Detailed information and requirements for apparatus, containers, tests, transportation, packaging, and safety of sealed source
- Classification, identification, and test procedures of sealed sources
- Site inspection, source exchange, and source container maintenance service

5.1 Specific considerations

Any type of work involving ionizing radiation should be carried out in a manner such that the radiation exposure to individuals (radiation personnel and members of the public) is As Low As Reasonably Achievable (ALARA), without exceeding the limits. This is because radiation exposure can cause harmful effects to the health of exposed individuals, and in normal radiation work such effects may not be readily recognizable.

5.2 Types of sources used in industrial gamma radiography

5.2.1 Radiography sources

Gamma-emitting radio nuclides as well as X-ray generating equipment are used as radiation sources for industrial radiography work. The choice of a particular radiation source is decided on the basis of the material and thickness of the object to be radiographed.

Personnel Protection and Safety Equipment for the Oil and Gas Industries.
DOI: http://dx.doi.org/10.1016/B978-0-12-802814-8.00005-5
© 2015 Elsevier Inc. All rights reserved.

5.2.2 Source pellets and source pencils

The isotopic sources used for gamma radiography are either in metallic form (^{60}Co, ^{192}Ir) or in metallic salt form (^{137}Cs, ^{170}Tm). These sources are sealed in stainless-steel capsules to avoid any damage to actual source during use. These capsules are incorporated inside a source assembly, for easy handling, which in turn is loaded in a radiography camera. The source assembly may be a rigid one or a flexible one depending on the design of the radiography camera. The source assembly when rigid is called a source pencil and when flexible is called a "pig tail."

5.3 Selection of radiography site

An ideal site for radiography is one that is away from any occupied areas, away from storage of explosives and inflammable materials, and one that is situated in a corner area with minimum occupancy. In situations where there is little choice in selection of a site because of the nature of thermal plants, for cross-country pipelines, etc., it must be ensured that the radiography work is carried out only during the time when there is no one around. Planning of the site must be done for a specific radiation source.

When radiography work is to be carried out long-term, the radiography site should be in an area provided with suitable fencing such as with ropes and radiation symbols to prevent unauthorized entry. Continued occupancy outside the fenced area must be restricted. Temporary shielding can also be improvised by stacking heavy steel or concrete objects around the area. The storage room for gamma radiography equipment and sources should be as close to the working site as possible.

5.4 Radiation exposure to human beings

If the human body is exposed to radiation externally or internally, the damage to tissue depends on the type, strength, and duration of the dose. The following types of radiation exposure may occur.

5.4.1 External radiation exposure

External radiation exposure can be expected when handling sealed radioactive substances in technical facilities. Protection against this type of exposure is possible if the equipment is handled properly and radiation-protection regulations are followed. In exceptional cases, external exposure is also possible if skin or clothing are contaminated, but only if the source is leaking, or when handling open radioactive substances. In these cases, special precautions must be taken.

Table 5.1 **Table of allowed dose**

Dose in mSv per year	Classification		
	A	B	N
Whole body, head and trunk	50	15	5
Hands, forearms, feet and ankles	500	105	50

N = not classified people; A = classified people category A; B = classified people category B.

Depending on the degree of occupational exposure, personnel is classified in different categories. These dose-rate values, recommended by international agencies, are the basis of the Radiation Protection Regulations.

- Persons not exposed occupationally: The general public must not be exposed to an annual dose exceeding 5 mJ/kg = 5 MSv (0.5 rem).
- Persons exposed occupationally-category B: Persons who are exposed to an annual dose of more than 5 mJ/kg = 5 mSv (0.5 rem) but less than 15 mJ/kg = 15 mSv (1.5 rem) belong to the category B. The body doses are recorded but medical examination is only required when handling open radioactive sources. During any quarter the body dose must not exceed fifty percent of the annual dose.
- Persons exposed occupationally-category A: Persons who are exposed to an annual dose exceeding 15 mJ/kg = 15 mSv (1.5 rem) must be classified in category A. The maximum permissible radiation dose for these persons is 50 mJ/kg = 50 mSv (5 rem) per year.

The person doses are to be determined by means of officially evaluated dosemeters. A medical examination once a year is mandatory. In this case, too, the body dose per quarter must not exceed 50% of the annual dose. Occupationally exposed persons of category A must be examined by an authorized physician; occupationally exposed persons of category B only if they handle open radioactive substances. This examination is repeated once a year. Further employment in the control area is only permitted after a certificate of authority has been granted. Table 5.1 shows the allowable doses for different classifications of people.

5.5 Biological basics of radiation

5.5.1 Dangers of radiation

If live-tissue is exposed to radiation, chemical and biological processes occur in the individual cells, which may change, damage, or destroy the cells. Alpha, beta, and gamma-radiation interact with the electrons of the atom shell and neutrons are retarded by the nuclei. Thus, electrons may be separated and as a result the atoms will become ionized. These ions are unstable and therefore react with adjacent atoms. Thus, undesired combinations may result, which can be harmful or even toxic.

Table 5.2 **Relationship between dose and effect**

Dose	Effect
Up to 0.2 Sv (20 rem)	No effect evident
Up to 1 Sv (100 rem)	Slight changes of the blood structure, but no serious damage is likely to occur
Up to 2 Sv (200 rem)	Radiation hang-over, vomiting, serious illness possible, good change of recuperation
2–6 Sv (200–600 rem)	Increase in mortality
More than 6 Sv (600 rem)	No chance of survival

Somatic Radiation Damage: Somatic radiation damage can occur as a result of short-term as well as long-term radiation exposure. Short-term exposure of the whole body may cause the following damage:
- Radiation hang-over
- Retardation of blood formation
- Sterility
- Inflammatory diseases of the skin

Table 5.2 shows the relationship between dose and effect in the human body when exposed to short-term radiation.

Permanent exposure to radiation with even distribution will cause much less damage, due to the regenerative capacity of living organisms, but may nevertheless lead to chronic illnesses, such as leukemia or cancer. This is also the case if the body is exposed only once to a high dose of radiation.

Genetic Radiation Damage: Genetic radiation damage is caused by changes in the reproductive cells and can lead to mutations. A lower limit for the probability of mutations cannot be specified. In assessing this limit it is necessary to take into consideration the natural radiation (cosmic and terrestrial radiation) to which human beings are exposed and which may be quite high in certain areas.

5.6 Radiation-measuring techniques

5.6.1 Measuring systems for dose-rate measurements

The human body cannot sense nuclear radiation. To detect radiation it is necessary to use suitable measuring instruments, with appropriate detectors for the different types of radiation. The most common detectors are ionization chambers, counter tubes (Geiger-Muller counter tubes or halogen counter tubes, respectively), and scintillation counters.

5.6.2 Ionization chamber

An ionization chamber is basically a gas-filled plate capacitor in which the radiation received triggers positive and negative charge particles (ions and electrons), which

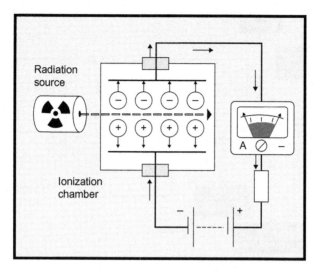

Figure 5.1 An Isonation chamber.

generate an electric current directly proportional to the dose rate (see Figure 5.1). Since this current is very small, it is necessary to amplify it substantially.

5.6.3 Counter tubes

Counter tubes (GM tubes) are built in such a way that the radiation received triggers a flow of electrons by means of ionization of the special gas filling (gas-amplification effect), as shown in Figure 5.2. Thus, strong pulses are generated that can then be counted by simple means. The number of pulses per unit of time is a measure of the dose rate. Although the resolution and the energy independence are not very high, counter tubes are well suited for applications in certain ranges of photon energy and are also preferred because of their low cost. (This is also true for the associated electronics.)

5.6.4 Scintillation counter

With a scintillation counter flashes of light are generated in a crystal by the radiation received, and these flashes are registered at the photo cathode of a photo multiplier and transformed into electrical pulses (Figure 5.3). The average of these pulses is a measure of the dose rate. Scintillation counters have high detection efficiency for radiation, but the technical expenses are quite high.

Construction and arrangement of detectors for nuclear radiation have to be suited for the detection of the different types and energies of radiation with their individual characteristics. Since the penetrative capacity of alpha and beta radiation is low, the windows of the detectors have to be thin. Alpha and beta radiation ionize the gas filling in the ionization chamber or the counter tube, or stimulate a scintillator to emit light. With gamma radiation evidence is produced by secondary electrons being

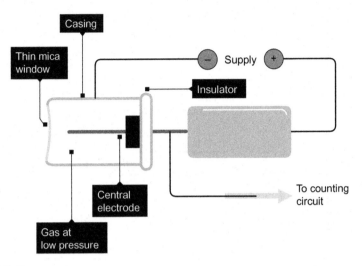

Figure 5.2 Counter tubes.

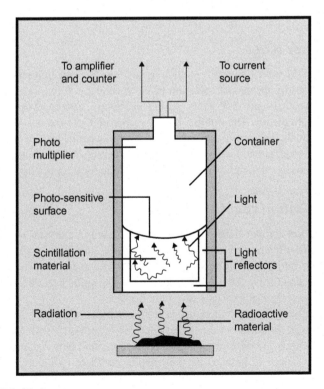

Figure 5.3 Scintillation counter.

directly released across the detector walls, or by stimulating a scintillator. Exposure rate meters, warning assemblies, and monitors for X and or gamma radiation of energy between 80 Kev and 3 MeV should comply with the BS 5566 standard.

5.6.5 Personal-dose measurement

Personal dosimetry instruments are used to measure not only the intensity of a radiation field (dose rate), but also the accumulated radiation dose while taking into consideration the duration of the reaction.

- Film dosimetry: The darkening of photographic emulsion is utilized to determine the radiation dose. Film dosemeters contain film in a cartridge protected from the light; this cartridge contains numerous metal filters. From the darkening produced behind the filters, the exposure dose and the "radiation energy" can be determined. The advantage of film dosemeters is their robustness and their small dimensions (see Table 5.3).
- Glass dosemeters: Contain a silver-activated phosphate glass enclosed in a capsule. Depending on the degree of radiation, strongly fluorescent spots are generated, which, upon evaluation, are stimulated by UV-light. The intensity of the fluorescent light is proportional to the dose received by the glass. Glass dosemeters are sensitive to dirt and therefore have to be protected (see Table 5.3).
- Pocket dosemeters: Provide direct reading of the dose received. They consist of a small electrometer whose cross-wire is made visible on a scale through a magnifying glass. Pocket dosemeters are charged by a brief connection to a voltage supply and discharged according to the radiation dose to which they are exposed. To prevent error caused by self-discharge, which is unavoidable, a daily reading of the pocket dosemeters is recommend (see Table 5.3).
- Thermoluminescent dosimetry: A thermoluminescent dosimeter (TLD) contains lithium fluoride (LIF), commonly used because of its negligible fading at room temperature and its low average atomic number. The TLD monitoring badge is comprised of a TLD holder with a clip, filters, and TLD chip. Thermoluminescence is the emission of light from previously irradiated material after gentle heating.

5.7   Radiation-protection techniques

5.7.1   Basic principles of radiation protection

To avoid damage to the human body with near certainty, an annual dose for the people classified in the different categories is fixed internationally. The aim of radiation protection is to adhere strictly to the specified permissible dose values and furthermore to avoid unnecessary radiation to keep the radiation dose for personnel as low as possible.

The formula for calculation of radiation dose shows what kind of radiation protections can be carried out. The radiation dose (D) depends on the activity (A) of the source, its specific gamma radiation constant (k), the distance (a) from the source, the radiation time (T), and the weakening factor (s) of the available shielding.

$$D = \frac{A \cdot K \cdot T}{a^2 \cdot s} \tag{5.1}$$

Table 5.3 The different kinds of personal dose meter systems

	Measurement	Evaluation	Characteristic features
Film	Translucency of film	Evaluation only by official authority and only at fixed intervals. Measuring value is primarily used for subsequent dose balancing; useless for immediate measurement; info is available to user only after a long time (weeks).	Universally applicable; yields much information when evaluated properly. disadvantages: Subject to many uncontrollable error influences; limited measuring range; can not be stored.
Pocket	Discharge of capacitory chamber	Read-out immediately and at any time by the user. Automatic evaluation systems with computer connection to the data evaluation system are being developed.	Fast, accurate information. Disadvantages: short measuring range; regular charging required.
Glass	Fluorescence intensity with unstimulation 6200 A	Evaluation any time and, due to storage of measuring values, as often as desired by the user and the official authority.	Ideal long-term dosemeter with high reproducibility, due to storage of measured values. Disadvantage:Too inaccurate in the low energy range(roentgen).Attempts to improve the measuring sensitivity and the energy dependence are being made.
Thermolu-minescent	Measurement of the light photons emitted after heating the previously irradiated materials	The system provides sensitivity with a high degree of accuracy at low exposure levels, detecting 10mrem (0.1 msv) with a standard deviation of 10 percent or less. Results are easily computerized.	There is no necessity for a development laboratory. TLDs may be reused many times. Relatively unaffected by moisture common solvents and minor physical abrasion.

Note: The results of all measurements and calculation have to be recorded and kept on file for 30 years, and have to be submitted for inspection by the supervisory authority on request.

Table 5.4 The thickness of lead required to reduce gamma radiation intensity by various transmission factors

Radionuclide	Transmission factors: lead		
	0.5	0.1	0.01
Cobalt-60	15 mm	43 mm	86 mm
Iridium-192	3.5 mm	12 mm	28 mm
Thulium-170	0.8 mm	5 mm	19 mm
Ytterblun-169	0.8 mm	2.9 mm	8 mm

The activity of a source and the corresponding specific gamma-radiation constant are determined by the measuring task. When designing a measuring system, the designer should try to keep the required source activity as low as possible by selecting suitable detectors and evaluation instruments.

From the above formula, the following radiation protection measures, which illustrate some important basic principles of radiation protection, can be derived:

- Increasing the Distance: To the radiation source, i.e., the distance between the source and the body. Since the dose rate (just as light) follows the square law, doubling the distance means reducing the radiation intensity to a quarter. This is the most efficient as well as the easiest method of radiation protection. It is important, therefore, to keep the largest distance possible when operating in the proximity of radioactive substances, especially those people who are not involved with the operation. Even weak sources generate a substantial dose rate if the distance is short. Test sources with very low activities must not be touched by hand, but only with pliers or tweezers.
- Shortening the Duration of Exposure: The time (T) has a linear effect, i.e., doubling the period of exposure gives twice the radiation dose. Operations close to the source should be well planned, so that the time of exposure in the immediate vicinity of the source is kept as short as possible.
- Use of Shielding: With a high weakening factor (s), which depends in an exponential function on the product of thickness and density of the material. Apart from a few exceptions, radioactive substances used in industry are already installed in a suitable shielding when delivered. The shielding is effective only when the shielding functions properly and is handled safely.
- Use Shielding Between Sources and Personnel if Possible: Dense, high atomic number materials, such as lead are preferred. Materials such as concrete must be much thicker to provide the same effective shielding. Table 5.4 shows the thickness of lead required to reduce gamma-radiation intensity by various transmission factors.

Table 5.5 shows transmission factors for concrete.

Transport containers alone do not normally provide sufficient screening for permanent storage. A storage cell or container should be used.

5.7.2 Radiation-protection areas

- Restricted areas: Areas with a dose rate higher than 3 mSv/h (300 mrem/h) must be secure, so that no one can enter unchecked. Access is only permitted under specific conditions and

Table 5.5 **Transmission factors for concrete**

Radionuclide	Transmission factors: concrete		
	0.5	0.1	0.01
Cobalt-60	165 mm	335 mm	545 mm
Iridium-192	130 mm	245 mm	390 mm
Thulium-170	No Data		
Ytterblun-169	Available		

Figure 5.4 The symbol for ionizing radiation.

if there is an absolute need for it, the body dose should be calculated and the personal dose measured. These areas are normally only the useful beam at the surface of the shielding. If it is possible for any part of the body to enter this area, it must be screened accordingly.
- Control areas: These are areas with dose rates that are equivalent to or higher than 7.5 μSv/h (0.75 mrem/h). Control areas must be marked off and provided with a radiation warning symbol and marked "control area." For these control areas the following warning signs should be displayed:
  1. The basic symbol shown in Figure 5.4 indicates the potential or actual presence of ionizing radiation.
  2. Such additional inscriptions, colors, or symbols as may be required to indicate in a manner understandable to all concerned the magnitude and particular nature of the exposure risk:
  3. Entry to the control areas is only permitted for people carrying out specific operations. The body dose must be determined or the personal dose measured. The authority concerned may grant exceptions if it can be proved that the whole body dose will not exceed 15 mSv/year (1.5 rem/year).
- Monitored areas
  Plant Monitoring: The plant-monitoring area starts in the control area with a dose limit of 15 mSv per year (1.5 rem/year), if a person stays in the area for 40 hours per week (which is equivalent to a dose rate of 7.5 μSv/h or 0.75 mrem/h), and reaches to a dose rate of 5 mSv per year. This will result in a dose rate of 2.5 μSv/h (0.25 mrem/h). Steps must be taken to ensure that people will not be exposed to a higher dose than 5 mSv per year (0.5 rem/year) taking into account the actual visits to this area.

The general dose-rate limits are 0.3 mSv per year (30 mrem per year). People in this area must not be exposed to a higher annual dose than 1.5 mSv (150 mrem).

5.8 Classification designation of sealed source

The classification of a sealed source should be designated by the code ISO/followed by a letter and five digits. The letter should be either C or E.C., which designates that the activity level of the sealed source does not exceed the limit established in standards and that the activity level of the sealed source exceeds the limit established in widely accepted standards, respectively. The first digit should be the class number, which describes the performance for temperature. The second digit should be the class number, which describes the performance for external pressure. The third digit should be the class number, which describes the performance for impact. The fourth digit should be the class number, which describes the performance for vibration. The fifth digit should be the class number, which describes the performance for puncture.

5.9 General requirements for radiography equipment

5.9.1 X-ray radiography equipment

X-ray equipment used in industrial radiography must be provided with adequate built-in safety features so as to minimize radiation hazards to its users and to the public. X-ray tubes are provided with lead-lined housing so as to reduce the leakage radiation below the level specified by the determining authority. In the case of conventional X-ray units, the leakage radiation at 1 m from the target should not exceed 1 R in 1 hour or 10 mGy in 1 hour. Further, the use of beam-limiting devices is also recommended. For rod-anode tubes that are used for panoramic exposure, and for which no tube shielding exists, this specification does not apply. The control console must have a lock to prevent the use of the X-ray machine by unauthorized persons. A red light must be provided on the control console as well as on the tube head to indicate the beam "on" position. In the case of conventional X-ray units, the cable length between the tube head and the control console should not be less than 20 m and the operator must make use of the full length of the cable while energizing the X-ray tube.

In the case of a linear accelerator the maximum leakage radiation at 1 m from the target (other than the primary beam direction) should not exceed 0.1% of the primary radiation measured at 1 m from the target.

5.9.2 Gamma-radiography equipment

Each model of radiography camera is designed to house a particular source of specified maximum strength. Gamma-ray sources of appropriate strength are used in radiography cameras. Depending upon the nature and strength of the source used and the type of radiography work, radiography equipment either of the stationary type or of the mobile type can be used. Higher-source strengths of (i.e., more than 20 Ci of ^{192}Ir and 5 Ci of ^{60}Co) are permitted only for radiography cameras provided with remote-handling mechanisms such as flexible cable operation or electrical or pneumatic operation so that all operations can be done from a safe distance, thereby minimizing radiation exposure to the operator.

The source housing of gamma-radiography equipment must have adequate shielding to restrict the leakage radiation within limits stipulated. The leakage radiation levels in the "off" position of industrial gamma-radiography equipment stipulated by the determining authority are:

At a distance of one m from the surface of the source housing when the source is in its fully shielded position:
The maximum radiation level in any direction must not exceed 0.1 mSv/h (10 mrem/h).
The average radiation level must not exceed 0.02 mSv/h (2 mrem/h).
At a distance of 5 cm from any point on the surface of the source housing, in any direction:
The maximum radiation level must not exceed 1 mSv/h (100 mrem/h).
The average radiation level must not exceed 0.2 mSv/h (20 mrem/h).
In the case of transport container (lead pots) used for temporary storage/transport, the leakage radiation levels in the full-shielded position of the source should not exceed:
2 mSv/h (200 mrem/h) at a distance of 1 m from the surface of the container.
0.1 mSv/h (10 mrem/h) on the average of radiation level.

The cameras and containers must be provided with built-in locking arrangements so as to prevent unauthorized handling. The radiation-warning symbol must be conspicuously visible on the camera/container.

5.10   Apparatus for gamma radiography

An apparatus for gamma radiography should be designed for the conditions that may be encountered in use and that may adversely affect safe operation. Designers and manufacturers should give particular consideration to the following:

- Durability and resistance to corrosion of components and their surface finishes, particularly where the functioning or moving parts may be affected
- Need to prevent the entry of water, mud, sand, or other foreign matter into the controls or moving parts, or the facility with which the apparatus may safely be cleaned out using, e.g., a hose and water
- Effect of temperatures that may be encountered in use
- Possibly damaging effects of gamma radiation on any nonmetallic components such as rubbers, plastics, jointing, sealing compounds or lubricants in close proximity to the sealed source
- Provision of appropriate accessories designed for the secure mounting of the exposure
- Container or exposure head in different positions of use
- Interchangeability of source holders and other replacement components

- Provision of instructions for use, periodic inspection, and maintenance

Where depleted uranium is used as the shielding material of an exposure container, it should be clad with a nonradioactive material of sufficient thickness to attenuate or absorb the beta radiation. If the nonradioactive cladding is liable to react with the depleted uranium at elevated temperatures, then the depleted uranium should be given a suitable surface treatment to inhibit this effect.

The sealed-source capsule should be:

- Free from surface radioactive contamination
- Leak-free
- Physically and chemically compatible with its contents
- Not contribute significantly to the activity of the radioactive material in the case of a sealed source produced by direct irradiation

5.10.1 Sealed-source certificate

The manufacturer should provide a certificate with every sealed source or sealed-source batch. The certificate should in every case state:

- Name of manufacturer
- Classification designated by the code established in 5.8
- Serial number and brief description, including chemical symbol and mass number of the radionuclide
- Equivalent activity and/or radiation output in terms of fluence rate, as appropriate, on a specified date
- Method used and result of test for freedom from surface contamination
- Leak test method used and test result

Note: In addition, the certificate may include, as appropriate, a detailed description of the source, in particular:

- For the capsule: Dimensions, material, thickness, and method of sealing
- For the active contents: Chemical and physical form, dimensions, mass or volume; percentage of undesirable radionuclides from the point of view of the use to which the sealed source is to be used

5.11   Source marking

Whenever physically possible, the capsule should be durably and legibly marked with the following information, which is given in order of priority:

- Mass number and chemical symbol of the radionuclide
- Serial number
- For neutron sources, the target element
- Manufacturer's name or symbol

The marking of the capsule should be done before the sealed source is tested.

5.12　Container marking

Each exposure container or a metal plate permanently fixed to the container should be permanently and indelibly marked by engraving, stamping, or other means with the following:

> Basic ionizing radiation symbol, complying with ISO-361
> Word "RADIOACTIVE" in letters not less than 10 mm in height
>> Maximum rating of the container:
>> For a cobalt 60 source, shown as "Rating × Bq ^{60}Co (y Ci ^{60}Co)"
>> For a caesium 137 source, shown as "Rating × Bq ^{137}Cs (y Ci ^{137}Cs)"
>> For an iridium 192 source, shown as "Rating × Bq ^{192}Ir (y Ci ^{192}Ir)"
> ISO marking indicates the manufacturer's claim that the exposure container and its accessories conform to international standard; this claim should be stated in the manufacturer's literature
> Manufacturer's type and serial number
> •　Class M and F containers: A class M or F exposure container should be marked with the mass of the container without removable accessories

5.13　Radiotoxicity and solubility

Except as required by standards, radiotoxicity of the radionuclide should be considered only when the activity of the sealed source exceeds the value shown in relevant standards. If the activity exceeds this value, the specifications of the sealed source have to be considered on an individual basis. If the activity does not exceed the values shown in the standards the aforesaid specification may be used without further consideration of either radiotoxicity or solubility. A quality-control program is essential and should be operated in both the design and manufacture of sealed sources that are to be classified.

5.14　General consideration of exposure container

5.14.1　Safety devices

- Locks: On all exposure containers, a series of beam emissions of source projections should be possible only after a manual unlocking operation. An exposure container should be provided either with an integral lock and key or with hasps through which a separate padlock can be fitted. The lock should be either of the safety type, i.e., lockable without the key, or an integral lock from which the key cannot be withdrawn when the container is in the working position. The lock should retain the sealed source in the secured position and should not, if the lock is damaged, prevent the sealed source when it is in the working position from being returned to the secured position. If separate padlock is used, there should be an additional device to provide a positive means of retaining the sealed source in the secured position.
- Source-position indicators: An apparatus for gamma radiography should clearly indicate whether the sealed source is in the secured or the working position. If colors are used, green

Table 5.6 Exposure rate limits

1	2		3	4
	Maximum exposure rate, mSv/h (mR/h)			
Class	**On external surface of container**		**50 mm from external surface of container**	**1 m from external surface of container**
P	2 (200)	or	0.5 (50)	0.2 (2)
M	2 (200)	or	1 (100)	0.05 (5)
F	2 (200)	or	1 (100)	0.1(10)

should only indicate that the source is in the secured position and red should indicate that the source is not in the secured position, but colors should not be the sole means of indication.

- System failure: remote control system, which is not manually operated, should either:
  Be designed so that a failure of this system causes shutter closure or the return of the sealed source to the secured position, or
  Be accompanied by a safety device, preferably manual, permitting shutter closure or the return of the sealed source to the secured position without unduly exposing personnel to radiation.
- A remote control system, which is manually operated, should be designed so that it is impossible for the sealed source to be withdrawn from the rear of the exposure container whilst operating, connecting or disconnecting the remote control cable.
- Unauthorized operation: Where a remote control is incorporated, there should be provision to prevent its unauthorized operation when the operator is not in immediate attendance, e.g., by a removable winding handle. The source holder should be designed in such a way that it cannot release the sealed source accidentally and should provide it with positive retention and mechanical protection.

5.14.2 Exposure rate in the vicinity of the containers

An exposure container should be made in such a way that when locked in the secured position and equipped with sealed sources corresponding to the maximum rating, the exposure rate, when tested, does not exceed the limit in column (4) and one of the other limits in columns (2) and (3) of Table 5.6.

5.15 Handling facilities of exposure containers

5.15.1 Portability

A class P exposure container should be provided with a carrying handle. A class M container should be provided with a lifting device. Such a handle or device should be adequate for its purpose and so secured that it cannot be accidentally removed from the container. (Such an adjunct is optional for a class F container.)

5.15.2 Mobility

The equipment provided for moving a class M exposure container should have a turning circle of 3 m or less, and should be fitted with an immobilizing device.

5.16 Manufacturing and production tests for exposure containers

Manufacturers should provide the necessary production quality-control inspections and tests. The program should include at least the tests specified in the following standards:

Resistance to normal conditions of service covering the tests for:
Vibration
Shock
Endurance
Kinking, crushing tensile
Accidental dropping

in accordance with standard ISO 2855.

Shielding efficiency test:

in accordance with standard ISO 3999.

Tests for mechanical remote control devices:

in accordance with standard ISO 3999.

Production test:

in accordance with standard ISO 3999.

Leak test:

in accordance with standard ISO/TR 4825.

Testing procedures:
Temperature test
External pressure test
Impact test
Puncture test

in accordance with standard ISO 3999

5.17 Packing and transportation of radioactive substances

5.17.1 Transportation

Transportation of radioactive substances on public transport is only permitted if approval has been granted by the supervisory authority. The transportation approval stipulates that delivery can only be done if the following points have been considered:

- Packing Type A
- Marking of package
- Marking of the shipping documents
- Observe maximum permissible activity

The transportation of radioactive material can take place only according the regulations for transportation of dangerous commodities. For transport regulation reference is given in the International Atomic Agency Safety series No. 6, No. 37.

5.17.2 Identification of the sealed source in the container

The user should ensure that the following information is displayed in a durable form, attached to the exposure container:

- Chemical symbol and mass number of the radionuclide
- Activity and the date on which this activity was measured
- Identification number of the sealed source

5.17.3 Packaging

Packaging of radioactive materials is generally controlled by the International Atomic Energy Agency transport (IAEA) regulations (Safety Series No. 6, 1973 revised edition). There are two categories of packaging:

1. Type A: Designed to retain the integrity of containment and shielding under the normal conditions of transport.
2. Type B: Designed to withstand the damaging effects of a transport accident. Packages containing radioactive materials must conform to certain maximum external radiation levels according to the method of transport being used. Under IAEA Category III-yellow and exposure rate not exceeding 2 mSv/hr at any point on the surface of the package is allowed.

The package must also have a Transport Index (TI) within the limits prescribed for the particular method of transport. (The Transport Index is the number expressing the exposure rate in 2 mSv/hr at 1 m from the surface of the package.) For most passenger-carrying aircraft the TI must not exceed 3 or 4, but for freighter aircraft a TI up to 10 is generally allowed. The IAEA transport regulations prescribe for each nuclide, maximum activities that may be transported in a type A package. The activity levels (A1 and A2 limits) depend on the toxicity, etc., of the nuclide and may vary according to whether or not the nuclide is contained in a capsule for which a "Special Form" certificate has been issued.

- Exempt packaging: Packages are exempt from the packaging regulations if they conform to the following requirements:
  Maximum activity of the contents is less than:
  10^{-3} A1 (special form approval), or
  10^{-3} A2 (for other sources).
  The exposure rate at any point on the surface of the package does not exceed 0.5 mrem/h.

Transport containers are designed for maximum safety and economy in transport and conform to the appropriate international regulations. Wherever possible light-weight nonreturnable containers should be used but shipments requiring more shielding should be packed in returnable containers.

- Nonreturnable containers: These usually consist of a lead shield in a sealed can packed either in a cardboard box or in an expanded polystyrene casing. Packages of this type meet the IAEA requirements for Type A. Modified packaging to conform to Type B is also available. The sealed may be opened with a domestic can opener. Refer to unpacking instructions where applicable.
- Returnable containers: Returnable containers usually consist of a heavy lead shield inside a steel drum. In the steel drum there may also be a cork or fiber liner to give protection in case the container is involved in a fire. Specially designed containers should be used for loading high-activity and high-energy gamma and neutron sources.
- Customers' containers: Customers' own containers must meet the relevant transport regulations; formal evidence of this is required before shipments can be arranged.
- Receipt of source package: The package should be inspected on arrival and if any damage is observed that could have resulted in damage to the product then it should not be opened. Measure the surface dose rate on the container. It should not exceed 2 mSv/h. A higher reading may indicate that the source is not in a safe position or that the shielding is damaged. Check that the documentation and label description agrees with the order acknowledgment. Notify the Radiological Protection Supervisor that the package has arrived. Update the official record for radioactive substances accounting purposes, noting the identification, activity and date.

The shielding provided by transport containers is adequate to comply with the maximum dose-rate levels specified in the IAEA transport regulations. However, these levels are normally too high to allow storage of the package in workplaces.

If the package is not opened immediately, a suitable and secure store must be provided. This store should be reserved for radioactive materials only and must be adequately shielded, correctly labeled, and fully secured against intrusion by unauthorized persons. The external dose rate should not normally exceed 2.5 Sv/h (0.25 mR/h).

- Unpacking: Sources must only be unpacked in a controlled area by trained, competent, and authorized personnel. Radiation levels should be checked using a dose-rate meter at each stage of unpacking. The exposure rate at the outer surface of the package may be as high as 2 mSv/h and dose levels at each stage of unpacking will increase. Various packing combinations are used depending on the type of sources. Steel drums are the most commonly used form of packaging. For these types proceed as follows:
  1. Remove steel closing band and lid
  2. Check the enclosed documents
  3. Remove cork lid and spacer, if fitted
  4. Lift the lead pot out of the drum, leaving the cork liner in place

Notes of caution. The lead pot is heavy. Use assistance if necessary. Place the lead pot on firm, level ground. The dose rate on the lead pot may be as high as 15 mSv/h, so contact time should be minimal.

5.18   Procedure to establish classification and performance requirements for sealed source

If the desired quantity does not exceed the allowable quantity of standard, an evaluation of fire, explosion, and corrosion hazards should be made. If no significant hazard exists, the sealed source's classification may be taken directly from the standards. If a significant hazard exists, the factors should be evaluated with particular attention to the temperature and impact requirements. If the desired quantity exceeds the allowable quantity of standard, an evaluation of fire, explosion, and corrosion hazards and a separate evaluation of the specific sealed-source use and sealed-source design should be made.

5.18.1   Identification of sealed sources

The classification designation should be marked on the sealed-source certificate and, where practicable, on the sealed-source capsule and the sealed-source container.

• Classification of Sealed-Source Performance: This is a list of environmental test conditions to which a sealed source may be subjected. The tests are arranged in order of increasing severity. The classification of each sealed-source type should be determined by actual testing of two sources (sealed, prototype, dummy of simulated) of that type for each test, or by derivation from previous tests, which demonstrate that the source would pass the test if the test were performed. Different specimens may be used for each of the tests. Compliance with the tests should be determined by the ability of the sealed source to maintain its integrity after each test is performed. A source with more than one encapsulation should be considered to have complied with a test if it can be demonstrated that at least one encapsulation has maintained its integrity after the test. Leak test methods for sealed radioactive sources are given in ISO/TR 4825. When leak testing a simulated source, the sensitivity of the chosen method has to be adequate.

5.19   Sealed-source application and exchange procedures

5.19.1   Choice of sealed sources

A source used to produce radiation field should be sealed in a suitable container or prepared in a form providing equivalent protection from mechanical disruption. The following characteristics are desirable, if consistent with the work being carried out:

The activity of the source used should be a minimum.
The energy or penetrating power of the emitted radiation should not be greater than that necessary to accomplish the task with a minimum total exposure.
If possible, the radioactive material in the source should be of low toxicity and in such a chemical and physical form as to minimize dispersion and ingestion in case the container should be broken.

Sealed sources should be permanently marked to permit individual identification and facilitate determination of nature and quantity of radioactivity without undue exposure of the worker.

Sealed sources or appropriate containers should be regularly examined for contamination or leakage (smear tests, and/or electrostatic collection may be used). The interval between examinations should be determined by the nature of the source in question.

Mechanically damaged or corroded sources should not be used and should immediately be placed in sealed containers. They should be repaired only by a technically skilled person, using suitable facilities.

5.19.2   Methods of use of sources

Sources should always be handled in such a way that proper location is possible at all times. Inventories should be kept. If any person has reason to believe that a source has been lost or mislaid, he should notify the "safety officer" immediately. If the loss is confirmed, the designated authority should be notified immediately.

Sources should be handled in such a way that the radiation dose to personnel is reduced to a minimum by such methods as shielding, distance, and limited working time. Sources should be handled in such a way as to avoid hazards to all personnel including those not involved in the operations. Attention should be paid to people in adjacent areas including rooms above and below. Areas subject to high radiation levels should be clearly marked, roped off, and evacuated if occupied.

Beams of radiation arising from a partially shielded source should be clearly indicated. Care should be taken to ensure that such a beam is stopped at the minimum practical distance by suitable absorbing material. Monitoring procedures should be planned to take into account the sharp collimation of radiation fields that may occur.

When practical, sealed sources should be used in enclosed installations form, which all persons are excluded during irradiation. Sources should not be touched by hands. Appropriate tools should be used, e.g., long-handled, lightweight forceps with a firm grip. If needed, even more elaborate means of protection have to be considered, such as master slave manipulation, etc.

Work with radioactive materials should be planned to permit as short an exposure as possible. The extent of protection provided by limiting working time can easily be lost if unexpected difficulties occur in the work, so that dummy runs should preferably be performed whenever it is possible. Although work should be planned to limit exposure time to a safe figure, if sufficient shielding cannot be provided and time of exposure must be controlled, this should be carried out in a systematic way, preferably with time keeping and warning services outside the responsibility of the actual personnel.

5.19.3   Special use of sealed sources

• Industrial gamma radiography: The controlled area should be clearly marked with easily recognizable signs. Such an area should be made inaccessible to unauthorized personnel.

Light or audible signals, or both, should be provided to give adequate warning before and during irradiation. The radiographic setup should be completed before starting the irradiation. For radiography, which requires the removal of the sealed source from its shielding container, a clearly identifiable dummy capsule should be used during any preliminary adjustments that may be necessary. If the sealed source must be handled outside the container, this should be done automatically or by remote means, so as to give adequate protection to all personnel concerned with the operation. A radiation detection instrument should be used to verify that the radiographic source has been correctly returned to its shielding container at the end of radiographic exposure. When an industrial gamma-radiography source is used away from the premises of normal use, notices consisting of diagrams and/ or photographs with dimensions and identifying features of the radioactive source and the steps to be taken by any person finding such a source should be prepared. These notices should be displayed at the area where the source is being used until removal of it from the area has been verified.

- Thickness gauges, static eliminators, and similar devices using sealed sources: Radioactive materials used for thickness gauges, static eliminators, and similar devices should be in the form of sealed sources conforming to the general provisions for sealed sources. Whenever practicable, the normally unshielded portion of the sealed source should be protected against mechanical damage and provided with a cover plate, shutter, or shield that can be readily secured so as to effectively intercept the useful beam. Wherever possible, such devices should be installed or shielded to ensure that the levels of irradiation of all people, including those installing or maintaining the sealed source or any machinery or plant in close proximity to it, should conform to the allowable doses for the general public (so avoiding the need for personnel-monitoring procedures and special medical examinations). Such devices should be conspicuously and permanently marked so as to warn personnel of the presence of radioactive material and the need to avoid unnecessary exposure. In case of a breakage of the source, the "safety officer" or other designated persons should be notified at once.

5.19.4   Source-exchange procedure

Source changers are used to transport new sources from the manufacturer to the user. The changer is coupled to a source projector and the old source is transferred from the projector to an empty channel in the changer. (This allows an opportunity to service the projector if necessary.) Then the new source is transferred from the changer to the projector. Finally, the old source is returned to the manufacturer in the changer.

The general procedures described in this section must be read in conjunction with the specific procedures relevant to particular container types. It is important that the complete procedure is thoroughly understood before any source unload is attempted.

- Layout of equipment: Source exchanges must only be carried out in a Controlled Area. Use any available radiation shielding, e.g., a wall, if possible.
  Arrange the source changer (or lead pot) and the projector so that one length of guide tube will fit between them without any sharp bends or kinks in the tube. Caution: Any bends in the guide tube must have a radius of not less than 500 mm.
  Lay out the drive cable between the projector and the drive cable control unit. Caution: Any bends in the control cable must have a radius of not less than 1 meter.

Locate the drive cable control unit as far away as possible from the projector and source changer. Preferably the control point will be outside the Controlled Area.

- Equipment assembly: Connect the guide tube to an empty hole in the source changer, using an adapter as appropriate (if a clamp is fitted, lift lever). Caution: The source changer must remain upright at all times. Do not lay it on its side.
- Connect the drive cable to the projector as per manufacturer's instructions. Caution: Do not unlock at this stage.

Connect the other end of the guide tube to the projector as described in the manufacturer's instructions. Caution: Minimize time spent near to the projector.

Position a dose-rate meter close to the control point to continuously monitor dose rate to which the operator is exposed.

Check that any personnel remaining in the vicinity are wearing monitoring equipment as specified by the rules (film badge, TLD, dosimeter, QFE, etc.).

- Transfer of decayed source into source changer

Set the radiography projector for exposure.

Sound and/or indicate visually the appropriate warning device(s) for imminent source exposure. Check area is clear of personnel and all access points are secure.

Crank the decayed source rapidly from the projector to the source changer.

Note: The radiation intensity will increase greatly as the source is first exposed, decrease slightly as the source is cranked out, and then drop to background level when the source is correctly loaded in the source changer.

Check the dose-rate meter reading. Caution: Do not move towards the projector or changer units if the reading remains high.

When satisfied that the source is located in the source changer, approach the equipment with a dose-rate meter. The dose rate at 1 m should be about 0.75 mSv/h (75 mR/h) for a lead pot, or 100 Sv/h (10 mR/h). Caution: If significantly higher dose rates are measured as you approach the equipment, STOP, check the operations and return to a low dose area. Check with your supplier.

Check the dose rates on all sides of the projector, on the guide tube and on all sides of the source changer. Caution: The maximum dose rate at the surface of a source changer should be 15 mSv/h (1.5 R/h) for a lead pot, or 2 mSv/h (200 Mr/h).

- Disconnecting the source

When satisfied that the source is properly loaded, uncouple the guide tube from the source changer. For a lead pot type, carefully unscrew the guide tube, taking care not to pull it away from the source changer as this may dislodge the source just transferred from its shielded position. For 650 units, open the latched source guide.

Caution: Do not move the source more than 10 mm from its stored position. Monitor the dose rate during this operation to ensure that you are warned if the source becomes exposed. Disconnect the drive cable from the source holder assembly taking precautions not to move the source. Amertest sources are disconnected by moving the lock pin of the connector toward the source and sliding the drive cable out through the keyway. Do not bend or twist. For other equipment, see appropriate manufacturer's instructions.

Replace the closure nut on the source changer (if fitted) using firm finger pressure or close the clamp, disconnect the guide tube.

Wire the old source identity plaque to the source changer so that the position of the source can be traced.

- Projector maintenance: The opportunity should be taken at this stage to inspect the empty projector. Routine maintenance work in accordance with the manufacturer's recommendations may conveniently be scheduled to coincide with the source replacement. To enable the drive cable to be disconnected from a projector, it is necessary to fit a test connector to the drive cable before withdrawing it into the projector. Test connectors (jumpers) are normally fitted in the drive connector dust cap of the projector. The drive cable connector should be checked for wear using a "no go" gauge.

- Transfer new source into projector

Identify the position of the required new source. Each source position is shown by marker tape and a source identification plaque or by a loading chart.

Remove the appropriate closure nut (if fitted) and connect the drive cable to the source holder (lift clamp lever if fitted). Attach the guide tube.

Couple the source connector to the drive cable. Amertest sources are connected by depressing the lock pin with a thumbnail, sliding the drive cable connector into the keyway, then releasing the lock pin. Make sure that the connection is secure. For other equipment-use equivalent methods, see manufacturer's instructions. Caution: Do not move the source more than 10 mm from its stored position. Monitor the dose rate during this operation to ensure that you are warned if the source becomes exposed.

For the 650-source changer, close and latch the source guide after attaching the guide tube and the drive cable to the source holder as above.

Retire to the control point, sound warning devices and take precautions as for source unload. Crank the new source rapidly from the source changer to its storage position in the projector. Caution: Observe the dose-rate meter during the operation. The radiation intensity should increase as the source exits the source changer, increase as it approaches the projector, and drop to a low level when the source is properly stored in the projector.

Survey the projector and the guide tube with the dose-rate meter to ensure that the transfer has been properly completed the dose rate at the surface of the projector should be less than 2 mSv/h (200 mR/h) and less than 100 Sv/h (10 mR/h) at 1 meter.

When satisfied that the source is properly stored, lock the projector and remove all guide tubes and controls. Attach the new source identification plaque to the projector.

Remove the guide tube from the source changer. Replace the lock nut (close clamp if fitted) or hold down cap. Ensure adequate means of identifying loaded positions.

5.19.5 Specific container types

Specific features of three main source changer types are described in this section. Details of their construction and operation should be read in conjunction with the previous section on general methods of source exchange.

- TEN650 source changer

To open, remove the cover from the TEN650 by unlocking the padlock, breaking the wire seal, and removing the bolts. Caution: Monitor for radiation beams from the source position. Remove the shielded source and hold down cap from the top of the unit by breaking the wire seal and unbolting. The loaded position will have the source-identification plaque attached. The unloaded position will be unlabeled. Caution: Check visually that the position chosen for loading into is empty.

Connect the guide-tube extension to the fitting above the empty chamber of the TEN650 source changer. Close and latch the source guides to secure the tube.

To seal the container for returning sources, bolt down the source hold-down cap in place and close with the seal and wire. Caution: The cap must be bolted firmly in position over the source connectors.

- Cable-type source holders

Follow this general procedure for source transfers. Caution: Follow all the safety precautions and monitoring procedures, using a dose-rate meter at every stage to check that the source is correctly located.

To open, unscrew the two nuts and remove the cover plate. Caution: The shielding insert containing the source holder is now loose. Do not remove the insert from the pot, since this could give rise to very high dose rates 1 Sv/h (100 R/h).

Select an empty storage position in the pot-identified by not having "radioactive" label tape on the closure nut or a source identification plaque attached. Remove the closure nut (or clamp lever, if fitted). Caution: Monitor for radiation beams from source positions. Check visually that the loading position chosen is empty.

Connect the adaptor onto the source tube. If a clamp is fitted, connect the adaptor to the clamp and lift the lever. Connect the source guide extension tube (open at both ends) to the adaptor.

When the old source is fully wound into the lead pot, remove the adaptor and uncouple the drive cable, taking care not to pull the source out of the source storage tube. Replace the closure nut, using firm finger pressure only, or close the clamp and insert the wire seal.

Determine the position of the new source by reference to the loading chart. Remove the corresponding closure nut, connect the drive cable to the source holder and lift the clamp lever.

Move away to the drive control unit and wind the new source into the projector. Secure the source in the projector.

Replace the closure nut or close the clamp. Replace the cover plate and tighten the nuts to clamp it securely in position. Replace the lead pot in the transport drum.

- Back-shielded source-holder assemblies (Teletron, Gammamat)

Follow the general procedure for source transfers. Caution: Follow all the safety precautions and monitoring procedures, using a dose-rate meter at every stage to check that the source is correctly located.

To open, slacken the two nuts securing the cross bar. Raise, turn and lift off the lid. Caution: The shielding insert containing the source holder is now loose. Do not remove the insert from the pot, this would give rise to very high dose rates >1 Sv/h (100 R/h).

Remove any top shielding to reveal the shielded ends of the source holders. Caution: Monitor for radiation beams from source positions.

Select an empty storage position in the pot by monitoring for minimal radiation and then confirm by visual examination.

Connect the source guide extension tube (open at both ends) to the empty position, using an adaptor if necessary.

When the old source is fully wound into the lead pot, remove the adaptor and uncouple the drive cable, taking care not to pull the source out of the source storage tube. Replace the closure nut, using firm finger pressure only, or close the clamp and insert the wire seal.

Determine the position of the new source by reference to the loading chart on the back page. Remove the corresponding closure nut, connect the drive cable to the source holder and lift the clamp lever.

Move away to the drive control unit and wind the new source into the projector. Secure the source in the projector.

Disconnect the guide tube, keeping any adaptor in a safe place.

Replace top shielding (if fitted). Replace the lid, turn until it drops into position, and bolt securely.

5.19.6   Returning the container

Before attempting to dispatch radioactive material the user should be familiar with IAEA Standards Safety Series No. 5. In particular the user must ensure that:

The container and any other packaging are fully approved for the radioactive material that the user intend to dispatch. Designated Authority approval certificates (Types A, B or Special Form) must be valid.

The container and any other packaging is undamaged and complete.

The description on the paperwork matches the material in the container.

Dose-rate and contamination measurement have been carried out and recorded and that these comply with the statutory limits for transportation. The maximum surface dose rate must be less than 2 mSv/h (200 mR/h) and the T.I. must be less than 10. The maximum removable surface contamination must be <1 Bq/cm² averaged over 300 cm².

The outside of the container and/or package is correctly labeled.

The consignor's certificate "Shipper's Declaration for dangerous goods" is completed.

Caution: Containers are designed for the shipment of specific types of sources. They must not be used for returning a source if the source, or any part of it or the package is damaged, modified or incomplete or if the source assembly being returned is not identical to the one received.

· Repacking: Place the lead pot in the drum. Replace the cork spacer and lid. Place the steel lid on the drum and position the closing band so that it covers the joint between the drum body and lid. Tighten the closure until the lid is securely held on the drum. Tap the band gently all round with a rubber hammer or a block of wood. Retighten the closure. Repeat as necessary until the band is securely in position without any chance of working loose in transit. Fit sealing wire to the closing band to ensure that the seal must be broken if the band is removed.

· Monitoring the container: When the source has been loaded into the container, measure radiation dose rates to ensure conformance with Regulations:

Surface dose rate: Measure the radiation dose rate as close as possible to all surfaces of the container, including the base. It must not be greater than 2 mSv/h (200 mrem/h) at any point. Usually the source to be returned will be of much lower activity than the new one, so the surface dose will be well within this limit. Caution: If a high dose rate is detected anywhere on the surface of the container, check that the sources are correctly positioned in the lead pot or source changer and that all shielding components are properly in place.

Transport Index (TI): Measure the dose rate at one m from all surfaces of the drum. The maximum value found in millirem per hour (mR/h) is the Transport Index (TI). Where the dose rate is measured in units of microsieverts per hour (Sv/h), divide the value by 10 to obtain the TI. The TI value must be quoted on the labels and on the Consignor's Certificate (Shipper's Declaration for Dangerous Goods). For most passenger-carrying aircraft the TI must not exceed 3 or 4, but for freight aircraft or sea or road transport, a TI of up to 10 is generally allowed.

· Labeling the container

Remove all old labels

Fill in the return address label and attach it to the container

Attach two "radioactive" labels on opposite sides of the container

Fill in the contents (e.g., Iridium-192), activity (e.g., 0.74 TBq, 20 curies), and TI (e.g., 3.0) on both labels. Radiation exposure is described in Figure 5.5.

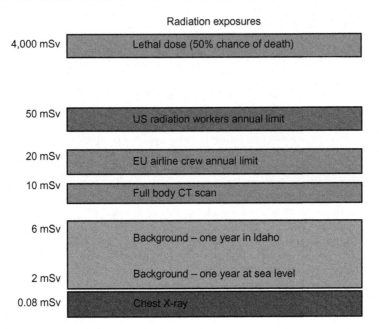

Figure 5.5 Maximum radiation level.

Consignor's certificate: (Shipper's Declaration for Dangerous Goods): A consignor's certificate is required in addition to the usual shipping documents. For shipments by air, the consignor's certificate must be in the form specified by the relevant authorized people. For other shipments, the certificate may be in any form providing it describes the radioactive contents and concludes with the signed declaration. This is to certify that the contents of this consignment are fully described above, are safely packed in a proper condition for transport, and the container is properly marked and labeled, in accordance with applicable standards. To complete the consignor's certificate, the following details must be given:

Shipper's name and address

Consignee's name and address

Description of the source either:

- Special form material described as "UN 2974 Radioactive Material, Special Form, N.O.S Class 7" or,
- Other sealed sources, which are described as "UN 2982 Radioactive Material, N.O.S. Class 7."

Nuclide

Activity on dispatch quoted in Bq (see decay chart, or calculated from source data)

State Physical Form: Solid or Special Form

Designated authority package certificate number (if Type B)

Designated authority special form certificate number (if applicable)

Category of radioactive label (e.g., yellow II, III etc.)

Type of package (A, B, or Excepted)

Container identification/serial number

TI (Transport Index)

Important: If returning a container with radioactive material is a legal responsibility, ensure that the container is properly packaged and labeled and ensure that all necessary paperwork is done – in particular, the shipper's or dangerous goods declaration.

To ensure the assembly conforms with the Types A or B approval certificate and to assist in the tracking of the containers, the inner container must be returned in the original outer container.

- Depleted uranium containers: For containers using depleted uranium as shielding (e.g., 650 source changer), special documents are required to describe the goods for transport purposes.
  Complete a Consignor's Certificate as for a loaded container, but use the following descriptions:
  Shipper's name and address
  Consignee's name and address
  Description – either:
  1. For surface dose rates less than 5 µSv/h (0.5 mR/h) state "UN 2910 Radioactive Material, Excepted Package, Class 7. Articles Manufactured from Depleted Uranium." Category and type are Excepted, TI not applicable, or,
  2. For surface dose rate greater than 5 µSv/h (0.5 mR/h) states "UN 2912 Radioactive Material, Low Specific Activity, Class 7 LSA N.O.S."
  3. "Physical Form: Solid"
  4. Group Notation: LSA-1
  5. Category is "Yellow II"
  6. Type is "Industrial"
  7. TI as appropriate
  8. Container number as appropriate
  9. Sign Declaration
  Attach two radioactive labels as described in the loaded container shipping instructions.
- Returning empty containers: The following procedure should be used to return empty containers:
  Remove all old adhesive labels
  Fill in the return address label and attach it to the container
  For empty lead shielded pots attach two "empty" labels ensuring that one covers over the metal "radioactive" label on the lid of the drum (if present).
- Loading Chart: Different shielding inserts are used for the different types of source holders. The way sources are arranged in these inserts is shown below. The source positions are identified by numbers stamped on the inserts. The inserts can contain up to 10 individual source positions.

5.20 Waste disposal

Radioactive waste must be disposed of properly, i.e., it has to be deposited at the collection site for nuclear waste. This obligation must be observed strictly. It is illegal to dilute or reduce the concentration of radioactive substances, so that the radiation falls below the permissible limits, and therefore the regulations are no longer applicable. If the sealed radioactive substances are no longer contained, the designated authority must be notified immediately and steps taken to ensure that the contamination cannot be dispersed. Proper handling and disposal of possibly leaking sources or contaminated parts of the equipment must be coordinated with the designated authority.

Note: Radioactive substances that are no longer used are not necessarily radioactive waste. These substances must be properly disposed of as well, e.g., by depositing them at the governmental collection site or by returning them to the manufacturer. In the latter case, the manufacturer will fill out a certificate acknowledging receipt, which will also serve as proof that the substances have been disposed of properly.

5.21　Specific safety procedures for radiography

5.21.1　Work practices

Radiography cameras must be operated only by certified radiographers. As far as possible, field radiography should be carried out during the night when there is little or no occupancy around. Field radiography during the daytime may be permitted on a restricted scale when occupancy is minimum, e.g., during lunch or on holidays. If field radiography is carried out at the same location repeatedly, it is advisable to provide either wire fencing or a temporary brick enclosure.

An appropriate area around the radiation source must be cordoned off during field radiography so that the radiation levels outside the area do not exceed the reference radiation levels for members of the public. The distance to be cordoned off is determined by the type and strength of the radiation source used, the type of exposures given, the nature of occupancy and the total exposure time per week. The radiation levels along the cordon must be monitored by a suitable and calibrated radiation survey meter to confirm the cordon distance is indeed adequate. Radiation warning symbols must be conspicuously posted along the cordon. Placards displaying the appropriate legend must be posted at the cordon. The placard and the radiation symbol should be readable from a distance of 6 to 7 m under normal illumination.

Entry into the restricted area by unauthorized persons must be strictly prohibited during exposure. When the radiography work is carried out at night-time, the radiography site up to the boundary of the cordon on all sides must be adequately illuminated throughout the duration of radiography work. Red warning lights must be conspicuously displayed during night along the cordon and especially at the point of entry.

The concerned radiographer must be available at the site very near the cordoned area throughout the exposure. Wherever practicable, field radiography work should be limited to collimated exposures. Wherever collimated exposures would suffice but cannot be given with the available equipment, suitably improvised collimators should be used. During collimated exposures the primary beam should be directed toward areas of minimum occupancy. All panoramic exposures with the sources of activities greater than 20 Ci of ^{192}Ir must be carried out using remote control system.

Manipulating devices must be used for handling the sources of activities up to 8 Ci of ^{192}Ir during panoramic exposures. While using an X-ray machine, the full length of the cable connecting the X-ray tube and control console must be used. The full length of the manipulator rod should be made use of during panoramic exposures so that maximum possible distance is maintained between the radiation source and the operator. The source pencil must never be touched or handled directly with hands.

Radiography work at elevated places and at locations where accessibility is restricted or limited should be carried out preferably by remotely operated cameras. Suitable supporting and fastening devices must be used for hoisting and positioning the radiography cameras/Xray machines in order to avoid mishaps such as accidental fall etc.

The radiography equipment should always be operated by positioning oneself behind the camera/X-ray machine making use of the shielding provided by the body of the equipment. All operations should be planned in advance and executed in minimum possible time. The radiography work must be carried out only under the supervision and guidance of the designated authority.

After termination of each exposure, it must be verified by means of a radiation survey meter in proper working condition that the source has indeed returned to its safe position inside the camera. After completion of each exposure, the source pencil must be securely locked in the camera. Prior to on-site transport of the camera with source from one place to another in a road vehicle/trolley it must be ensured that the source pencil is securely immobilized and locked in the camera. This would avoid any accidental opening of the shutter and falling of source pencil from the camera. A log book must be maintained at every site to record the following details regarding the use of the source:

- Date and time of taking out the camera from storage
- Model and serial number of the camera
- Nature and strength of the source
- Name of the radiographer
- Location of use
- Type and total number of exposures given
- Duration of use
- Date and time of return of the source with the camera to storage

5.21.2 Additional safety measures in the use of teleflex-type radiography cameras during field radiography

A wide-range radiation survey meter must be available at the radiography site. Coupling between the teleflex cable and the source cable and also between the camera and guide tube must be physically verified prior to the operation of the unit. Any sharp bend either in the driving cable or in the guide tube must be avoided in order to facilitate smooth and trouble-free movement of the source through the guide tube. While operating the camera, the operator must make use of the full length of the cable so that maximum distance is always maintained between the operator and the source.

The smooth functioning of the driving system must be periodically inspected. Prescribed lubricating agents should be applied at periodic intervals in order to facilitate smooth functioning of the camera. Any defect, however trivial, must immediately be attended to. A checklist may be prepared and used for such periodical inspection.

The source must never be driven under force. Should there be any obstruction to the smooth movement of the source through the guide tube during operation, the source must be immediately retrieved to the camera, the fault must be investigated and rectified before the radiography work with the camera is carried-out. The radiation survey meter must always be used during such occasions to verify the position and also its proper storage inside the camera. If the source is stuck in the guide tube and cannot be retrieved, radiography work must be suspended, and emergency procedures must be followed.

5.22   Radiation-protection safety

5.22.1   Safety measures

When designing the installation of radioactive system, the possibility that a fire could start must be considered. Flammable substances must not be stored in the proximity of radioactive substances. They should be covered and protected properly, so that a possible spreading of a fire to the radioactive sources will be prevented. It is mandatory to coordinate all preventive measures against fire with the appropriate fire authorities. They must be informed about the type, scope, and place of application of the radioactive substances used, in order to be prepared in the event of fire. When designing alarm plans, possible special features of the radiometric measuring system have to be mentioned, and the safety officer to be notified in the event of an emergency.

5.22.2   Malfunctions and accidents

The Radiation Protection Regulations define malfunction as even which for safety reasons prohibits continuation of the operation of the facility. Malfunctioning means a device necessary to guarantee safe operation of the facility, e.g., the seal of the active radiation beam of the shielding, no longer functions properly. An accident is an event that could expose people to a radiation dose that exceeds the permissible limits, or could cause contamination by radioactive substances. In terms of safety, malfunctioning and accidents are very serious events and appropriate steps must be taken immediately to prevent dangers for people as well as for facilities, or to reduce them as much as possible. It is therefore important that personnel be aware of preventive measures and be prepared for possible malfunctions of facilities or accidents, so that dangerous consequences can be avoided as far as possible by a proper reaction of the personnel.

In any case, the safety officer who checks the situation at site and takes all necessary steps to prevent unnecessary radiation exposure of the personnel has to be notified immediately. All official authorities listed in the emergency procedure including local suppliers should be informed. The necessary steps should be taken in the following order:

1. Locate the source.
2. Measure the dose rate.
3. Guard and mark the control area.
4. Secure the source and shielding.
5. Check the function and efficiency of the shielding.

6. Record the event and assess possible radiation exposure of the personnel concerned. In case the source capsule is damaged, the following points have to be considered:
7. Avoid contamination.
8. Handle source with tools (e.g., tweezers) and put both in a plastic bag.
9. Stay behind an auxiliary shielding (e.g., concrete, steel, or lead plate).
10. Check if vicinity is free of contamination.
11. Secure the radioactive waste properly (deposit at designated collection site or return to manufacturer).

5.23 Fire, explosion, and corrosion

In the evaluation of sealed sources and source-device combinations, the manufacturer and user have to consider the probability of fire, explosion, and corrosion and the possible consequences. Factors that should be considered in determining the need for actual testing are:

- Consequences of loss of activity
- Quantity of active material contained in the sealed source
- Radiotoxicity
- Chemical and physical form of the material and the geometrical shape
- Environment in which it is used
- Protection afforded the sealed source or source-device combination

5.24 Health requirements

Health and safety rules (conforming to widely accepted standards) should be prepared for the areas in which radioactive material is to be handled. All necessary operating instructions should be provided as well as suitable installation and equipment. Provisions should be made for necessary medical supervision of personnel and for suitable medical casualty service. Only persons medically suitable and adequately trained or experienced should be allowed to work with radioactive material. All personnel liable to exposure to ionizing radiation in the course of their work should be instructed about the health hazards involved in their duties. Suitable training with reference to health and safety should be provided for all staff. A person technically qualified to advise on all points of radiation safety should be available, and the authority in charge of the installation should consult on all points of radiation safety. Procedures should be provided to handle persons who are exposed to radiation hazards.

5.24.1 Medical casualty service

The form of medical casualty service provided will depend on the availability of medical staff within the establishment. First-aid advice and equipment should be immediately available throughout the working area. The scope of first aid and treatment attempted should be based on medical advice. Arrangements for referring casualties and personnel contamination problems to medical services at an appropriate stage should be clearly defined and known.

5.24.2   Personnel monitoring

- External radiation monitoring in which radiation measuring devices are worn by personnel
- Internal contamination monitoring in which suitable instruments may be used or body wastes may be sampled and analyzed to determine the presence and quantity of radioactive material within the body

5.24.3   Area monitoring

- Determination of radiation levels and air contamination in the working area
- Measurement by the use of radiation-measuring instruments and devices
- Calculation based on the amount of radioactive material present, its form, and the nature of the processes in which the workers will be exposed

5.24.4   Determination by personnel monitoring

- Monitoring for external radiation exposure with personnel dosimeters: This simple and convenient method should be used for the measurement of external radiation exposure of all personnel in the controlled area. The preferred device is the film dosimeter, which permits measurement of the accumulated radiation dose over a period. This film also provides a permanent means of checking the accumulated external radiation exposure record, which should be kept for every individual. Similar film dosimeters should be used on the hands, wrists, or other extremities when these are exposed to higher radiation fields than the trunk of the body. Pocket ionization chambers, luminescent individual radiation detectors, and thimble chambers supplement these film dosimeters and are particularly useful where an immediate and sensitive measurement is needed in connection with a specific task. In the use of both film dosimeters and ionization chambers for personnel monitoring, serious errors may occur unless standard procedures are adopted.

5.25   Inspection

5.25.1   Site inspection

A trained safety engineer should visit the work site or equipment store to perform equipment inspection and overhaul. The source projector maintenance service and gamma-radiography source projector should be regularly inspected and maintained. The booklet radiation safety for site radiography recommends that this be carried out annually.

A complete inspection should be made to check for any weakness in moving components or safety interlocks. Any maintenance that is required must be carried out by skilled people. Repainting and provision of new labels should be included when appropriate. Certificates should be issued after each inspection and maintenance work is done. An inspection should normally be carried out during a source exchange. Caution: Source exchanges must only be carried out by a person who knows the precautions to be taken when working with radiation and is fully informed about the operations he is about to perform. A radiation monitor (dose-rate meter) must be used at all times during source movements and particularly after each source movement (for source exchange or routine radiography) to check that the source is fully retracted and is in the fully shielded position.

5.25.2 Periodic inspection and maintenance of radiography equipment

The shielding integrity of radiography equipment must be regularly checked once a month. The cameras and the associated accessories must be inspected periodically and any defect rectified immediately by a person duly authorized for the purpose by the designated authority. The licensee must not undertake any repairs/modifications to radiographic cameras. However, some minor repairs such as mending hinges of the shutter, the source pencil arrester lock, etc., can be done. Cameras must not be repaired with the radiography source in it. The source must be unloaded into a temporary source container prior to undertaking the repair.

Radiography cameras with the source inside must never be taken out of an approved site without obtaining permission from the appropriate authority. However, source/camera movement is permitted for urgent radiography jobs, provided all the relevant information is transmitted to the Division of Radiological Protection simultaneously.

5.25.3 Service instructions

When writing up service instruction, which should include the necessary rules of behavior, the following should be taken into consideration:

- Assembly and disassembly (the radiation path of the shielding must remain locked)
- Operations in the immediate proximity of the shielding.
- Ensure that the shutter of the shielding is locked if the vessel on which the source is mounted has to be entered or, with a density gauge, removed from the pipe
- Responsibility for the key of the lock of the shield in the container

5.25.4 Inspection and testing

All guide tubes, cable connectors, and other associated equipment must be inspected before use as described in the equipment manufacturer's operating and maintenance manual. Direct inspection of the source assembly is not possible (or safe) without specialized equipment. If normal source movements are difficult or impeded, this suggests damage. Seek further advice from the local radiological protection service. Leakage tests must be carried out at intervals as specified in local regulations. A test according to BS 5288 on the source itself is not possible unless shielded remote-handling facilities are available.

Table 5.7 Lists of "Recommended Working life" (RWL) for radiography sources

Nuclide		RWL
Iridium-192		1 Year
Cobalt-60	(Portable unit, Fixed installation)	15 Years
Thulium-170		1 Year
Ytterbium-169		1 Year

To test for source leakage, the exit port of the projector or storage/transport container should be wipe tested and the result recorded including:

- Source identity method used and date of test result (numerical)
- Pass/fail statement (limits) and reason for test
- Remedial action if failure
- Testing organization and signature

5.25.5 Recommended working life

The Recommended Working Life (RWL) is the period within which the source should be replaced. The period has been determined on the basis of factors such as toxicity of nuclide, total initial activity, source construction, half-life of nuclide, typical application environment, operation service experience, and test performance data. Table 5.7 lists the RWLs for some radiography sources.

The assessment of the RWL is based on the assumption that the source is not used in adverse environments. It is the user's responsibility to inspect and test the source regularly in order to assess at what point during the RWL the source should be replaced and sent for disposal. Advise should be sought regarding the RWL for sources used in adverse environments or for sources that, having completed the RWL, appear satisfactory and, subject to full inspection by a competent laboratory, may be suitable for an extended period of use.

5.25.6 X-ray lead-rubber protective apron

This section of standards applies to X-ray lead-rubber protective aprons with or without back panel, used by personnel during medical X-ray diagnostic examination with X-ray generated and voltage up to 150 Kev peak, and intended to give to the body of the operator or patient a measure of protection against scattered radiation.

The aprons should be designed to give the minimum protective value specified by the manufacturer and in no case less than 0.25 mm lead equivalent for X-rays generated at a voltage of 150 kV peak. All joints or seams in the apron should offer at least the same protection. The protective material should be either natural or synthetic rubber compound incorporating lead or a compound of lead. It should comply with the test requirements, and together with its attached waterproof fabric, should be freely

Table 5.8 **The general dimensions of X-Ray lead-rubber protective aprons**

Size designation	Length (Measured from the center of the shoulder)	Width
33 inch	33 ± ¼ in (83.8 ± 0.6 cm)	24 ± ¼ in (61 ± 0.6 cm)
36 inch	36 ± ¼ in (91.4 ± 0.6 cm)	24 ± ¼ in (61 ± 0.6 cm)
38 inch	38 ± ¼ in (95.5 ± 0.6 cm)	24 ± ¼ in (61 ± 0.6 cm)

flexible. The apron should consist of an X-ray protective layer(s) or covered on all exposed surfaces by a waterproof fabric integrated with the protective layer. The edges should be protected by binding, and the binding should be of such a nature as to remain flexible during the life of the apron. The fixings used for the supporting harness, and for attaching the back panel to the shoulder pieces, should not reduce the life of the apron under normal conditions of use.

The apron should be so designed that the width of the protective material on each shoulder should not be less than (76 mm), and the shoulder pieces should extend over the back of the shoulder by not less than 152 mm. When specified by the purchaser, the shoulder pieces should incorporate a suitably covered padding of soft resilient material 3 in (76 mm) wide and extending a distance of 6 in (152 mm) on either side of the center of the shoulder. Table 5.8 shows the dimensions of X-ray lead-rubber protective aprons. Figure 5.6 shows the general dimensions of X-ray lead-rubber protective aprons.

Note: Urethane foam, 25 mm thick, when uncompressed and of a density of 27 to 30 kg/m3 is one suitable material.

The illustration in Figure 5.6 is diagrammatic only; it is given to define the essential dimensions but is not intended to indicate details of design.

5.25.7 Activity level

This section establishes the maximum activity of sealed sources, for each of the four radiotoxicity groups discussed in the following, for which a separate evaluation of the specific use and design is not required. Sealed sources containing more than the maximum activity should be subject to further evaluation of the specific use and design. The activity level of a sealed source for purposes of classification is shown in Table 5.9, and of course, it should be based on the time of its manufacture. The table also defines the physical, chemical, and geometrical forms of the radionuclide used to determine these properties; they should be the same as the physical, chemical, and geometrical forms of the radioactive material within the sealed source.

5.25.8 Sealed-source performance requirements for typical use

This section discusses a list of typical applications in which a sealed source or source-device is used, together with an estimate of their minimum performance requirements.

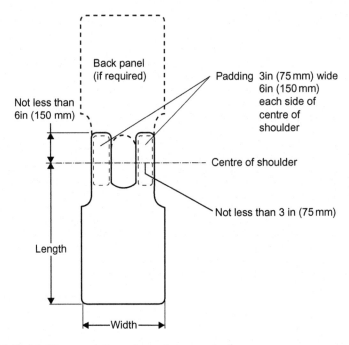

Figure 5.6 Sketch (1) general dimensions of x-ray protective aprons.

Table 5.9 **Activity level**

Radionuclide group (From Annex A)	Maximum activity, TBq (Ci)	
	Leachable* and/or	Non-leachable† and not
	Highly reactive‡	Highly reactive§
A	0,01 (about 0,31)	0,1 (about 3)
B1	1 (about 30)	10 (about 300)
B2	10 (about 300)	100 (about 3000)
C	20 (about 500)	200 (about 5000)

Notes:
*Leachable-greater than 0.01% of the total activity in 100 ml in still H2O at 20°C in 48 hrs
†Non-leachable-less than 0.01% of the total activity in 100 ml in still H_2O at 20°C in 48 hrs.
‡Highly reactive-highly reactive in ordinary atmosphere or water (metallic, Na, K, U and Cs, etc.).
§Not highly reactive-not highly reactive in ordinary atmosphere or water (Au, Ir, ceramics, etc.).

This estimate takes into account normal use and reasonable accidental risks but does not include exposure to fire or explosion. For sealed sources normally mounted in devices, consideration is given to the additional protection afforded the sealed source by the device when the class number for a particular use was assigned. Thus, for all uses, the class numbers specify the tests to which the sealed source should be subjected, except that for the ion-generators category, the complete source-device combination may be tested.

Obviously, this section does not cover all sealed-source use situations. If the particular use or accidental risks are likely to differ from the values suggested in the estimate, or if the sealed-source use is not shown, the specifications of the sealed source should be considered on an individual basis by the supplier, the user, and the regulating authority. The numbers shown in Table 5.10 refer to the class numbers. Attention is called to the International Atomic Energy Agency tests for special form radioactive material. These are not of general application but may be relevant when formulating special tests.

5.26   Technical information

- SPECIFICATION: The strength of a radiation source may be specified either by its radiation output or by stating the radioactivity of its contents. For most applications, the user is mainly interested in the source's radiation output, and requires information about content only for licensing or commercial reasons. For many sources, the radiation output is not simply related to the activity content because factors such as self-absorption and attenuation by the capsule cause a nonisotropic output distribution. For this reason, the radiation output in the given direction, using the most appropriate method, does not rely on an estimate of the amount of radioactive material in the source. A real source emits anisotropically, so it is necessary to specify the direction in which a measurement has been made, as well as the distance from the source. The distance taken from the center of the source and the direction is normally radial for cylindrical sources and axial for disc sources. The relationship between equivalent activity and exposure rate is different for each radionuclide, depending on the type and quantity of radiation emitted in each nuclear transformation. The accepted values for the most commonly used high-energy gamma-emitting nuclides are given in the Table 5.11. In the SI system, source strengths may be expressed in terms of air kerma rate at 1 m in grays per hour. The equivalence between the old units and the new is shown in Table 5.11.

5.27   Test device

The test device should be used for checking the dosimeters for reliable operation; it should accommodate a maximum of 8 dosimeters. The dosimeters should be set to zero and then subjected in the test to many hours of radiation with a known total dose that will cause a corresponding dosimeter indication. In order to operate the test device, a separate radiation source, type Automess 6706, is required. This radiation source contains Cs 137 as the radioactive substance, with an activity of $9.9\,\mu Ci$. It is mandatory to register the procurement and utilization of the test device.

The test device, including the radiation source, should be tested and approved for use in accordance with the calibration validity requirements. The calibration validity period of 2 years for calibrated dosimeters can therefore be extended to 6 years when a test device is used, if test measurements are made at least semi-annually. Figure 5.7 shows Personal dosimeter for radiation protection. Ten indicating ranges, from 0.1 R

Table 5.10 Sealed source performance requirements for typical usage

Sealed source usage	Sealed source test & class				
	Temperature	Pressure	Impact	Vibration	Puncture
Radiography–Industrial					
Unprotected source	4	3	5	1	5
Source in device	4	3	3	1	3
Medical					
Radiography	3	2	3	1	2
Gamma teletherapy	5	3	5	2	4
Interstitial & interacavitary appliances*	5	3	2	1	1
Surface applicators	4	3	3	1	2
Gamma gauges (medium & high energy)					
Unprotected source	4	3	3	3	3
Source in device	4	3	2	3	2
Beta gauges & sources for low-energy gamma gauges or X-ray fluorescence analysis (excluding gas-filled sources)	3	3	2	2	2
Oil-well logging	5	6	5	2	2
Portable moisture & density gauge (including hand-held or dolly-transported)	4	3	3	3	3
General neutron source application (excluding reactor start-up)	4	3	3	2	3
Calibration sources-Activity greater than 1 MBq	2	2	2	1	2
Gamma irradiation sources					
Unprotected source	4	3	4	2	4
Source in device	4	3	3	2	3
Ion generators†					
Chromatography	3	2	2	1	1
Static eliminators	2	2	2	2	2
Smoke detectors	3	2	2	2	2

Notes:

*Sources of this nature may be subject to severe deformation in use. Manufacturers and users may wish to formulate additional or special test procedures.

†Source-device combination may be tested.

Table 5.11 **Technical information**

Nuclide	Equivalent activity	Exposure rate at 1 meter	Air kerma rate (K) in air at 1 meter (approx.)
137 Cs	1 Ci	0.32 R/h	2.9 mGy/h
60 CO	1 Ci	1.30 R/h	11 mGy/h

Figure 5.7 Personal dosimeter for radiation protection. * Ten indicating ranges, from 0.1 R to 200 R. Energy range can be: 40 keV to 3 MeV 18 keV to 3 MeV. *Should be low self-discharge and vacuum-sealed metal-glass fusion. *Rugged and drop-proof construction.

Figure 5.8 Direct reading alarm dosimeter.

to 200 R. Energy range can be: 40 keV to 3 MeV 18 keV to 3 MeV. Figure 5.8 illustrates a Direct reading alarm dosimeter.

5.27.1 *Application and design*

The Alarm-DOSimeter (ADOS) is a portable, battery-operated dosimeter for measuring photon radiation (gamma and X-rays). A GM tube is used as the radiation detector. The radiation values are processed by a microprocessor, and the digital, Liquid-Crystal Display (LCD) has four digits. Ten adjustable dose-alarm thresholds and one fixed dose-rate alarm threshold provide audible and visual alarms whenever a certain level of radiation is exceeded.

The ADOS is designed primarily for use as a personal dosimeter in all areas in which people are exposed to the hazards of increased photon radiation, e.g., X-ray diagnostics, radiotherapy, nuclear medicine, technical applications, and transportation of radioactive nuclides. The ADOS can also be used for local dose measurements.

The robust, waterproof housing is made of diecast aluminum. Power is supplied by a commercially available 9 V battery; due to the low power consumption at low radiation levels, an alkaline battery gives a life of approximately 2000 hours continuous service.

All the necessary operations, such as switching the dosimeter on and off, setting the dose-alarm threshold, resetting the dose and acknowledging alarms are effected with a single push-button. The device is thus completely autonomous, and no additional equipment or tools are required to operate it.

For special applications, the radiation values can be read and the dosimeter programmed automatically using an analyzer (optional). The data transfer between the dosimeter and the analyzer is contactless, and takes place via an inductive sensor inside the cover of the battery compartment.

5.27.2 Dose storage, dose resetting

The dose is displayed continuously and is stored in nonvolatile form. The term "nonvolatile" means that the dose is not lost even if the dosimeter loses power; in other words, it still remains stored when the dosimeter is switched off or when the battery is replaced. When the dosimeter is switched on again or a new battery is inserted, the dose that thereafter has accumulated will be added to the dose that is already stored. The dose can only be reset directly after the meter is switched on (or optionally using an analyzer).

5.27.3 Dose alarm

The dose-alarm threshold can be set to one of the fixed values when the dosimeter is switched on; it is set to 0.2 mSv by default. If the dose reaches or exceeds the alarm threshold, a dose alarm is output; the dose display flashes and an intermittent alarm tone begins. The dose alarm can be reset by pressing a button. If the dose rises still further by a fixed amount, a new dose alarm is output as a reminder that the permitted dose has already been exceeded – a so-called "post-alarm."

This dose-alarm method has been selected for the following reasons: On the one hand, it must be possible to reset the alarm, since a tone that cannot be reset is likely to annoy the user. Yet on the other hand, it is essential to ensure that an alarm that has been reset is not forgotten if the dose continues to rise. The user is thus given an opportunity to bring his work to some sort of conclusion before leaving the radiation field. For example, if he is working in a radiation field in which the first dose alarm is issued after one hour, a new dose alarm will be output roughly every minute. The currently adjusted dose alarm threshold can be displayed at any time by pressing a button.

5.27.4 Dose-rate alarm

The dose-rate alarm threshold is permanently set to 1 mSv/h. If this value is reached or exceeded, an audible and visual alarm is issued and the dose rate is displayed as a digital value. The dose-rate alarm can likewise be reset by pressing a button; it is reset automatically when the dose rate falls below the alarm threshold again. If a dose alarm and a dose-rate alarm occur simultaneously, the dose alarm has priority. The dose-rate alarm is not displayed until the dose alarm has been reset.

5.27.5 Remaining time

The dosimeter continuously calculates the time remaining up to the adjusted dose alarm threshold in hours and minutes, taking account of the current dose and dose-rate values. If the dose alarm threshold is reached or exceeded, the remaining time is set to 0. The remaining time can be displayed any time by pressing a button (maximum value 9 hours 59 minutes).

5.27.6 Maximum dose rate

The dosimeter remembers the maximum dose rate that has been measured since the last time it was switched on or since the dose was last reset. This value cannot be read off directly on the dosimeter, but can only be read from the dosimeter and displayed with the help of an optional analyzer.

5.27.7 GM-tube monitoring

The dosimeter continuously checks the GM tube with regard to the expected, minimum pulse rate, even if there is currently no external radiation. If no pulses are detected for a fixed period of time, the GM-tube circuit is assumed to be defective and a resettable audible/visual alarm is output.

5.27.8 Battery monitoring

The battery voltage is measured by the dosimeter at intervals of 5 min, and can be displayed by pressing the button. If the battery voltage falls below 5.0 V, a resettable audible/visual alarm is output.

5.27.9 Analyzer

The dosimeter incorporates an inductive sensor that enables data to be exchanged between it and an analyzer. Simply inserting the dosimeter in the slot in the analyzer permits the following functions to be activated:

- Readout and resetting of the dose and the maximum dose rate
- Readout and programming of the dose alarm threshold, the dose-rate alarm threshold, and a personal identification number for the user
- Readout of the serial number of the dosimeter

It should be stressed again at this point that an analyzer is not essential for operating the ADOS; it merely offers an additional degree of sophistication and extra functions.

5.27.10   Check device

The check device 704.1 ADOS can be used for a quantitive check of the radiological functions of the dosimeter. The device is microprocessor-controlled and – apart from inserting the dosimeter and extending the Cs137 radiation source – it performs all the necessary steps automatically; namely, resetting the dose, selecting the radiation time, reading the dose at the end of the radiation time, and printing out the results. A measurement process that simultaneously exposes 15 dosimeters to radiation takes 7 to 8 min using a radiation source with a rated activity of 37 MBq (1 mCi).

First aid and sanitation

6

Vast complexes and varieties in nature and locations of operations within the oil, gas, and petrochemical industries require specific requirements related to sanitary and first-aid measures. This chapter outlines these specific requirements.

Sanitation and first aid are the two key requirements for keeping plants/machinery, working places, and personnel in healthy conditions. The dangerous consequences of undesirable sanitary conditions and poor first-aid procedures in the oil, gas, and petrochemical industries are briefly categorized as follows:

- Unsafe working conditions
- Malfunctioning machineries
- Poor health of personnel

6.1 Sanitation and hygiene of the plants and workshops

- Cleanliness: Not less than 8 hours a day are spent at work; if the working surrounding are dirty and depressing the worker tends to become dirty and depressed. The standard of behavior, especially in young workers, is set by standard of cleanliness and hygiene of working places.

6.1.1 Lavatories

Filthy lavatories invite disgusting habits. The general appearance of the factory, benches, tools, floors, and walls must be spotless, and the addition of proper cloakrooms and the adequacy washing and lavatory accommodation make a special impression.

Lavatories should be of the water-carriage type and separate lavatories for each sex provided at the rate of 1 for every 25 women and 25 men up to the first 100 men and then at the rate of 1 for every 40 men in privacy and ventilated. In women's lavatories, sanitary towels should be available from automatic machines, and bins provided to receive soiled towels. The bins must be emptied and the soiled towels burned daily. Wash basins with taps and wash plugs are unsatisfactory as they are seldom kept clean; the fountain type with a spray of water controlled by a foot pedal is most satisfactory. Soap should be dispensed from containers either in the form of liquid or powder. Hot-air dryers are more satisfactory than towels, and though it seems to take much longer to dry the hands on a hot-air drier than with a towel, the extra time taken is only a few seconds if the drier is in frequent use so that hot air is delivered immediately.

From a medical point of view, employees who are disabled or unfit should use standard urinals or European lavatories. The rate of these lavatories depends on the number of men in a factory or local requirements.

Personnel Protection and Safety Equipment for the Oil and Gas Industries.
DOI: http://dx.doi.org/10.1016/B978-0-12-802814-8.00006-7
© 2015 Elsevier Inc. All rights reserved.

6.1.2 Colors

Colors used to paint walls and machineries should be cheerful and pleasing; in cloak-rooms and lavatories the walls should be covered with tiles and the floors with tiles or terrazzo that are easily washed.

6.1.3 Gulley

A gulley running the whole length of one side of the room with the floor sloping slightly toward its washing facilities should be provided.

6.1.4 Lockers

Separate lockers should be provided for each worker. Means of drying wet clothes and shoes should be provided; hot pipes under the lockers are satisfactory but if the clothes are crowded together in the lockers, drying may be delayed. The best locker is that in which clothes are hung on coat hungers. A separate shelf should be provided for hats handbags. Mirrors should be fixed away from washing fountain or basin and in women's cloakrooms a wide shelf for handbags should be provided beneath the mirrors. Adequate space is essential in all cloakrooms and lavatories if a high standard of behavior is to be achieved.

6.1.5 Lighting

The intensity of lighting in lavatory should be greater than in the factories so that on entering a lavatory a sense of cleanliness is engendered. As a rule, the intensity of lighting should not be less than 26.9 Lux (25-foot candles).

6.1.6 Living quarters

The most important sanitary factors that should be observed are:

- Daily cleaning of rooms, bathrooms, and lavatories
- Weekly change of bed linen
- Bed linen should be changed before the shift change

6.2 Mechanical and electronic equipment and radiation

6.2.1 Machinery and plant

Machines should be so designed to facilitate the movement of employees and not only to design the space required for the size of the machines. Seating required for the prevention of fatigue. Foot pedals are often inconveniently situated, both in height and distance from the center of gravity to the site of operation, thus causing an undue strain on the muscles of the back, pelvis and opposite leg. Shock absorbers should be fitted on all pedals whether foot or hand operated. The convenience and conversely fatigue of the employees should always be considered in the design of a machine. The layout of control and recording instruments should be studied so that the "pattern" of the operation is quickly observed. If control movements are required these movements should be geared to secure an optimum relation between speed of control and precision of operation.

6.2.2 Electronic equipment

Electronic equipment that is utilized in the control rooms, keyboards, and computer rooms, etc., should be regularly checked for any malfunctions and kept sanitized. Monitors should be equipped with special filters for prevention against radiation.

6.3 Radiation

6.3.1 Biological effects

Radioactivity causes ionization of proteins. This effect is best shown by disintegration of chromosomes at the point of passage of rays and particles. Tissues undergoing mitosis are selectively sensitive to ionizing radiation. The most important effects of radioactivity are as follows:

- Germinal epithelium is damaged
- Hematopoietic system
- Gastro-intestinal membranes
- Rarefying osteitis and bone neoplasms
- Tumor formation
- Post cortical cataract
- Skin disorders

6.4 Chemical substances

Chemical substances may be absorbed into the body in three ways:

- By ingestion and absorption from the alimentary tract
- By absorption through the skin
- By inhalation and absorption through the lungs

The first way is rare and often a semi-accidental cause of industrial poisoning, but the last is the most common way in which industrial poisons are absorbed. The lungs are a great mode of entry for most industrial toxic substances, first because of the nature of industrial processes, and second because substances absorbed through the lungs enter the systemic bloodstream. Substances absorbed through the intestines pass through the portal system and elimination or detoxification by the liver may occur and a toxic quantity can reach the systemic blood.

6.4.1 Prevention

Prevention and avoiding of contamination of the skin consists of the following:

- Chemical substances should not be handled more than is essential
- Adequate ventilation of workshops
- Good nutrition
- Periodic medical examination
- Selection of workers for employment
- Adequate washing facilities
- Barrier substances may be applied in the form of powder, cream, or varnish

6.5   Personal sanitation

6.5.1 Personal hygiene

In addition to cleanliness in the workplaces, personal cleanliness is of the upmost importance. Cloakrooms, washing rooms, mess rooms, bath, nailbrushes, towels, and soap must be provided. The hands should always be washed before eating, and the workers urged to take a warm bath. Food and drink should not be brought into the workrooms and smoking at work must not be allowed.

- Skin: Facilities for washing and taking showers after work should be available and facilities for complete change of clothing before work should be provided.
- Dirty clothing should be washed daily and a good neutral soap should be in hand.
- Barrier creams should be available wherever recommended by the medical authorities. It should be applied to the skin to prevent irritating of substances from coming in contact with the skin.
- Hair: Long hair should be kept covered by some hygienic protective cloth when working at workshops.
- Dental examination: Regular dental examination can achieve the hygiene of the mouth.

6.5.2 Protection against infection

Protection against infection requires the following precautions:

- Provision of wholesome water and food
- Sanitary and washing accommodation
- Maintenance of individual cleanliness and cleanliness of surrounding
- Detection and treatment of carriers of vermin and diseases

6.5.3 Prevention against accident and fire

Below is the list of prevention against accident and fire:

- Adequate floor space and cubic space for each worker 11/327 cubic meter (400 cubic feet per person as the minimum permissible space)
- Control of toxic hazards
- Elimination of electric shock and burn
- Control of fire
- Precautions against hazards of traffic movement

6.6 Medical examinations

A program of medical examination should be established as following:

- Examination prior to placement
- Periodical examination when the employee is exposed to special risk
- Special examination
- Termination examination

The details and techniques of medical examination will vary according to the industrial hazards.

- Pre-placement medical examination: A properly conducted pre-placement examination protects employer and employee. Industrial Medical Officer must have full knowledge of two factors: First, the work to be undertaken, including any special risks. Second, an exact knowledge of patient's physical and mental state acquired by a full investigation of family and personal history.
- Periodic medical examination: Periodic examination is meant a medical inspection of an employee at specified intervals to prevent risk either to the employee or to the fellow employees, or detect hazards before any permanent harm has resulted.
- Special examination: Employees having on-the-job difficulties that may be health related are benefited by special examination. Job transfer also often requires medical evaluation.
- Termination examination: Upon termination of employment of an employee, a medical examination should be conducted.

6.7 Site conditions

Work environments should provide the following.

6.7.1 Physiological requirements

- Maintenance of thermal environment (including appropriate humidity) sufficiently warm to prevent excessive heat loss and not too hot to prevent adequate heat loss from the body
- Atmosphere of reasonable purity
- Admission of adequate daylight, the provision of sufficient artificial light and the avoidance of glare
- Protection against excessive noise
- Adequate space for work and movement
- Nutritional needs

6.7.2 Psychological requirements

- Work within (but near to) limit of mental capabilities
- Work without undue time stress and with reasonable hours of employment
- Hygiene and social standards not less than those prevailing in the community
- Opportunities to find out satisfaction in work and in membership of a group
- Recreational facilities

The general environmental conditions in which work is performed, though frequently overlooked, are the most important of all factors affecting well-being at work. Another important factor is the relationship between the employee and the management. Unsatisfactory environmental conditions give rise to a fall in output, a lowering of health, an increase in accidents, and a host of real or imaginary grievances.

6.7.3 Heating

The temperature of workplaces after the first hour of work should not be less than 15.50°C (60°F), though even this is chilly for light work. For light work most people are comfortable at approximately 19°C (67°F) with an air velocity of 33 m (100 feet) per minute, though the range of comfortable temperature may extend from 12 to 24°C (54 to 76°F). The output of an employee decreases as the temperature rises above the comfortable level. At least one thermometer should be provided in each workroom. The thermometer should be about 1.65 m (5 feet) above the floor with the bulb freely exposed and so situated as to record fairly the conditions to which employees are exposed.

6.7.4 Ventilation

Ventilation is related to heating, since it is common knowledge that in the absence of proper heating, ventilation of a room is reduced by the occupants. The object of good ventilation is to avoid these conditions, to create an adequate change of air, and to supply clean air, so that the air is kept comfortable and body odors are removed.

6.7.5 Humidity

The amount of sweat evaporated from the skin, and loss of heat by this means, is influenced by the humidity of the air. High atmospheric temperatures prevent loss of heat by radiation. Dry air is unpleasant over long period, since dryness and soreness of the nose and pharynx are produced together with cracking of the lips. However, the humidity should be between 60 to 70%.

6.7.6 Lighting

The best light is daylight, and the proper arrangement of adequate window space or roof lights directed toward the north will always provide better illumination than artificial light. The characteristics of daylight are its great intensity (on a dull day equivalent to 200 or more foot candles and on a bright sunny day 1000 or more foot candles) and the diffuseness of the light.

Recommended values for illumination:

- Fine assembly work – 106.60 Lux (100-foot candles)
- Drawing office – 32.28 Lux (30-foot candles)
- Ordinary bench and machine work – 10.75 Lux (10-foot candles)
- Corridors – 5.38 Lux (5-foot candles).

Others:

- Work of simple character not involving close attention to detail, 4.30–6.45 Lux (4–6 foot candles)
- Casual observation where no specific work is performed, 2.15–4.30 Lux (2–4 foot candles)

6.8 Dust, spray, gases, and vapors

Dust, spray, gases, and vapors may all find access to the body through the respiratory tract. Chlorine has an immediate action on the respiratory tract; phosgene has delayed action; lead, hydrogen, and sulphide act when absorbed into the bloodstream, and others may show their effect many years later by their action on the lung such as with silica, or on the body as with manganese.

6.8.1 Dust

Dust hazard is the most difficult substance to control. Construction of dust-proof apparatus or the reduction of dust by ventilation is often a matter of extreme difficulty. Liquids and gases are easily confined. Pneumoconiosis is defined as all forms of pulmonary reaction to inhaled dust. Legally, "pneumoconiosis" is associated with fibrosis of the lungs consequent on the inhalation of dust.

6.8.2 Silicosis

Silicosis is defined as a pathological condition of the lungs due to inhalation of silicon dioxide. Silica occurs in various states of purity in earths, ores and stones. Quartz, granite, schist, and sandstone consist of pure or nearly pure silica. The classical lesion produced by silica is nodular fibrosis.

6.8.3 Asbestos

Asbestos is varying composition and consists of silicates, several base metals, principally magnesium and iron and to a less extent, calcium, sodium, and aluminum combined in a fibrous form. Long fibers are used for weaving into cloth, belts, safety curtains, and brake linings, but asbestos board, paper and insulating materials are prepared from short-fibered material.

Asbestos penetrates into the alvedi, whereas asbestos fibers tend to remain in the fine brochioles. After months or years of increasing shortness of breath the patient usually dies. Asbestos is one of the exogenous causes of lung cancer and mesothelioma of the pleura and peritoneum.

6.8.4 Prevention of silicosis and asbestoses

- Substitution of less dangerous materials
- Suppression of dust at the source
- Segregation of dusty processes
- Protection of workers
- Medical examination

6.9 Infectious disease

Infectious disease is also known as contagious or communicable disease due to parasitic organisms capable of transmission from some reservoir of infection to susceptible human recipients. High-infectivity disease, demonstrated by virus infections, attacks many people in a short period of time; low infectivity disease, e.g., bacterial and fungal diseases, attacks few people.

Control and prevention of infectious diseases:

Detection of source of infection and elimination of such sources
Disinfection:
     Concurrent disinfection
     Terminal disinfection
Immunization of exposed people
Education of community in risk of infection and methods of avoiding infection; advice on general hygiene

6.10 First aid and rehabilitation

Proper medical and surgical treatment from the moment of the accident to full recovery does much to lessen the consequence of an industrial accident. Facilities needed for proper first aid vary with different localities. Whatever is provided should be freely accessible to workers. The importance of the early and proper attention on the most trivial injuries cannot be overstressed in the prevention of sepsis and other complications leading to lost time. The extent to which a first-aid room is used depends on its siting.

For reduction of infection and disability and in line with the requirement of work insurance compensation law, regardless of the extend of injury, all employees must report accidents, so that immediate attention is given to injured employee thus investigation of accident to be carried out for future preventive measures.

6.10.1 First-aiders and teaching

In every industry or workplace first-aiders have a real part to play. But the emphasis of their work is different. Conventional first-aid course devote much space to the control of serious hemorrhage, fractures, and other form of severe injury. When a serious accident occurs the first-aider must know what to do and how far to go. But if expert help is quickly available heroic first aid is seldom needed.

The day-to-day picture of first aid in the factory is rather a stream of minor injuries and minor ailments, small cuts and burns, colds, and headaches. Many first-aiders will have to treat the workmen themselves, and they will never be seen by a trained nurse or a doctor.

In undertaking full treatment of minor injuries and ailments, the first-aider is shouldering a serious responsibility. He must know his job and his limits, and when to call for help. Given his knowledge, given the tools he needs for the job, he is the real first line of defense in the health care of his workmates.

Training in first aid has been carried out by the different organizations. Certificates are issued and competitions promoted, and a great network of enthusiastic voluntary workers have pursued these activities with an almost religious fervor. Though the scope of their teaching may change in certain respects, those who have been basically trained by these organizations are good as industrial first-aiders. However, they have to acquire a rather different approach if they are to play a proper part in modern industrial medicine and must be full of enthusiasm and intense interest in accepting this responsibility.

There must be one first-aider responsible for every first-aid box on each shift, and at least one deputy ready to take over in case of absence or illness of the first-aider. Key first-aiders should have some general training before they learn industrial first-aid treatment.

6.10.2　First-aider tools

The first-aider must be provided with tools for the job. Depending on the number of employees in the plant, the minimum contents of the first-aid box should be specified.

6.10.3　First-aid boxes

There are three box types A, B, and C (not the official nomenclature).

Box	A	is for workplace up to 10 workers.
Box	B	is for units with 11–50 workers.
Box	C	is for factories with more than 50 workers.

Boxes could be improved by certain additions and omissions, and official revision of first-aid box contents has recently been completed. Nevertheless, most of the basic items are essential, in particular, the official sterilized individual dressing is still the best emergency dressing as a prelude to removal for treatment elsewhere.

The classification of box size on the bases of the number of the workers at risk is a guide to minimum needs only. Some workplaces sustain very few minor injuries. Other machine shops or the like have a heavy minor causality rate. In consequence, the latter will use first-aid supplies far quicker than the former.

Most boxes supplied commercially have fronts to provide a workspace for the first-aider. This is a good arrangement, but the supporting chains are usually far too flimsy. Wooden boxes are superior to those made of tin, as the metal is more inclined to warp. Internally, every box should have a space in which bottles can be kept upright. This space must be at least 254 mm (10 in) in height if it is to contain 567 gm (20 oz) bottles. A minimum space of 254 mm (10 in) by 152 mm (6 in) by 102 mm (4 in) for 567 gm (20 oz) bottles, and 217 mm (8½ in) by 127 mm (5 in) by 102 mm (4 in) for 57 gm (10 oz) bottles should be considered. Boxes supplied commercially seldom have this space; consequently, in most small factories, dust-covered bottles decorate the top of the first-aid box. The box must be plainly marked "First Aid." The first-aid kit should be inspected periodically and maintained in complete readiness.

6.10.4 Sterilized individual dressing

The official sterilized dressing is made in three sizes:

- **Small:** for injured fingers
- **Medium:** for injured hands or feet
- **Large:** for other injured parts

The dressing consists of a thick absorbent pad, with a layer of lint, or preferably gauze, on the side to be applied to the wound, and a roller bandage stitched to the other side. The whole forms a small roll, which is wrapped in paper and enclosed in a cardboard box. The dressing itself, inside the paper, is sterilized. Sometimes the pad is medicated; this is unnecessary and undesirable. When the paper covering has been turnoff, the bandage will be found to be rolled in such a way that the pad can be applied to a wound without being touched by the hand, and so remains germ-free.

This is an excellent true first-aid dressing for any wound that is extensive or bleeding much, and the treatment should be done by a trained nurse, doctor or at hospital. As a dressing for small injuries to keep on while at work it is much too bulky.

A special version of the sterilized individual dressing is the sterilized burn dressing; in this the pad is impregnated with picric acid. Knowledge of the proper treatment of burns is rapidly advancing, and it is now clear that the application of picric acid or any other antiseptic or crust-forming chemical as a first-aid dressing does harm rather than cure. For burns, the use of the nonmedicated simple sterilized individual dressing is more beneficial rather than the sterilized burn dressing. The minimum supplies of sterilized individual dressing in first-aid boxes are as follows:

	Box A	Box B	Box C
Small Individual Dressing	6	12	24
Medium Individual Dressing	3	6	12
Large Individual Dressing	3	6	12

6.10.5 Cotton wool

Boxes have to contain a "sufficient" supply of sterilized absorbent cotton wool in 14.17 gm (½ oz) packets. Cotton wool in substantial quantity is occasionally needed by the first-aider for padding a splint, or mopping up a lot of blood. For such purposes, the 14.17 gm (½ oz) packets have the great merit of cleanliness and convenience.

Each type of first-aid box should contain six of these packets. The disadvantage of the 14.17 gm packet is that this quantity is far too much for most single use such as cleaning a wound and the remainder of a package is left about opened; which is no longer sterile, and soon gets physically dirty.

Small pledgets or pieces of cotton wool are essential for wound cleansing. For this purpose a cotton wool strip dispenser, as used in barber's saloons, is very useful. A screw-top jam jar with 12.7 mm (½ in) hole cut in the metal top. Clean cotton wool is cut into 19 mm (¾ in) strip, and packed neatly into the jar, the end being threaded through the hole in the top, pledgets can then be pulled off as required. Every type of first-aid box should contain clean cotton wool in a strip dispenser. The regular stocking up of the dispenser should be done by the responsible authority on a clean table, in a clean room, with clean hands, preferably a trained nurse.

If a first-aider is to attempt to remove foreign bodies from the eye, cotton wool is needed in one other form. The "individual applicator" consists of a wisp of clean cotton wool wound round an orange stick and stored in an envelope. This is a permissible alternative to the corner of the one-too-clean packet handkerchief, which is still in use in many workplaces. To discourage the use of handkerchief such disposable applicators are included in all types of first-aid boxes.

6.10.6 Adhesive tape

A sufficient supply of adhesive tape should be put in all boxes. Adhesive tape is used in two forms:

- The individual small tape, with a gauze-dressing attached. Many excellent proprietary varieties are available, with the gauze plain or medicated. Plain nonmedicated gauze is preferable.
- Strips on the reel of tape are cut as required. These strips are not normally applied directly to wounds, but are used to hold other dressing in place.

Every type of first-aid box must contain a large tin of individual adhesive tape dressing, preferably in three sizes. These are useful in the treatment of small cuts. The gauze dressing may extend from edge to edge of the tape, or it can be centered only with a complete surrounding of adhesive. For most purposes, it is preferable that the dressing that stretches from edge to edge to be applied that permits the escape of skin moisture, and so prevents the development changing. Rather than do this, it may be better to cover the tape with a short length of ordinary bandage, which can then be changed as often as necessary. A reel of tape is worth its place in the first-aid box, though it should never be applied directly to a wound without some kind of dressing between it and the injury. Its great value is in securing ordinary bandage ends in place.

6.10.7 Protection from oil

In many jobs, it is necessary to protect a wound from oil, particularly cutting oil. Oil is not necessarily germ-infected; indeed some cutting oils contain an added antiseptic. Nevertheless, oil must be kept away from wounds to prevent their affecting the raw tissues, for this may lead to the development of skin sensitivity later. Oil also delays healing.

The obvious step is to cover the wound or the dressing with some oil-and water-proof barrier. The barrier has to be water-proof, as many lubricating fluids have a watery basis. To achieve this, rubber finger-stalls and gloves, water-proof tapes, and self-sealing crepe-rubber dressing covers have been tried. With one or two possible exceptions, these all have the serious disadvantage that they retain perspiration, producing a soggy skin around the wound, and so delay healing.

At present, all first-aid cabinets contain one 76 mm (3 in) roll of self-sealing crepe rubber, which can be used to make an individual fitting finger-stall that is regarded as obsolescent. A water-tight occlusive should remain on only when the patient is actually at work. It should be removed on leaving work in the evening, and preferably also at the lunch break, and reapplied at the start of work in the morning or afternoon.

Provided it is properly applied and frequently changed, the best protection against oil is an ordinary roller bandage applied over some other dressing. It will need changing at least three times a day, at start of work and at the end of the morning and afternoon shifts. An oily bandage left in contact with damaged skin overnight predisposes to oil acne and dermatitis.

6.10.8 Roller bandage

The proper use of the roller bandage can be taught only by demonstration and practice. The allocation of roller bandage in boxes should be as follows:

	Box A	Box B	Box C
Roller Bandage 25.40 mm (1 in)	6	9	12
Roller Bandage 51 mm (2 in)	6	9	12

The 25.40 mm bandage is suitable for fingers and hands and 51 mm for limbs. In using a roller bandage, the first-aider should observe the following points:

- Clean the hands before breaking the paper seal.
- Break the paper by grasping in both hands and contra-rotating.
- Always work with the bandage rolled. Attempts to apply an unrolled bandage soon results confusion.
- Keep the coil of unused bandage close to the part being bandaged and pull firm after each turn. There is a difference between pulling firm and pulling tight that can be taught only by demonstration and practice.
- Do not apply too much bandage.
- The correct way of tying a bandage must also be taught by demonstration. The bandage end should be nicked with scissors split for about 30 cm and knotted once to prevent further splitting. Bandages round the fingers, hands, or forearms should usually be tied. Bandages round the arm or leg should be fixed with a safety pin.
- The split bandage ends should be tied with a reef knot and the ends cut short to prevent their catching in machinery.
- The knot and ends are then best covered with a piece of adhesive tape. Some advocate fixing the bandage with strapping alone, without a knot, which is not safe.
- Used roller bandage should be fixed with a pin and carefully preserved for future use.
- There are special methods of bandaging the knee, elbow, shoulder, ankle, scalp, ear, and eye. The first-aider should never have to undertake these complicated maneuvers. With these types of injury sterilized individual dressing should be used and then refer the injured for further treatment to a nurse or doctor. The idea that bandages must be applied from the extremities of the body working toward the heart is an outdated myth.

6.10.9 Triangular bandage

This 965 mm (38 in) triangular bandage is 965 mm (38 in) along each of its two shorter sides. It is made by cutting a square piece of linen or calico diagonally. The present requirement is 6 triangular bandages as in Box C only. The recent revision specifies 2 triangular bandage in Box A, 4 in Box B, and 8 in Box C.

The triangular bandage may be used either as a bandage for holding a dressing or a splint or for covering a large burn. The first-aider with a supply of individual sterilized dressing will not need to use it for holding a dressing in a place. To hold splints in place, it is folded on itself three times to produce a stout narrow binder. Further details are covered under the section on fractures.

When used as a sling, the right angle of the triangle should point outward behind and beyond the elbow, and the front layer of the sling should pass over the shoulder on the injured side (Figure 6.1). To sling the arm at an angle of 45 degrees, the triangular bandage folded narrow may be used as a "collar-and- cuff" sling. This is essentially no more than a clove hitch round the wrist. Slings can be improvised with safety pins, a neck-tie, or simply by using a jacket.

6.10.10 Tulle-gras dressing

First-aiders often ask for a small soothing dressing that can safely be applied to burns and that will not stick, especially since the use of acriflavine emulsion has been discouraged. Individual sterilized tulle-gras dressing, contained between two slips of transparent paper, stored in a small tin should be used. Tulle gras is curtain netting, impregnated with petroleum jelly. Twelve such dressings should be included in Boxes A, B, and C.

6.10.11 Splints

"Suitable splints" with cotton wool or other padding used to be included in Box C but are now obsolcte. Splints are easy to improvise, and often the best splint is the human body itself. Nevertheless, on some occasions, a few pieces of wood are useful.

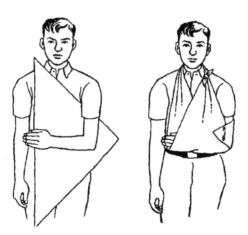

Figure 6.1 The triangular bandage.

6.10.12   Other statutory requirements

Every Box of A, B, and C should contain a tin of safety pins of assorted sizes, as well as a copy of special form, a single-sheet to give the outline of first aid. The following items, which still have to be stocked for statutory reasons, will soon lapse into harmless disuse:

In Boxes A and B:	Iodine (2%) in alcoholic solution
In Boxes B and C:	Oily cocaine eye drop with camel hair brush in the cork
In Box C:	Tourniquet

To carry out the treatments as intended here certain other items are needed. In compiling this list, careful study has been made and experienced. First-aiders should already be aware and are expected to utilize them in their workplaces.

- Cetrimide (1%), either two 280 gm (10 oz) bottles; these bottles should have plastic screw cap, not corks
- Gallipot, 57 gm (2 oz)
- Kidney dish, 152 mm (6 in)
- Proprietary noninflammable tape remover, 28–113 gm (1–4 oz) bottle; this is also useful for cleaning oil from the skin around wounds
- Small unbreakable tumbler
- Eye-bath unbreakable
- Blunt-nosed surgical scissors with chain attached; the length of chain helps to prevent the scissors getting lost
- Splinter forceps
- Clinical thermometer
- Magnesium trisilicate tablets, 50
- Aspirin and phenacetin tablets, 50
- Formalin throat tablets, 50

The gallipot, kidney dish, tumbler, and eye bath should always be washed thoroughly with soap and hot water and dried on a clean towel after use. If this is not accomplished, infection could be spread from patient to patient.

6.10.13   Siting the first-aid box

Ideally, the first-aid box should meet the following:

The box should be fixed on the wall over a small enamel-topped table, which should be kept clear and clean.

There should be a strong chair close at hand, on which the patient can sit while being treated.

There should be sink, with running water, soap, and towel close by, for the use of both the patient and first-aider. A drinking fountain is an advantage; it has special value as it can be used for washing out the eye after chemical splashes.

Beneath the table, there should be a pedal controlled bucket for the disposal of used dressings. It is part of the first-aider's job to see that it is emptied regularly and kept clean.

It is particularly important to try to preserve a small clear workspace to handle first-aid treatments.

6.10.14 Replenishment

Regular inspection and replenishment of first-aid boxes is part of duty of the trained nursing staff. The frequency with which this has to be done will depend on the number of causalities to be treated. In the industrial premises, plants are divided into the following categories for periodical inspections:

- Weekly
- Monthly
- Quarterly

These visits for inspection and replenishing help to build a useful link between first-aiders and trained industrial nurses. If stocks run low between visit, first-aiders are responsible for letting this be known.

First-aid boxes should never be kept locked; first-aid that is delayed while a key is searched for is a travesty. Normally, only a trained first-aider must be responsible for the box and its stocks. His name, and that of his deputy, should be on the outside of it. Where a first-aid box is supplementary to a factory first-aid room or medical department, it may be used only when the room or department is shut. In such a case, it is helpful if instructions to patients needing treatment are also displayed on the box.

6.11 Principles of wound treatment

A wound is defined as any break in the skin with or without injury to the deeper tissues. Thus, the term "wound" covers every type of skin break, from the trivial scratch to the severe crush injury.

The skin is the body tissue most liable to injury, and it is estimated that every day there are a million skin injuries of sufficient size to meet at least a first-aid dressing. Of this one in every ten needs attention at a plant surgery of industrial health center. It follows that wounds are by far the most common reason for first aid. The following are some typical industrial wounds:

- A straight cut from a chisel or sharp metal edge. This is an incised wound.
- A treating wound with ragged edges where flesh is caught in a machine. This is a lacerated wound.
- A crushing wound with the flesh around bruised and injured, from a hammer-blow, or injury from a spanner or rollers. This is a contused wound.
- A deep stab, from stepping on a nail. This is a puncture wound. Incidentally, a severe puncture wound may bleed very little or even not at all.
- A scraping wound or graze, where the skin surface is torn by a file or sandpaper, for example. This is an abrasion.

6.11.1 Major and minor wounds

A wound is considered either a minor or simple wound or a major wound.

- Minor or simple wounds: The ordinary everyday small skin cut of workplaces that can properly be treated by the first-aider.
- Major wounds: Everything more severe than the minor wound. In these the first-aider gives true first-aid treatment only, pending the arrival of, or referral to, a trained nurse or a doctor.

This division of wounds emphasizes the most important decision the first-aider has to make. He must never feel reluctant about passing on the patient to more skilled hands. In the case of obviously severe wounds, there is no difficulty in making a decision; nor is there any with a 12.7 mm (half-inch) long, shallow graze on the hand. Between these two there are many types of wounds where the first-aider will have to make judgments.

There are three points to be considered:

1. Position of wound: Any wound around the eye or involving the skin of the face is serious. Any wound, other than a small shallow cut, of the finger, hand or wrist is to be treated as serious; even a small scar on a finger may reduce the skill and affect the livelihood of a manual worker. Any wound of the abdomen is serious.
2. Type of wound: Any wound with ragged edges or with the flesh around it bruised is serious, because the damaged tissue is more liable to infection. Any deep wound or stab or puncture wound is serious, because infection carried in by the wounding object is more likely to gain a foothold, because there may be unseen damage to deeper tissues. Any gaping wound the edges of which do not easily come together is serious, because the exposed raw area is more likely to get infected, and the scar will be wide and disabling.
3. Complications of wound: Any wound from which the blood pumps out in jerks is serious, because this means an artery has been cut. Any wound from which the blood gushes out in steady stream is serious because this means a vein has been cut. Any wound more than 3 mm (one eighth-of-an-in) deep may involve damage to muscles, tendons, nerves, or other structures. This risk is greatest in the wrist, hand and fingers. The first-aider cannot tell if these structures have been injured. Therefore, any cut more than 3 mm (one eight-of-an-in) deep, especially in the wrist, hands, or fingers, is serious.

6.11.2   Infection

Infection means the entry of harmful germs into a wound where that start to grow and multiply. Clearly, the prevention of infection in first aid is just as important as the control of bleeding.

6.11.3   Cleaning the wound

A major wound needs thorough cleaning by a trained nurse or doctor. An extensive major wound may need opening and cleaning thoroughly by a surgeon, with the patient or at least the wounded part anesthetized. Delay in getting a major wound properly cleaned increases the likelihood of the germs gaining a foothold in the tissues. The first-aider's job is to cover the major wound with a sterile pad as quickly as possible.

A minor wound is best cleaned by washing thoroughly with clean water under a running tap. If there is any visible dirt present around the minor wound, it may be washed away with soap and water. Better even than soap is the recognized detergent cetrimide (cetavlon); this has an antiseptic action as well, but does not injure the tissue.

When running water is not available at the first-aid point; cleaning of the wound and surrounding skin should be done with cotton wood dipped in cetrimide. Finally, first the wound, then the surrounding skin, should be thoroughly dried with fresh dry pieces of cotton wool.

6.11.4 Closing and covering the wound

Any wound that is left gaping is more liable to become infected. Even if not infected, a gaping wound will heal much more slowly, and will leave behind a wide and perhaps disabling scar. The first-aider must regard any gaping wound as a major wound, to be covered with a clean or sterile dressing and passed on at once to a trained nurse or doctor. Many gaping wounds will require stitching (suturing) to bring the edges together, for which most doctors use a local anesthetic.

In covering a major wound, the first-aider must take all reasonable steps to keep germs away from the cleaned wound and the dressing. The hands of first-aider should be clean, and he must be careful not to cough, sneeze or talk over the wound. Even more important is to keep his own skin germs away from the wound or anything that is going to touch the wound surface. This means no touching of the wound with the fingers, and no touching of the surface of the dressing placed neat to the wound. Every first-aider must practice this simplified "no-touch" method of dressing wounds until he does it quite automatically.

The wound that has been properly cleaned, closed and covered will heal in the shortest possible time, almost without pain, and with the smallest possible scar.

6.11.5 Re-dressing minor wounds

A minor wound should be re-dressed as seldom as possible. If there is no pain, it is only necessary to change the outer dressing when it is soiled; the dressing immediately over the wound should be left in position, if possible for 48 hours. Exactly the same care must be used in changing a dressing as when the dressing is first applied. If the patient complains of pain or discomfort in a minor wound on the day after injury or thereafter, the first-aider must refer the patient at once to trained nurse or doctor, as infection is likely to have occurred.

6.11.6 Foreign bodies in wound

A large foreign body, such as a piece of metal or glass, if sticking out of the wound, should be removed gently, provided this can be done without putting the fingers into the wound. If the foreign body does not come out easily, or if there is projecting bone, rolled bandages, with the paper removed, may be placed on each side of the projecting object; the wound, object and rolled bandages are then covered with a large individual first-aid dressing; this should be bandaged in place firmly but not tightly. Any elaborate building up, while the wound is left uncovered, increases the chance of infection. An alternative method of bandaging over a foreign body without pressing on it is to use individual sterilized dressing on either side of the wound as shown in Figure 6.2; this method is specially useful where the wound is large.

If there is sever bleeding from a wound in which there is a foreign body, control of the bleeding must take precedence over treatment of the foreign body. Small foreign bodies should not be touched, but a note that they have been seen should be sent with with the patient. In a severe injury, there may be a rare occasion that a piece of bone is projecting through the wound or the skin; this should be left alone and not touched.

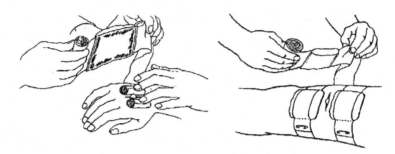

Figure 6.2 Sterilized dressing on either side of the wound.

6.11.7   Special wounds

- Small crush, graze, or laceration: Any crush, grace and laceration, other than a very small one, is to be treated as a major wound and referred to trained nurse or doctor. First-aiders often ask to be allowed to use acriflavine or some other oily dressing for small crushes, grazes and lacerations, since it prevents sticking when the dressing is changed. But acriflavine has the disadvantages of iodine and other chemical antiseptics, and oily preparations delay healing. Frequent changing of the dressing should not be needed unless the injury has become infected. The proper treatment for a really small crush, graze, or laceration is thorough cleaning with cetrimide, followed by a dry dressing, protected by a bandage to keep it clean. Anything large should be covered with a sterilized first-aid dressing and referred to a trained nurse or doctor.
- Puncture wound: This may be caused by a nail through the boot, a drill that slips, a glass splinter, a wire brush, or any other thin pointed object. All such wounds should be treated as serious, because germs, particularly tetanus germs, may be carried deep into the tissues where they cannot be reached by ordinary cleaning. There is no point in cleaning the wound and the skin around unless the skin is dirty, by application of a small dressing; the patient should be referred as soon as possible to a trained nurse or doctor.
- Animal or human bite: The mouth is full of germs, so bites are usually badly infected, because they are often lacerated or punctured wounds. Even a small bite should be treated as a major wound.
- Puncture wound of the chest cavity: Such a wound may damage the lungs. The patient may cough up blood and find it hard to breath. He may breath easier if propped up in the semi-sitting position with comfort for the patient. A puncture wound of the chest is a rare occurrence. Any patient with a chest wound should be taken to the hospital as quickly as possible.
- Wound of the abdomen: Because of the risk that an abdominal wound may have punctured the stomach or bowels, it is very important that the patient should be given nothing to eat or drink. He should be taken to a hospital without delay.

6.12   Bleeding

Bleeding (hemorrhage) is part of the body's natural response to injury so it should not cause alarm in the first-aider or the patient. Bleeding is nature's means of wound cleansing, because it washes dirt out from the bottom of the wound. Too much bleeding is a danger, simply because, beyond a certain point, the body cannot swiftly make up for blood loss. But bleeding from most wounds will stop spontaneously without any treatment at all.

The body has two very effective methods of stopping bleeding:

The clotting of blood, as a result of its coming into contact with cut and injured tissue.
The pulling back and shrinking of the cut end of blood vessels, so that the holes from which the blood coming out get smaller and may close entirely.

6.12.1 Bleeding from minor wounds

This will occur during the cleansing of the wound; it helps to make the cleansing more thorough. As soon as the wound is covered and the edges drawn together by the dressing, clotting of the blood will take place and the bleeding will stop.

6.12.2 Bleeding from major wounds

This will also usually stop on its own when a dressing is applied. The first-aider can apply three ways to help the body to stop such bleeding.

6.12.3 Rest

Make the patient lie down quietly, and keep the wounded part still. This lowers the blood pressure and slows the pulse, so that the volume of the blood flowing through the injured part is reduced.

6.12.4 Raising the injured part

If the injured part is raised above the level of the rest of the body, the amount of blood reaching it will be less for simple hydraulic reasons. A wounded arm or leg may be raised and put on pillows, but the stomach or chest cannot be effectively raised.

6.12.5 Pressure on the part that is bleeding

This is the most important and most effective way of controlling bleeding. It can be stated that if enough pressure is applied, hemorrhage can always be controlled.

• Applying pressure: Place a clean pad over the wound and bandage it firmly in place. If blood quickly comes through the first pad, put another pad on top, and bandage this firmblood still comes through, ly in place. If blood comes through the second pad, apply a third pad. If press firmly with the hands on the third pad, and hold in position until a doctor can take over.
As already stressed, sterilized individual dressing is ideal for the control of bleeding, since it has a built-in pad attached to a bandage and the whole dressing is sterilized.
If an appropriate first-aid dressing is not available, a rolled-up bandage may be used as a pad, or a clean folded handkerchief. If necessary, a clean handkerchief may also be used as a bandage.

Every first-aider, especially in the oil, gas, and petroleum industries should have been trained in the use of a pad and bandage to control bleeding, so that he can effectively handle when faced with his first major wound.

- Pressure points: Certain points between the heart and the site of bleeding where, by pressing hard against an underlying bone, the arterial flow can be stopped is wrongly exercised and should be abandoned. It is not expected the first-aider to risk the patient's life by hunting for a pressure point, instead of applying direct pressure to the place that is actually bleeding.
- Tourniquet: First-aid boxes must contain a rubber or pressure bandage for use as a tourniquet. It should never be used, as it is not a first-aid measure. It is often ineffective and frequently harmful. If improperly applied, it can cause death of a limb. If improperly applied, it can increase bleeding by obstructing the veins but not the arteries. Finally, it is never necessary, as bleeding can always be stopped by the safe simple method of direct pressure.
- Importance of blood loss: About one-eleventh of the weight of the body is blood. There are about 5.67 L of blood in the average adult. A normal adult can lose a 0.47 L of blood without ill effect; many people give this much blood twice a year to the blood transfusion service. Most bleeding is not serious, and the first-aider need never be frightened by it. The loss of a large amount of blood produces a very dangerous state. As the bleeding continues, it leads to pallor and weakness, then unconsciousness and finally death. If life is to be saved, after the bleeding has been controlled by firm pressure, it is vital at the earliest possible moment to replace the blood that has been lost, using a blood transfusion. A patient who is believed to have lost a large amount of blood must be moved as swiftly as possible to a hospital where a blood transfusion can be started at once. If transfusion can be started within a half-hour, life will probably be saved; delay of over an hour may prove fatal. By making arrangements quickly and calmly, the first-aider is acting in a life-saving role. It will help the doctor at the hospital to estimate the amount of blood that has been lost, and the amount of blood the patient needs, if the blood lost can be mopped or scooped up, and the blood and stained dressing, cotton wool and clothing put in an enamel basin and sent with the patient to hospital. But do not waste time on this if it means delay in taking the patient to hospital. Cover the patient with two blankets or a coat. Apart from lifting out of danger or on to a stretcher, keep movement to a minimum.
- Nose bleeding: Epistaxis or nose bleeding may follow on the nose, nose picking a bad cold; such nose bleeding will usually stop quickly. Or it may follow a severe head injury, which means usually that the skull is fractured. Often nose bleeding is spontaneous and has no obvious external cause; this type is more likely to last for some time and can be serious. It is not part of first-aid to attempt to diagnose the cause of spontaneous nose bleeding.

First-aid treatment in the absence of major injury is as follows:

Sit the patient up, with the head slightly forward, so that any blood that runs down the back of the nose can escape from the mouth instead of being swallowed.
Make him breath through his mouth, and pin the nose firmly so that the nostrils are closed. Thereafter, he must be warned not to sniff.
Apply cold water to the bridge of the nose, using handkerchief or cotton wool soaked in it.

If the bleeding continues or recurs, the patient should be seen by a doctor. The first-aider should never attempt to plug the nose.

6.13   Sting, insect bites, and blisters

6.13.1   Bee and wasp stings

The types of injuries occur both indoors and outdoors.

- Bees: The bee leaves both its sting and poison-bag behind. If the sting is grasped with a pair of forceps, in order to pull it out, the contents of the poison-bag may be pumped into the patient. The sting is best lifted or scraped off the skin with one blade of a pair of forceps or with a pin. The patient should then suck the wound and spit out. The only other local treatments of any value are: The application of a proprietary "antihistamine" ointment sold in a collapsible metal tube; failing this, a cold compress or an ice-pack may help. These are described in the next section. Bee's venom is not acid and treatment with mild alkali is useless. If the sting is in the mouth, skilled nursing or medical help is required at once. While help is coming, the patient should be given a piece of ice to suck.
- Wasp: The wasp leaves no sting behind, so the patient should suck the wound and spit out forthwith. Further local treatment is exactly the same as for a bee sting (antihistamine ointment, a cold compress or an ice-pack). Like bee venom, wasp venom is a complicated mixture of organic compounds and it is not alkaline so vinegar or lemon juice are valueless as methods of treatment. If the wasp sting is in the mouth, skilled help should be sought at once, and ice given to suck. With any sting, the patient may start to swell up either around the injury or generally, show signs of shock. If this happens skilled nursing or medical help is needed immediately.
- Spider and snake: Spider and snake bites can occur in the working areas. Those at risk are dockers and banana-ripening store operatives. The creatures are imported in the banana bunches. The snake is most often seen in the different part of world.
  Treatment of snake bites: Wash the bites thoroughly, to remove any venom that the snake may have spot out into the skin. Suck the wound hard and spit out. Tie a bandage tightly round the limb, between the bite and the body. This will not stop the blood flow, but will cut down the flow of lymph (body tissue fluid) back to the body; it is in the lymph that the venom mainly travels. The bandage should be loosened for half a minute every quarter of an hour. Patient should be visited by skilled helper at once, or sent to hospital immediately. The snake should be killed and sent in a box with the patient for identification if possible.
- Mosquito bites: Patients sometimes arrive at work with painful swelling due to mosquito or other bites. These are not first-aid problems and need nursing or medical examination and care.

6.13.2   Treatment and care in hospital

- Systemic: This is the same as for any potentially necrotic and infected wound; antibacterial agents in adequate dosage. Tetanus antitoxin also should be given, since the snake's mouth may have transmitted tetanus bacilli or spores.
- Supportive: It is essential to prevent exertion, reassure the patient, prohibit alcoholic beverages, and order complete rest in bed. To relieve nervousness and pain, pentobarbital 100–200 mg orally may be given and repeated, if necessary, every 4–6 hours. In respiratory depressants morphine must be avoided. For treatment peripheral vascular failure, either strychnine 1–2 mg subcutaneous or by mouth. To combat collapse, normal saline with 10% dextrose, and either whole blood or human blood plasma should be given. The patient must be kept under observation at least for 24 hours.

- Specific: Antivenin (antitoxin) and polyvalent antivein serum for all snake poisoning is commercially available. Systemic administration, an injection of 2–3 cc or more around the wound should be applied to minimize tissue necrosis, subsequently, similar injections proximal to the wound as the tourniquet is shifted. After adequate doses of serum have been injected, the tourniquet can be removed. For systemic treatment, the dosage and route of administration will depend upon the age, size and clinical condition of the patient. If the patient is a child or in shock intravenous administration may be indicated, provided it has been demonstrated beyond doubt that he is not allergic to those serum. Otherwise, intramuscular injections are necessary. The injections should be repeated every 1–2 hours until symptoms are significantly diminished; they should be continued at the same rate as long as the swelling, paralysis, or other symptoms are progressing. It must be remembered that over-treatment is the lesser error in snake venenation; up to 100 cc (occasionally more) of antivenin may be required.

6.14   General effects of serious injury

6.14.1   General shock

Every severely injured patient soon becomes very ill. This illness is known as shock. Without proper treatment, shock is often fatal. With proper treatment applied quickly enough, the patient almost always recovers. Proper treatment of shock can be summed up in the words "blood transfusion." Even a half-hour delay decreases the patient's chance of recovery.

The first-aider's duty is simple: It is to speed the removal of the severely injured patient to a properly equipped hospital, doing only what is necessary meanwhile to prevent the shock from getting worse. If the severely injured patient is in a hospital within half-an-hour, the first-aider will have played a major part in saving his life.

There are six kinds of shock:

1. Primary
2. Secondary
3. Hemorrhagic
4. Traumatic
5. Toxic
6. Nervous

The word "shock" is used only for true wound shock, i.e., for items 2–3 and 4 above. The muddle over the shock is matched by the confusion and controversy over treatment. The following points are based largely on the valuable research carried out by hospitals:

- Shocked patient: The state and condition of the patient with shock is as follows:
  Facial expression is anxious and worried-looking, or staring in a vacant way
  Skin is pale-white, ashen-gray, or slightly blue
  Skin feels cold, yet in spite of this it may be soaked in sweat
  Patient is sometimes restless, fidgety, and even talkative, but may be dull, and sometimes even unconscious

Breathing is rapid and shallow, sometimes sighing

Pulse is usually rapid and feeble, though occasionally normal

Patient usually complains little of pain, but may complain greatly of thirst

External signs of the cause of shock, such as injury or bloody vomit

The first-aider cannot measure the blood pressure. If he could, he would usually find it low or even very low. Similarly, he would find the body temperature to be subnormal, though he must not waste time trying to take it. A shocked patient does not always show all of the above conditions at the same time. In medicine, there are exceptions to even the best word pictures. Thus a patient with shock due to heart attack or a bad fracture may be in great pain.

6.14.2 Shock

Shock is due to loss of body fluid. This happens in four different ways:

1. Bleeding: This may be:
   External, from the outer surface of the body
   Internal, from the inner surfaces of the body into stomach or gut, e.g., stomach ulcer or into the soft tissues of the body, (around the broken ends of bone)
2. Seeping away of plasma from the capillaries. Plasma is the fluid part of the blood. The capillaries are the small tubes that join the arteries to the veins. They are the finest blood vessels of all and they have the thinnest walls. Those that start to leak in shock are:
   At the site of injury, especially if the injury is a crush or a burn
   In the rest of the body, probably mainly in the muscles and the gut
3. Vomiting
4. Sweating

Each of these ways of losing body fluid must be looked at in rather more detail. But the fluid in each case comes either directly or indirectly from the blood. Moreover, the more rapid the blood loss, the smaller is the amount needed to produce shock. By contrast, a much greater blood loss can be born without symptoms of shock provided it occurs sufficiently slowly.

6.14.3 Bleeding as a cause of shock

Actual blood loss is now regarded as by far the most important factor in producing shock after severe injury. External blood loss can be seen, and the blood should be mopped up and collected and sent with the patient to hospital, to help the surgeon to judge how much has been lost.

Internal blood loss into the tissues themselves can be equally important as a cause of shock. If a large bone is broken, there will usually be a great deal of bleeding into the tissues around the broken ends, even though nothing shows from outside. A broken shin-bon (or tibia) will cause an internal bleeding or about a pint of blood, not really enough on its own to produce shock. But a broken thigh bone (or femur) will cause two-and-a-half to three pints of hidden internal bleeding, with quite considerable shock as a result.

6.14.4   Capillary leakage and results of fluid loss

When tissues are injured, they produce certain chemical substances that pass into the blood. These substances affect the capillaries in the immediate vicinity of the injury and also generally throughout the body. Their distant effect can be shown in the following ways:

> If the veins from the injured part are temporarily blocked, the degree of shock is reduced; and when the block is released, the shock gets worse.
> The capillaries in the burned area itself become very leaky, and considerable quantities of fluid can be lost from the burned surface.

Apart from severe burns, capillary leakage is a major factor in producing shock whenever there is substantial damage or destruction of living tissue. Such injury is usually due to crushing, from falls, collapsing building, pinning under vehicles, also limbs crushed in rollers or torn off by machinery.

• Results of fluid loss: Too little blood is reaching the brain.

6.14.5   Positioning of shocked patient

If the patient is vomiting or is semi-conscious or unconscious, and the injuries permit, he should be gently rolled into the semi-prone position (Figure 6.3); once again, there should be nothing placed under the head to raise it.

6.14.6   Heat

Patient feels cold when the blood vessels in the skin have all closed down as part of a deliberate move to force what little blood is available to the brain and other vital organs. If external heat is applied the skin is made to glow, blood will be drawn away from the vital places. The shocked patients who were not warmed did as well as or even better than those who were warmed. Therefore, the right first aid is to dispense entirely with artificial sources of heat.

6.14.7   Fluids

The shocked patient is often intensely thirsty because of blood and fluid loss. It follows that the only safe way is: "no fluids or sweets of any kind" to be given by mouth to the shocked patient.

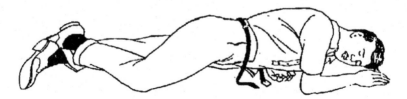

Figure 6.3 Positioning of shocked patient.

6.14.8 Morphia

Morphia is not a treatment for shock but a means of relieving pain. It is needed only if pain is continuous and severe as, for example, when a limb is trapped in machinery. Obviously it will not be necessary if the patient is unconscious.

Severe crush injuries, with much destruction of muscle tissue, involve an added risk besides shock. Debris and poisons from the crushed muscle released into the blood damage or even destroy the kidneys. The resulting condition is called the "crush syndrome." The same kind of kidney damage may follow severe burns. The general care of crush injury follows precisely the same lines as for shock.

6.14.9 Fainting

A person who has fainted looks "shocked." There is extreme pallor, with beads of cold sweat on the forehead. It may be impossible to feel the pulse. Breathing may be shallow and sighing. But in a few moments recovery starts and consciousness begins to return.

The person who faints is usually young and healthy. The cause may be mental, such as the sight of blood, fear of an injection, or sudden bad news or physical such as extreme pain or standing for a long time to attention.

The only treatment needed is to loosen any tight clothing around the neck; if consciousness does not return within two min, the patient should be rolled into the semi-prone position (Figure 6.3) and expert help should be gained.

The patient who feels he is about to faint can usually prevent this by pulling his stomach, buttock, and leg muscles tight, and holding them tight a minute or so.

6.14.10 Electric shock

Electric shock is the general bodily reaction to the passage of an electric current. It may vary from slight tingling to sudden unconsciousness looking just like death. But the first-aider must never presume death in electric shock, since the breathing may stop and the pulse vanish, but life can still be restored.

Direct current is less dangerous than alternating current for the following reasons:

Direct current produces a single violent muscular contraction, which tends to throw the patient away from the source of the shock. The resulting fall is as likely to cause injury as the shock itself. By contrast, alternating current produces continuous muscle spasm, which may cause the affected muscles of the arm and hand to grip involuntarily the source of electric supply. Therefore, a continuous prolonged shock is more likely.

The lowest fatal voltage ever recorded was 38. A great deal depends on the contact between the source of electricity and the skin, and between the skin and the ground. A metal floor will also increase conductivity. A person who is fatigued is at more risk of shock than one who is rested.

With very high voltage, the current usually does not penetrate the body deeply, because the electrical pressure is so great that the tissues and conductors are destroyed.

Ordinary domestic AC current alternates at 50 cycles per second. Such a current can just be felt if it is of 1 mA. By contrast a DC current should reach 5 mA before it is perceptible. One-hundred mA AC is the usual minimum fatal current, but as low a figure as 20 mA. AC has caused death. The length of duration of exposure to a current is very important; with exposures of over 5 sec, the danger of serious injury is great.

The skin has a very high electrical resistance about 3000 ohms if dry and healthy. Once this resistance is overcome, the current follows the internal water courses of the body. A current passing from head to leg, as in judicial execution, will travel via the fluid around the brain and spinal cord, damaging vital nerve system on its way. A current passing from leg to leg does less harm than one passing from arm to leg, since the latter will pass over and often damage the electrical mechanism of the heart.

Most electric shocks occur among electricians and one-third of all fatal electrical accidents are due to portable electrical apparatus and hand-tools. A severe electric shock may occur during electric welding, where a sweaty welder may come in contact with a metal sheet that could be live. A fatal shock may be caused by a jib-crane fouling an overhead cable; or a metal strip may touch the "live" overhead wires feeding an electric gantry.

- Symptoms: These may vary from muscle spasm and pain to unconsciousness and even deep coma. The muscle spasm may be momentary with a single direct-current shock, or continuous from alternative current. Pain in the affected muscles may be intense. In as much as the patient cannot overcome the spasm of the muscles by an effort of will, the muscles are effectively paralyzed, as long as the shock continues. If the spasm is strong enough, the electric current may paralyze the breathing muscles, or put the breathing control center in the brain out of action and such a paralysis is usually transistory. At the same time, the electric current may partially paralyze the heart muscle. As a result the heart beats rapidly but feebly, in a state of "flutter"; in this state, although the blood is still circulating, the pulse cannot be detected. It follows that the absence of both pulse and respiration in a patient unconscious from electric shock are not signs of death. Prolonged artificial respiration may yet save life.
- First aid and treatment: Speed and coolness are essential, and may be life-saving. The first move is to disconnect the patient from the source of the electricity:
  Switch off the current.
  If this is impossible, pull or push the patient away from the source of the electricity, while taking great care not to make electrical contact with either the ground or the patient.
  Stand or kneel on a dry nonconductor, such as a dry rug, mackintosh, or rubber mat.
  Pull or push the patient away from the source of the electricity, again using a dry nonconductor. Considerable force may be needed to get the patient free. If the patient has to be grasped, use special electrician rubber gloves, or dry sacking, a dry coat or several thicknesses of dry paper. If a crooked stick is available, this should be used.
  Avoid contact with any part of the patient that may be moist, e.g., the armpits or crotch, or the face, which may be wet with spittle.
  With very high voltages at electricity stations or overhead wires, the patient will be thrown clear. If not, the danger to a rescuer while the current is still on is very great, and all possible precautions should be taken. Effort should be made to get the electricians to switch off the current before rescue is attempted.

Once the patient has been rescued from contact with the electric source, if breathing has ceased or is very feeble, artificial respiration should be started at once, using the methods described in standards. At the same time, the standard treatment for shock in an unconscious patient should be applied, but this definitely takes second place to artificial respiration. Since artificial respiration may have to continue for half an hour or more, a resuscitator or a rocking stretcher is of the greatest value.

In about half of all electrocution cases with cessation of breathing, there is recovery with artificial respiration; nine out of ten patients who start breathing again do so within half-an-hour of artificial respiration being started. Delay in starting artificial respiration can prove disastrous. If it is started at once, 70% of the patients recover. If there is more than three min' delay only 20% recover.

First-aiders should get to know the position of the electrical switches in the part of the factory for which they are responsible.

6.15   Fractures

A fracture is a broken or cracked bone. Broken bones in industries are the small bones of hands and feet, usually happen as a result of object falling. Safety protective boots are means of preventing fractures.

6.15.1   Role of first-aider

In the oil and gas industries because skilled help can almost always be quickly obtained, the first-aider's role in fracture treatment is to look after a patient with fracture of thigh until expert helpers arrive. But with a suspected fracture of the arm, hand, or foot, the first-aider may well have to get the patient ready for transport as a sitting case to the hospital or industrial health center. Fracture of the thigh is a task for experienced ambulance service people to transport the patient.

The serious injured patient will often have one or more fractures. The treatment of the patient's general condition must have priority; care of the fracture will be limited to making the patient as comfortable as possible.

With the patient who has sustained a moderate and local injury, the first-aider must always foresee the possibility of a fracture. In such cases, he should call for help or refer the patient to the industrial medical department or hospital.

Transport of severe fractures is thoroughly explained in the first-aid manuals. The ambulance worker must know all these situations and first-aid basic principles. The industrial first-aider needs to know only certain basic principles and how to apply them if the need arise.

6.15.2   Types and signs of fracture

Many varieties of fracture are described. For the industrial first-aider only two are important:

- Closed or simple
- Open or compound

Most fractures are closed. Open fracture is so rare that many first-aiders will never see one. An open or compound fracture is one where there is an outside wound as well as a fracture, and a communication between the skin, air and the broken bone-ends. This greatly increases the risk of germs getting into bones. The first-aider can observe a compound fracture if there is a broken bone-end sticking out from a wound or through the skin, or if broken bone is visibly in a wound. But in most compound fractures, the bone cannot be seen in the wound. The first-aider can explain that there is a wound outside and a broken bone inside; whether they can be put back together is a matter for the surgeon to investigate.

The safe way to treat the wound is cover it as quickly as possible with a large individual dressing in order to keep out any infection. Once this is done, the patient's general condition and the fracture itself can be attended to. It is particularly important to handle any such injury extremely gently. One rough movement may link together an outside wound and an inside fracture and so convert a closed into an open fracture.

For the first-aider there are only two certain signs of a fracture:

1. If the patient is conscious, he claims that he heard or felt a bone snap.
2. The limb or injured part is often bent in a way that could happen only if the bone was broken. This is called the "deformity"; it can usually be detected without removing the clothes. Deformity is best appreciated by comparing the injured and uninjured limbs.

6.15.3　Fracture of individual bones

Certain bones are particularly liable to get broken. Often the deformity or change in shape produced is so characteristic that by simply looking at the injured part it is clear there is a fracture.

- Collar bone or clavicle: The cause is usually a fall on the outstretched hand. The arm is held tight against the side of the chest, and any movement gives pain over the collar bone.
- Upper-arm bone or humerus
- Again the arm is held tight against the side of the chest, but this time pain on movement is over the broken humerus (Figure 6.4)

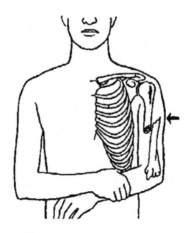

Figure 6.4 Upper-arm bone or humerus.

- Forearm bones: The radius and ulna (Figure 6.5). The injured forearm is supported with the other hand. There will be pain at the site of break. The amount of deformity depends on the extent of the fracture. A young person may crack one of the forearm bones only part of the way through; this is called a green-stick fracture. If one bone alone is broken, the other will act as a splint.
- Forearm bones at the wrist: The common cause is a fall on the wrist, particularly in an elderly woman. The fracture is called colles's fracture and the deformity, seen from the side, is like a dinner fork.
- Small bones of the wrist and hand: The usual causes are jerks, falls and blows. Like same fracture when cranking a diesel engine and "Kick's Back."

6.15.4 Thigh bone or femur at the hip

The femur is the largest bone in the body and when it breaks accompanies by shock. In old people, the femur is fragile and a simple fall will snap the "neck" of the femur close to the hip joint. The deformity is quite characteristic. The leg is held rolled outward, so that the toes point away from the other foot. Sometimes it can be seen that the injured leg is shorter (Figure 6.6).

6.15.5 Fracture of thigh bone

Because it is so strong, the femur will be broken only by great violence, such as a fall from a height or motor vehicle accidents. Pain and uneasiness will be extreme; the leg will be held quite still; there may be shortening.

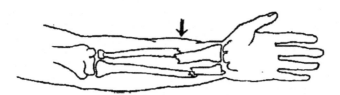

Figure 6.5 The radius and ulna bones.

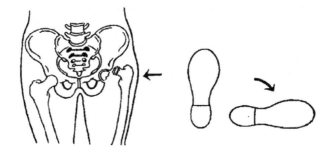

Figure 6.6 Fracture of thigh bone.

6.15.6   Shin bone

The tibia and fibula-the large shin, the tibia, is just under the skin, so a fracture can be felt quite easily by running a finger along it. Generally, the thin little fibula is also broken. The common causes are road accident, falls and football injuries.

6.15.7   Shin bone at the ankle

It is usually impossible for the first-aider to distinguish between a badly strained and broken ankle. The cause is usually a twist or a slight fall. Occasionally the whole foot is pushed backward on the leg and in addition to fracture of bone the ankle is dislocated.

6.15.8   Ribs

Rib fractures are common. They may be caused by sudden compression of the chest, or by falls, e.g., on the corner of a workbench. There is usually no deformity, but sharp pain on breathing or coughing.

6.15.9   Skull

With head injury, the general condition of the patient matters much more than the local damage. Falls, blows and road accidents are the usual causes. Often the patient will be drowsy or unconscious. Blood from the nose or ear, following a blow on the head, suggests a broken skull.

A bad bruise on the scalp may feel like a fracture of the skull; there is a raised circular swelling with an apparent deep or hole in the center. Usually there is no fracture, but this is a matter for a trained nurse or a doctor to decide.

6.16   Care of fractures, strains, and sprains

6.16.1   Principle of first-aid care

The principle of first-aid care of any fracture is to steady the broken bone-ends so that the patient can move or be moved without added pain or further injury. The injured part should be steadied and supported to prevent movement of the broken bone-ends. This means that the joints at each end of the broken bone must be held still.

If the limb is in a very unnatural position, it should be moved with great care and without force that patient can lay down as natural position as possible. If the position of limb has not much changed should not be moved. If the patient is to move, or to be moved without further expert help, the injured part should be fixed in a comfortable natural position. The patient's clothes should not be taken off as this may harm the broken bone-ends.

6.16.2 Fractures of hip, thigh and shin

Patients with fractures of the hip, thigh, and shin will normally be transported to hospital by ambulance as quickly as possible. Any splinting needed should be done by medical professionals.

If, for any reason, the first-aider has to splint a fracture of hip, thigh or shin, the safest way is to tie the two damaged and undamaged limbs together with four to six folded triangular bandages. Should an assistant be available, he may at the same time exert a steady pull on the injured foot, without bending or turning it in any way. This pull is to overcome, or at least reduce, the muscle spasm around the fracture that is the main cause of the pain (Figure 6.7).

Plenty of cotton wool should be placed round the injured limb, before the two bandages are applied on either side of the fracture. Never tie bandage directly over a fracture. On no account should any attempt be made to remove the clothes. It is reasonable, however, to roll up the trousers or pull down the stockings to see if a fractured shin bone has penetrated the skin. Once the limb is properly immobilized, the patient may be lifted carefully on to a stretcher.

6.16.3 First aid of other fractures

Patients with severe head injuries will go straight to hospital under expert care; they will usually be unconscious and the fractured skull as such needs no first aid. Patients without sustaining shock, but with fractures or suspected of the arm, forearm, wrist, ankle, hand and foot, collar bone, and ribs should be transported as a sitting case to the industrial medical department or hospital. For such patients, a firm bandaging should be done before being moved.

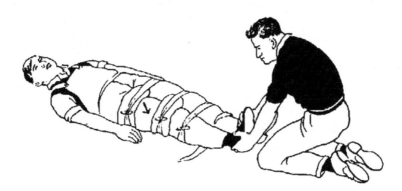

Figure 6.7 Fracture of hip, thigh and shin.

6.16.4   Procedure for firm bandaging of other individual fractures

With suspected fractures around the shoulder, in the arm or forearm, it is usually enough to apply carefully and gently an ordinary right-angle sling, without taking off the clothing. The conventional methods of splinting are as follows:

- Collar bone: Cotton-wool pad should be applied in the armpit. The upper arm to the side of the chest should be bonded with two triangular bandages. The forearm to be supported in a sling at an angle of 45 degrees. A large cotton-wool pad should be placed under the sling end that passes over the injured collar bone.
- Humerus: The side of the chest should be used as a splint. A large cotton-wool pad should be placed between the arm and the chest, and arm to the side of chest bonded with two triangular bandages. The forearm in a sling, at a right-angle, should be supported.
- Radius and ulna: Pad with cotton wool a splint long enough to extend from the elbow to the junction of the fingers and hand. Fix the splint to the forearm and hand along the palm surface, with a bandage at either end. Place cotton-wool pad on each side of the fracture and bandage over them. This treatment also applies to fractures of radius and ulna at the wrist, or other doubtful wrist injuries.
- Ankle: Pad all round with cotton-wool and bandage firmly. No weight should be borne on the injured ankle.
- Hands and feet: Fractures of the small bones of the hand and foot, fingers and toes require no first-aid splinting. The injured hand should be rested in a sling. No weight should be borne on the injured foot.
- Ribs: Fracture of the ribs require no first-aid splinting. If pain is extreme, this may be eased by propping up with several pillows. Whenever bandages or slings are used for fixing fractures, these should be secured firmly enough but not too tight as the tight bandage will cause the part below it to start to swell.

6.16.5   Fractured spine

Fracture of the spine may happen in the neck or the back. A broken neck may follow when diving into a pond that is too shallow. A sudden stop of a car, motorcycle, plane, or train are common injuries of the neck. The head jerks forward or backward and snaps the neck.

A broken back is due to a fall from a height, such as scaffolding; it may happen regardless of whether the head or feet, buttocks or back strikes the ground first. The back may also fracture by direct violence, e.g., when a heavy weight material falls on the back.

The damage to the bone is comparatively unimportant but it matters to the spinal cord inside the bone. Any damage to the spinal cord is absolutely permanent. There can be no recovery from the paralysis (loss of movement of muscles) and loss of sensation below the level of the damage. Because movement of broken spine may itself cause damage to the spinal cord, the first-aider should take absolutely no measures unless he has to.

The first-aider will suspect or recognize a broken spine by the following signs:

1. Story of accident
2. Pain at the place of injury
3. Patient feels "afraid to move" and may be unable to move if he tries

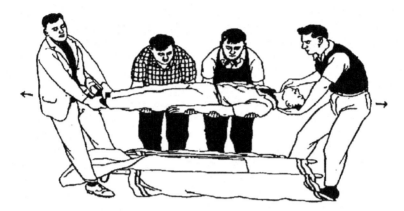

Figure 6.8 Fractured spine.

If it is absolutely necessary to move the patient or adjust his position, it should be done very gently and slowly. The greatest care should be taken not to bend the back or neck or twist the spine. For anything more than the slightest movement, head and foot traction should be used, preferably with four people helping (Figure 6.8). But it is emphasized that this is a job for expert first-aider who have practiced the maneuver carefully. If lifting is absolutely necessary, then the opportunity should be taken to put the patient on to a flat hard stretcher without pillows, or on to a door. But the proper course is always to wait for the expert medical professionals to handle it unless there is an overwhelming reason for not doing so.

If patient with a broken back is found lying on his face, he may with advantage be transported on his face. With a broken neck, the patient should be moved on his back, with his head supported between two rolled blankets, sand-bags, or bricks wrapped in cotton-wool.

6.16.6   Recovering fracture

Patients should be encouraged to go back to work; this is the way of keeping a patient generally fit. First-aiders can play a valuable part and offer certain practical advice.

A tape splint should not be covered with a rubber glove; the retained sweat softens the tape. For similar reason, it is important not to rub a tape or to get suds or water on it. The patient with crutches or in tape should be encouraged to move around from time to time and not remain standstill. The rubbers on the ends of crutches should be in good repair.

6.16.7   Strains and sprains

- **Strains**: A strain is an injury to a muscle or tendon. A sprain is an injury to a joint. With both strains and sprains, the first-aider's prime duty is to make sure that other serious injury is not undetected. The decision is beyond his responsibility, therefore, if there is the least doubt, the patient should be referred to a trained nurse or doctor. Signs of strain are sharp pain in a muscle or tendon, the affected part is held stiff. The muscles most commonly strained are those of the back. A severe strain may involve the complete rupture of a muscle or tendon. The pain is more severe, there may be great swelling, and the affected part cannot be moved. Such cases will probably need surgical treatment. No rest is required for simple strain, since active movement from the start hastens recovery. To relieve pain, a cold compress can be applied. Many industrial strains particularly those of the back, can and should be prevented. Modern mechanical handling methods can get rid of much "back breaking" incidents. When manual labor cannot be avoided, its proper technique should be learnt; the motive power should come from the hip and thigh muscles with bending at the hip and knees, rather than from the back muscles with bending at the spine (Figure 6.9). The two figures show the right and wrong way to carry heavy objects.

- Sprains: Sprains will happen by the same kinds of injury that cause fractures. Taking x-ray pictures will show if bones have been fractured. In a sprain, the ligaments and other soft parts around the joint are either stretched or actually torn. It is usually either a twist or a wrench. There is pain at the point of injury and the joint is held stiff. Swelling may be considerable.

6.16.8 Dislocations

A dislocation is the displacement of one or more bones at a joint. Dislocations are much less common than either fractures or sprains. There is loss of movement in the dislocated joint and the joint looks peculiar. The pain is often described as "sickening." Often the patient can tell what has happened. Dislocation of a joint may be repeated again.

The shoulder is most commonly dislocated owing to a fall on the outstretched hand. Jaw, usually dislocates because of a big yawn. Dislocation of other parts of body include ankle, thumb, and finger joints. It takes great force to dislocate either the elbow or the knee.

The first-aid treatment is to support the part beyond the dislocation in the position of greatest comfort and to get expert help. The first-aider should never himself try to put back a dislocation, as by doing so he may cause a fracture. Quite often, a dislocation and fracture occur together. Diagnosis of these double injuries is beyond the first-aider control.

Figure 6.9 Wrong right strains and sprains.

6.17 Burns and scalds, electrical and heat injuries

A burn is tissue damage caused by dry heat; and scald is damage by wet heat. Tissue damage in direct contact with strong chemical is referred to as a chemical burn. The seriousness of any burn depends on four factors:

1. Area
2. Depth
3. Part of body affected
4. Age of the patient

6.17.1 Area of burn

The skin area involved in a burn is more important than the depth. Even a superficial burn involving more than 5% of the body surface is serious; if more than 15% of the surface is involved, the condition is extremely dangerous, and the patient may die of shock, unless blood transfusion is started within an hour or so.

In all large burns, there is severe shock, due to great quantities of body fluid lost from the raw surface of the damaged tissues, or by the swelling of the burned part. Naturally, the larger the burned area, the greater the shock.

Burning sterilizes the tissues, but the damage and the exposure of a large raw area greatly increase the chances of subsequent infection. The greater the area, the greater the infection risk. The first-aider should help to keep the burn clean and infection-free; but wrong action will introduce infection.

6.17.2 Depth of burns

For practical purposes two "depths" of burns have to be recognized:

1. Superficial burn: Only the outer layers of the skin are affected. The burned area goes red, and blisters may or may not form. Pain is considerable, but the burn usually heals rapidly and there is little scarring. Large superficial burns produce considerable shock.
2. Deep burns: All the layers of the skin are damaged, and the fat and muscle beneath the skin, and even the bone, may be involved. The burned area is yellowish-white or completely charred. If the skin is completely destroyed, there will be less pain than in superficial burns, because the surface skin nerves have also been destroyed. Deep burns often become infected. They heal very slowly and scarring is often serious.
   - Part of body: Burns of the face and hands are more serious than burns of corresponding size elsewhere because quite small scarring may disable affected part of body and cause ugliness.
   - Age of patient: Children and old people react severely to burns, and often have extensive burns.

6.18 Varieties of burns

Dry hot burn may be caused by contact with hot metal such as soldering filler rod or an unprotected hot bag. The burn is localized and may be superficial or deep.

Dry cold burns may be caused by contact with liquid gases such as liquid oxygen or carbon dioxide. The burned area is marked, localized and paled.

Fire burn may be caused by furnace back-fire, flammable oil products, solvents, or a burning building. The clothes usually catch fire and the burn often covers a large area; parts of it may be superficial and other parts deep. Charred clothing may stick to the burn. The patient is usually very shocked.

Sunburn will be caused by exposure to sunbeams or artificial lights. It is very superficial but there is often considerable reddening and blistering.

Friction burn is a rare type of burn. It is caused friction of a fast moving rope.

Wet burns or scalds may be caused by steam, hot water, hot oils, cooking fat, hot solvents or tar. They are usually superficial but are often extensive and therefore serious.

6.19 First-aid treatment of burns

In treating a burn or scald the objectives are to:

1. Prevent shock
2. Avoid infection
3. Relieve pain

The first-aider must not allow anything bactriologically "dirty" to be put on the burn, e.g., grease or ointment from an old pot. He must not touch the burn with his hands, and should speak as little as possible until the burn has been covered with a clean or sterile dressing.

For treatment purposes, burns are considered as trivial, medium, and serious. The first-aider can safely treat the trivial burn himself, but any medium burn, larger than a coin or an average cigarette burn, should receive expert treatment from a trained nurse or a doctor, so that the chances of infection may be kept to a minimum. The serious burn involving more than a few square cm of skin will be sent direct to hospital.

6.19.1 Trivial burns

Trivial burns are often very painful. The pain is quickly relieved by holding the burn part under running cold water. If after this, there is any sign of injury to the skin, the burn should be carefully cleaned with cetrimide or soap and water, and cotton wool, in the same way that minor wounds are dealt with. After cleaning, the burn and surrounding skin should be dried with clean cotton wool, and covered with an individual sterilized tulle-gras dressing. The dressing is contained between two slips of transparent paper. One slip is pulled off, and the dressing, still attached to the other slip, the second slip is then quite easily removed, leaving the sterilized tulle gras in place. In applying the tulle gras the first-aider must take care not to touch the dressing, except at the corners of edges separating it from the slip of paper, forceps will be helpful if the tulle gras is too large and it should be cut to the right size before the slips of paper are removed. The tulle gras is covered with a small individual sterilized dressing, an individual plaster or clean cotton wool and a roller bandage. If there is a blister, it should not be pricked, and the first-aider should not try to remove dead skin.

6.19.2 Medium burns

Since thorough cleansing of the burned area will be undertaken by the doctor or nurse, the first-aider's duty is simply to cover the burn with one or more individual sterilized dressing, and to get the patient to the expert as quickly as possible. There is no point in putting tulle gras on any burn that is efficiently cleaned.

6.19.3 Serious burns

No attempt should be made to clean the burn or to take off the clothes, or to pull away any charred clothing that has stuck to the burn; the burning will itself have sterilized the whole area. The burn area should be quickly covered with one or more large sterilized individual dressing. If the burn area is extensive, a clean towel or sheet should be used as a covering. At the hospital, cleaning will be done with full surgical precautions in an operating theatre.

Rapid replacement of body fluid lost is the life-saving treatment; in such cases, quick transport of patient to hospital is literally vital. Attempts to carry out blood transfusion or even intravenous saline infusion outside hospital cause more harm by delaying full-scale controlled fluid replacement. They are now resorted to only when the patient is trapped and cannot be quickly released, or where the distance to hospital is considerable.

The general treatment of shock should be followed. If the patient is thirsty, he may wash out his mouth and spit out. Only if there is considerable delay in getting the patient to hospital can small sips of water be given. Larger quantities of fluid taken suddenly may cause vomiting. With small burns, hot sweet tea is harmless.

6.19.4 Electric burns

Electricity can cause burns, "electric shock," or both. Burns occur at the points of entry of an electric current, that is, the points of contact with a live conductor. A common cause is electric short circuit of portable hand tools, especially if there is inadequate earthing. Severe burn with extensive charring of the tissues will occur when body comes in contact with high-tension lines. The heat and destruction from a high-tension contact are so great that the part of conduction is broken and the injured man runs away with his clothes on fire.

A mild electric current can produce a pattern on the skin like the branches of a tree or the meshes of wire netting. This is probably because the electricity flows along the trickles of sweat on the skin. A moderate current will produce a dry, shriveled burn, with little pain, less than from a heat burn of the same size. There is little or no reddening around the burn, and the burned tissue takes the form of cone with the point inward, extending down from the skin into the deeper structures. Quite a small burn may involve tendons and other important structures, and this may not be apparent for 3 or 4 days.

Sometimes the point of entry of an electric current may be similar burn at the point of exit. Where the entry burn is on the hand, there may be an exit burn on the foot.

• Treatment: Even the smallest electric burn should be covered with a clean dry dressing and referred to a nurse or doctor. The "devitalization" of the tissue around the burn will delay healing and increase the risk of infection. The best treatment is a small skin-graft, usually applied in the out-patient department.

6.20   Heat stroke

The heat stroke is a rare and somewhat dangerous condition that occurs when the overheated patient has neglected treatment and continued for some time in a very hot environment. The first, and much more common, effect of too much heat is heat exhaustion. This is also known as miners' or stokers' cramp.

6.20.1   Cause

The essential cause is loss of too much body water and body salt as a result of too little replacement of what has been lost by sweating. Sweating is part of the natural mechanism of cooling the body. It is not the production of sweat, but its evaporation from the body surface lowers the body temperature. If the body is getting too little water and salt to replace what is lost as sweat, or if the surrounding air is so full of water-vapor that the sweat cannot evaporate, the body cuts down on further sweating and the internal temperature starts to rise. If this is allowed to continue, true heat stroke develops.

6.20.2   Symptoms

Heat exhaustion, the skin is clammy, and the patient irritable; he complains of severe cramps in the limbs.
Early heat stroke, the skin is hot and dry, and the irritability and cramps are much more severe. In the second stage of heat stroke, the patient may be found unconscious, breathing hard, and sometimes twitching a little. The skin is dry, red, and burning hot.

6.20.3   Prevention

Among those specially liable to heat exhaustion and heat stroke are especially marine strokers in the tropics. Workers at furnaces and in foundries and other very hot places should be provided with special salt drinks that may be flavored with orange or lemon and glucose. The workers concerned soon learn for themselves how much salt drink they require to meet their differing individual needs.

Working in air-tight rubber protective clothing may produce heat exhaustion and heat stroke, especially if the weather is warm. The layer of air between the skin and the protective clothing soon becomes saturated with sweat, and an artificial humid atmosphere is produced. If such clothing is essential for heavy work, its outside should be soaked in cold water. The evaporation of this water will cool down the worker inside. In very hot condition 0.24 to 0.48 L of sweat may be lost per hour, and this should be recouped by fluid intake.

6.20.4 Treatment

The patient should be removed from the heat, stripped to the waist, and bathed or sprinkled with cold water. He should then be fanned with towels to encourage the evaporation of the water, which will cool the patient further. This cooling process must be stopped when the patient's temperature has fallen to 36.8°C. If the patient is conscious, or as soon as he becomes conscious, he should be given copious droughts of cold water, with a salt spoonful of common salt per tumbler and orange or lemon to improve the taste. On recovery, the patient should rest (the length of rest depends on severity of the attack). All cases of heat exhaustion or heat stroke should be seen by a doctor or a nurse before returning to work.

6.20.5 Sunstroke

This is usually a combination of heat exhaustion and ordinary fainting. It is particularly liable to occur in those who are suddenly exposed to the heat in unsuitable clothing. Its treatment is the same as for heat exhaustion.

6.21 Chemical burns, injuries, and poisons

Chemical substances may harm the human body in three ways:

1. By direct burning the skin or eyes
2. By irritation of the skin, so that dermatitis is produced
3. By entering the body and causing rapid or slow poisoning

Almost all chemical substances can cause trouble if misused. If used with proper care, they can be handled with complete safety. In this type of work, prevention is the target.

6.21.1 Chemical burns

Chemical burns may be caused by acids or alkalis. In either case, speedy treatment is vital. The acid or alkali should be washed off at once, or at least greatly diluted by flooding the affected part with large volumes of water. Thus if no special antidotes are available, a chemical splash in the eye should be treated by holding the eye open under a running cold tap, or by plunging the upper part of the face into a bucket of cold water and blinking hard. Similarly, an acid or alkali splash on the skin should immediately be held under a running tap.

Antidote has an almost great significance, but there is always the danger that with chemical burns precious minutes may be lost hunting for an antidote when speedy treatment with water will fulfill the same result. Only after this has been done should time be given to finding and applying the correct chemical antidote unless, of course, a large volume of antidote is immediately at hand.

6.21.2   Acid treatment and prevention

Acid may be quick-acting or slow-acting. The chief risk come from filling, transporting, and emptying carboys, and from accidental spilling and splashing. Those without technical training e.g., cleaners in laboratories, run special risk, and should be carefully instructed in the necessary precautions.

- Quick-acting acids: With quick-acting acids the patient feels irritation and burning almost at once. These kind of acids are: hydrochloric acid that is used in pickling- vats, metal wire drawing and miscellaneous usage. It produces a dark brown blister that later turns black. The other quick-acting acids are nitric acid, nitro-hydrochloric acid, sulphuric acid, etc.
- Slow-acting acids: With slow-acting acids there is no immediate pain, so that the patient may not know that he has been in contact with the acid for a period of half-an-hour to four hours. By then, the acid has penetrated deep into the tissues. Hydrofluoric acid, hydrobromic acid, carbolic acid, and oxalic acid are in this group.

6.21.3   Treatment of acid splashes

Quick action is required with either type of acids as given below:

Wash off the acid immediately with a large volume of water from a tap, shower, or bath. Continue washing until the neutralizing antidote is available.

If water is not available, the acid should be dabbed off the skin with cotton wool, a clean rag or a handkerchief. Any wiping movement must be avoided for this tends to spread the acid.

If an antidote is immediately available in large quantity, it should be used instead of water, but it must be used freely and copiously. If there is only a small supply, it should be applied as soon as the affected part has been completely flooded and douched with water. The antidote here recommended is "buffered phosphate solution" that has the valuable property of neutralizing both acids and alkalis. If this is not available, a solution of bicarbonate of soda (2 tablespoons for 0.48 L (one pint) of water) may be used.

If the clothes are contaminated with acid, they should be removed at once if possible. If not immediately possible, the affected area of clothing should be flooded with water or antidote. If in doubt swill everywhere.

Slow-acting acids should be dealt with as above, but special treatment by a trained nurse or doctor will be needed to neutralize any acid that has penetrated into the tissues, e.g., calcium gluconate may have to be injected under a hydrofluoric-acid burn.

Every suspected cause of a slow-acting acid burn should be seen by a trained nurse or doctor as soon as possible after initial first-aid treatment. With quick-acting acids, the same applies if, after initial treatment, the skin shows any change or the patient feels any adverse effects, or if the quantity of acid involved was considerable.

6.21.4   Prevention

The prevention of chemical burns should be considered by the management and in consultation with the plant/complex medical officer. This will include the provision of first-aid facilities at all danger points. The industrial medical officer should make sure that first-aiders and those concerned know how to use these facilities.

6.21.5 Alkalis and treatment

Alkali burns are more serious than acid burns, because the alkalis tend to penetrate quickly into the tissues, and to go on acting even after thorough washing and neutralization. Thus alkalis closely resemble the more dangerous slow-acting acids. An alkali burn is therefore usually worse than it appears at first. Once the alkali has penetrated, the skin appears pallid and sodden, and later a deep slow-healing ulcer may develop. The main alkalis used in industry are caustic soda, caustic potash, ammonia, bleaching powder, lime, and cement.

6.21.6 Treatment of alkali splash

First-aid treatment is exactly the same as for acids, with the first emphasis on speedy complete washing with a large volume of water; this may be followed by buffered phosphate solution. If the solution is available in a large quantity, it may be used instead of water from the start.

If buffered phosphate is not available, dilute vinegar (two tablespoons to 0.48 L (a pint) of water) or citric acid tablets dissolved in water may be used, but these are unlikely to be available in industry, they add little to the benefit of the water douche.

With lime, bleaching powder, or cement, solid particles should be removed from the skin before the part is flooded with water, as water makes them stick. Removal is best done with a piece of cotton wool or a soft brush.

All alkali injuries should be seen by a trained nurse or doctor at the earliest possible moment. The provision of first-aid facilities at danger point is even more important with alkalis than with acids.

6.21.7 Tar burns

Burns caused by tar should be covered with a dry dressing and the patient referred to a trained nurse or doctor. Solidified tar is itself a good dressing, so no attempt should be made to remove it.

6.22 Chemical skin irritation

Dermatitis or inflammation of the skin is of great importance in industry. Almost any chemical substance can produce dermatitis in a person whose skin is sensitive, yet others can handle the same substances with complete immunity. A good example is dermatitis produced by water in some washerwomen. Strong alkaline soap may also produce dermatitis. Some substances are particularly liable to cause trouble, e.g., acids and alkalis, solvents and degreasers, detergents, oils and tars, glues, synthetic resins, plastics, accelerators, and metallic irritants, such as mercury and arsenic, nickel and cyanide, and sugar, flour, and certain woods.

The first-aider should never attempt to deal with a case of industrial dermatitis, or any other skin condition. The treatment should be carried out by an expert at the earliest possible stage. Delay makes treatment far more difficult, and exposes others to the same risk.

Here again, prevention proper planning should include personal cleanliness of employees, the use of a carefully selected barrier cream or other physical protection, proper hygiene of wash-places and lavatories, changing and cleaning of protective clothing, and special duties for first-aiders.

6.23 Chemical poisons

Chemical substances may enter the body through the skin, the lungs or through the stomach and digestive system. The subject of industrial poisoning is of great extent, most of it being outside the range of the first-aider. He should however, know how to deal with such emergencies as may arise and he should be aware of the existence of certain possibilities.

The direct action of chemicals on the skin has been dealt with in standard, but certain chemicals, e.g., chrome and nickel, may produce ulcers in the skin or in the membrane lining the nose. Such ulcers are known as "trade holes." Fortunately these are now extremely rare. Certain other chemicals can penetrate the skin without damaging it. As a consequence, they have to be handled with great care.

6.23.1 Gases, fumes, and dusts

Gases, fumes, and dusts are important hazards in certain industries. Many dusts, though unpleasant, are not poisonous. But dust-containing particles of silica of certain size are liable, over the years, to produce severe lung damage. These risks are now well known and general preventive measures should be taken.

Chemicals entering via the mouth, stomach, and digestive system are comparatively low in industry. Poisoning may happen accidentally or by attempted suicide. Pollution of hands will contaminate food. This emphasizes the importance of washing the hands before food is eaten, and no food or drink should be served in places where poisonous chemical processes are involved.

6.23.2 Notifiable Industrial Diseases

The well-known industrial poisons have been very largely brought under control. These poisons mainly cause symptoms of very slow onset, and are therefore seldom seen by the industrial first-aider.

There are 14 different industrial diseases and conditions are known by doctors, these are: Lead, Phosphorus, Manganese, Arsenic, Mercury, Carbonbisulphide, Aniline, Benzen, Anthrax, Compressed air illness, Toxic jaundice, Toxic anemia, Chemical skin cancer, Ulceration due to chromium.

Note: Industrial dermatitis is not a "notifiable industrial disease."

6.24 Unconsciousness, gassing, and asphyxia

When a patient has become unconscious the first-aider should make an immediate assessment of what has happened and check to see if the patient is or is not breathing. Most unconscious patients will breath, but if breathing has ceased, the patient is in immediate danger of asphyxia and urgently needs artificial respiration. The unconsciousness and asphyxia are different medical cases, even though they may both be present at the same time. With unconsciousness, there may or may not be asphyxia. But with asphyxia, there is always unconsciousness.

6.24.1 Ascertaining the Cause

There are three kinds of situations:

Where the cause is obvious
Where it is probable
Where the first-aider can see no obvious cause

It is vitally important to make this assessment, since the first step in first-aid is to remove the unconscious person from danger area, and this can be done only after a broad decision about the probable cause has been made.

- Where the cause is obvious: Some circumstances in which the patient is found, show fairly clearly what has happened, e.g., where unconsciousness is due to partial drowning, electric shock, head injuries, or attempted suicide. The patient who has attempted suicide may be found hanging, or with his head in a gas-oven and pillow under the head, or in bed with an open bottle of tablets beside him. In these circumstances the first-aider should waste no time to save life if he can.
- Where the cause is probable: In situation where the cause of unconsciousness is probable by accidental gassing, (domestic or industrial). Gassing in industry may have many different causes. The first-aider should know of the existence of risks in any particular processes in the plant area. Some common industrial processes always have certain risk. When stacks and boilers develop defects, it will cause gas or vapor to blow back and result in asphyxia or unconsciousness of men working nearby. Similarly, men working in deep holes, wells, closed tank are subject to special risks.
- No obvious outside cause: The first-aider will not be able to make an accurate diagnosis in cases where there is no obvious external cause, though he may have his suspicions. It will help him to remember that there are six common causes as follows:
Fainting
Fits
Strokes
Diabetes
Alcohol
Hysteria

6.24.2 Care of unconscious patient

When the patient has been moved off the danger zone there are certain general lines of care, whatever the cause of unconsciousness, which should always be followed as indicated.

The unconscious person should be moved out from danger area. If he is not in danger, no attempt should be made to move him. The patient should be rolled over into the prone or semi-prone position. An unconscious patient may suffocate, if left lying on his back. The tongue falls back into the throat and may block the entry of the windpipe. Suffocation will happen if the patient has false teeth. In addition, saliva or vomited material may enter breathing passage and cause serious results. Often the unconscious patient will be in shocking state, struggling for breath, and the color of his skin will turn blue. This is due to the patient lying on his back with obstruction of the air passage.

Many lives have been lost because patients have not been turned over into the prone or semi-prone position. Prone means face downward and the elbows bent, so the forearms and hands are under the forehead (Figure 6.10). Semi-prone means that the patient's body is on its side, his face turned toward the ground. To stop the body from rolling right over, both arms should be bent naturally at the elbows, and the upper leg bent slightly at the hip and knee, so that it falls forward over the lower leg and acts as a supporting strut (Figure 6.3). If there is retching or vomiting, the semi-prone position is preferred as the mouth and nose are more easily kept clear.

Before rolling the patient over, no obvious fractures should be present. If there is a fracture roll the patient over but support the fractured part. Rolling should be done firmly but gently, moving the whole body into what looks like a natural and easy position.

Take any false teeth out of the mouth gently. If the jaw is tightly closed, do not try to force it open. Raise the point of the chin with the hand, so that the neck is bent slightly back. This helps to open up the air passage at the back of the mouth. Loosen any tight clothing, especially round the neck or waist. If the patient has to be moved, he should be lifted carefully on to a stretcher, still in the prone or semi-prone position, and carried in this way.

6.24.3 Don't do these things

- Don't force fluid into an unconscious patient's mouth. He cannot swallow and will probably inhale it and may get pneumonia.
- Don't slap or throw water over him.
- Don't try to transport him sitting up. He must be moved lying down in the prone or semi-prone position. Attempts to sit up an unconscious person, e.g., in the back of a car, have proved fatal.

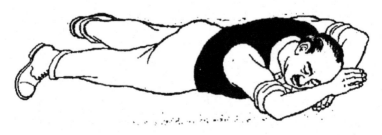

Figure 6.10 Patient in prone position.

6.24.4 Internal causes of unconsciousness

Reference should be made to the following causes:

Fainting
- Fits: Fits are alarming but are usually quickly over. They are almost always due to the condition of epilepsy, and the patient will often have and had previous attacks. As a result of new drugs used to control them, epileptic fits are much more rare than they used to be; but they may occur if the patient forgets to take his medicine, or to bring it with him to work. At the start of the fit, the person utters a cry and then falls over. The limbs stiffen and then start to jerk. The patient may froth at the mouth, bite the tongue, pass urine, or pass a motion. All the time he is quite unconscious, and when the violent phase is over he falls into what appears to be a deep sleep. This usually lasts only a short time. He may hurt himself in falling. Pillows, coats and other soft objects placed around him are safer and more effective than human strength. Never force the jaws apart in order to prevent tongue biting; it is possible to knock out teeth and fracture the jaw. If the mouth is open, it is reasonable and safe to put in a gag. This is no more than wedge to keep the jaw apart. Such wedge is a piece of firewood with a clean handkerchief wrapped round it. Another is a stout pencil, not less than 12 cm (5 in) long. Never tell an epileptic patient what his fit is like, as he may be quite unnecessarily distressed. Unconsciousness during the fit is one natural blessing of the disease. After the attack is over, the patient should be advised to report to his own doctor as soon as possible.
- Strokes: Strokes are caused by a bursting artery or a blood clot in the brain. Though a stroke is sometimes fatal, but many patients have recovered. Good first-aid care, as already described in 19.3 may save life. The patient is usually elderly. He may feel giddy and may, or may not, pass out completely. As a result of the injury to the brain, he usually loses the ability to move one side of the body wholly or in part. This involves most obviously the arm or leg. At the same time, the other side of the face is also paralyzed. In an unconscious patient who has had a stroke, the paralyzed cheek may be seen flapping in and out each time the patient breathes. The facts leading to the conclusion that the patient has had a stroke are as follows: his age in his 50 s or 60 s; the color of his skin, which is usually blue; loud harsh breathing, called stertorous breathing; the flapping cheek; and dribbling from the corner of the mouth. Treatment is generally as set out above of course, the first-aider must send for skilled help without delay.
- Diabetes: Some employees hide the fact that they are diabetics from their workmates or doctors, which has disastrous results. It is in their own best interest that their condition be known by those around them. The most usual cause of trouble in a diabetic is over-action of a normal dose of insulin, as a result of physical fatigue, excessive work or worry, or missing a meal. The patient may become giddy, confused, and even apparently mentally disordered. The treatment is to give sugar at once, preferably in the form of a sweet drink. The physician should be called at once.
- Hysteria: The first-aider should never assume that an unconscious patient is hysterical. Hysteria hardly ever causes complete unconsciousness. Occasionally, however, a patient, usually a young girl but sometimes an older woman or a man, will become typically "hysterical." The situation usually occur when an unrest or anxiety, or natural disaster arise. Hysterical involve bad behavior, like screaming or violent weeping, may lead to panic. In such circumstances, firm physical measure is justified to prevent panic from breaking out. More occasionally, hysterical behavior can follow serious injury or disease of the brain. In these cases, it appears to be caused by lack of oxygen to the brain tissue, and other than the hysteria of panic. Patients should be treated with gentle but firm kindness rather than the traditional slap.

6.25   Rescue operation of gas casualty

As gas casualties are of considerable industrial importance, the first-aider should be fully trained and practiced in rescue work. The following are the general principles to be observed:

Before entering outdoor or indoor gas-filled areas the doors and windows should be opened so as to blow gas or fumes away.

A damp cloth or towel tied round the face gives no protection against gas.

If two or more people are present, one should stay outside in case the rescuer himself needs rescuing. A lifeline tied round the rescuer's waist should always be used to pull a man along the ground.

If the rescue worker has to make a dash into a gas-laden atmosphere, he should take slowly six really deep breaths, then hold his breath and dash in. He will be able to hold his breath for three-quarters of a minute to one minute at the most.

In gas-filled places, the light is often poor. Some gases, e.g., carbon monoxide and methane, are inflammable. The first-aider engaged in rescuing a gas casualty should never use a naked flame.

Respirators should not be used by the inexperienced rescue worker or untrained first-aider. The proper use of respirators requires a good deal of practice. The first-aider who puts one on for the first time in a real action situation may easily panic.

6.26   Industrial gases

There are four types of gas encountered in the oil, gas, and petroleum industries:

• Irritant gases
• Asphyxiating or smothering gases
• Issue-poisoning gases
• Narcotizing gases

6.26.1   Irritant gases

Irritant gases are immediately detected by their effects, particularly on the nose and eyes. The smell is powerful, and the eyes start to water. Those exposed will run away for their lives. These gases are less dangerous than those that are nonirritant. The common irritant gases are as follows:

• Sulphur dioxide (SO_2) is used in the manufacture of sulphuric acid, and in fumigation and refrigeration; and in ordinary smoke.
• Ammonia (NH_3) is utilized in refrigeration and ice-making, and number of other industrial processes.
• Chlorine (Cl_2) is used in bleaching, paper-making, etc. Phosgene ($COCl_2$), mainly used as a war gas, is produced during the manufacture of some aniline dyes. It is also produced when trichlorethylene is inhaled through a lighted cigarette; hence the instruction that those using trichlorethylen should not smoke at work.

6.26.2   Simple asphyxiating gases

The air we breath consists of about four-fifths nitrogen and one-fifth oxygen. The nitrogen is inert; the oxygen is absorbed by the blood and carried throughout the body to enable the tissues to live; without oxygen the tissues die. Asphyxiating gases work simply by replacing the oxygen in the air. It follows that they should be present in very large quantities to get rid of enough oxygen to do harm. Most of them do no smell; this makes them the more dangerous. The following are the common asphyxiating or smothering gases:

- Nitrogen (N_2) is important for practical purposes only in wells, mines, and other deep holes where all the oxygen have been used. Absence of oxygen is shown when a safety-lamp flame, lowered into the hole, goes out.
- Methane (CH_4) is the gas most commonly found in mines, where it is called "fire damp" because it explodes if exposed to flame or spark.
- Carbon dioxide (CO_2) is produced by the living tissues of the body as a waste product and breathed out by the lungs. Large quantities are produced in brewing, aerating and fermenting. It may also be found in mines, tunnels, cellars, and boilers.

6.26.3   Tissue-poisoning gases

Small quantities of tissue-poisoning gases exert a disproportionate poisonous effect. They are absorbed quickly into the blood from the lungs (or even from the mouth) and quickly poison the living tissues by preventing their intake of oxygen. The common gases under this heading are in the following clauses.

6.26.4   Carbon monoxide (CO)

CO is perhaps the most important industrial gas poisoning. It is produced when coke, coal or gasoline is burned. In consequence a black flue that causes the combustible products to leak out into a workplace that is carbon monoxide poisoning gas. The same result may be brought about by gasoline engine when working in a closed space. Exhaust from this engine contains 7% carbon monoxide.

6.26.5   Hydrogen cyanide (HCN)

HCN is so poisonous that it is usually used only in the open air. Sometimes, however, it is used for fumigation of premises or dirty fabrics. It has a smell of bitter almonds and is almost instantly fatal.

Wherever it is used, the maker's precaution card should be exhibited and antidotes should be immediately available.

6.26.6   Hydrogen sulphide (H₂S)

H_2S is evolved in glue making, tanning, mines and oil industry. In small concentration it is violently irritating and has a foul smell. In large concentration a man inhaling it may drop down dead.

The symptoms of gassing depend on the nature of the gas, the amount inhaled and the length of exposure. With the irritant gases, coughing and watering of the eyes and nose are immediately apparent. With the tissue-poisoning or narcotizing gases, the patient quickly becomes unconscious but may retain a good color. With the simple asphyxiating gases there are usually two stages:

1. Partial asphyxia: The patient feels dizzy and weak and may stagger and collapse. There may be difficulty in breathing, with panting and gasping. Occasionally there is convulsion, especially as the patient breathes out.
2. Full asphyxia: The patient is unconscious and blue, especially at the "tips" of the body, nose, ears, lips and fingers. Breathing is first intermittent and then absent. The pulse is first weak and then absent. The absent pulse does not necessarily mean, however, that the heart stopped.
   * Treatment of gassing: The treatment of gassing is briefly summarized:
   remove from danger area artificial respiration if breathing has ceased administration of oxygen treatment of shock and general care of the unconscious patient.

6.27   Artificial respiration

Artificial respiration, or artificial breathing, is required when breathing has stopped, but life is not extinct. Patients who need artificial respiration are always unconscious; but most unconscious patients have not stopped breathing and artificial respiration is not required. The most usual causes of cessation of breathing are electric shock, drowning, carbon monoxide poisoning, and pressure on the chest; like person left under debris.

In such cases, the time between the cessation of breathing and the stopping of the heartbeat is short. The purpose of artificial respiration is to give the heart and other tissues the oxygen they need and get rid of unwanted carbon dioxide from the body to encourage the lungs to start work again. Artificial respiration should be started on the spot, unless the patient has to be moved out of contaminated air.

6.27.1   Methods of Artificial Respiration

Many ways of artificial respiration have been devised. There are five main headings as follows:

* Push methods: In "push" methods the operator pushes on the outside of the chest to force air out, relying on the natural recoil of the ribs to suck air in. This method cannot be used if the ribs have been fractured.
* Pull methods: In "pull" methods the operator moves the arms so as to stretch and expand the chest causing an intake of air. The best known "pull" method is called the Silvester, but experience has proven it to be unsatisfactory.
* Rocking methods: In rocking methods the principle is to use the diaphragm and the contents of the abdomen as a piston, first to compress and then to inflate the lungs. This is more efficient than any of the manual methods, but it requires special apparatus.
* It is possible to improve a rocking stretcher, but this is not without risk. In a fully equipped industrial medical department or industrial health service, a proper rocking stretcher should be available.

- Suck-and-blow methods: The lungs may be expanded and contracted in a natural way by first applying a positive pressure, then a negative pressure, either outside the walls of the chest or directly down the windpipe. An outside pressure can be applied only with an elaborate mechanical apparatus. Direct inflation and deflation of the lungs by air or oxygen is achieved by alternately blowing and sucking through the nose and mouth and the air passage. Provided there is a clear "air way" such a method is usually effective. It is used by all modern anesthetists during operation when chest and other muscles have been temporarily paralyzed by medication. For the first-aider there are two possible "suck-and-blow" methods:
  Mouth-to-mouth
  Resuscitator
- Mouth-to-mouth: The operator has to blow hard into the mouth of the patient, making sure that the patient's chin is well up, the mouth open, the tongue out of the way, the nostrils closed, and above all there is a good fit of lips to lips. The method is very effective and far easier to carry out than might be expected. Every first-aider should know how to perform mouth-to-mouth artificial respiration. It can be learned from books or lectures, but the most valuable method is to practice on a human dummy.

The details of the mouth-to-mouth are as follows (see Figure 6.11):

The patient lies on his back (the so-called supine position). Compare this with the Holger-Nielson method where the patient lies on his face, in the prone position.
Sweep the finger round inside the mouth to remove any pond or seaweed or false teeth and to make sure that the tongue is forward.
Kneel comfortably on one side of the patient's head, so that one's mouth can come naturally over his.
Bend the patient's head right back, as far as it will go. This will open the air-passage behind the tongue.
With one hand, hold his chin up and back with the other, pin his nose closed.
Take a deep breath. Apply your mouth to his, getting as good a fit as you can. Then blow, until out of the corner of your eye, you see his chest rise.
Take away your mouth, and watch his chest sink back.
Give the first six blows quickly, leaving in between each blow just time enough for the chest to sink back.
Thereafter, blow at the rate of 10 blows a minute.
- Holger-Nielsen method: Before starting the Holger-Nielson method, the following steps should be taken as quickly as possible:
  Roll the patient into the prone position.

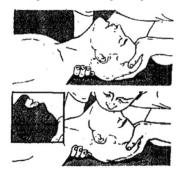

Figure 6.11 Mouth-to-mouth respiration.

Put the finger inside the mouth and sweep it around to remove any obstruction, e.g., sea or pond weed or false teeth.

Make sure that the tongue is hanging in its normal forward position.

Loosen the collar.

Move out any "lumps" in the clothing from the front of the chest such as, a tin in a pocket. An object may harm the ribs when artificial respiration is started.

If the patient has been submerged or has been vomiting, the first-aider should stand astride the patient, clasp the hands underneath his stomach and raise him quickly a short distance from the ground. Repeat twice. This helps to empty the air passages. Wet clothing should be taken off immediately.

The steps referred to above should not take more than a minute. Artificial respiration should then be started. Doctor and ambulance to be called and arrangements for a rocking stretcher or a resuscitator to be made.

Artificial respiration should be continued rhythmically without stopping, until natural breathing starts again, or until the doctor pronounces the patient dead.

- Artificial respiration and the Holger-Nielsen method (see Figures 6.12 and 6.13): This method is also known as the "back pressure arm-lift" method, which is a good descriptive title:

  Position: The patient should be placed in the prone position, with the elbows bent and projecting out sideways and the hands crossed under the head. The head will be turned slightly on one side so that the cheek rests on the hands. The nose and the mouth must be clear of any obstruction.

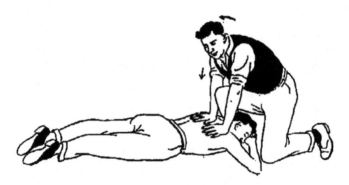

Figure 6.12 Holger nelson method (a).

Figure 6.13 Holger nelson method (b).

The operator kneels on one knee at the head of the patient and facing him. The knee is placed in the angle between the patient's head and his forearm. The opposite foot is placed near the patient's other elbow. Alternatively, the operator may kneel on both knees, one on either side of the head. If the one-knee position is used, he will find it an advantage to change knees from time to time. The operator places his hand on the flat of the patient's back. The tips of the thumbs should be just touching, with the fingers pointing downward and the wrists on a level with the armpits.

Movements: In making the movements, the operator's arms should be kept straight and the body weight is used to show the effects. All the movements should be made steadily, slowly and rhythmically, the operator counting out loud slowly as he proceeds.

Movement	Time	Count
First Movement: Compression of patient's chest	2 sec	"one, two"
Second Movement: Slide hands to patient's elbow	1 sec	"three"
Third Movement: Raise patient's elbow	2 sec	"four, five"
Fourth Movement: Lower elbows and slide hands to patient's back	1 sec	"six"

Compression of a patient's chest causes breathing out or "expiration." Raising the patient's elbows causes breathing in or "inspiration." The state of the cycle of movements on the patient is as follows:

Breathing out	2 sec
Relaxation	1 sec
Breathing in	2 sec
Relaxation	1 sec

The full cycle takes 6 sec, giving a rate of artificial breathing of 10 to 1 minute.

• Oxygen therapy: Oxygen should be obtained and administered to the patient during artificial respiration.

6.28 External heart massage

If the heart has ceased to move the blood around the body, artificial respiration is no use. It is sometimes possible to re-start the heart by pressing over the lower half of the sternum or breast bone, which is called an external heart massage.

If after 12 breathings with the "mouth-to-mouth" method the patient still looks dead, and there is no change in the color of the skin or lips and no signs of

Figure 6.14 External heart massage.

spontaneous breathing, it is worth trying external heart massage. Place the ball of one hand over the lower half of the breast bone. This will be found at the top of the inverted V made by the lower ribs. Place the second hand over the first. Give six sharp presses at one-second intervals (Figure 6.14). Then give mouth-to-mouth lung inflation, and repeat the whole cycle. Stop the external heart massage as soon as the color improves. But continue with the mouth-to-mouth artificial respiration. If two first-aiders are available one can give heart massage while the other gives mouth-to-mouth artificial respiration.

6.29   Eye injuries

More than one in ten of all accidents involve the eye, and the most common of these is a "foreign body" in the eye. It is therefore eye injury that constitutes one of the most important parts of the first-aider's work.

Most industrial eye casualties never reach the hospital but are dealt with by first-aiders, nurses, medical officers, or general practitioners. Fifty percent of casualties are caused by working with an emery wheel when a foreign body separates from the surface of the wheel, travels slowly, and does not penetrate too deeply into the eye. It is usually composed of metallic dust, abrasive, and bonding material, and is nonmagnetic. Other industrial eye casualties are related to turning, milling, spinning, boring, hammering, and chipping.

6.29.1   Examining the eye

Any first-aider who may be called upon to deal with a colleague's eye an injury should know how to examine it. The sequence is as follows:

1. The patient should be seated, with a good light shining on his face and eye. The first-aider should stand behind the patient and support the head against his own body.

2. The patient's head should be tipped well back and the eye then held open with two fingers. It will make much easier to examine the bad eye if the patient is asked to keep both eyes open. The patient should be asked to look slowly and in turn at each of the four points of the compass, so that the whole of the exposed eye may be carefully inspected. There should be no hurry. It is particularly important to inspect the front of the cornea (the transparent curved surface covering the pupil and iris).
3. If nothing can be seen, the lower lid should be pulled away from the globe while at the same time the patient looks upward. This enables foreign body under the lower lid to be seen. It is possible to turn up the upper lid in such a way that its under surface can be seen, but as this maneuver needs considerable experience if it is to be satisfactorily performed, it is best not done by first-aiders.

A foreign body may be seen on the front of the cornea, on the white of the eye, or on the red inside lid. It may be black or glistening, and may be fixed or moving.

6.29.2 Removal of foreign bodies

A foreign body that moves will probably come out of the eye very easily. Nature's method is to flush the eye with tears from the tear gland. Flushing can be exercised by getting the patient to blow his nose strongly, and blinking several times. On no account should the eye be rubbed. Should this fail, the eye should be washed out with an eye-bath, using ordinary water. The bath is completely filled and the eye lowered until in contact with it. The bath is then raised and the patient blink under water; this is easier if both eyes are blinked at once. If this fails to remove the foreign body, it is certainly stuck to the surface of the cornea. The eye-bath should be washed and dried after use before putting it away.

The first-aider who feels confident and has been properly instructed may make only one more attempt, to remove a foreign body if failed to get it out by washing. But if expert medical or nursing help is readily available, it is better to pass the patient on for skilled attention.

The attempt to remove the foreign body should be made with clean cotton wool on an applicator. When one has been used it should be thrown away immediately. The eye should be held open and a single sweep with the cotton wool should be made over the foreign body. If it is loose, it will be seen attached to the tip of the cotton wool.

The first-aider should never use a matchstick or the corner of a handkerchief. Neither will be sufficiently clean to be safe. A camel-hair brush is also unsatisfactory because it is too soft; moreover, if not sterilized after use, it will carry germs from eye to eye. Unless the first-aider is completely certain that a foreign body has been removed, the patient should be referred to a nurse or doctor at once. If the patient complains of any pain at all after removal, this also is an indication of referral. In any case, to make sure that the foreign body has been taken out, the eye should be carefully inspected under a good light.

A foreign body so well embedded in the surface of the cornea that it does not project may at first, cause no pain. Any patient who complains of pain and thinks that the foreign body entered at some earlier time should be referred straight away to a nurse or doctor.

It is essential that the first-aider should take no risk in dealing with eye injuries. If there is slightest doubt, the patient should be sent to a nurse or doctor at once. Always cover the eye with a medium-size dressing or eye-pad before referral.

All first-aid boxes must contain "an approved eye ointment." Small magnets are sometimes used by first-aiders. In practice, these are virtually useless, as almost all easily-removed foreign bodies are nonmagnetic.

6.29.3 Glass in the eyes

It may be very difficult to see glass in the eye. Moreover, a piece of glass is liable to cut the surface of the eye, sometimes severely. The first-aider must on no account wash out the eye, for fear of the washing-fluid getting into the glob through the cut. Nor must any drops, ointment, or liquid paraffin be put into the eye by the first-aider, and he should not attempt to remove glass from the eye. The eye should be covered with a medium-sized individual sterilized dressing, and the patient sent for expert treatment as quickly as possible.

6.29.4 Dust in the eyes

Dust may blow in through the open doorways of a factory or be blown up following the use of compressed-air hose for cleaning debris. The eye should be washed out using water and an eye bath. If the irritation is not speedily relieved and there is a scratch of the cornea the patient should be sent to a nurse or doctor.

6.29.5 Foreign Bodies in the Eyes

A foreign body that penetrates the globe of the eye will not be visible when the eye is examined, through a small cut in the cornea or the white of the eye may be seen. Such an accident usually happens by hammering or chipping with a mushroom-headed chisel. A mushroom-headed hammer is equally dangerous. An eye accident following the use of such a chisel or hammer must be assumed to be serious, and should be sent for immediate treatment. The eye should be covered with a medium-sized individual sterilized dressing and then the patient should be transported by ambulance; movement of the head and upper part of the body must be kept to a minimum for fear of starting bleeding within the eyeball.

6.29.6 Welding and the eyes

Exposure of the unprotected eye to gas or electric welding or cutting is the most common cause of conjunctivitis in industry. There are three common types of welding:

- Spot welding: The operator should wear goggles or have the eyes protected with a mica or other transparent shield. The only risk to the eye is from sparks. Eye injuries from spot welding should be referred for expert treatment, as tiny pieces of metal are usually stuck to the burned conjunctiva. Gas welding oxygen 2204°C (4000°F) and acetylene 3315°C (6000°F) are the common flames used.
- Electric arc welding: The temperatures here are similar to those with gas welding.

Welding or cutting places should be well-ventilated, to be certain that harmful gases are not present; they should also be screened to prevent exposure to the strong ultra-violet rays that are produced, particularly with gas and electric welding.

6.29.7 Arc-eye or welder's flash

Eye injury of welder's flash is due to exposure of the unprotected eye to gas or electric welding or cutting. The operator should use dark goggles or a dark shield. For the sake of safety, all cases of arc-eye should be referred for treatment. The first-aid treatment is to wash out the eye with water or a simple solution, but a special "arc-eye-lotion." Exposure of unprotected eye to infra-red rays from furnaces, molten glass or white-hot metal can, over many years, damage both the lens and the cornea.

6.29.8 Chemical splashes in the eye

In dealing with chemical splashes, first aid is of utmost importance, since it can, if done promptly and efficiently, save sight. As with chemical burns of the skin alkalis are more dangerous even than acids. Unless the alkali is removed at once, it combines with tissues of the eye and goes on acting on the tissues long after the eye has been thoroughly washed out. A neglected alkali burn of the eye will in consequence continue to increase in size and depth despite washing with antidote, and this may cause loss of vision.

6.29.9 Treatment and prevention

- Treatment: Unless antidote is immediately available, the head should be held under a tap, or plunged into a bucket to clean water; the victim should blink vigorously. Eye wash fountain gives quite a good eye-irrigating jet. The patient may have difficulty in opening the eye. He should be told to try to hold both eyes open. If the first-aider is trying to irrigate the eye with water or antidote, the patient should sit or lie with the head tilted right back, and an assistant should hold the eye open; if he is not available, the first-aider may use the first and second fingers of the left hand. The jet of water or antidote should not be directed right on the front of the eye; instead, the patient should be told to look outward and the jet directed on to the inner angle of the eye (Figure 6.15). Every industrial first-aider should have experience irrigating an eye. Wherever there is a high risk of alkali or acid splashes, buffered phosphate can be supplied with an irrigating bottle or strong canister. Irrigation should be continued, with short rest pauses for 5 to 10 min. The patient should then be transferred as swiftly as possible to expert nursing or medical care. After alkali splashes, this irrigation may have to continue for up to an hour.
- Prevention: If goggles or protective face-shields were more widely worn in industry, eye injuries whether from chemical splashes or foreign bodies would be far fewer.

Here are some practical points about the relative dangers of different processes and materials:

Grinding wheels should be eye-guarded
Transparent plastic shields to be fitted
Wear goggles

The first-aider has a real part to play in encouraging the use of eye protectors whenever there is a risk of eye injury.

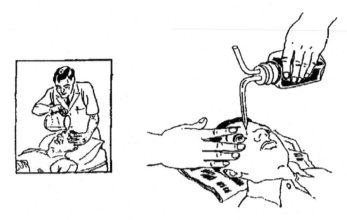

Figure 6.15 Treatment of eye injury.

6.30   Aches and pains, transport, and records

In dealing with illness as a result of injury, the first-aider should make some simple practical decisions as follows:

- Is it a minor condition that will get better quickly at the work site?
- Is the patient sufficiently ill to be sent to the industrial medical center?
- Is the patient so ill that skilled help should be called at once?

6.30.1   Care of minor aches and pains

The first-aid box contains four items for use in appropriate cases of aches and pains: a clinical thermometer, magnesium trisilicate, sedatives, and aspirin tablets.

Every first-aider should take a patient's temperature and read a clinical thermometer. If the temperature is above normal the patient should be referred to a doctor and if patient looks ill referral is essential.

Magnesium trisilicate (1–2 tablets) may be safely and beneficially given to the patient with a hangover, or to the regular gastric sufferer who is under medical treatment. Severe stomach pain should never be treated by the first-aider.

Sedatives (1–2 tablets) relieve the mild headaches of ordinary or a headache accompanied by any other symptoms, should always be referred to a trained nurse or doctor.

6.30.2   Moving an Injured Person

Any severely injured or ill person should be moved as little as possible until experienced ambulance personnel, nurse or a doctor are available. The transport of the injured is a specialized branch of first aid, calling for considerable practical training and experience. The industrial first-aider may occasionally have to move injured person out of a position of immediate and continuing danger, and in emergency may have to transport him to an ambulance or clinic center. To meet these emergencies, some practical experiences, on the lines set out below, are essential. For the demonstration of work described hereunder a stretcher, two blankets and a strong scarf are the only equipment needed.

It is difficult but not impossible to move an injured person safely without a stretcher. It is easier to move a patient without a stretcher with two bearers than with one. It is easier both to load and to carry a stretcher with four bearers than with two. But in emergency one person can move another provided that the proper techniques have been carried out.

6.30.3   Preparing a stretcher

If two blankets are available, they should be arranged on the stretcher that is known as the "fish-tail" position. The patient's feet and legs should be covered with the "fish-tail" and the body and head wrapped in the lower blanket, tucking in firmly with the longer side. If only one blanket is available, it should be arranged on the stretcher diagonally. The patient should be folded into the blanket, with the longer angle being turned over on top and tucked in.

6.30.4   Loading a stretcher

There should be four loaders, one of whom must give orders so that all act together. Three men lift the patient; the fourth pushes the stretcher, with blanket or blankets, under the lifted patient, so that he can be gently lowered in the right position of the stretcher.

The "three man lift" is an art to be implemented by practice. Its object is to lift the patient while keeping the head, body and legs in straight line. All three men must be on the same side of patient. They all kneel on one knee, in each case the knee near the patient's feet. Their other knees are near the patient's head from a shelf on which the patient can be rested. Hands and arms are gently but firmly insinuated right under the patient. The first man has to raise the head and shoulders. The second man, who should be the strongest, has to raise the chest and abdomen. The third man has to raise the legs, with one arm under the thighs and the other under the calves; he should take care not to let the feet sag and the knees bend (Figure 6.16). When all are ready, the leader gives the command to lift, and the patient is raised and rested on the lifter's bent knees, so that the stretcher can be slipped into position. Again, the leader must give the command to lower, so that all three move as one.

The "three-man lift" may also be used for carrying a patient a short distance. It is then spoken of as the "human stretcher." If only three loaders are available, they will be a "three-man lift" and carry the patient on the stretcher.

If there are only two loaders available, both should stand astride the patient facing his head. The first passes his arms under the patient's shoulders; the second passes one arm under the buttocks and the other under the calves. When both are in position, the man in the rear gives the command to lift. With short steps, they then walk

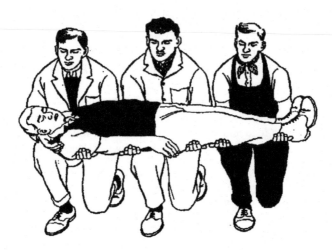

Figure 6.16 Loading a stretcher.

over the stretcher, and lower the patient on to it. This procedure is known as the "straddle-walk."

It is sometimes necessary to load an unconscious patient on to a stretcher. In that case lift in the prone position and carry in the semi-prone position. Attempts to lift in the semi-prone position are dangerous, as the unconscious patient may roll out of the lifter's arms. Carrying in the prone position is difficult because of the position of the patient's arms, and also the airway may be obstructed. The transport of the patient with a broken back or neck has already been discussed under the care of fractures.

6.30.5   Carrying a stretcher

Carrying a stretcher is more difficult than it looks, practical experience is needed both as a bearer and as a "patient." It is easy for the unskilled to tip a patient off a stretcher. There has been much discussion as to whether it is better to carry a patient head first or feet first. The conventional method of carrying feet first should be considered, though there are exceptions to this, such as lifting into an ambulance. The strongest man or men should be at the head. This is because the upper half of the body is heavier than the lower half. The command to "lift," "move forward," and "stop" should be given by one of the men at the rear and of the stretcher. The smoothest carry is achieved by all four people finding a rhythm.

6.30.6   Blanket lift

Four men can carry a severely injured man, making use of a single blanket. The blanket must first be inserted under the injured man. This is done by rolling up the blanket longways and placing the roll beside the patient. Three people pull the patient toward them and a fourth inserts the roll under the patient (Figure 6.17(a)). The patient is

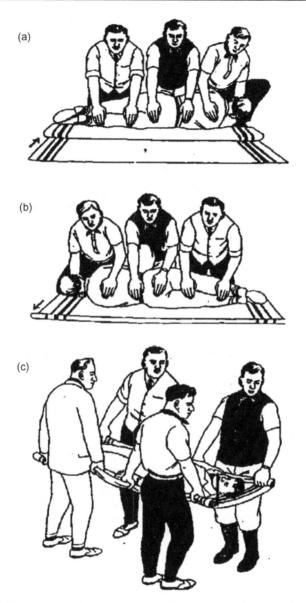

Figure 6.17 a) Three people pull blanket under the patient b) The patient is lowered on to the roll (c) Blanket is lifted and moved like a stretcher.

lowered on to the roll, then pulled or pushed up to other way (Figure 6.17(b)); this enables the roll to be pulled through. The patient is then lowered on to the blanket. At the outset, the blanket should be so placed that, when the patient is in position, a small roll can be made along each side of the patient. For lifting, one man takes hold of a half each of these small rolls, and the blanket is lifted and moved like a stretcher

(Figure 6.17(c)). Note particularly the position of the bearer's hands. The hands in the middle of each roll must be closed together. Otherwise it is impossible to maintain the tension of the blanket needed to keep it flat. An efficient blanket lift is impossible with fewer than four bearers.

6.30.7　Chair lift

If a patient can stand or sit but cannot walk, two men can move him using a "chair lift." The familiar "bandy chair" needs no equipment, but the patient must be able to use his arms to grip the neck of his carriers. A real chair is much better. This is carried by two men facing each other; each man grasps the back of the chair and a front leg, close to the point where it joins the seat. Care has to be taken not to tip the patient forward. The "real chair lift" makes it comparatively easy to carry a patient up or down stairs.

6.30.8　Single-handed lifts

If a patient can just stand and has the use of his arms, the familiar "pick-a-back" is useful. For the pick-a-back, the rescuer must use both his hands and cannot therefore climb a ladder.

The "fireman's lift" leaves a hand free, and so makes ladder-climbing possible. It requires considerable strength on the part of the rescuer and good balance; it cannot therefore be used if the patient is very heavy, unless the rescuer is proportionately strong.

The patient must be helped to stand upright, facing the rescuer. The rescuer grasps the patient's right wrist with his left hand, then bends down until his head is just under the patient's right hand. This brings the rescuer's right shoulder level with the lower part of the patient's abdomen. He then puts his right arm between the patient's legs, and grasps the leg firmly. The weight of the patient is then taken on his right shoulder. As he rises to stand upright, the patient is pulled across both shoulders. The patient's right wrist is then transferred to the rescuer's right hand, thus leaving his left hand free.

A little practice will soon demonstrate the value and the limitation of the fireman's lift.

6.30.9　Transport by helicopter

In off-shore platforms that are located at sea or other sites that are far from hospitals or medical centers a medically equipped helicopter should be available for transporting patients who are suffering from serious accidents or illnesses.

6.31　Record-keeping

Certain records of industrial accidents should be kept, but for the most part, minor accidents are excluded. The first-aider in charge of a first-aid box or post should keep a complete record of all that have happened. Such a record may be kept in an ordinary calendar book, appropriately ruled, sometimes called a "day book." The day book should give the date and the name of the patient, the nature of the injury or condition, the cause if this can be stated, the treatment given, and the disposal (back to work, to industrial nurse or doctor, or to own doctor or hospital as the case may be). Simple abbreviations will soon be devised. Writing must be kept to a minimum, or it will soon be neglected. The day book should be kept in the first-aid box.

The accident book is a special book. It is the statutory duty of the employer to provide this book and safety officials should investigate circumstances of accidents and make whatever arrangements are necessary in accordance with procedures in hand.

Safety belts

7

Safety belts and harnesses are means of protective equipment that are worn around the upper parts of the body protecting the user against falls and creating self-confidence when used in the correct manner. In designing and selecting a belt or harness for any particular work, care should be taken to ensure that the equipment gives the user, as far as it is compatible with safety, the maximum degree of comfort and freedom of movement, and also in the event of a user falling, the greatest possible security against injury.

Self-locking anchorage, lanyards, and other components are safety protective devices against falls. In assessing the performance of safety belts and harnesses the focus should be placed on maintenance, inspection, and storage of equipment. This chapter applies to protective equipment against falls and covers the minimum requirements for:

> Design, adjustment, use of equipment, material specifications, manufacture, tests, performance requirements, and inspection of safety belts and harnesses.
> Design, materials for lanyards (chain and webbing), requirements for braided ropes, and test of self-locking safety anchorages.
> Safety belts are not expected to conform to the designs illustrated in Figures 7.1 through 7.4; compliance is only required with respect to the dimensions specified (dimensions are in mm).

The types of belts and harnesses are as follows:

- Type (A): Pole belts.
- Type (B): General-purpose safety belts.
- Type (C): Chest harness.
- Type (D): General-purpose safety harnesses.
- Type (E): Safety rescue harnesses.

7.1 Design

7.1.1 Type A

The belt should be capable of being firmly attached round the user and at the same time firmly secured round the structure. It should be made to one of the following general patterns:

> Separate waist belt and pole strap connected with snap hooks and "D" rings or other suitable fittings.
> Waist belt and pole strap permanently connected on one side, and connected on the other side with snap hook and "D" ring or other suitable fitting.
> Either (a) or (b) but having a waist belt with suspended breeching belt and pole strap designed so that the load is taken by the user's buttocks.

Personnel Protection and Safety Equipment for the Oil and Gas Industries.
DOI: http://dx.doi.org/10.1016/B978-0-12-802814-8.00007-9
© 2015 Elsevier Inc. All rights reserved.

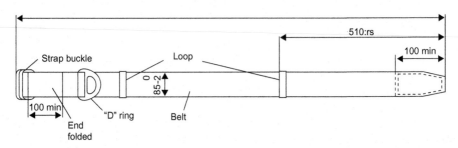

Figure 7.1 Safety belt with one "D" ring and without double – spiked buckle.

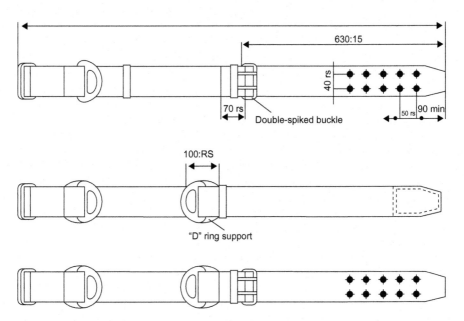

Figure 7.2–7.4 Other dimensions and nomenclature as of Figure 7.1.

7.1.2   Types B and C

The belt should be designed to comply with the requirements of standard when tested on any safety lanyard attachment point on the belt.

7.1.3   Type D

The harness should be designed to comply with the requirements of standard when tested on any safety lanyard attachment point on the harness. The harness should provide support for the body around the lower chest, over the shoulders, and around the thighs.

7.1.4 Type E

Rescue harnesses should be designed to comply with the requirements of the standard when tested on any of the rescue-line attachment points on the harness.

7.2 Use of belts and harnesses

- Type A: Used by linesmen, not intended in situations permitting a drop of more than 60 cm.
- Type B: Used in conjunction with lanyard where mobility is limited. Length of drop to within a max. of 60 cm.
- Type C: Chest harness is used with lanyard. Combined effects of anchorage and length of lanyard limits drop to max. of 2 m.
- Type D: Used in conjunction with lanyard where freedom of movement is required. Limit of drop to max. of 2 m.

7.3 Means of adjustment

7.3.1 Types A, B, C, and D

- Means of adjustment for length to fit the user.
- Self-lock adjusters securely lock on belt or strap, and do not present roughened surfaces and sharp edges (kNurled bars are permitted).
- Retain pole strap at the extremity of adjustment.
- "D" rings should be attached to waist or breeching belt.
- Connection and disconnection of the snap hook to be of single-handed operations.
- When "D" ring is fastened to the waist or breeching belt, the belt should pass through "D" ring and secured by reinforcement.
- If it is possible for snap hook to pass through "D" ring, it should be easy to remove it after any degree of rotation.

Notes:

1. If pole belt is used with breeching attachment, the droppers should be located on waist belt or on pole strap so that they can not be moved laterally more than 10 cm.
2. "D" ring or rings or other equivalent facility should be provided on the waist belt for the attachment of the safety lanyard and should be capable of accepting two such lanyards.
3. Where the "D" ring is secured to waist belt by a loop, the loop should be as strong as the belt and should pass through the "D" ring and capable of easy withdrawal after any degree of rotation.
4. The "D" ring or other equivalent facility provided for the attachment of lanyard should be located in the upper part of the harness so that the angle formed between the spine of suspended user and lanyard does not exceed 45°.

7.4 Materials for belts and harnesses

7.4.1 Webbing for belts and harnesses

- Quality: The yarn used for man-made fiber webbing should be of virgin, bright, high-tenacity polyamide, nylon, or polyester fibers having a uniform breaking strength or any other suitable man-made material. Natural fiber webbing should be made from flax or cotton yarn, well-spun, and evenly twisted, and with a uniform breaking force. It should be suitably processed at an appropriate stage of manufacture to render it rot-proof by suitable processing in accordance with BS 2087.

- Strength: Webbing used for primary straps should have a minimum breaking force of 9 kN (902 kgf) per 25 mm width. Webbing used for secondary straps should have a minimum breaking force of 4.5 kN (451 kgf) per 25 mm width. This ensures adequate thickness to prevent roping of the webbing.

- Leather: Best quality butt leather only should be used. The leather should be free from flaws that would reduce its strength and from soft and loose fibered leather. There should be no blind warbles in those parts where buckle holes are punched. The leather should not be treated as to obscure defects, and should not be treated with a nonpermeable surface finish. It should not be stained with compounds of iron. The pH value of an aqueous extract from the leather should not be below 3.3. The tensile strength of the leather should not be less than 20.7 N/mm² (211 kgf/cm²). For this test one sample is to be cut from each butt, the sample being cut parallel and adjacent to the backbone with one end of the sample within 50 mm of the root of the tail. The minimum thickness in the restricted portion of the test piece should be used for determining the area of cross section. The leather should not crack on the grain side when bent grain outwards through an angle of 180° around a mandrel of diameter 19 mm when tested in accordance with method 7 of BS 3144: 1987. This test should only be applied to parts of leather where there are no buckle holes or stitching.

7.4.2 Threads for sewing

- Color: Threads should be of a different color from the sewn material.
- For hand-sewing leather: The threads should preferably be of best quality flax or hemp, and should be of 6-cord No. 12 white flax or thread of comparable strength. The cords of strands used in making up the threads should be well twisted and thoroughly waxed. Alternatively, equivalent and suitable synthetic threads should be used.
- For machine-sewing leather: Stout linen or similar thread, well-waxed, and of suitable thickness, should preferably be used. Alternatively, equivalent and suitable threads of man-made fibers should be used.
- For natural fiber webbing: Best-quality linen with thread of size appropriate to the thickness of folded webbing to be sewn should preferably be used. Alternatively, equivalent and suitable thread of man-made fibers should be used.
- For man-made fiber webbing: Best-quality man-made fiber with thread appropriate to the thickness of folded webbing to be sewn should be used, and it should be compatible with the chemical resistance of the main fabric.

7.4.3 Rivets and washers

- For leather: Tinned solid copper rivets of best quality should be used with tinned copper washers.
- For webbing: Rivets and washers for leather, or other suitable rivets of comparable quality, may be used in addition to stitching. The riveted strength should not be less than the unriveted strength.

7.4.4 Metal components

- Materials: Metal components should be constructed either of stainless steel or of one of the metals specified in Table 7.1.

- Finishing: All metal components should be smoothly finished, free from any defects due to faulty material or manufacture, and those made other than of stainless steel should comply with the requirements specified in Table 7.1 that are appropriate to the finish used. When, in a multi-part component, more than one finish is present, each finish should be assessed separately.
- Hooks: Hooks should be of the self-closing variety and of such a type of design that pressure exerted accidentally on the tongue or latch will not permit disengagement; this should be achieved using a locking device to prevent the accidental opening of the tongue or latch. The springs of the hooks should preferably be so loaded that, when the hooks are closed,

Table 7.1 Coatings of metal components

	Coating	British Standard	Grade	Assessed for
Steel	Electroplated zinc[a]	1706	Zn_3	Appearance, adhesion, coating thickness
	Electroplated cadmium[a]	1706	Cd_3	Appearance, adhesion, coating thickness
	Hot dip galvanized	729	NA	Appearance, adhesion, coating thickness
	Sherardized	4921	Class 2	Coating thickness
	Electroplated nickel	1224	Medium application grade	Appearance, adhesion, coating thickness and, for nickel plus chromium
	Electroplated nickel and chromium	1224	Service condition No. 2	Corrosion resistance
Copper or brass	Electroplated nickel	1224	Medium application grade	Appearance, adhesion, coating thickness and, for nickel plus
	Electroplated nickel and chromium	1224	Service condition No. 2	Chromium corrosion resistance
Aluminum	Anodized	1615	AA10	Appearance, coating thickness, sealing
	Electro-Plated nickel	1224	Medium application grade	Appearance, adhesion, coating thickness and, for nickel plus
Aluminum alloy	Electroplated nickel and chromium	1224	Service condition No. 2	Chromium corrosion resistance
Screw threads	Any of the above coatings covered by BS 3382	3382	NA	Appearance, adhesion, plating thickness porosity (where appropriate)

[a]Denotes preferred finishes. Zinc coatings are more suitable for general use including use in industrial atmospheres, and cadmium is more suitable for use in marine environments.

the springs rest tightly in position and are free from any movement until pressure is applied to engage or release. Alternatively, hooks or main connectors should be so designed that, when intended to be affixed only to a mating fitting, they cannot be accidentally released from such a fitting.

• Coatings: Metal components, other than those constructed of stainless steel and threaded components, should be coated. Threaded components, other than those constructed of stainless steel, should be coated in accordance with BS 3382: Parts 1, 2, 3, 4, or 7 as appropriate to the metal and coating to be applied. When in a multipart component more than one coating is present, each coating should comply separately with the requirement of standards. If metal components have been coated with a plastics material, the plastics coating should be removed before performing corrosion tests in accordance with Table 7.1.
Note: Where there are dissimilar metals in contact, attention should be given to the possibility of galvanic action.

Notes:

1. Where tolerances are important it is recommended that the manufacturer should also specify a maximum coating thickness (which should not be less than twice the minimum thickness).
2. Attention is drawn to the clauses concerning hydrogen embrittlement in the relevant standards. It is necessary that the platter be informed of the specification of the steel to be plated.
3. The cleaning and preparation of parts before coating should be carried out to the highest standard (see BS 7773).

7.5   Lanyards

7.5.1   Design of safety lanyards

Safety lanyards are an essential component of general-purpose safety belts and harnesses, and it is essential that they are always attached to a "D" ring or other equivalent facility provided on the primary straps.

• Length: Safety lanyards should be designed so that their length cannot be extended beyond their intended length by any arrangement of their components. The effective length of the lanyard, including attachment devices and any shock absorber should not exceed 1.2 m for Type B belts and 2 m for Types C and D harnesses.
• Design of rope safety lanyards: Lanyards made from man-made fibers should have a spliced eye at each end for attaching to the belt or harness and for attachment of the safety snap hook or other means of attachment to the permanent structure. The splice for laid ropes or loops should consist of four full tucks using all the yarns in the strands and two tapered tucks. The length of the splicing tails emerging after the last tuck should be at least one rope diameter. Tails should be whipped to the rope with a sealing compound compatible with the rope fiber and protected with a rubber or plastics sleeve. Eyes should be formed round a plastics or metal thimble of appropriate size and strength. Eight-strand (plaited) polyamide or nylon ropes should be spliced by making one double-strand full tuck and four single-strand full tucks. The tails emerging after the last tuck should be at least two rope diameters in length and should be whipped to the ropes as described above.

- Design of chain safety lanyards: Lanyards should consist of an arrangement of length(s) of chain, terminal, and intermediate links such that one end can be secured with a self-closing hook to the harness or belt and the other end to an anchorage point.
- Design of webbing lanyards: Lanyards should comprise the webbing and any thimbles, fairing or protective piece, hooks, and other fittings necessary for the lanyard to comply with the requirements of standards whether it is permanently attached to a safety belt/harness or is supplied as an accessory. The webbing should be secured with a compatible synthetic thread. It should be noted that the shock-absorbing properties of the lanyard will be reduced if it is formed of two or more lengths of webbing sewn together lengthwise.

7.5.2 Materials for safety lanyards

A lanyard is protective equipment against falls consisting of rope or webbing with an attachment device (eye, snap hook) at its ends that connects a belt (safety belt, safety harness, rescue harness) to an anchorage point (see Figure 7.5).

Warning: In choosing a safety lanyard it is important to keep in mind that if there is a requirement to protect a user against a drop of between 60 cm and 2 m, chain and natural fiber rope are unsuitable unless adequate shock absorbing properties are built into the harness or lanyard. These materials should only be used when overriding considerations render unsuitable the use of polyamide, or nylon, and polyester ropes. Safety lanyards should never be knotted. Polyamide, or nylon, degrades in direct contact with acids. Polyester degrades in direct contact with alkalis and may swell in contact with certain chlorinated solvents. Attacks by concentrated phenols are severe and should be avoided.

7.5.3 Materials for rope lanyards

Ropes should have a minimum diameter of 12 mm and a minimum breaking force of 29.4 kN (3000 kgf). Ropes should be made of virgin, bright, high-tenacity continuous polyamide, or nylon, or polyester filament and should comply with the requirements of BS EN 696, 697, 699, 700, 701 (1995). For the purposes of standards, Table 7.2 covers 8-strand (plaited) polyamide, or nylon, filament ropes.

Note: The reference or size number of an 8-strand plaited rope corresponds to the circumference in inches of a 3-strand rope of the same fiber having an equivalent mass per 100 m and breaking strength. This number may be derived from the 3-strand rope diameter in millimeters by dividing by 7.

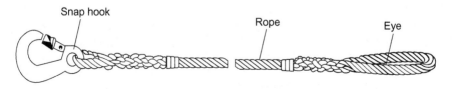

Figure 7.5 Example of lanyard (illustrating a rope lanyard with snap hook and eye).

Table 7.2 Requirements for 8-strand plaited polyamide (nylon) and polyester filament ropes

Size or Reference Number	Polyamide (Nylon) ropes						Polyester ropes	
	Nominal Mass Per 100 m	Minimum Breaking Load	Minimum Breaking Strength	Maximum Length of 20 Plait Pitches	Nominal Mass per 100 m	Minimum Breaking Load	Minimum Breaking Strength	Maximum Length of 20 Plait Pitches
	kg	kg	kN	m	kg	t	kN	m
1	4.2	1.4	14	0.30	5.1	1.0	9.8	0.30
1½	9.4	3.0	29	0.45	11.6	2.3	23	0.45
2	16.6	5.3	52	0.60	20.5	4.1	40	0.60
2½	26.0	7.3	81	0.75	31.9	6.3	62	0.75
3	27.3	12.0	118	0.90	46.0	9.1	89	0.90
3½	51.0	15.8	155	1.05	62.8	12.2	120	1.05
4	66.4	20.0	196	1.20	81.9	15.7	154	1.20
5	104	30.0	294	1.50	128	23.9	234	1.50
6	150	42.0	412	1.80	185	33.5	329	1.80
7	203	56.0	549	2.10	251	44.7	438	2.10
8	265	72	706	2.40	327	57.0	559	2.40
9	336	90	883	2.70	414	72	706	2.70
10	415	110	1080	3.00	511	88	863	3.00
11	501	131	1285	3.30	619	106	1040	3.30
12	597	154	1510	3.60	736	125	1225	3.60
13	700	180	1765	3.90	860	145	1420	3.90
14	810	210	2060	4.20	1000	165	1620	4.20
15	930	240	2355	4.50	1150	190	1865	4.50
16	1060	270	2650	4.80	1310	215	2110	4.80
17	1200	305	2990	5.10	1480	245	2400	5.10
18	1340	340	3335	5.40	1660	270	2650	5.40

Note: The reference or size number of an 8-strand plaited rope corresponds to the circumference in inches of a 3-strand rope of the same fiber having an equivalent mass per 100 m and breaking strength. This number may be derived from the 3-strand rope diameter in millimeters by dividing by 7.

7.5.4   Strand (plain or hawser laid) ropes

• Direction of lay: For 3-strand (plain or hawser laid) ropes the direction of lay, should be "Z" or right-hand lay (see Figure 7.6).
• Length of lay: The length of 10 lays should be as specified, when tested in accordance with A.3 of BS 4928: 1985.

Notes:

1. The length of one lay is illustrated in Figure 7.6.
2. 3-strand ropes made from polyamide (nylon) and polyester may be heat treated to set the lay and obtain dimensional stability.

7.5.5   8-Strand (plaited) ropes

• Construction and twist of strands: 8-strand plaited ropes should consist of four pairs of strands, each alternative pair consisting of two "S" twist strands and two "Z" twist strands, respectively (see Figure 7.7).
• Length of plait pitches: The length of 20 plait pitches should be as specified, when tested in accordance with BS 4928: 1985.

Note: The length of one plait pitch is illustrated in Figure 7.7. Table 7.3 covers 8-strand (plaited) polyamide, or nylon, filament ropes having referenced and size Nos. 1½ and 1¾.

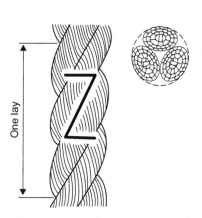
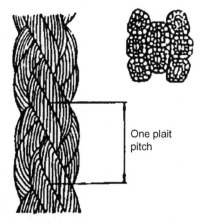

Figure 7.6 3 – Strand plain or hawser laid rope.   **Figure 7.7** 8-Strand plaited rope.

Table 7.3 **8-strand (plaited) polyamide, or nylon, filament ropes having referenced and size numbers**

Reference or Size no.	Nominal mass Per 100 m	Min. breaking force		Maximum length of 10 stitches
	kg	kN	kgf	mm
1½	9.37	29.4	3000	420
1¾	12.80	40.2	4100	490

Figure 7.8 shows an example of a self-locking safety anchorage with ladder. Figure 7.9 illustrates an example of a self-locking safety anchorage on a lanyard. Figure 7.10 shows an example of a self-locking safety anchorage attached directly to the belt. Figure 7.11 illustrates an example of a self-locking safety anchorage with a lanyard.

7.5.6 Double-braided ropes

Double-braided ropes should comprise a braided core covered by a braided sheath. Half the strands in both the core and sheath should have "S" twist and half should have "Z" twist. With the exception of the requirements of standards each strand of the sheath should be of the same construction and each strand of the core should be of the same construction.

Note: The constructions of the sheath strands and the core strands need not be the same.

Figure 7.8 Example of a self-locking safety anchorage with ladder.

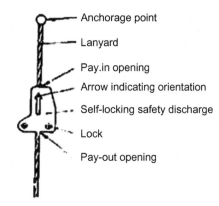

Figure 7.9 Example of a self-locking safety anchorage on a lanyard.

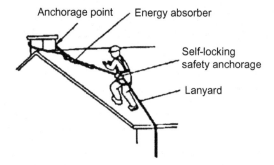

Figure 7.10 Example of a self-locking safety anchorage attached directly to the belt.

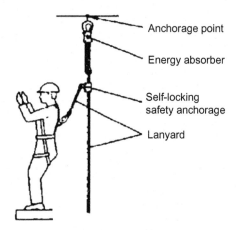

Figure 7.11 Example of a self-locking safety anchorage with a lanyard.

7.5.7   Materials for chain lanyards

The chain should comply with the requirements of the 6.3 mm chain size given in BS 4942: Part 3. The terminal egg links and intermediate links should comply with the requirements given in BS 2902 for those links used with ¼ in chain. If special intermediate links are used, the dimensions should be agreed upon by the buyer and the manufacturer, and the links should comply with the appropriate performance requirements of BS 2902.

The materials used for terminal egg links and intermediate links in the lanyard should be compatible with the chain for heat treatment purposes. After fitting any terminal egg links and/or intermediate links, the lanyard should be adequately heat treated with a final process of hardening and tempering. This requirement is waived if the chain itself has been hardened and tempered previously and if link heaters are used to treat the additional links.

- Material for webbing lanyards: The yarn used for webbing lanyards should be of virgin, bright, high-tenacity polyamide, or nylon, or polyester fibers having a uniform breaking strength. Webbing lanyards should be tested in accordance with BS 1397 and should have a minimum breaking force of 20 kN (2040 kgf) and a maximum width of 50 mm.

7.6   Manufacture

7.6.1   Webbing safety belts, harnesses, and lanyards

The attachment of metal load-bearing components and the making of splices and joints in the material of the belt or harness should be such that the finished assembly will comply with the requirements of standards. All machine sewing should be carried out with even tension on a suitable lockstitch machine and securely finished off by back sewing for at least 13 mm, except where sewn by an automatic lock stitching machine when the first and last stitches should be sewn in such a way as not to provide a natural starting point for a break in the stitching. Sewing should not be carried out within 2.5 mm of any edge of the webbing. This does not exclude over-sewing of sealed ends. Heat-sealed edges should not be over-sewn unless the over-sewing stitching is protected.

Where a prong buckle is used, the belt should be provided with effective reinforcements that will prevent the disengagement of the prong if the webbing fails. Metal load-bearing components should be designed or protected to prevent abrasion of the webbing passing round them. All securing buckles (i.e., buckles other than those used primarily for adjustment of fit) should be designed so that either they can only be assembled in the correct manner or, if they are capable of assembly in more than one way, each method of assembly should comply with the requirements of the standards.

7.6.2   Leather safety belts and harnesses

- Thickness and width: Leather safety belts and harnesses should be at least 4.75 mm in thickness and load-bearing straps should not be less than 50 mm wide.
- Quality: They should be free from cuts or other flaws due to manufacturing processes and should be well and smoothly finished at the edges.
- Position of holes: Strips cut, for safety belts and harnesses, in which holes for buckles are punched should be placed so that the shoulder ends are either used to receive the buckles or attached to other portions of the belt or harness, and so that the holes are punched in the butt portion.
- Splices: Splices should be used only where the design of a pole belt renders their use unavoidable and they should, where possible, be placed where they are reinforced by the body belt.

7.6.3   Beveling of ends

The ends of all overlaps, joints, or splices, or straps of reinforcements, where attached to the belts and harnesses, should be satisfactorily beveled (skived) and fitted to ensure sound joints and to avoid abrupt terminal points.

7.6.4   Punching

At the end of a strap to which a buckle is fitted, a crew punch should be used to make the hole through which the tongue is to pass. At the other end, the holes for the buckle tongue should be made with an oval punch.

7.6.5   Buckles

Buckles should be inserted so that when the leather is bent around the shoulder of the buckle, it fits over a core of not less than 13 mm in diameter. Beveled leather reinforcements should be passed around the shoulders of buckles before attaching the belt or strap.

7.6.6   Sewing

All hand-sewing of leather should be carried out by the "double-hand" method, with not less than six and no more than seven stitches to 25 mm. Sewing should be continuous and back-sewn with at least two drop-stitches before fastening off the threads. For leather, if not hand-sewn, all sewing should be carried out on a heavy lockstitch machine and as stated.

Sewing should not be carried nearer than 6 mm to the edges of the leather on strips 38 mm wide or more, and not nearer than 5 mm on strips less than 38 mm wide. In no case should stitching be placed at right angles to the length of the belt or strap. Belts or straps 50 mm wide or more should have three rows of sewing; those less than 50 mm wide should have two rows of sewing.

7.7   Performance requirements

7.7.1   Tests

- Self-locking safety anchorages: A self-locking safety anchorage is an item of protective equipment consisting of an anchorage line (e.g., rigid or flexible) and a movable arrestor attached to it, to which a safety belt or safety harness can be fixed using a connector. An arrestor is a device that is fitted to an anchorage line to move in the direction specified, is designed to permit the attachment of a connector, responds to loading, and, thus, keeps the person to be protected attached to the anchorage. The connector may be part of the arrestor. An anchorage line is a device that permits the arrestor to move in the direction specified. An anchorage line may be fitted with a bracket to fix it to a ladder or to a structure. A connector (e.g., rope, strap, fittings, chains) is the part of the protective equipment that connects the arrestor to the "D" rings and has the safety belt or safety harness. Points are a device permitting the arrestor to be moved from one anchorage line to another. An attachment/detachment point is the point on the anchorage line where the arrestor is fitted or detached.

When type "A" belts are tested the provision given in BS 1397 (1979) should be followed. When Types B, C, D, and E belts and harnesses are tested the provision given in BS EN 354, 355, 358, 361, 362, 363, 364, 365 (1993) should be followed.

7.7.2   Strength and tests

- Hooks, safety lanyard, and pole-belt attachment fittings: Any hooks should be of the self-closing variety of such a type or design that pressure exerted accidentally on the tongue or latch will not permit disengagement. This should be achieved using a locking arrangement to prevent the accidental opening of the tongue or latch. The springs of the hooks should be free from any movement when the hooks are closed, and engaged or released when pressure is applied on them. Each attachment or components of hooks, safety lanyard should be tested to 11 kN (1120 kgf). The application of the load should reproduce as closely as practicable the direction in which the component is stressed in service. After testing the component should be free from flaw, defect, or distortion. When tested to destruction components should have a minimum breaking force of 22 kN. In the case of hooks, tests should be carried out with latches closed by the self-closing arrangement but with the securing device in the open position.
- Other load-bearing components: Each of these components should be tested to 50% of the ultimate tensile strength but with a maximum of 11 Kn (1120 kgf) without showing signs of flaw, defect or permanent distortion. The application of the load should reproduce as closely as practicable the manner in which the component is stressed in service.
- Additional provisions for castings: Where castings are used for steel load-bearing components they should be made by the investment casting process and should be in accordance with BS 3146. Where castings are used for aluminum load-bearing components they should be made by gravity die-casting.

Note: If safety belts, harnesses, and lanyards are to be used in potentially flammable or explosive atmospheres, buyers should specify that no metal components should be made of aluminum, magnesium or titanium; neither should any alloy containing one or more of these constituents be used unless both the total content of these three constituents does not exceed 15% by mass, and the content of magnesium and titanium together does not exceed 6% by mass. These limitations have been imposed to avoid the hazards of spark due to friction between rusted steel or iron and the metals described.

7.7.3 Method of Use

The following points are applicable to any work:

Inspection of the appliance before use.
Correct fitting and adjustment.
Selection and inspection of suitable anchorage points. A suitable anchorage point is one that is strong enough and allows free movement of the attachment and is as high as possible to reduce the amount of fall. It should also be as nearly vertical as possible above the place of work to reduce the liability to swing.
If the appliance has been used to arrest an accidental fall it should be withdrawn from use. It is strongly recommended that consideration be given to it being destroyed.
Where work is such that the position of anchorages cannot be used above the point of work the use of double lanyards or self-locking safety devices, or both, is recommended. Where double lanyards are used it is essential that one safety lanyard always be attached to an anchorage.
· Instruction for Use: Clear instructions for fitting, adjustment, and use should be supplied with each safety belt and harness. The instructions should include a warning in general terms and also susceptibility of material to any kind of chemicals.

7.8   Inspection

The user should make a visual inspection at least daily before using the appliance to ensure that the appliance is in a serviceable condition. A record should be kept of all examinations. Each belt should be marked with a serial number for identification purposes. The appliance should be withdrawn from use if found to be damaged and it should not be returned to service until the necessary repairs have been effected.

7.8.1 Frequency of examination and inspection

Users should establish their own routine inspections in accordance with Appendices A, B, and C, and the following procedure is also recommended:

All rope used as a component of, or in conjunction with, safety belts or harnesses should be examined immediately before being taken into use.
When rope has been brought into use the user should also inspect it every week, or more frequently if used under adverse conditions or subjected to very hard wear.
All rope used as a component of, or in conjunction with, safety belts or a competent person should also examine harnesses every 3 months, or more frequently if used under adverse conditions or subjected to very hard wear.

7.9 Markings

7.9.1 Markings on belts and harnesses

Safety belts and harnesses should be clearly and indelibly marked or permanently labeled by any suitable method not having a harmful effect on materials with the following information:

Number of national standard.
Name, trademark, or other means of identification of the manufacturer.
Year of manufacture.
The words "maximum safe drop 2 m (or 60 cm)" as appropriate, together with details of recommended safety lanyards for use with the safety belt or harness.
Type of belt or harness, i.e., A, B, C, D, or E.
Manufacturer's serial number.

7.9.2 Markings on lanyards

Lanyards not permanently attached to belts or harnesses should be clearly and indelibly marked or permanently labeled by any suitable method not having a harmful effect on materials with the following information:

The manufacturer's model number and the type of belt or harness i.e., A. B, C, D, or E, with which the lanyard is designed to be used. Chain lanyards should be additionally marked in a manner to enable the manufacturer to identify the batch from which the chain and any intermediate link, terminal link, and special intermediate link were selected.

7.9.3 Labels attached to lanyards

Lanyards should be supplied with an attached label bearing the words: "For maximum safety, attach the free end to a point as high as possible above and avoid looping the lanyard around small joists and angles with narrow edges."

Each belt and harness should be supplied wrapped, but not sealed, in moisture-proof material and should bear in a clearly visible manner appropriate instructions for storage in accordance with standards.

7.10 Storage, examination, and maintenance of safety belts and harnesses

- Records: A card or history sheet should be kept for each belt and harness and particulars of all examinations and other items of interest recorded. Each belt and harness should be marked with a serial number for identification purposes.
- Storage: Belts and harnesses should be stored in a cool dry place and not subjected to direct sunlight.

- Examination: To provide the maximum degree of safety for users, it is essential that all belts and harnesses are thoroughly examined periodically by a competent person, and any showing any defect should be withdrawn from service immediately.

During the examination particular attention should be directed to the following points:

- Webbing and leather: Examine for cuts, cracks, tears, or abrasions and undue stretching and damage due to deterioration contract with heat, acids, or other corrosives.
- Snap hooks: Examine for damaged or distorted hooks or faulty springs.
- Buckles: The tongues should be carefully examined where fitted to the shoulder of buckles; inspect for open or distorted rollers.
- Sewing: Examine for broken, cut, or worn threads.
- Ropes and chains: Examine for any damage or signs of wear, and in, the case of ropes, interstrand wears, unraveling extension, and fusion.

7.11 Maintenance of equipment

7.11.1 Examination of man-made fiber rope lanyards

It is very important that rope used in conjunction with a safety belt or harness is periodically and carefully examined by a competent person. The following are the main causes of weakness in ropes:

External wear due to dragging over rough surfaces causes a general reduction of the cross section of the strands.
Local abrasion as distinct from general wear may be caused by the passage of the rope over sharp edges while under tension and may cause serious loss of strength.
Cuts, contusions, etc., that cause internal as well as external damage.
Internal wear caused by repeated flexing of the rope, particularly when wet.
Heavy loading may result in permanent stretching so that the extension available in an emergency is reduced.

Note: The extension should not exceed 10% for polyamide and polyester fiber ropes.

Chemical attack may be of many forms. All rope is susceptible to attack by acids even in fine spray or mist as may be given off in some industrial processes, and alkalis may be harmful if concentrated.
Strong sunlight may cause degradation, although only after prolonged exposure.
Heat may, in extreme cases, cause charring, singeing of fusing. Any singes would obviously result in rejection of the rope, but the rope may be damaged by heat without any such obvious sign. Never dry a rope in front of a fire or store it near a source of heat.

7.11.2 Maintenance of belts and harnesses

- Maintenance of leather belts and harnesses: Leather belts and harnesses should be cleaned and dressed and examined regularly. The frequency will depend upon the conditions under which the equipment is used, but in any even should not be less often than every 3 months.

- Maintenance of man-made fiber webbing belts or harnesses: Belts and harnesses made of man-made fiber webbing should be cleaned and examined regularly. The frequency will depend upon the conditions under which the equipment is used, but in any event should not be less than every 3 months.

7.11.3 Examination of man-made fiber webbing

The following are the main causes of weakness in man-made fiber webbings and the signs by which they may be recognized. If, after examination, there is any doubt as to the safety of the belt or harness it should be withdrawn from service:

- General external wear: External wear due to contact with rough surfaces causes filamentation, this will be shown by a fluffiness of the surface. Local abrasion as distinct from general wear may be caused by the passage of the webbing over sharp edges or protrusions while under tension and may cause serious loss of strength.
- Cuts, contusions, etc. Cuts on the edges of the webbing exceeding 6 mm in length, or holes cut or burnt in the webbing are potentially dangerous and should lead to rejection.
- Chemical attack: Oil, grease, creosote, or paints stains are harmless, but other forms of chemical attack of a sufficient degree may be indicated by local weakening of softening of the webbing so that the surface fibers can be plucked or rubbed off, as a powder in extreme cases.
- Heat may, in extreme cases, cause fusing: Any signs, other than the heat-sealing of webbing edges carried out during the manufacturing processes, should obviously lead to rejection.

7.12 Recommendations for the selection and use of appropriate appliances

- Selection: It is strongly recommended that where a choice of appliance is possible a harness is used as opposed to a belt.
- Method of use.
  Inspection of the appliance before use.
  Correct fitting and adjustment.
  Selection and inspection of suitable anchorage points.
  After use, the appliance should be stored.
- Inspection: The user should make a visual inspection at least daily before using the appliance to ensure that the appliance is in a serviceable condition. No less often than at quarterly intervals the appliance should be examined by a competent person other than the user. A record should be kept of this examination. Each belt should be marked with a serial number for identification purposes.
- Storage: While on site and when not being worn appliances should be stored in accordance with A. 2. Dry the appliances naturally away from an open fire or other source of heat.

Safety portable ladders

8

This chapter describes the minimum requirements for the construction, testing, care, and use of the industrial types of portable aluminum and wood ladders in order to ensure safety under normal conditions of use.

Accidents caused by falls from ladders happen frequently, and many of them have occurred from misuse, especially overloading. Safe handling, storage, transport, maintenance, and inspections are briefly discussed in this chapter, and it is emphasized that these guidelines should be thoroughly observed when ladders are handled. For wood ladders, the risks associated with different types of tests are significant, therefore, it is strongly recommended that these tests are conducted by manufacturers.

This chapter specifies the minimum requirements for the materials, dimensions, and workmanship for aluminum and wood ladders. It also describes rules for construction, testing, and care of ladders under normal conditions of use.

8.1 Types and classes

- Types.

  In this chapter ladders fall into the following categories:

  Aluminum ladders.
  Wooden ladders.
  Specific types of ladders.

  Classes: According to general condition and frequency of use portable ladders are in three different classes:

- Class 1: Industrial. For heavy duty where relatively high frequency and onerous conditions of use, carriage and storage occur. Suitable for industrial purposes.
- Class 2: Light trades. For medium duty where relatively low frequency and reasonably good conditions of use, storage, and carriage occur. Suitable for light trades.
- Class 3: Domestic. For light duty where frequency of use is low and good storage and carriage conditions occur. Suitable for domestic and household purposes.

Notes:

1. Single-section and extension ladders of Class (2) are designed for ease of handling if the load does not exceed 105 kg. They are suitable for light trades and domestic applications and therefore are not recommended for use in construction or other heavy industry.
2. Class 3 as stated above is for domestic use.

Personnel Protection and Safety Equipment for the Oil and Gas Industries.
DOI: http://dx.doi.org/10.1016/B978-0-12-802814-8.00008-0
© 2015 Elsevier Inc. All rights reserved.

8.2 Materials (Industrial)

8.2.1 Aluminum ladders

The materials from which the component parts are made should be in accordance with relevant standards. Guide brackets and fixed and latching hooks should be made from the materials given in either (a), (b) or (c) as follows:

Aluminum alloys
Mild steel
Whiteheart malleable cast iron complying with BS 6681
- Hinges: Hinges should be made from the materials given in either (a), (b), or (c) as follows:
Aluminum alloys
Forged steel or steel strip
Whiteheart malleable cast iron complying with BS 6681

Feet of stiles and capping for upper ends of stiles or ends of treads should be made from the materials given in either (a), (b), or (c) as follows:

Plastics (see note below)
Rubber (see note below)
Wood
Decking of lightweight stagings should be made from the materials given in either (a), (b), or (c) as follows:
Aluminum alloys
Plastics (see note below)
Wood, as specified
Other components should be made from aluminum alloys as given in (a) to (f) as follows:
- Drawn tube: designation 6063 (HT9) and 6082 (HT30) of BS 1471, or equivalent
- Extruded sections: designation 6063 (HE9), 6082 (HE30), 6063A, 1200 (E1C), and 6061 (HE20) of BS 1474 or equivalent
- Longitudinally welded tube: designation 5251 (NJ4) of BS 4300: Part 1 or equivalent
- Castings: LM6 and LM25 of BS 1490 or equivalent
- Components formed from sheet and strip: designation 1200 (S1C), 3103 (NS3),5154A (NS5), 5251 (NS4), and 6082 (HS30) of BS 1470, or equivalent
- Forgings: 6082 (HF30) of BS 1472, or equivalent

Note: Plastic materials and rubber should be selected with regard to the stresses to which they may be subjected and their resistance to environmental deterioration, especially that due to ultraviolet light.

8.2.2 Wood ladders

- Species of wood: The wood used should be selected by the manufacturer. All like components, e.g., stiles, rungs, etc., in any single piece should be of the same species of wood, as far as can be achieved by visual examination.
- Stiles for builder pole ladders: Stiles should be made from European whitewood (picea abies, Abies alba) or European redwood (Pinus sylvestris), or equivalent.

- Stiles for other ladders, trestles, and stagings: Stiles for other ladders should be made from European redwood, Douglas fir, imported Sitka spruce, Eastern Canadian spruce, European whitewood, western hemlock or hembal, or equivalent. Laminated stiles should be permitted provided these are made from the above woods and glued with phenolic or resorcinol resin adhesives to comply with the weather-proof and boil-proof (WBP) resistance requirements of BS 1204: Part 1.
- Rungs for builder pole ladders, single-section, and extension ladders: Rungs should be made from European oak, American white Oak, European ash, American white ash, hornbeam, yellow birch, hickory, robinia, keruing, pau marfim, ramin, or equivalent. Treads should be made from any of the woods with the addition of Parana pine, ramin and keruing.
- Cross-bearers for stagings and trestles: Cross-bearers should be made from any of the woods.
- Decking for stagings: Deckings should be made from any of the woods. Plywood of moderately durable rating or better with WBP bending, aluminum, or other man-made material should be permitted provided that the deadweight of the staging does not exceed that for a wood slatted staging and that the material used for decking is capable of supporting a load of 90 kg when supported at 380 mm centers.
- Quality of wood: Sapwood should be permitted in all species of wood except oak. All wood should be free from apparent damage, fungal decay, and insect attack, except for an occasional ambrosia beetle hole. Kiln dried wood should be free from case-hardening and honeycombing. Pieces abnormally light in weight should be excluded, as should wood that is abnormally heavy; like components should be reasonably matched. Edges of sawn faces should be so finished that there is no rough surface that might constitute a hazard for the hands of the user.
  Note: Ambrosia beetle is a forest beetle that cannot live in wood other than freshly felled trees; the hole is circular and not more than 2 mm diameter and the lining of the hole is stained a blue color.
- Rate of growth: For stiles, treads, rails, and bracings the number of growth rings per 25 mm should not be less than 6. In addition, in the case of oak and ash the number of growth rings per 25 mm should not exceed 16.
- Knots: Rungs of builder pole ladders, single-section ladders, and extension ladders should be free from knots.
- Slope of grain: The combined slope of grain should not be steeper than 1 in 10, when determined in accordance with the standard methods.
- Stiles of builder pole ladders
- Knots: The total diameter of the knots within 30 mm of each side of the center of a rung hole should not exceed, when measured on the convex side of the stiles, one-quarter of the depth of the flat face of the stiles at this point (see Figure 8.1). The total diameter of knots allowed in the remaining region between the 30 mm bands should not be more than the depth of the flat face of the stile measured at a point midway between adjacent rung holes. No knot should exceed 12 mm diameter, in its greater axis, in the top 4 m of a stile and 20 mm between that point and the base.
- Surface checks: Surface checks should be permitted provided that the checks do not deviate more than 1 in 10 to the axis of the stile and should not be more than 1.5 mm in width or 200 mm in length at the time of manufacture.
- Spiral grain: Spiral grain having a deviation not steeper than 1 in 10 should be permitted. Tests should be carried out on the convex face of the stile.
- Stiles of single-section ladders, extension ladders and lightweight stagings.
- Knots: Sound knots not over 7 mm in class 1 on their greater axis should be permitted provided they do not occur in the outer quarters of the face.

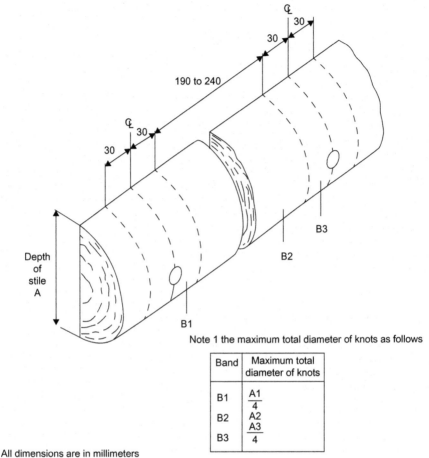

Note 1 the maximum total diameter of knots as follows

Band	Maximum total diameter of knots
B1	$\dfrac{A1}{4}$
B2	$\dfrac{A2}{A3}$
B3	$\dfrac{A3}{4}$

All dimensions are in millimeters

Note 2 A1 A2 and A3 are the values of A at the centers of the bands B1 B2 and B3

Figure 8.1 Knots in stiles of builders pole ladders.

- Surface checks: Surface checks not exceeding 130 mm long or 1.5 mm wide should be permitted provided they do not run out to the edge of the face in which they occur.
- Selection of stiles (single-section and extension ladders): Each stile should be inspected visually for quality and finish before assembly.
- Resin pockets: Resin pockets should be permitted if their length is not more than 1½ times the width of the face in which they occur and that the width does not exceed 3 mm and the depth does not extend to the opposite face.

8.2.3 All component parts of shelf ladders, step ladders, trestles, and stagings (other than stiles of stagings)

- Knots: Knots other than staging slats. Sound knots not larger than one-sixth of the width of the face in which they appear should be allowed providing the edge of the knot is not closer to the edge of the component than the diameter of the knot. No knot should be nearer than 25 mm to any joint.

- Staging slats: Sound knots not larger than 12 mm in diameter should be allowed providing they are at least 50 mm apart. Edge knots should be allowed providing they extend no further than 6 mm from the edge. Splayed knots should not be allowed.
- Surface checks: Surface checks not exceeding 1.5 mm wide and 60 mm long should be permissible, providing the check does not deviate by more than 1 in 10 and does not run out to the edge of the component.
- Slope of grain: Combined slope of grain on components other than treads, when determined in accordance with the method given in Appendix A should not exceed 1 in 8 for Douglas fir and 1 in 10 for other species. On treads, the grain starting from the top face at one end should not run out on the underside before entering the housing at the other end.

8.2.4 Moisture content

The moisture content at the time of manufacture should be in the range of 16% to 22% (see BS 4471).

8.3 Other components and material

8.3.1 Tie rods

Tie rods should be made of mild steel and be as specified in either (a) or (b):
Plain tie rods should be no less than 3.9 mm in diameter, except in the case of folding trestles, when the rods should not be less than 5.9 mm in diameter.
The ends should be passed through mild steel washers of 15 mm minimum outside diameter and no less than 1.4 mm thick, and should be securely riveted over them and smoothly finished. Threaded tie rods should not be less than 2.9 mm diameter with 4 mm rolled threads and buttons, except in the case of tie rods for folding trestles, which should have 6 mm rolled threads on rods no less than 4.9 mm diameter.

- Ladder buttons: Ladder buttons should be of malleable cast iron or pressed steel.
- Reinforcing wire (for rung reinforcement): Rung reinforcement is optional but when used the reinforcing wire should consist of two galvanized mild steel wires, of no less than 3.5 mm diameter twisted together. Wires should be properly tensioned and anchored into the stiles or retained in position by buttons or other equally efficient tensioning devices.
- Reinforcing wire for ladder stiles: Reinforcing wire for ladder stiles should consist of a seven wire strand, of mild steel wire having a characteristic strength of 425 N/mm² minimum, and of the appropriate overall diameter as given in Tables 8.3, 8.5, and 8.6 of any other suitable reinforcement of equivalent strength. When such reinforcement is specified in these tables it should be fitted under tension into grooves in the stiles and anchored securely so as to pretension the stiles. The material should pass through the stiles beyond the end rungs and be no less than 100 mm from the stile ends and should be anchored so that the stiles are not cut into when under load.
- Nails: Nails for securing treads of steps and the tops of swing-back steps should be either wire, twisted, or screw nails nominally 60 mm long and 2.5 mm diameter, or annular ring shank nails nominally 60 mm long with a 2 mm diameter shank.
- Screws: Screws should be of steel and should comply with the strength requirements of BS 1210.

8.3.2 Fittings for extension ladders

All metal fittings should be of steel, wrought iron, malleable cast iron, or aluminum of suitable strength, and should be well finished and securely fitted. On rope-operated ladders the pulley wheel may be of cast iron or nylon, or other material of adequate strength and durability to provide a factor of safety of no less than eight times the weight of the extending section or sections.

All bearing and rubbing surfaces of the hooks should be finished smooth and should be free from sharp edges liable to cause indentation of the stiles or rungs.

- Fixed and latching hooks: Fixed and latching hooks should be such that they bear evenly over a length of no less than 12 mm along each end of the engaged rung.
  Note: Where aluminum alloy fixed hooks or latching hooks are used in conjunction with rungs or treads treated with a copper-containing preservative, the intermittent contact between the hooks and the preservative is not considered to be harmful to the aluminum.
- Guide brackets: The brackets should enclose one side and 75% of the back of the stile to which they are fixed and provide a bearing on the front of the sliding stile of no less than 75% of the stile width. The brackets should be properly formed with no tool marks that could affect their strength or performance and should have all sharp corners removed. The internal radius of any bend should not be less than the thickness of the material.
- When tested in accordance with relevant standards a bracket should show no sign of distortion or permanent deflection and the bracket fixings should not have become loosened or be damaged. Each bracket should be securely fitted by at least one bolt and one countersunk wood screw. A pair of guide brackets should be fitted to each section other than the top section.
- Guide groove: Where a guide groove is employed and it is necessary for a cut out to be formed adjacent to the groove, the openings so formed by the cut away portion, to facilitate the removal of an upper section from the lower section, should be protected by a mild steel plate, designed to protect the sides and edges of the openings.
- Latching device: The design of the latching device in ladders of the rope-operated type should be such that in the event of the rope breaking or being released accidentally, it engages and arrests the descent of the ladder section.

8.4 Specific types of ladders

The following types of ladders are discussed in this chapter:

Single-section and extension (aluminum)
Single-section and extension (wood)
Shelf ladders (aluminum)
Shelf ladders (wood)
Swing-back steps (aluminum)
Swing-back steps (wood)
Backed steps (aluminum)
Backed steps (wood)
Folding trestles (aluminum)
Folding trestles (wood)
Lightweight stagings (aluminum)
Lightweight stagings (wood)

8.4.1   Single-section and extension (aluminum)

- Lengths: The lengths of single-section ladders and extension ladders when fully extended should not exceed the 17 m in class 1.
- Feet: The lower end of each stile should be closed by blocks of hardwood, plastics or rubber or fitted with an articulated foot soled with slip resistant material. The feet should project to form a wearing surface and should be securely fixed but easily removable for renewal. The length of projection of the blocks or fittings should be taken into account when determining the overall length of the ladder.
- Rungs: Each rung should have a textured surface on the working face to reduce slipping. All rungs should be securely fixed so as not to rotate in their supports and if the ends of the rungs protrude through the stiles they should be smoothly finished so as not to injure a user's hands.
- Spacing of rungs: Rungs should be uniformly spaced at 250 mm to 300 mm centers. The distance from the ends of the stiles to the nearest rungs center should be 125 mm to 300 mm.
- Fittings for extension ladders: Fittings for extension ladders should be such that the width of parts bearing on the rungs is no less than that given in Table 8.1. The shape of hooks should be such as to require upward movement of the upper section to disengage them from the rungs. The fittings should have no sharp edges liable to cause indentation of the stiles or rungs. Latching devices, if fitted, should not be dependent for their operation on springs and, if of the rope-operated type, they should be such that in the event of the rope breaking or being released they engage automatically and prevent uncontrolled closure of the ladder, where the latches are acting as a pair, they should be connected to ensure movement in unison. The latches or fixed hooks should bear equally on the rungs. Guide brackets should be formed with no tool marks that would affect their strength or performance. Any sharp corners should be removed.
- Ropes: The strength of rope attachments should be such as to provide a factor of safety of no less than 8 times the mass of the extension section or sections. Ropes should be hemp sash cord, made from yarn, complying with Table 1 of BS 6125: 1982 or other material of equivalent strength. The nominal diameter of the ropes and breaking load should not be less than the appropriate values given in Table 8.2.
- Overlap of sections of extension ladders: When fully extended, the effective overlap between adjacent sections of the ladder should not be less than the following:

  1.5 t for closed lengths up to and including 5 m

  2.5 t for closed lengths over 5 m and up to and including 6 m

  3.5 t for closed lengths over 6 m (see also Figure 8.3)

where t is the spacing of rungs (in mm).

Table 8.1   Fittings for extending ladders

Major horizontal rung dimension or diameter	Minimum width of fitting bearing on surface of rungs
mm	mm
Up to and including 31 Over 31 and under 39 39 and greater	12.0 8.5 6.0

Table 8.2 **Ropes for extending ladders**

Ladder duty rating	Minimum nominal diameter of ropes	Minimum breaking load
	mm	kg
Class 1	10	410

Table 8.3 **Stile sizes of class 1 single-section ladders**

Length of ladder		Minimum cross section	Stile reinforcement overall diameter
Over	Up to and including		
m	m	mm	mm
—	5	69 × 31	None
5	6	82 × 31	5.38
6	7.3	89 × 35	5.9

8.5   Single-section and extension sections (Wood-class I)

8.5.1   Single-section

- Stiles: The finished sizes of stiles should be in accordance with Table 8.3. The ends of the stiles should be suitably chamfered or rounded, and the edges should have a small radius to remove the sharp corners.
  Note: Ladders designed for use where there may be electrical hazards may have the stile reinforcement omitted provided the width of the stile is increased by 6 mm.
- Distance between stiles: The minimum width between the inner surface of the stiles at any point should not be less than 235 mm and at the bottom no less than the appropriate value in Table 8.4.

8.5.2   Rungs

Rungs should be either rectangular or circular. Typical rung patterns are shown in Figure 8.2. All rungs should be a drive fit into their holes and should not rotate. The end of the rungs or the holes in the stiles, or both, should be coated with the adhesive specified in BS 1204

- Rectangular rungs: Rectangular rungs should not be less than 36 mm × 22 mm and should be housed full section 5 mm to 8 mm deep into the stile, through tenoned and double-wedged. The tenons should be the full depth of the rungs and no less than 12 mm thick.

Table 8.4 **Minimum inside width of ladders**

Length of ladder		Minimum inside width at bottom
Over	Up to and including	
m	m	mm
—	3.0	242
3	3.5	247
3.5	4	252
4	4.5	257
4.5	5	262
5	5.5	267
5.5	6	272
6	6.5	277
6.5	7	282
7	7.4	287

• Circular rungs: Circular rungs should be as given in either (a) or (b):
Tapered, with a minimum center diameter of 35 mm
Parallel, with a minimum diameter of 31 mm

The ends of circular rungs should be as given in (1) to (3):

1)	Cylindrical	See Figure 8.2, (b), (d), and (e)
2)	Tapered	See Figure 8.2, (b), (d), and (e)
3)	Shouldered	See Figure 8.2, (b), (d), and (e)

A cylindrical end should have a diameter at the point of entry into the stile between 25 mm and 28 mm and should butt against the end of a blind hole, terminating between 9 mm and 12 mm from the outer face of the stile. A tapered end should have a diameter at the point of entry into the stile of between 25 mm and 31 mm and reduce to between 16 mm and 19 mm.

A shouldered end should have a diameter at a point of entry into the stile of between 28 mm and 35 mm. It should be housed between 5 mm and 8 mm full section into the stile reducing thereafter to between 16 mm and 22 mm. In no case should a shoulder butt closely against the inner face of the stile without housing.

Tapered and shouldered ends of rungs should be finished flush with the outer surfaces of the stiles, or if in blind holes, between 6 mm and 12 mm from the outer surface.

• Spacing of rungs: Rungs should be uniformly spaced at centers of 250 mm to 300 mm, except in the case of the top and bottom rung, which may be positioned at a distance of between 125 mm and 300 mm measured from the end of the stile to the center of the nearest rung.

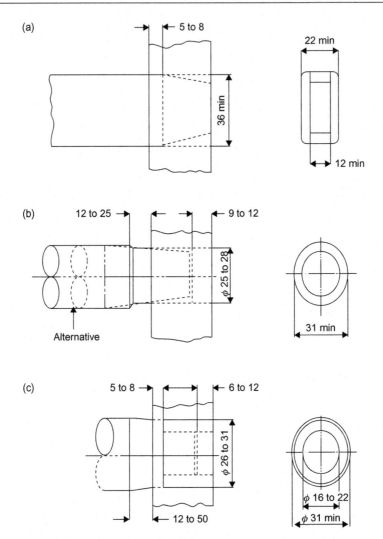

Figure 8.2 Typical rung patterns. (a) Rectangular rung housed full section into stile through tenoned and double wedged (b) Parallel rung cylindrical end (c) Paralle rung shouldered end housed full section into stile finish flush with outside surface of stile or bilnd hole.

8.5.3 Tie-Rod and rung reinforcement

The ladders should be provided with tie-rods or reinforcing wires. Any projection of tie-rods, reinforcing wires or washers above the surface of the stiles should be smoothed off to prevent injury to the hands of the user.

- Tie-Rods: Tie-rods, if used, should be fitted immediately below the first or second rung from each end of the ladder and below intermediate rungs at points not more than nine rungs apart.
- Reinforcing wires: Reinforcing wires, if used, should be fitted adjacent to and underneath each rung centrally in the width of the stile.

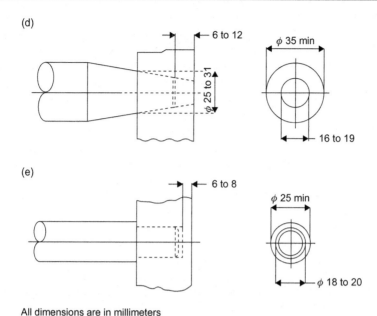

All dimensions are in millimeters

Figure 8.2 (Continued) (d) Tapered rung tapered end finish flush with outside surface of stile or blind hole (e) Parallel rung shouldered end for class 2 ladders only.

8.5.4   Extension sections

- Stiles: The finished sizes of stiles should be in accordance with Table 8.5. The ends of the stiles should be chamfered or rounded and edges should have a small radius to remove the sharp corners.
- Distance between stiles: The width between the inner surfaces of the stiles of the top section should be no less than 235 mm and not more than 360 mm. The stiles should be parallel. The width of the sections of extension ladders should be such as to provide a minimum clearance consistent with the operation of the ladder.
- Overlap of sections, rungs, rung spacing, and tie-rods: Items of the above headline (see Figure 8.3) should comply with relevant standards.

8.6   Shelf ladders (Aluminum)

8.6.1   Construction

- Lengths: The lengths of shelf ladders should be as specified in standards.
- Distance between stiles.
  Parallel-sided ladders: The working width between the inner edges of the stiles should be no less than 355 mm.
  Tapered ladders: The working width between the inner edges of the stiles at the level of the uppermost tread should be no less than 250 mm. This dimension should be increased by 12 mm to 25 mm per tread for each successive tread below the uppermost tread.

Table 8.5 Principal dimensions for class 1 extending ladders

Ladder type	Length of ladder closed		Minimum cross section stiles	Stiles reinforcement overall diameter	Rope diameter (nominal)
	Over	Up to and including			
	m	m	mm	mm	mm
Doubles					
(a) ropeless	—	3.0	69 × 28	—	—
(b) with ropes or ropeless[a]	3.0	5.0	69 × 31	—	Rope optional 10
	5.0	6.25	82 × 31	5.38	10
(c) with ropes	6.25	7.3	89 × 35	5.38	10
Trebles					
(a) ropeless	—	3.0	69 × 31	—	—
(b) with ropes	—	3.0	69 × 31	—	10
	3.0	4.5	69 × 31	5.38	10
	4.5	6.0	93 × 31	6.40	10

[a]It is strongly recommended that extension ladders longer than 4.5 m should be rope-operated and that care be taken to avoid compression damage by bumping against a wall or support.

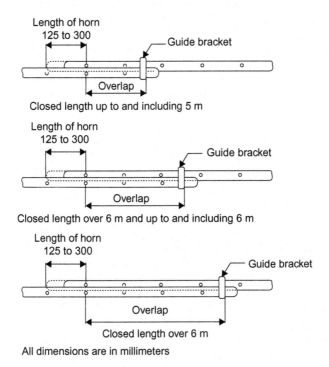

All dimensions are in millimeters

Figure 8.3 Overlaps of extending ladders.

Table 8.6 Dimensions of stiles for class 1 shelf ladders

Length of shelf ladders		Minimum cross-sectional area
Over	Up to and including	
m	m	mm
—	2.5	69 × 28
2.5	4.0	69 × 31

- Feet: The lower ends of the stiles should be fitted with feet of hardwood, plastic, or rubber. The feet should project to form a wearing surface, and should be securely fixed but easily removable for renewal.
- Treads: Treads should be no less than 75 mm wide from back to front and should have textured upper surfaces. When the ladder is inclined at any one angle within 65° to 77° to the horizontal the treads should be horizontal.
- Spacing of treads: Treads should be uniformly spaced at 225 mm to 300 mm. The distance from the bottom of the stiles to the upper surface of the lowest tread should be 125 mm to 300 mm.

8.7 Shelf ladders (Wood class 1)

Wood-type shelf ladders should comply with the appropriate standards and the following:

- Stiles: The finished sizes of stiles should be in accordance with Table 8.6. The bottom ends of the stiles should be cut parallel with the treads and suitably chamfered.
- Distance between stiles: The width at the top of the ladder between the inner faces of the stiles should not be less than 250 mm and not more than 375 mm. When the stiles are not parallel, the distance between them at any point on the ladder should not exceed 550 mm.
- Treads: Treads should consist of machined wood 89 mm × 22 mm minimum section. They should be housed 5 mm to 6 mm full section into the stile and should be secured in position with two nails or two 50 mm no. 8 gauge screws. A pair of corner blocks or brackets should be fitted underneath of the bottom tread. Shelf ladders should be so designed that when the treads are horizontal the ladder is inclined at an angle of 75 ± 2° to the horizontal.
- Spacing of treads: If a cross-bar is fitted then the distance from the top tread to the top of the stiles should be a maximum of 600 mm.

8.7.1 Tie-rods and tread reinforcement

Shelf ladders should be fitted with tie- rods or reinforcing wires. Any projection of the rods, reinforcing wires or washers above the surface of the stiles should be smoothed off to prevent injury to the hands of the user. Tie-rods, if used, should be fitted, at least immediately below the second tread from the bottom of the ladder and below other treads at points not more than four treads apart. Reinforcing wire, if used, should be fitted adjacent to and underneath each tread centrally in the width of the stiles.

8.8 Swing-back steps (Aluminum)

8.8.1 Construction

- Stiles: Stiles should be of sufficient width to provide secure bearing for the treads. The steps should be designed so that when fully open the inclination of the front stiles to the horizontal is within the following limits:
  Steps of heights up to 1675 mm: no less than 65° and not more than 70°
  Steps of heights over 1675 mm: no less than 65° and not more than 75°

  For Class 1 steps the stiles should be at least 75 mm from front to back.

- Back: A back should be hinged to the top using:
  Single hinge extension across the full width of the steps or
  Pair of hinges of wrought or forged aluminum alloy, steel, or malleable cast iron or
  Pin hinges

  The back should be constructed of either:

  Stiles and rails or
  Stiles and rungs

  Rungs should be spaced such that the top of the rungs and treads are at the same level when the steps are opened.

- Feet: The four feet of the steps should all be on the same plane when the steps are in the open position and should be soled with hardwood, plastics, or rubber. The soling material should be securely fixed but easily removable for renewal.
- Treads: Treads should be no less than 75 mm wide from back to front and should have textured upper surfaces. The steps should be so designed that when they are in use on a level surface the treads are horizontal ± 2°. For Class 1 steps the whole of the tread section should be within the outline of the stiles.
- Number of treads: Swing-back steps should have any number of treads up to maximum of 16 steps in Class 1.
- Top: The top should not be less than 100 mm wide from back to front and may overhang at the back, front, or sides, except that any projection at the front should not exceed 30 mm. The upper face should have a textured surface.
- Restriction of opening: The degree of opening of the steps should be limited using a locking bar on each side between the front stile and the back so that when fully extended the inclination of the front stiles for steps of the appropriate height, and that of the back no less than 72° and not more than 80°. The locking bar or device should engage positively in the open position to form a rigid connection between the front and back sections. Folding stay bars should positively engage in the open position by locking over center.
  Note: One-piece type locking bars (or tie bars) may be used provided that they are designed and fixed higher on the front of the step and lower on the back to ensure that forces are transmitted to the lower part of the back leg assembly.

Table 8.7 **Dimensions of stiles for class 1 swing back steps**

Length of steps		Minimum cross section dimensions	
Over	Up to and including	Front	Back
m	m	mm	mm
—	1.8	69 × 28	64 × 25
1.8	4.0	69 × 31	69 × 31

8.9 Swing-back steps (Wood)

8.9.1 Wood-type swing-back steps class 1

Wood-type swing-back steps Class 1 should comply with the appropriate requirements of 6.2 and of the following:

- Front stiles: The finished sizes of front stiles should be those specified in Table 8.7. The bottom ends of the stiles should either be cut parallel with the treads and suitably chamfered, or should be rounded.
- Distance between front stiles: The width between the inner surfaces of the front stiles should not be less than 250 mm and not more than 375 mm at the top of the steps. The width should be greater by between 25 mm and 50 mm for each 500 mm of stile length below the top of the steps such that the longest unsupported tread does not exceed 550 mm.
- Treads: Treads should consist of wrought rectangular wood 89 mm × 22 mm minimum section. They should be housed 5 mm to 6 mm full section into the stiles and should be secured in position with two nails or two 50 mm no. 8 gauge screws. A pair of corner blocks or brackets should be fitted underneath the bottom tread. Steps should be designed so that the treads are horizontal when the steps are fully opened. The top of the steps should be included in the number of treads for ordering purposes.

8.9.2 Back-hanging board

A back-hanging board, 120 mm minimum wide and of not less thick than the back stile, should be secured to each front stile by at least one 6 mm bolt and at least one countersunk wood screw, no less than 50 mm long and of no. 10 gauge. The width of the hanging board may be made from two pieces, glued with the adhesive specified in BS 1204.

8.9.3 Top

The front stiles should be housed 5 mm to 6 mm full section into a top which should be secured to each stile and the back-hanging board by glue and either nails or screws. The top should not be less than 125 mm × 28 mm thick and should be either a single piece or two pieces glued with the adhesive specified in BS 1204.

8.9.4   Back

The back should consist of stiles of the minimum dimension. The length of the back stiles should be such that when the steps are fully open the front stiles are inclined at an angle to the horizontal of:

For heights up to and including 1375 mm: no less than 65° and not more than 70°
For heights over 1375 mm: no less than 65° and not more than 75°

At the same time the back should be inclined at an angle to the horizontal of no less than 72° and not more than 80°. The distance between the back stiles should vary in the same proportion as the front stiles.

8.9.5   Top rail

A top rail, no less than 69 mm wide and of the same thickness as the back stiles, should be either:

Through tenoned into the back stiles and double-wedged with tenons of 10 ± 1 mm or
Lapped at both ends to a depth of 6 mm and at an angle suitable for a snug fit over the back stiles to which the top rail is secured by gluing with the adhesive specified in BS 1204 and, after assembly, by screwing with two 38 mm no. 10 gauge C/sunk head steel screws staggered diagonally.

8.9.6   Lower rails

Steps not exceeding 2.28 m long should have one lower rail: steps over 2280 mm but not over 3800 mm should have two lower rails. The bottom rail should be so positioned that its center line should not be less than 250 mm and not more than 500 mm from the ends of the stiles. Lower rails should not be less than 69 mm wide, of the same thickness as the back stiles and should be either:

Lapped at both ends to a depth of 6 mm, each joint being secured by gluing with adhesive specified in BS 1204 part 1, 2 and screwing with two 38 mm no. 10 gauge screws staggered diagonally or
Through tenoned into back stiles, double-wedged and glued, having tenons of 10 mm + 1 mm thick

8.9.7   Hinges

The back should be hinged to the back-hanging board by two mild steel back flaps or strap hinges having no less than 50 mm length of joint for steps up to 2.5 m long, and no less than 63 mm length of joint for steps over 2.5 m long. Each hinge flap should be secured with one bolt or rivet and at least two steel countersunk headed screws no less than 19 mm long. The end of the bolt should be riveted over the nut unless self-locking nuts are used.

8.9.8 Cords

The steps should be fitted with two plaited or braided cords of equal length no less than 6 mm diameter, in accordance with BS 6125, or material of equivalent strength. The length of the cords should be such that when fully extended the front and back stiles are at the angles. Cords should be fixed by passing them through the sides of the front and back stiles or lower back rail and should be either knotted at both ends or knotted at one end and stapled at the other.

8.10   Backed steps (Aluminum)

- Back: The back should be constructed of stiles and rungs and should comply with standard. In addition, the rungs should be spaced so that the top of the rungs and treads are at the same level when the steps are open. The hinge device joining the back to the front should be of a type that will limit the extent of opening.
- Working height: The maximum working height provided for a scaffold board should be 1785 mm above floor level.

8.10.1 Wood-type backed steps class 1

Wood-type ladder backed steps should comply with the appropriate requirements of the standard and the following:

- Back-hanging board: A back-hanging board, no less than 66 mm deep and 28 mm thick, should be secured to each front stile, and immediately beneath the top, by one countersunk wood screw no less than 50 mm long, and one nail.
- Back: The back should be constructed of stiles and rungs. The stiles should consist of wrought rectangular wood. The rungs should be either rectangular or circular as specified for standing ladders. The length of the back stiles should be such that when the steps are fully opened the front and back stiles are inclined at the appropriate angles and that when the steps are fully closed and used as a shelf ladder they should stand on the front stile. The width between the back stiles should be the same as between the front stiles or they may be parallel with the width determined at the top.
- Spacing of rungs: The rungs in the back stiles should be so spaced that when the steps are fully opened the top surfaces of the rungs are level with the top surfaces of the treads in the front stiles.
- Tie-rods and rung reinforcement: The back should be provided with tie-rods or reinforcing wires fitted adjacent to and underneath the center of each rung. Any projection of the tie-rods, reinforcing wires or washers above the surfaces of the stiles should be smoothed off to prevent injury to the hands of the user.
- Hinges: The front and the back of the steps should be connected using shouldered or lipped trestle hinges of steel, wrought iron or malleable cast iron which limit the extent of opening of that specified in standards. The minimum length of hinges should be as given in Table 8.8.
- Check blocks: To relieve the hinges of strain when ladder backed steps are being moved in the closed position, check blocks should be fitted to the inside faces of both front stiles, each secured by one bolt and one countersunk wood screw, or they may be glued and fixed by two countersunk wood screws.

Table 8.8 **Dimensions of stiles for class 1 folding trestles**

Length of stiles		Minimum cross section
Over	Up to and including	
m	m	mm
—	3.0	69 × 31
3	4.6	69 × 35

8.11 Folding trestles (Aluminum)

- Stiles: The stiles of both halves should be of equal length and should be adequate to provide secure anchorage and enable the cross-bearers to support the test load. The inside width at the top of the trestle should not be less than 500 mm and should be increased by no less than 30 mm in each 300 mm of length of the stiles.
- Feet: The 4 feet of the trestles should all be on the same plane when the trestles are in the open position and should be soled with hardwood, plastics or rubber. The soling material should be securely fixed but easily removable for renewal.
- Side plates: Side plates should be fitted to keep the stiles in register when the trestle is closed. Hinges. The hinges should be of:

Cast or forged aluminum or

Forged steel or steel strip complying with type CR4 of BS 1449: Part 1 or

Whiteheart malleable cast iron complying with BS 308

They should be trestle hinges of the locking type and should limit opening to a contained angle of no less than 30° and not more than 40°.

Note: Ropes or locking bars or similar devices as used for swing-back steps or platform steps may be fitted to reduce concentration of load on hinges.

- Cross-Bearers: Cross-Bearers should be spaced not more than 610 mm apart and should be staggered alternately on each half of the trestle at half of this distance except that there should be a top cross-bearer at the same level on each half.

8.12 Wood-type folding trestles (Class (1))

Wood-type folding trestles Class 1 should comply with the requirements of the following clauses.

- Stiles: The finished sizes of stiles should be no less than those given in Table 8.8.
- Distance between stiles: The width between the inner surfaces of the stiles should be no less than 500 mm at the level of the top cross-bearer and should be increased by no less than 50 mm for each 500 mm length of stile. The taper of the two halves of the trestle should be identical.

Table 8.9 **Length of hinges for class 1 folding trestles**

Stile length		Hinge length	No. of screws	No. of bolts
Over	Up to and including			
m	m	mm		
—	1.8	200	2	1
1.8	2.5	250	2	1
2.5	3.3	300	2	2
3.3	4.6	375	2	2

- Cross-bearers: The cross-bearers should consist of wood no less than 69 mm × 28 mm. The tenons should be 15 mm to 16 mm thick and the full width of the cross-bearer; the tenons should pass through the stile and be double-wedged. The tenons and/or the mortises should be glued with the adhesive specified in BS 1204.
- Spacing of cross-bearers: The top of the top cross-bearer on each half of the trestle should be no less than 110 mm from the top of the stiles. The other cross-bearers should be spaced so that those on one half of the trestle lie midway between those on the other half of the trestle. Except for the spacing of the top two cross-bearers on one half, the spacing of the cross-bearers on each half of the trestle should be 500 mm to 610 mm. Where uniform spacing as specified would bring the bottom cross-bearer less than half of the bearing spacing from the bottom of the stiles, the bottom cross-bearer should be positioned on the same level as the lowest cross-bearer of the other half of the trestle.
- Contained angle between halves: The combined angle between the halves of the trestle when fully open should not be less than 24° or more than 36°.
- Tie-rods: Tie-rods should be provided at a frequency of at least two on each half of trestles with stiles under 3.0 m long, and three on each half of trestles with stile length over 3.0 m. The tie-rods should be fitted immediately below the second cross-bearer from the top and either under the bottom cross-bearer or under the second cross-bearer from the bottom. When fitted, the third tie-rod should be fitted under the cross-bearer nearest to the center of the length of the stile.
- Hinges: The two halves of the trestle should be connected using shouldered or lipped trestle hinges of steel, wrought iron or malleable cast iron which limit the extent of opening. The minimum length of the hinges should be in accordance with Table 8.9.
- Check blocks: To relieve the hinges of strain when the trestles are being moved in the closed position, check blocks should be fitted to the inside faces of both stiles of one half of the trestle, each should either be secured by one bolt and one countersunk wood screw, or should be glued and fixed by two countersunk wood screws.

8.13   Aluminum-type lightweight stagings

When tested in accordance with Appendix L the residual deflection should not exceed 1/500 of the span or 3 mm, whichever is greater.

8.13.1   Cross-bearers

Cross-bearers should be of aluminum or wood; where wood is used the cross-bearers should comply with the appropriate requirements.

8.13.2   Decking

When subjected to a mass of 90 kg applied to an area 50 mm × 50 mm mid-way between two adjacent bearers, the decking should not fracture and there should be no permanent distortion.

The decking should consist of either:

Aluminum or wood slats no less than 60 mm wide with a maximum gap between the slats of 7 mm. The gap between the slats and stiles should not be more than 10 mm.

Plywood of minimum thickness 9 mm bonded with an adhesive complying with type WBP of BS 1204:

Part 1. The face veneer of the plywood should run longitudinally with the staging. Where plywood is scarf jointed to form a continuous length, any such scarf should be at least 90 mm long.

The top of the decking if metal or plastics should be textured to provide a slip-resistant surface.

8.14   Wood-type lightweight stagings class 1

Lightweight stagings should comply with the appropriate requirements of section 6.2 and the following.

Note: The stagings covered by this chapter are intended to be capable of supporting three workmen of average weight, reasonably spaced apart, and their hand tools. The load should not exceed 270 kg if distributed over the length of the staging or 180 kg if concentrated in the middle third. The length and width should comply with paragraph.

8.14.1   Stiles

The stiles should be planed on the outer face and the two edges and may be fine sawn on the inner face. The finished sizes should be in accordance with Table 8.10.

• Jointed stiles: Stiles for stagings 4.3 m and over may be scarf jointed in which case they should comply with (a) to (g).

Where jointed stiles are used, the joint should be a scarf and should have the slope visible along the edges not steeper than 1 in 18, and there should not be more than one joint per stile. Test evidence should be available to show that at least a 95% efficiency in bending strength can be obtained from the joint when manufactured under normal production conditions in the relevant wood species (see BS 1129 for test procedure).

Both portions of a jointed stile should be of the same species of wood, and should match, i.e. quarter sawn to quarter sawn or flat sawn to flat sawn. Joints should be positioned so that

Table 8.10 Dimensions of stiles for class 1 lightweight stagings

Length of staging (nominal		Minimum cross section dimension of stiles
Over	Up to and including	
m	m	mm
–	4.3	69 × 31
4.3	5.5	93 × 31
5.5	7.3	93 × 35

they have no part nearer to the center of the stile than 450 mm. Where both stiles are jointed, the centers of the joints should be separated by at least one quarter of the stile length. The moisture content of the pieces for jointing should not vary by more than 3% from each other, and should not exceed 17% moisture content.

The faces of the scarfs should be cleanly cut by fine set saw or planing to give a flat surface without tearing or crushing the fibers. The slope of the scarf should run in the general direction of any slope in the grain. Care should be taken to keep the cut surfaces clean to assist in this and also to avoid distortion joints should be assembled as soon as possible on the same day.

The adhesive used should be of type WBP in accordance with BS 1204: Part 1. The method used in mixing the adhesive should be in accordance with the manufacturer's instructions, as should the spreading, open and assemble times, curing temperature, cramping and conditioning times. Any cramping arrangement should be such that the pressure is uniform over all the glue line, so causing a continuous film of adhesive. The pressure should be as specified in the adhesive manufacturer's written instructions, but in no case should be less than 0.69 N/mm2 and there should be a continuous squeeze out along the full length of the glue line.

After release from the cramp and subsequent machining, a visual inspection should take place to ensure that all glue lines are continuous, and that there is perfect bonding of the fibers on the faces where the joints appear.

A sample joint should be made in each batch for testing to destruction, to ensure that the glue mix for the batch is correct (see BS 1129 for procedure). Should the glue joint fail under testing but the wood remain intact, the glue mix should be regarded as unsatisfactory and the batch rejected.

8.14.2 Cross-bearers

The cross-bearers should be of wood and of cross-section no less than 31 mm deep and no less than 22 mm wide; they should have tenons or pins on each end no less than 22 mm × 22 mm or no less than 25 mm diameter in cross-section and no less than 22 mm long. They should be spaced at centers not exceeding 381 mm except in the case of the space between the end three cross-bearers when a maximum space of 475 mm is permissible. The distance between the center of the end cross-bearer and the end of the stile should not be more than 75 mm.

8.14.3 Decking

The decking staging should be constructed using one of the following materials and methods:

Wood slats having a width no less than 60 mm and not more than 150 mm and a thickness no less than 12 mm. The finish should be either wrought or fine sawn. The gaps between the slats should not exceed 10 mm and the gaps between the slats and the stiles should not exceed 15 mm.

Both ends of each slat should be fixed by two countersunk wood screws 32 mm long and of no. 8 gauge when fixing to softwood cross-bearers and 25 mm and of no. 8 gauge when fixing to hardwood cross-bearers. When fixing slats wider than 75 mm then three screws at each end are required. Fixing of all other cross-bearers should be by two nails, staples, or wood screws no less than 32 mm long and 16 swg. No fixing should protrude above the surface of the slats. Joined slats and decking are permitted provided that each joint occurs over a bearer no less than 34 mm wide and not more than two joints occur on any one cross-bearer (see Figure 8.4).

Wood laminates bonded with a type WBP adhesive in accordance with BS 1203: Part 1. The face veneer of plywood should run longitudinally with the staging. The plywood should be a minimum of 9 mm thick and be capable of sustaining a load of 90 kg when supported by the cross-bearers at 380 mm spacing. The wood laminates may be scarf jointed to form a continuous length, and any such scarf should be 90 mm long. Joints required to form a continuous length of decking should be either: 30 mm half-lap located centrally over a bearer fixed with at least 4 countersunk no. 8 screws 32 mm long through the laps into the bearer, or butted and screwed with 4 countersunk no. 8 screws 32 mm long as in Figure 8.4(a).

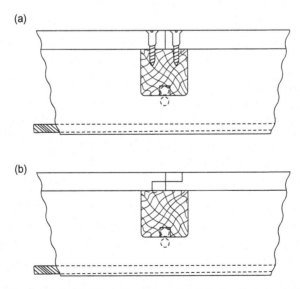

(a)

(b)

Figure 8.4 Decking joints for lightweight stagings. (a) Butted and screwed (For Screwes length see clause 7.12.4(a)) (b) Halved and glued (nailed or stapled)

8.14.4 Tie-rods

A tie-rod should be fitted immediately below, immediately alongside, or into a groove beneath, each cross-bearer.

8.14.5 Stile reinforcement

The reinforcement should be fitted in grooves running centrally along the lower edge of the stiles and should be properly tensioned and secured in position. Reinforcement may pass over ends of stiles or pass through each stile 100 mm to 150 mm from the end of the stile.

Stiles for lengths up to and including 5.4 m should be reinforced with a 7 wire mild steel strand of 5.38 mm overall diameter having a minimum characteristic strength of 425 N/mm^2 or a single wire, strip or different strand of equivalent strength. Stiles for lengths over 5.4 m should be reinforced with a 7 wire mild steel strand of 6.4 mm overall diameter having a minimum characteristic strength of 460 N/mm^2 or a single wire, strip, or different strand of equivalent strength.

8.15 Performance tests

The performance tests described below are for specific types of wood and aluminum ladders.

8.15.1 Performance test for aluminum-type of single-section standing and extension ladders

- Deflection under load: When tested in accordance with Appendix B the deflection of the loaded stiles should not exceed the limit determined from the graph shown in Figure 8.5. In addition, after removal of the test load there should be no permanent damage and the residual deflection should not exceed 1 mm per m of test span.
- Torsional rigidity: When tested, the difference between the deflections of the two stiles should not exceed the limit determined from the graph shown in Figure 8.6.
- Strength: When tested in accordance with Appendix C, after removal of the test load the residual deflection should not exceed 1 mm per meter of length between the supports plus 1 mm.
- Twist: When tested in accordance with Appendix D, the angle of twist should not exceed the value given in Table 8.11.
- Sideways bending: When tested in accordance with Appendix E, the deflection measured midway between supports should not exceed (0.0033 L + 18 mm), where L is the effective span (in mm) and the residual deflection should not exceed 1 mm per meter.
- Cantilever bending: When tested in accordance with Appendix F, the residual deflection of either stile should not exceed 6 mm.
- Rungs: When tested in accordance with Appendix G, the rungs should support the load. In addition, after removal of the test load there should be no damage or permanent deflection.

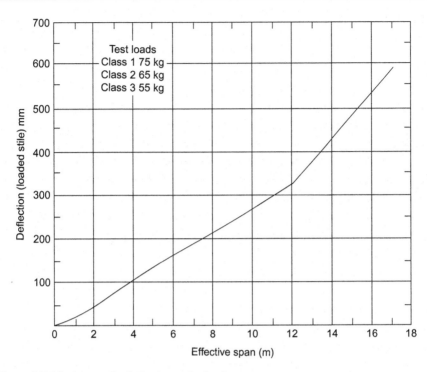

Figure 8.5 Maximum stile deflection under load.

8.16   Performance test of aluminum-type shelf ladders

Test for treads: When tested in accordance with Appendix H the tread should support the load. In addition after removal of the test load the residual deflection of the tread should not exceed 1 mm.

8.17   Performance test of aluminum swing-back steps

8.17.1   Rigidity

When tested in accordance with Appendix I.1, the steps should show no damage or permanent deflection on removal of the load except that a residual spread of up to 8 mm, measured between the ends of the front and rear stiles, is acceptable.

8.17.2   Test for treads

When tested, the tread should support the load. In addition, upon removal of the test load the residual deflection of the tread should not exceed 1.0 mm. Tests for deflection

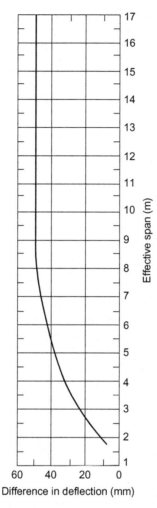

Figure 8.6 Torsional rigidity: maximum difference in deflection between stiles.

Table 8.11 **Maximum allowable angle of twist**

Duty rating	Type	Maximum allowable angle of twist
		Degrees (°)
Class 1	Industrial	18

under load, strength, sideways bending, and cantilever bending should be as specified in standards.

8.17.3 Performance test of aluminum ladder backed steps

Tests for rigidity, front, and back are:

> For rigidity
> For front
> For back

8.18 Tolerances on sizes

The overall length of wood-type ladders should be subjected to the following tolerances:

- 50 mm for sections up to and including 3 m
- 75 mm for sections over 3 m and up to and including 6 m
- 100 mm for section over 6 m
- 5 mm for rung and tread centers on a single product

Where a size is specified as a nominal dimension a tolerance of ± 5% applies when determining the acceptable actual size. When checking the dimensional requirements of ladders, any cut outs for joining members, fitting guide brackets, or similar purposes should be discounted. Where treads have a rectangular cross-section the dimensions specified in standards are overall dimensions. For aluminum types of ladder a tolerance of 25 mm should be permitted on the nominal length of all ladders. Spacing of rungs and treads should not vary by more than 2.0 mm from the nominal spacing selected by the manufacturer.

8.18.1 Markings

Equipment should be clearly and durably marked. There should be no reduction in legibility at the conclusion of the test. Adhesive labels, where used, should not have worked loose or become curled at the edges.

Ladders, steps, trestles, and lightweight stagings should be marked with the following:

- Name, trademark, or other means of identification of the manufacturer or supplier
- Number and date of standard followed for manufacturing this product
- Class and duty rating

Ladders, steps, and trestles should be provided with a separate label displaying the following wording on a background colored blue for Class 1, yellow for Class 2, and red for Class 3.

- Inspect for damage before use
- Lean ladder at approximately 75° from horizontal (1 m out for each 4 m height)
- Ensure firm level base

- Check safety at top
- Avoid electrical hazards
- Avoid overreaching (do not push or pull from the ladder or climb higher than the third rung from the top)
- Keep a secure grip
- Never stand on top of swing-back step or step ladders
- Secure at top and bottom wherever possible

Lightweight stagings should bear a label fitted to the outside of one stile incorporating in letters no less than 4 mm high with the following: "Maximum load, three men spaced apart, and hand tools or 270 kg uniformly distributed."

8.19 Care and use of ladders

8.19.1 Handling

Equipment should be handled with care and not subjected to unnecessary dropping, jarring, or misuse. If it has fallen or received a heavy blow, it should be examined immediately and any damage should be eliminated and repaired by a competent person before putting back into service.

8.19.2 Storage

Equipment should be stored in such a manner as to provide ease of access and inspection and to prevent danger of accident when withdrawing for use. Ladders should be stored horizontally on racks designed for their protection when not in use. These racks should have supporting points at every 2 m to prevent any possibility of excessive sagging. At no time should any material be placed on the ladder while in storage. Wood plants should be stored in a location where they will not be exposed to the elements but where there is good ventilation; it should not be stored near radiators, stoves, steam pipes, or other places subject to excessive heat or dampness.

8.19.3 Transport

Ladders carried on vehicles should be adequately supported to avoid sagging and there should be minimum overhang beyond supporting points, which should be of resilient material. Ladders should be tied to each support point to minimize rubbing and the effects of road shock. Ladders should also be carefully loaded so that they are not be subject to shock or abrasion.

8.19.4 Maintenance

Equipment should be maintained in good condition at all times. Hardware, fittings, and accessories should be checked frequently to ensure that they are securely attached and in proper working condition. Moving parts, such as pulleys, locks, hinges, and wheels, should operate freely without binding or undue play and should be oiled frequently, and kept in good working order. All bolts and rivets should be in place and tight before use. Ropes or cables should be inspected frequently and those frayed or badly worn or defective should be replaced.

8.19.5 Inspection

Equipment should be inspected before and after use and periodically by a trained person. Those items found to be defective should be suitably labeled or marked, and should be withdrawn from service. The inspection should include checking the rungs, treads, cross-bars, and stiles for damage, defects and dents, checking the rung to stile connections, checking ropes and cables and all fittings, locks, wheels, pulleys, connections, rivets, screws, and hinges.

8.19.6 Painting

Wood equipment other than inserts may be coated with a transparent non-conductive finish such as varnish, shellac or a clear preservative, but should not be coated with any opaque covering. Preservatives for the treatment of wood components in aluminum plant should not contain copper salts. Aluminum should not be used in corrosive conditions. Aluminum ladders should not be used where any electrical hazard exists. The ladder should always be inspected carefully before use.

8.19.7 Angle of ladder

The ladder should be erected at an angle of 75° from the horizontal, i.e., the distance in feet from the vertical surface should be as near as possible one-quarter of the height reached by the top of the ladder.

8.19.8 Support

Equipment should be placed on a secure footing on a firm level base. It should not be used on ice, snow, or slippery surfaces, unless suitable means to prevent slipping are employed. It should not be placed on boxes, barrels, or other unstable bases to obtain additional height.

8.19.9 Fixing of the ladder

The point on which the top of a ladder rests should be reasonably rigid and have ample strength to support the applied load. The ladder should be securely fixed to this point. If such a fixing is impracticable the ladder should be securely fixed at or near the lower end either by staking or by roping.

8.19.10 Overlap

On extension ladders there should be an overlap of at least:

- 1½ rung spacings for ladders with closed lengths up to 5 mm
- 2½ rung spacings for ladders with closed lengths over 5 m and up to 6 m 3½ rung spacings for ladders with closed lengths over 6 m

The ladder should be raised and lowered by the user at the base of the ladder so that it can be seen that the locks are properly engaged. When extension ladders have been used previously as single ladders, care should be taken to ensure that the re-assembly of the ladder is properly carried out and that interlocking brackets and guides are correctly engaged.

8.20 Determination of slope of grain

To ascertain the slop of grain in wood it is necessary to study both faces and edges over the length looking for any deviations. If any seasoning checks are present, these will indicate the slop of grain, as will resin ducts. Slope of grain can be determined using a tool called a grain detector, which consists of a handle that swivels on a cranked rod with a gramophone needle set at a trailing angle at the tip of the rod, as shown in Figure 8.7. The needle is pressed into the wood between 1 mm and 2 mm deep and the scribe is pulled along with a steady action parallel to the edge of the wood: the needle will deviate in the direction of the grain. If there are any steps in the groove produced by the needle this indicates that the needle is climbing over the grain, it is then essential that another scribe is made to ensure the two grooves are parallel to each other.

The inclination of grain on a face is measured as shown in Figure 8.8 and Figure 8.9 in which AB is the line indicating grain direction, AC is a line drawn parallel to the edge of the member, BC is of length one unit (any convenient unit may be used), and

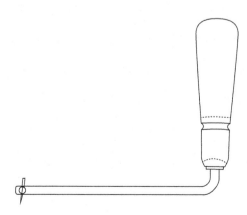

Figure 8.7 Swivel-handled scribe for determination of slope of grain in wood.

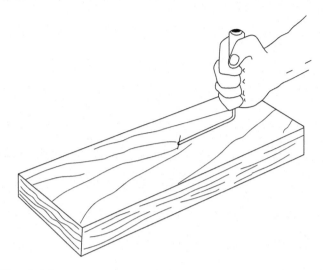

Figure 8.8 Use of scribe.

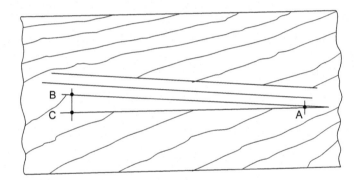

Figure 8.9 Measurement of slope of grain.

is at right angles to AC. Grain inclination is expressed as "one in x" where x is the length of AC measured in terms of BC.

When sloping grain occurs on two adjacent surfaces, the slope may be determined on both. The acceptance limits of both slopes considered together are shown in Figure 8.10.

8.21   Test for deflection of ladders and step fronts

If this test is being applied to an extension ladder, extend it fully to the appropriate overlap before the test is commenced. With the climbing face uppermost, support the ladder or steps horizontally under the stiles at each end rung or tread, or in the case of steps, where the hand or knee rail is an integral part of the step front, under the bottom

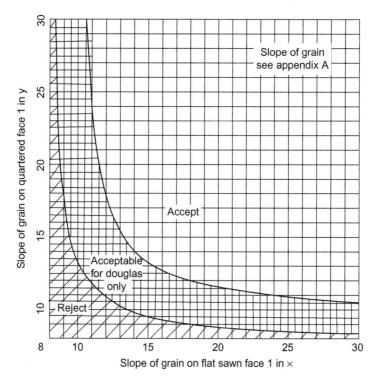

Figure 8.10 Acceptance limits of slope of grain on adjacent surfaces.

tread and at a point 200 mm in from the hinge point. Measure the clear span between the supports. This is regarded as the effective span for the purpose of this test. Apply a preload as given in Table 8.12, according to the class, to the center point of both stiles distributed over 50 mm for a duration of 30 s. Remove this load and establish datum. Then apply a test load as given in Table 8.12, according to the class, to the center point of one stile distributed over 50 mm. By any convenient means measure the vertical deflection at the center of the effective span of both stiles between the unloaded condition and after a period of no less than 30 s from the application of the full test load.

8.22 Strength test for ladders, step fronts, and trestle frames

If this test is being applied to an extension ladder, extend it fully to the appropriate overlap before the test is commenced. With the climbing face uppermost, support the ladder or steps horizontally under the stiles at each end rung or tread, or in the case of steps, where the hand or knee rail is an integral part of the step front, under the bottom tread and at a point 200 mm in from the hinge point. Apply a preload, as given in Table 8.13,

Table **8.12** **Loads for deflection test**

Duty rating	Preload	Test load
	kg	kg
Class 1: industrial	55	75

Table **8.13** **Loads for strength test**

Duty rating	Preload	Test load
	kg	kg
Class 1: industrial	95	130

according to the class, vertically at the middle of the ladder, distributed over a length of 50 mm for a duration of 1 min., so that the stiles are loaded equally. Remove this load and establish a datum point. Then apply a test load, as given in Table 8.13, according to the class, in the same way as the preload for a duration of 1 min. Remove the load. By any convenient means measure the residual deflection at the datum point.

8.23 Ladder twist test

The test unit should consist of a ladder base section of any length, supported over a 2 m test span. Place the ladder in a flat horizontal position and support it at each end, as shown in Figure 8.11. The distance between the pivot point center and the plane of the centerline of the rungs should not be more than 50 mm. Apply a preload torque of 6.5 kg m gently and then remove. The residual angle of pivotal support should be noted as datum position to establish a reference for angular deflection. Apply a test torque of 13 kg m in the same direction as the preload by either using a torque wrench or by applying a test load at the end of the arm. Measure the angle of twist from the datum position. Apply a second load of the same torque as the preload in the opposite direction and then remove. The residual angle of pivotal support should be noted as datum position. Apply a second test load in the opposite direction to the first test load. Measure the angle of twist from the second datum position.

8.24 Sideways bending (sway) test for ladders, step fronts, and trestle frames

Place the ladder, step or trestle on its side with the rungs or treads vertical. Measure each section of the ladder individually, the bottom stile being supported as for the

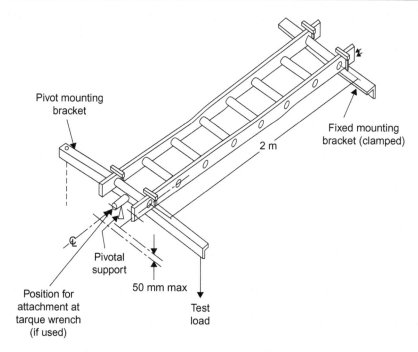

Figure 8.11 Arrangement for twist test on single or extension ladder. **Note:** The test span is 2 m, but any ladder section that is at least 2 m in length may be tested.

strength test of Appendix C under the end rungs (see Figure 8.12, or in the case of steps, where the hand or knee rail is an integral part of the step front, under the bottom tread and at a point 200 mm in from the hinge point. Apply a preload of 15 kg for 1 min. and then remove it to determine the datum for measurement on the lower edge of the lower stile. Apply a test load as given in Table 8.14, according to the class, at the center points of the span of the stiles distributed over 50 mm. By any convenient means measure the vertical deflection at the datum point on the lower edge of the ladder and then remove the test load. After 1 min. measure the residual deflection at the same point.

8.25   Cantilever bending (horn end strength) test for ladders, step fronts, and trestle frames

The test unit should consist of a step front, trestle frame, single ladder section, or the base section of an extension ladder. Any safety shoes or spikes affixed to the section should be removed before the test is conducted. The test unit should be placed on edge with the rungs or treads in a vertical plane. The lower stile should be clamped to a support and should be unsupported from the bottom end to the midpoint of the

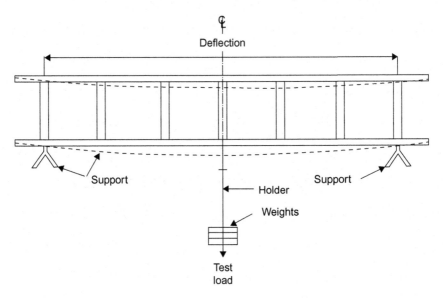

Figure 8.12 Arrangement for sideways bending (sway) test. **Note:** The deflection is the difference between the height of the lower edge of the ladder side when unloaded (solid line) and when loaded (dotted line). L is the effective span (in mm).

Table 8.14 Test load

Duty rating	Preload	Test load
	kg	kg
Class 1: industrial	15	27
Class 2: light trades	15	25
Class 3: domestic	15	23

lowest rung or tread. If the rung has a flat surface, that surface should be parallel to the end of the support (see Figure 8.13). Establish a datum point on the end of the stile to which the test load is to be applied. Apply a test load, as given in Table 8.15, according to the class, for a minimum period of 1 min., to the extreme bottom end of the upper stile. Apply the load to a block resting on the full width of the stile web and hold in place by a clamp.

The block should be 25 mm thick, 50 mm long measured along the stile and of width equal to the clear distance between flanges. Ensure that the load is suspended so that it is acting through the vertical neutral axis of the stile (see Figure 8.13(a)). Remove the test load and after 1 min. measure the residual deflection at the datum point. Repeat the loading and measurement procedure on the lower stile (see Figure 8.13(b)).

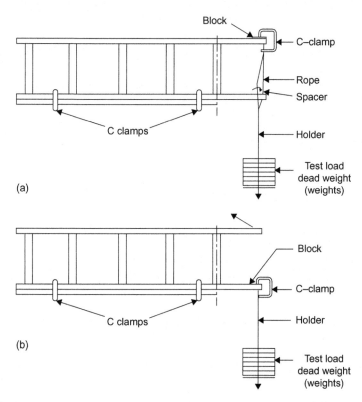

Figure 8.13 Arrangement for cantilever bending (horn end strength) test. (a) contilever in test (b) contilever out test.

Table 8.15 **Test loads for cantilever bending (horn end strength) test**

Duty rating	Test load
	kg
Class 1: industrial	125

8.26 Test for rungs

Support the ladder at an angle of 75° from the horizontal and with continuous support for both stiles over a length equal to three rung spacings. To the center rung of the three, apply a vertical load as given in Table 8.16, appropriate to the class of the ladder, distributed over a length of 50 mm for 1 min., as follows:

- At the center of the rung
- Close to one end

Remove the load and examine the rungs for permanent deflection or visible damage.

Table 8.16 **Test loads for rungs, treads and platforms**

Duty rating	Test load
	kg
Class 1: industrial	225

8.27   Test for treads of shelf ladders

Place the shelf ladder in the normal working position, supported so as to prevent movement of the feet or deflection of the stiles. To a typical tread apply a vertical load, as given in Table 8.16 over a length of 50 mm for 1 min. as follows:

- At the center of the tread
- Close to one end

Remove the load and examine the ladder for permanent deflection or visible damage.

8.28   Tests for steps

8.28.1   Test for rigidity

Attach securely, by any convenient means, a wheel (or roller) to the outer side of one of the back stiles. The wheel should be of metal and have a diameter of 50 mm; it should be mounted so that it can rotate freely, with its axis parallel to the treads, and raises the foot of the stile by 10 mm (see Figure 8.14).

Place the steps in the fully open position on a smooth level surface and apply the appropriate preload given in Table 8.17 to the top tread but one, adjacent to the stile on the same side as the wheel. In the case of platform steps, apply the preload to the tread immediately below the platform.

Maintain the load on the tread for 1 min., then remove the preload and then apply the appropriate test load given in Table 8.17 using the same procedure. After the 1 min. has elapsed remove the test load and the wheel before inspecting the steps for damage and deformation. Examine the steps for visible deformation or damage.

8.28.2   Test for treads

After completing the test given in I.1, apply to a typical tread a vertical load as given in Table 8.16 over a length of 50 mm for 1 min. as follows:

- At the center of the tread
- Close to one end

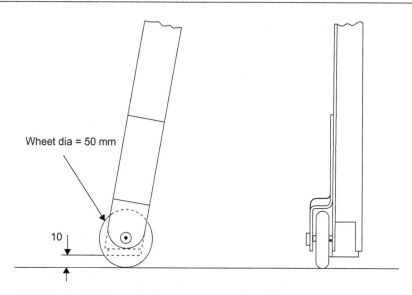

Wheel dia = 50 mm

10

Figure 8.14 Method of fixing wheel to bottom of backleg for step test.

Table 8.17 Test loads for steps and trestles

Duty rating	Preload	Test load
	kg	kg
Class 1: industrial	95	130

Remove the load and examine the ladder for visible damage. Place a 6 mm thick straight edge on the centerline of the tread-climbing surface so that it is symmetrically positioned and covers 95% of the length of the tread and the mid span point of the tread. Measure any space between the latter point and the straight edge.

8.28.3 Test for platforms

After completing the tests given in I.2 apply to the center of the platform a load as given in Table 8.16 over an area of 50 mm × 50 mm for 1 min. Remove the load and examine the platform and the ladder for permanent deflection or visible damage. Apply a load of 54 kg uniformly distributed as before. Take an initial reading for deflection measurements at the mid span position on each stile, (reading 1). Increase the load to a total of 540 kg and maintain it for 1 min. Reduce the load to 54 kg again and take a further deflection reading as above (reading 2). Calculate the residual deflection as reading 2 minus reading 1.

8.29   Test for trestles

Place the trestles in the fully opened position on a level surface. Apply the appropriate preload given in Table 8.9 at the load center of the cross-bearer and spread over a length of 50 mm. Maintain the preload for 1 min., then remove it. Apply a test load as given in Table 8.9 in three equal increments at the load center of the bearer, the load being spread over a length of 50 mm. Maintain the load for 1 min.

Remove the load and examine the cross-bearers and the whole trestle for permanent deflection or visible damage. Repeat the procedure for each cross-bearer.

8.30   Test for lightweight stagings

With the decking uppermost, support the lightweight staging, in a horizontal position, under both stiles at 150 ± 5 mm from each end. Apply a preload of 400 kg uniformly distributed over the area of the decking between the supports and maintain the load for 1 min. before removing the load. Apply a load of 54 kg uniformly distributed as before. Take an initial reading for deflection measurements at the mid span position on each stile (reading 1). Increase the load to a total of 540 kg and maintain it for 1 min.

Reduce the load to 54 kg again and take a further deflection reading as above, (reading 2). Calculate the residual deflection as reading 2 minus reading 1.

8.31   Test for extension ladder guide brackets

The sample to be tested should take the form of a length of stile to which a single guide bracket is attached using the normal method of fixing. Secure the stile using a clamping device or devices so that it is fixed rigidly in position. Apply the following load at the center of the overhang of the bracket for 1 min. (see Figure 8.15):

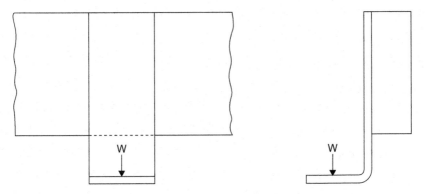

Figure 8.15 Application of load in test for extension brackets.

(a) (b)

Figure 8.16 Specific types of ladders. (**a**) Extension ladder (**b**) Swing back steps.

Class 1 extension ladders 130 kg.

Measure and record any movement and distortion of the bracket and its fixing and whether any remains after removal of the load.

Figure 8.16 shows different type of ladders.

Safety and firefighting equipment, part 1

9

Safety training is one of the important ways to motivate employees to reduce the number of accidents that occur but also to promote their general knowledge and self-confidence. To implement this training, a wide range of programs is needed that incorporates facilities for this purpose.

Provision of facilities in fire stations is also required and is a major concern of oil and gas processing plants to protect employees and property from the risk and thread of fire in residential homes and workplaces.

The equipment and facilities required should be envisaged by safety and fire and the management as authorities to determine the needs as given in the standards. In designing and specifying materials and equipment for fire stations, factors such as hazards involved in the process of oil and gas refineries, petrochemical complexes, production units, and pumping and compressor stations should be considered. Based on the information gained the size of building construction, facilities, equipment, and emergency response team as well as future development can be determined.

The goal of this chapter is to provide guidelines on the design and construction of facilities required for the control of accidents and fires. The standards are divided in two sections as follows:

Section 1: Safety- and Fire-Training Centers
Section 2: Fire-Station Facilities

9.1 Safety-and fire-training centers

9.1.1 Design and construction

- Assessment: In order to derive the maximum benefits from the resources available a comprehensive assessment of current and future needs must be done and the following subjects must be considered:
  Current and future needs
  Facilities currently available
  Adequate space required for indoor and outdoor training

In design and construction of training facilities the following courses should be considered:

Recruit training (basic course) for firefighters
Employees firefighting course

Personnel Protection and Safety Equipment for the Oil and Gas Industries.
DOI: http://dx.doi.org/10.1016/B978-0-12-802814-8.00009-2
© 2015 Elsevier Inc. All rights reserved.

Motor vehicle accident prevention defensive-driving course
First-aid course
Environmental pollution and hygiene control
Special course for supervisors
Off-the-job safety course

9.1.2   Layout of training center

The following are generally the building requirements:

Offices and administration facilities
Conference and classrooms including exhibition and library
Outdoors facilities for fire-service practical training

The following items should be considered in the planning:

Weather conditions: temperature, humidity, wind velocity, rain, and snow
Ample space between classrooms and outdoor practical facilities in training ground
Location of heating and air-conditioning equipment where regular maintenance can be easily performed. Avoid the installation of such units in classroom area
Space needed for visitors
Provision of storage for materials, equipment, and fuel
Communication requirements
Dining room, washroom, drinking water, and pantry
Parking lot

9.1.3   Location of training center

Some aspects to consider in determining the placement of the training facility are site, water supply, environment, security, support services, and access to utilities.

9.1.4   Site consideration

Proper drainage separated from oil installations drain system is a major consideration because of the use of large amounts of water for certain exercises. Also, effort should be made to ensure that drainage is sufficient for variable weather conditions as well as future expansion. The slope of the land may be advantageous for drainage but an excessive slope may be a negative safety factor for manpower and apparatus movement especially when the surface is wet.

The size of the site should be ample for planned buildings, parking, and future expansion. Allow adequate separation between buildings for safety. The site of the training facility should be located away from the center of community life to minimize negative impacts on adjacent land use.

9.1.5   Water supply

The maximum water supply required should be estimated so that an adequate system can be installed to deliver the necessary volume for training activities and domestic as well as off-shore desalinated water needs. A loop or grid system with properly placed valves would help to ensure an adequate water delivery. If possible, dead-end mains should be avoided. Two delivery hydrants with capacity of 120 m³/h of water supply and two water monitors should be installed, preferably up-wind.

9.1.6 Use guidelines

Rules should be developed regarding the use of the facility. The various components of the facility should be in use as much as possible, and the needs of users can be worked out with scheduling. Prevailing winds can be used to direct smoke away from neighbors. Shifting winds will have to be taken into consideration.

9.1.7 Administration and classroom building

- Offices: Office space should be provided for the head of training, instructors, and clerical personnel. Additional spaces are required for store laboratory, visual devices, and other material and equipment.
- Conference room: A conference room for meetings, lectures, and discussions with relative devices.
- Classrooms: Classroom size should be based by the number of trainees and the type of training to be conducted. The minimum space should be for 30 people. The instructor should be able to control room condition and audio-visual equipment. Good lighting is a must and the use of both individual controls and rheostats should be considered to vary the illumination. A podium light and separate white board illumination can make a presentation in a darkened room more effective. Electrical outlets in the floor and the walls should be spaced to eliminate the use of extension cords. Before the sound system is installed, the installer should eliminate dead spaces in the room. Classroom furniture has to be durable. Writing surfaces for use by the instructors and students should be provided. Folding tables, 45 cm wide, and stacking chairs permit greater flexibility in room utilization. Experience has shown that wider tables occupy space that can be better utilized. To lessen classroom disturbance the following features must be considered:
Doors to the room should open and close quietly, sanitary and refreshment facilities should be close to the room, and ceiling height must permit the hanging of wall screens or the placement of portable screens for good viewing. The ceiling height should be a minimum of 300 cm as dictated by experience.
A central air-conditioning system and heating units should be installed in the classroom because of noise.
- Audio-visual: To allow the instructor to take advantage of various media the following equipment should be available:
White board, magnetic board
Video with TV monitor
Slide projector, recorder, and video-editing machine
Computer, overhead, video projector
Amplifier and microphone
- To aid in the use of audio-visual equipment the following requirements should be provided:
Provide an extra electrical switch with a rheostat to control illumination
Place projector area near hallway so that the equipment can be easily moved
Provide heating, ventilation, and air conditioning

Provide projectors with permanent remote control wire
Install electrical receptacles in the floor to eliminate the use of extension cord
- Building Maintenance: The material used should be easy to maintain. Durable material would cut down supply of replacement and cost.

9.2 Firefighting training ground

The training ground including recruit section should consist typically of an area of about 1000 m² located in a safe position away from plant and storage facilities. A typical layout of a training ground is usually provided in the relevant standards. It consists of a concrete floor, sloping for drainage, surrounded by spillage walls of about 0.5 m high. The drain connection should be provided with an isolating valve, and a fire trap should be installed in the drain pit located outside the spillage wall. For Liquefied Natural Gases (LNG)/Liquefied Petroleum Gases (LPG) plant training grounds, an additional sump of 3 × 3 m is required with a total depth of 1 m (including a wall), both sump and wall should be lined with heat-resistant refractory bricks. Thermally insulated LNG/LPG filling and nitrogen cooling connections should be installed for the simulation of LNG/LPG fires.

A fuel tank is installed outside the spillage wall at a safe distance to provide the "mock-up" equipment with fuel for simulated plant equipment fires. The fuel tank may be pressurized with nitrogen to transfer the fuel to the equipment, or, when electrical power is available a pump can be used. For refinery installations, the mock-up equipment indicated in the layout should be installed as a minimum.

Locations with special fire hazard, additional equipment for simulating fires may be required. If gas cylinders are used, they should be included and stored on a concrete slab positioned outside the spillage wall at a safe distance, to feed gas to the mock-up equipment for a simulated gas fire.

A branch from the fire-water main with a capacity of 120 m³/h should run to the training ground. This branch should be fitted with two-way hydrants, two water monitors, and a full bore flushing connection for cleaning the branch line. The area should be accessible for a firefighting vehicle and, if the access road has a dead end, a turnaround should also be provided.

9.3 Other requirements

The following spaces and facilities are generally required for recruit training:

Ground for drill and practical exercises. The area should be approximately of 30 × 20 m.
Water sump to be provided for pump test. The sump to be of minimum of 2 × 2 m and 3 meter deep. The level of water to be at 2 m but a slop ramp should be made with 4 m wide and one meter high. The pump testing area should be fenced and the sump to be covered.
Construction of smoke building to acquaint the trainees with the skill and ability necessary for survival in an oxygen deficient atmosphere and learning to use breathing apparatus. The smoke building should have entry points and escape hatches and smoke used should be of

a controlled combustion with minimum toxicity. Hay-Straw-Cardboard boxes or similar combustibles are considerably safer Smoke Machine.

Floating-roof oil storage tank fire. Fire is extinguished by:

1. Foam pourer
2. Portable foam branch or fire extinguishers
   Ladder-training tower to be provided for ladder drill
   Flammable liquid burn area

9.3.1 First aid

Safety should be the foremost consideration in facility design. Accidents and illnesses do occur; therefore, properly designed first-aid room should be provided. Space should be provided so that temporary care can be administered to victims suffering from burns, cuts, cardiac distress, smoke inhalation, heat exhaustion, and other injuries or illnesses.

9.4 Fire-station facilities

This section covers the requirements for fire-station buildings. The fire stations should be built at safe locations away from any risk and as close as practicable to the fire-control area where additional personnel are available.

9.4.1 Categories

The size of fire stations should be determined by company management and safety and fire authorities, considering all risk factors, such as:

1. Size of the area
2. Combined fire and emergency factors
3. Fire-protection system installed
4. Availability of employees trained in firefighting operations
5. Availability of other sources of help
6. Fire-prevention techniques, design and enforcement; categories of fire stations are as follows:

A	Large	For the areas such as combining refineries, process, chemical, gas-treating plants, loading terminal, storage, and other service facilities including industrial and residential areas.
B	Medium	The same as above classified as lower risk.
C	Sub-Fire Station	For the area of high risk with fire station of Category A but close enough to reach within a 10-minute drive.
D	Retained	For the fire risk potential areas and plants covering approximately 2 to 5 km² and located far away from any fire stations (selected employees are trained and assigned as volunteer firefighter).

9.4.2 Category A: fire station

- Parking bays: The fire station should have six fire truck parking bays, for a minimum of six fire trucks (three first-aid emergency units and three auxiliary major emergency units). Firefighting units may be selected from the following, depending on the risk factors. The station should be designed and located such that future expansion of 25% will be possible:
General-purpose fire truck
Major fire truck
Dry powder or combined foam/dry-powder truck
Foam or water tanker
Hydraulic platform (boom)
Rescue tender
Fire-extinguishing trailer
Equipment trailer
Foam/water monitor trailer
Emergency ambulance
- Layout
The vehicles should be able to enter and leave the station parking bays at both sides, front and rear, of the fire station.
If the fire station is located at an authorized primary road, consideration should be given to the installation of traffic lights, operable from, and indicating when vehicles are leaving, the fire station.
The entrance and exits should be closed by doors, e.g., rolling shutters, counter weight, etc., designed for fast opening and constructed in such a way that the vehicles would be able to drive through without delay.
Open parking (in fire stations without doors) may be considered when climatic conditions allow, but a disadvantage will be entry of unauthorized people.
In locations where freezing can occur, the parking places should be protected accordingly. Each parking bay should be equipped with an electrical connection and cable for battery charging, and when required for heater in the engine cooling system of the vehicle, the plugs to be pulled out of the sockets, when the fire truck drives away.
The height above the parking places, including doors should not be less than 5 m. A free space of 1.5 m should be available between each vehicle, between vehicle and wall and all doors. The width of vehicle is approximately 2.5 m.
The length of a parking place should be based on the length of the longest firefighting vehicle pulling a mobile water/foam monitor that may be approximately 11 m.
An inspection pit should be available in the maintenance area of the fire station.
Parking area should be equipped with compressed air supply for pressurizing the brakes of the trucks when required.
- Workshop-office and other facilities: The following facilities should be available:
Workshop laboratory containing workbench, fixed drilling, and grinding machines with fire extinguishers testing equipment and tools for testing and servicing of other equipment such as breathing and emergency equipment.
Storage for spare parts.
Fire-extinguisher filling station.
Rooms for storage of CO_2, O_2, N_2, and transfer-charging compressor. An air compressor should also be provided in a separate room for charging cylinders of breathing apparatus.
A firefighting instruction room equipped with a white board-slide projector-video and other audio-visual media with screen, sized for about 30 people. Alternatively, where safety and fire training center is available, the training center should be used for such purpose.

Office accommodation for chief fire officer or head of safety and fire and staff administration should suit the requirements.

Lockers, rest room, dining room, and other facilities for fire-service personnel should be provided.

A control room with communication facilities and fire-alarm annunciator incorporating with panels should be provided.

Provision of storage and loading for foam compound and other materials used during fire and emergencies.

Hydrant(s) with a fresh water supply and facilities for flushing out of the piping systems of the firefighting vehicles should be provided at the rear of the fire station together with hose-cleaning equipment.

Firefighting training and exercise ground. All fire-station office buildings should be provided with air-conditioning systems.

Drying room for drying fire hoses, clothing, and other equipment.

Firefighters outfits for assigned assisting crew, if selected from plant employees.

9.4.3   Category B: fire station

Depending on availability on-site, fixed, or portable firefighting equipment, for a Category B fire station (24 hours manned) five fire trucks and trailers are the minimum requirements.

Types of trucks:
1. General-purpose fire truck, 1 unit
2. Major fire truck, 1 unit
3. Auxiliary fire trucks and trailer, 3 units
   - Layout

The fire station should be provided with parking accommodation of five bays including future expansion.

The following building spaces are required:

Offices for fire master and his staff
Workbench, fire extinguishers and emergency equipment testing and servicing facilities
CO_2, N_2, O_2 transfer charging units and air compressor
Training room for 20 men
Accommodation for rest room, locker room, dining room, etc.
Store room
Control room
Firefighting training ground with selected suitable equipment from Appendices A and B

Other facilities should be provided as follows:

Alarm annunciator and communication system
Visual aids for training
Connections and cable for charging batteries together with heater in the engine cooling system for all fire trucks where required
Firefighters outfit for assigned assisting crew if selected from plant workers or staff
Provisions for storage and loading of foam compound
Fire-water hydrant with hose-washing and cleaning rack

9.4.4   Category C: sub-fire station

For any fire risk area more than a 10-minute drive away from the main fire station, a sub-fire station is required. Truck and crew should attend to fight the fire before the main fire station's trucks and crews take over on their arrival.

- Parking bays: The building should consist of three parking bays one of them with inspection pit, two fire trucks, 1 unit of general-purpose and 1 unit of major fire trucks. The following are the station requirements:
  Parking bays should be equipped with connections for battery chargers and engine-heating system where required
  The bays should be of 11 m in length 45 m wide and no less than 5 m in height
  Locker room, a workbench, hose cleaning and washing rack, rest room, dining room, and drying room
  Offices, communication, and alarm annunciator
  Store room for fire equipment and foam liquid compound
  Fire-water hydrant with fire-hose washing and cleaning facility

9.4.5   Category D: retained fire station

Retained fire stations under control of area fire station Category A or B applies to those stations having three bays with two fire trucks selected for major emergencies and the manpower of two drivers for each shift. Another auxiliary fire truck may be used as spare.

In the case of a fire when the alarm is sounded the appropriate fire trucks will be driven by available drivers or reported for duty and retained selected trained personnel will be picked up on their way to the site of fire.

Retained fire stations can be totally unmanned. In that case, available trained firemen will attend the fire station when alarm has been given. The crew and the truck will proceed to the scene of fire. This system is entirely for major emergency cases. However, assistance should be given when needed.

Unmanned retained stations should be locked and will be opened only by means provided in main fire station or the keys left with selected people. Retained personnel will usually report to the station for checking and inspection of fire equipment once a week.

Requirements for this type of station are as follows:

Locker room
Direct communication with main fire station
Alarm annunciator
Store room
Hose-wash and cleaning facility
Office
Workbench

9.5   Fire-water distribution and storage facilities

This chapter specifies minimum requirements for water supply for firefighting purposes. It is important that all authorities concerned should work together to provide and maintain these minimum water supplies and discussions with municipality fire stations would include not only the water available from the hydrants but also help to assure the continuous and adequate flow of water for firefighting.

The following items are also included in standards:

Basics for a firefighting water system
Fire-water pumping facilities
Water tanks for fire protection
Fire-hose reel (water) for fixed installation
Water-spray fixed system

- Water is the most commonly used agent for controlling and fighting a fire, by cooling adjacent equipment and for controlling and/or extinguishing the fire either by itself or combined as a foam. It can also provide protection for firefighters and other personnel in the event of fire. Water should therefore be readily available at all the appropriate locations, at the proper pressure and in the required quantity.
- Fire water should not be used for any other purpose.
- Unless otherwise specified or agreed, the company requirements that are given for major installations such as refineries, petrochemical works, crude oil production areas where large facilities are provided, and for major storage areas should be applied.
- In determining the quantity of fire water, i.e., "required fire-water rate," protection of the following areas should also be considered:

General process
Storage (low pressure), including pump stations, manifolds, in line blenders, etc.
Pressure storage (LPG, etc.)
Refrigerated storage (LNG, etc.)
Jetties
Loading
Buildings
Warehouses

- Basically, the requirements consist of an independent fire-grid main or ring main fed by permanently installed fire pumps taking suction from a suitable large capacity source of water such as storage tank, cooling tower basin, river, sea, etc. The actual source will depend on local conditions and is to be agreed on with the company.
- The water will be used for direct application to fires and for the cooling of equipment. It will also be used for the production of foam.

9.6 Water supplies

The choice of water supplies should be made in cooperation with the relevant authorities.

9.6.1 Public water systems

One or more connections from a reliable public water system of proper pressure and adequate capacity usually provides a satisfactory supply. A high static water pressure should not, however, be the criterion by which the efficiency of the supply is determined. If this cannot be done, the post-indicator valves should be placed where they will be readily accessible in case of fire and not liable to injury. Where post-indicator valves cannot readily be used, as in a city block, underground valves should conform to these provisions and their locations and direction of turning to open should be clearly marked.

Adequacy of water supply should be determined by flow tests or other reliable means. Where flow tests are made, the flow in (L/min.) together with the static and residual pressures should be indicated on the plan.

Public mains should be of ample size, in no case smaller than 15 cm (6 in). No pressure-regulating valve should be used in water supply except by special permission of the authority concerned. Where meters are used they should be of an approved type. Where connections are made from public waterworks systems, it may be necessary to guard against possible contamination of the public supply. The requirements of the public health authority should be determined and followed. Connections larger than 50.8 mm to public water systems should be controlled by post-indicator valves of a standard type and located no less than 12.2 m from the buildings and units protected.

9.7   Basis for a firefighting water system

A ring-main system should be laid around processing areas or parts thereof, utility areas, loading and filling facilities, tank farms, and buildings while one single line should be provided for jetties and a firefighting training ground, complete with block valves and hydrants.

The water supply should be obtained from at least two centrifugal pumps of which one is electric motor-driven and one driven by a fully independent power source e.g. a diesel engine, the latter serving as a spare pump.

The water quantities required are based on the following considerations:

There will be only one major fire at a time.

As a recommendation in processing units the minimum water quantity is $200 \, dm^3/s$ or air foam making and exposure protection. It is assumed that approximately 30% of this quantity is blown away and evaporates; the balance of this quantity, which is $140 \, dm^3/s$ per processing unit, should be drained via a drainage system.

Note: The quantity of fire water required for a particular installation should be assessed in relation to fire incidents that could occur on that particular site, taking into account the fire hazard, the size, duties, and location of towers, vessels, etc. The fire-water quantity for installations having a high potential fire hazard should normally be no less than $820 \, m^3/h$ and no greater than $1360 \, m^3/h$.

For storage areas the quantity needed for making air foam for extinguishing the largest cone roof tank on fire and for exposure protection of adjacent tanks.

For pressure storage areas the quantity needed for exposure protection of spheres using sprinklers.

For jetties the quantity needed for fighting fires on jetty decks and ship manifolds with air foam as well as for exposure protection in these areas.

The policy for a single major fire or more to occur simultaneously should be decided on by the authorities concerned.

Note: The above specification is based on one major fire only.

For new installations the quantities required for items (a) to (f) above should be compared, and the largest figure should be adhered to for the design of the firefighting system.

The system pressure should be such that at the most remote location a pressure of 10 bar can be maintained during a water take-off required at that location.
Firefighting water lines should be provided with permanent hydrants.
Hydrants with 4 outlets should be located around processing units, loading facilities, storage facilities for flammable liquids, and on jetty heads and berths.
Hydrants with two outlets should be located around other areas, including jetty approaches.
Fire-hose reels should be located in each process unit, normally 31–47 m apart at certain strategic points.
The water will be applied using hose and branch pipes using jet, spray, or fog nozzles, or by fixed or portable monitors preferably with interchangeable nozzles for water or foam jets.

9.7.1 Fire-water ring main system

Fire-water ring mains of the required capacity should be laid to surround all processing units, storage facilities for flammable liquids, loading facilities for road vehicles and rail cars, bottle filling plants, warehouses, workshops, utilities, training centers, laboratories, and offices. Normally, these units will also be bounded by service roads. Large areas should be sub-divided into smaller sections, each enclosed by fire-water mains equipped with hydrants and block valves.

A single fire-water pipeline is only acceptable for a firefighting training ground. Fire water to jetties should be supplied by a single pipeline provided that it is interconnected with a separate pipeline for water-spray systems. The fire-water pipelines from the fire pumps to the jetty should be provided with isolating valves, for closing in the event of serious damage to the jetty. These valves should close without causing high surge pressures.

The fire-water mains should be provided with full bore valve flushing connections so that all sections and dead ends can be properly flushed out. The flushing connections should be sized for a fluid velocity in the relevant piping of no less than 80% of the velocity under normal design conditions but for no less than 2 m/s.

Fire-water mains should normally be laid underground in order to provide a safe and secure system, and which will give in addition, protection against freezing for areas where the ambient temperature can drop below 0°C. When in exceptional circumstances, fire-water mains are installed above ground they should be laid alongside roads and not in pipe tracks where they could be at risk from spill fires.

The basic requirements consist of an independent fire grid main or ring main fed by permanently installed fire pumps. The size of ring main and fire pumps should be such as to provide a quantity of water sufficient for the largest single risk identified within the overall installation.

Suction will be from a suitable large capacity source of water such as storage tank, cooling tower basin, river, sea, etc. The actual source will depend on local conditions and is to be investigated. Pump suction lines should be positioned in a safe and protected location and incorporate permanent, but easily cleanable strainers or screening equipment for the protection of fire pumps.

Advantage should be taken where available in obtaining additional emergency water supplies through a mutual aid scheme or by re-cycling but mandatory national or local authority requirements may modify these to a considerable extent.

9.7.2 Fire-water ring main/network design

The fire-water mains network pipe sizes should be calculated and based on design rates at a pressure of 10 bar gauge at the takeoff points of each appropriate section, and a check calculation should be made to prove that pressure drop is acceptable with a blocked section of piping in the network. The maximum allowable flow/velocity in the system should be 3.5 m/s.

Fire-water rates should, however, be realistic quantities since they determine the size of fire-water pumps, the fire-water ring main system and the drainage systems that have to cope with the discharged fire water. If the drainage system is too small or becomes blocked, major hazards such as burning hydrocarbons floating in flooded areas may occur to escalate the fire. Facilities for cleaning should therefore be provided. For large areas such as pump floors, and in pipe tracks, fire stops should be provided to minimize the spillage area. It is assumed that 30% of fire water evaporates or is blown away while extinguishing a fire. This figure should be taken into account for the design of drainage systems.

Under nonfire conditions, the system should be kept full of water and at a pressure of 2 to 3 bar gauge using a jockey pump, by a connection to the cooling water supply system, or by static head from a water storage tank. If a jockey pump is used, it should be 'spared' and both pumps should have a capacity of 15 m³/h to compensate for leakages.

The fire-water ring main systems should be equipped with hydrants. A typical arrangement of a fire-water distribution system is shown in Figure 9.1.

A single water line connected to the ring main system should run along the jetty approach to the jetty deck. This line should be fitted with a block valve located at a distance of about 50 m from the jetty deck.

For small chemical plants, depots, and minor production and treatment areas, etc., for which precise commensurate with the size of risk involved, requirements should be as specified or agreed with authorities.

9.8 Fire-water pumping facilities

Fire-water should be provided by at least two identical pumps, each pump should be able to supply the maximum required capacity for a fire-water ring main system. Fire-water pumps should be of the submerged vertical type when taking suction from open water, and of the horizontal type when suction is taken from a storage tank.

The fire-water pumps should be installed in a location that is considered to be safe from the effects of fire and clouds of combustible vapor, and from collision damage by vehicles and shipping. They should for example, be at least 100 m away from jetty loading points and from moored tankers or barges handling liquid hydrocarbons. They should be accessible to facilitate maintenance, and be provided with hoisting facilities.

The main fire-water pump should be driven by an electric motor and the second pump, of 100% stand-by capacity, by some other power source, preferably a diesel engine. Alternatively, three pumps, each capable of supplying 60% of the required

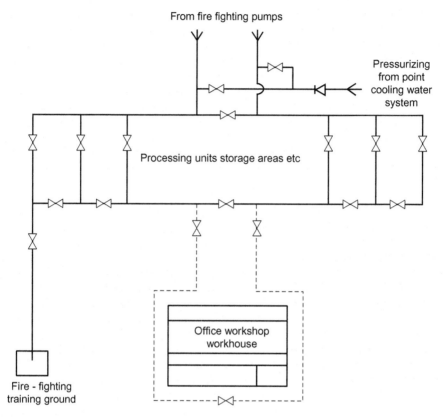

Figure 9.1 Typical sketch of fire-fighting water distribution system.

capacity may be installed, with one pump driven by an electric motor and the other two by diesel engines.

Note: Refinery with over 100,000 barrels a day capacity should have two electric and two diesel pumps.

When the required pump capacity should exceed $1000\,m^3$, two or more smaller pumps should be installed, together with an adequate number of spare pumps. The power of the drives, for both main and standby units should be so rated, that it will be possible to start the pumps against an open discharge with pressure in the fire-water ring main system under nonfire conditions, normally at 2 to 3 bar gauge unless otherwise agreed by the relevant authorities. The main firewater pump should be provided with automatic starting facilities that will function immediately the fire alarm system becomes operational due to one of the following actions:

When a fire call point is operated
When an automatic fire-detection system is operated
When the pressure in the fire-water ring main system drops below the minimum required static pressure which is normally 2 to 3 bar (ga)

The stand-by fire-water pumps should be provided with automatic starting facilities that will function:

If the main fire-water pumps do not start, or having started, fail to build up the required pressure in the firewater ring main system within 20 sec.

Manual starting of each pump unit (without the fire alarms coming into operation) should be possible at the pump, from the control center and, when necessary, from the gate house. Manual stopping of each pump unit should only be possible at the pump.

Fuel-tank Capacity. Fuel-supply tank(s) should have capacity at least equal to I gal per horsepower (5.07 L/kW), plus 5% volume for expansion and 5% volume for sump. Larger capacity tanks may be required and should be determined by prevailing conditions, such as refill cycle and fuel heating due to recirculation, and be subject to special conditions in each case. The fuel-supply tank and fuel should be reserved exclusively for the fire-pump diesel engine.

The pumps should have stable characteristic curves exhibiting a decrease in head with increasing capacity from zero flow to maximum flow; a relatively flat curve is preferred with a shut off pressure not exceeding the design pressure by more than 15%.

The total water supply within the refinery should be capable of supplying the maximum flow for a period of no less than 4 to 6 hours, consistent with projected fire scenario needs. Where the water system is supplied from a tank or reservoirs, the quantity of water required for fire protection. However, where the tank or reservoir is automatically filled by a line from a reliable, separate supply, such as from a public water system or wells, the total quantity in storage may be reduced by the incoming fill rate.

9.9 Plans

A layout plan should be prepared and approved in every case where a new private fire-service main is considered. The plan should be drawn to scale and should include all essential details such as:

- Size and location of all water supplies.
- Size and location of all piping, indicating, where possible, the class and type and depth of existing pipe, the class, and type of new pipe to be installed and the depth to which it is to be buried.
- Size, type, and location of valves indicate if located in pit or if operation is by post-indication key wrench through a curb box. Indicate the size, type, and location of meters, regulators, and check valves.
- Size and location of hydrants, showing size and number of outlets and if outlets should be equipped with independent gate valves. Indicate if hose boxes and equipment should be provided and by whom.
- Sprinkler and standpipe risers and monitor nozzles hose reels to be supplied by the system.
- Location of fire-department connections, if part of private fire-service main system, including detail of connections.
- Location, number, and size of fire-water pumps installed.

9.10   General winterizing

9.10.1   Water systems

When the lowest recorded ambient temperature is below 0°C, water mains should be buried below the frost line, but no less than 0.6 m. below ground level. Branches to equipment that should be shut down while the remainder of the unit is in operation should have the block valves protected using one of the following methods:

> Provide a bypass, just under the block valves, from the supply back to the return. This bypass should be 12.7 mm for lines 76.2 mm and smaller, 25.4 mm for lines 101.6 mm to 203 mm and 38 mm for lines larger than 203 mm. Provide a drain in the valve body just above the gate of the valve so that all water can be drained out of the line above the valve after it is closed, and another in the bonnet to drain the void around the gate and stem.

All portions of water lines above the frost line should be provided with drains.

In freezing climate where water lines must be above ground, branch lines should be taken from the top of horizontal main lines with the block valve in a horizontal position and drains should be provided on the dead leg side.

In climates where freezing occurs, provisions should be made to prevent stored water from freezing, e.g., by circulation or by heating; alternatively, storage capacity should be increased to compensate for the ice layer. The quality of the water should be monitored and treated to control the growth of algae and/or barnacles. The replenishment system should also include easy-to-clean strainer facilities.

In locations where freezing can occur, the fire-water pumps should be installed in a housing for protection; for other locations, a rain/sun cover only may be required. When the pump suction is taken from open water, a strainer system that is easy to clean should be provided. When the pump suction is taken from storage a strainer should be included in the replenishment supply to the storage tank. The discharge line from each pump should be fitted with a check valve, a test valve, a pressure gauge and a block valve with a locking device; the test valves should have a common return line with a flow metering unit. Each pump should be connected separately to a common manifold.

The pump common discharge manifold should normally be connected to the fire-water ring main system by two separate pipelines each with a block valve and of the same size as the ring main. Lines to jetty heads will normally take the form of dry mains due to the problems of providing frost protection.

9.11   Water tanks for fire protection

Standards cover elevated tanks on towers or building structures, grade or below grade water storage tanks, and pressure tanks.

9.11.1   Capacity and elevation

The size and elevation of the tank should be determined by conditions at each individual property after due consideration of all factors involved. Where tanks should supply sprinklers. Whenever possible, standard sizes of tanks and heights of towers should be used as given in standards.

The capacity of the tank is the number of cubic meters available above the outlet opening. The net capacity between the outlet opening of the discharge pipe and the inlet of the overflow should be at least equal to the rated capacity. For gravity tanks with large plate risers, the net capacity should be the number of cubic meters between the inlet of the overflow and the designated low-water level line. For suction tanks, the net capacity should be the number of cubic meters between the inlet of the overflow and the level of the vortex plate.

The standard sizes of steel tanks are: 18.93, 37.85, 56.78, 75.70, 94.63, 113.55, 151.40, 189.25, 227.10, 283.88, 378.50, 567.75, 757.00, 1135.50, and 1892.50 cubic meters net capacity. Tanks of other sizes may be built (according to NFPA-20).

The capacity of pressure tanks should be as approved by the authority concerned.

The standard sizes of wooden tanks are 18.93, 37.85, 56.78, 75.70, 94.63, 113.55, 151.40, 189.25, 227.10, 283.88, and 378.50 cubic meters net capacity. Tanks of other sizes may be built.

The standard capacities of COATED-fabric tanks are in increments of 378.5 to $3785\,m^3$ (According to NFPA 22). Figure 9.2 shows a typical fire-water storage tank.

9.11.2   Location of tanks

The location chosen should be such that the tank and structure will not be subject to fire exposure from adjacent units. If lack of yard room makes this impracticable, the exposed steel work should be suitably fireproofed or protected by open sprinklers.

Figure 9.2 A typical fire water storage tank.

Fireproofing where necessary should include steel work within 6.1 m of combustible buildings, windows, doors, and flammable liquid and gas from which fire might erupt.

When steel or iron is used for supports inside the building near combustible construction or occupancy, it should be fireproofed inside the building, 1.8 m above combustible roof coverings and within 6.1 m of windows and doors from which fire might erupt. Steel beams or braces joining two building columns that support a tank structure should also be suitably fireproofed when near combustible construction or occupancy. Interior timber should not be used to support or brace tank structures.

Fireproofing, where required, should have a fire-resistant rating of no less than 2 hours.

Foundations or footings should furnish adequate support and anchorage for the tower.

If the tank or supporting trestle is to be placed on a building, the building should be designed and built to carry the maximum loads.

9.11.3 Storage facilities

Fire water taken from open water is preferred, but if water of acceptable quality for firefighting in the required quantity, cannot be supplied from open water, or if it is not economically justified because of distance to install firewater pumps at an open source, water storage facilities should be provided.

Storage facilities may consist of an open tank of steel or concrete or a basin of sufficient capacity. The tank or basin should have two compartments to facilitate maintenance, each containing 60% of the total required capacity and there should be adequate replenishment facilities. A single compartment of 100% capacity is acceptable providing that an alternative source of water, e.g., from temporary storage will be available during maintenance periods. The replenishment rate should normally not be less than 60% of the total required fire-water pumping capacity.

If a 100% replenishment rate is available, the stored fire-water capacity may be reduced if agreed by the oil and gas companies. Authorities that may be considered for replenishment are plant cooling water, open water or below-ground water, provided that it is available at an acceptable distance and in sufficient quantity for a minimum of 6 hours uninterrupted firefighting at the maximum required rate.

9.11.4 Steel gravity and suction tanks

- Pressure Tanks: Pressure tanks may be used for limited private fire-protection services. Pressure tanks should not be used for any other purpose unless approved by the authority concerned.
- Air-pressure and water level: Unless otherwise approved by the relevant authority, the tank should be kept two-thirds full of water, and an air pressure of at least 5.2 bars by the gauge should be maintained. As the last of the water leaves the pressure tank, the residual pressure shown on the gauge should not be less than zero, and should be sufficient to give no less than 1.0 bars pressure at the highest sprinkler under the main roof of the building.
- Exception: Other pressures and water levels may be required for hydraulically designed systems.
- Location: Pressure tanks should be located above the top level of sprinklers.

9.11.5   Housing

Where subject to freezing, the tank should be located in a substantial noncombustible housing. The tank room should be large enough to provide free access to all connections, fittings, and manhole, with at least 457 mm around the rest of the tank. The distance between the floor and any part of the tank should be at least 0.91 m. The floor of the tank room should be watertight and arranged to drain outside of the enclosure. The tank room should be adequately heated to maintain a minimum temperature of 4.4°C and should be equipped with ample lighting facilities.

9.11.6   Buried tanks

Where lack of space or other conditions require it, pressure tanks may be buried if the following requirements are satisfied:

For protection against freezing the tank should be below frost line.
The end of the tank and at least 457 mm of its shell should project into the building basement or a pit in the ground, with protection against freezing. There should be adequate space for inspection and maintenance and use of the tank manhole for interior inspection.

The exterior surface of the tank should be fully coated as follows for protection against corrosion conditions indicated by a soil analysis:

An approved cathodic system of corrosion protection should be provided.
At least 305 mm of sand should be backfilled around the tank.
The tank should be above the maximum ground water level so that buoyancy of the tank where empty will not force it upward. An alternative would be to provide a concrete base and anchor the tank to it.
The tank should be designed with strength to resist the pressure of earth against it.
A manhole should be located preferably on the vertical centerline of the tank end to clear the knuckle while remaining as close as possible to it.
- Tank's material: Types of materials should be limited to steel, wood, concrete and coated fabric. The elevated wood and steel tanks should be supported on steel or reinforced concrete towers.

9.12   Fire-hose reels (water) for fixed installations

- Rotation: The hose reel should rotate around a spindle so that the hose can be withdrawn freely.
- Reel: The drum or hose support of the first coil of hose should be no less than 150 mm diameter. The fitting by which the hose is attached to the reel should be arranged in such a way that the hose is not restricted or flattened by additional layers of hose being placed on it.
- Manual inlet valve: The inlet valve of a manual reel should be a screwdown above-ground stop valve or a gate valve.
  Note: To facilitate ease of installation and maintenance a union should be fitted between the valve and the reel. The valve should be closed by turning the handle in a clockwise direction. The direction of opening should be permanently marked on the handle, preferably by an embossed arrow and the word OPEN.

- Reel Size: Reels should be of sufficient size to carry the length of hose fitted, excluding the nozzle, within the space defined by the end plates. The length of hose fitted should be not more than 45 m for 19 mm internal diameter hose or 35 m for 25 mm internal diameter hose.
- Range and water-flow rate: The water flow rate should be no less than 24 L/min. and the range of the jet should be no less than 6 m. The output of the nozzle, whether plain jet or jet/spray, should comply with the above flow rate except that the range of the spray should be less than 6 m (according to NFPA-22).

9.12.1   General considerations

- Limitations of hose in certain circumstances: Although standards permit up to 45 m of hose on hose reels, frequently there are circumstances in which there is a likelihood of the hose having to be handled by people of only moderate physical strength. In such cases, and also when the likely routes for the hose are tortuous, the length and size of hose on the reel should be limited, and the siting and provision of reels should be reviewed with these limitations in mind.

  Note: One hose reel should be provided to cover every 800 m² of floor space or part thereof.
- Siting: Hose reels should be sited in prominent and accessible positions at each floor level adjacent to exits in such a way that the nozzle of the hose can be taken into every area and within 6 m of each part of an area, having regard to any obstruction. Where heavy furniture or equipment is introduced into an area, the hose and nozzle should be capable additionally of directing a jet into the back of any recess formed. In exceptional circumstances consideration should also be needed as to the desirability of siting hose reels in such a way that if a fire prevents access to one hose reel site, the fire can be attacked from another hose reel in the vicinity.

9.12.2   Coordinating spaces for hose reels

The spaces required for most types of hose reels and their location in relation to floor or ground level for "horizontal" hose reels are not given as these are considered to be special installations. The range of acceptable choices from the point of view of dimensional co-ordination. First preferences are indicated by a thick blob and second preferences are indicated by a smaller blob.

Notes: The basic space accommodates:

Eeel and valve
Hanging loop of hose
Guide or necessary space for proper withdrawal of the hose
Component case (if any)

The space sizes have been based on the normal arrangement where the water supply is fed upwards. Downward or sideways feeds should be treated as special installations.

9.12.3   Water supply for hose reels

- Minimum requirement: As a minimum, the water supply to hose reels should be such that when the two topmost reels in a building or unit are in use simultaneously, each will provide a jet of approximately 6 m in length and will deliver no less than 0.4 liter/s (24 liters/min.).

For example, when a length of 30 m of hose-reel tubing is in use with a 6.5 mm nozzle, a minimum running pressure of 1.5 bar* is required at the entry to each reel and similarly for a 4.5 mm nozzle where a minimum running pressure of 4 bar is required.

$$*1\,bar = 10^5\,N/m^2 = 100\,kPa$$

9.13 Water-spray fixed systems for fire protection

9.13.1 Applicability

Water spray is applicable for protection of specific hazards and equipment, and may be installed independently of or supplementary to other forms of fire-protection systems or equipment.

- Hazards: Water-spray protection is acceptable for the protection of hazards involving:
  Gaseous, liquid flammable, and toxic materials
  Electrical hazards such as transformers, oil switches, motors, cable trays, and cable runs
  Ordinary combustibles such as paper, wood, and textiles
  Certain hazardous solids
- Uses: In general, water spray may be used effectively for any one or a combination of the following purposes:
  Extinguishment of fire
  Control of burning
  Exposure protection
  Prevention of fire
- Limitations: There are limitations to the use of water spray that should be recognized. Such limitations involve the nature of the equipment to be protected, the physical and chemical properties of the materials involved and the environment of the hazard. Other standards also consider limitations to the application of water (slopover, frothing, electrical clearances, etc.).
- Alarms: The location, purpose, and type of system should determine the alarm service to be provided. An alarm, actuated independently of water flow, to indicate operation of the detection system should be provided on each automatically controlled system. Electrical fittings and devices designed for use in hazardous locations should be used where required by standard.
- Flushing connections: A suitable flushing connection should be incorporated in the design of the system to facilitate routine flushing as required.

9.13.2 Water supplies

It is of vital importance that water supplies be selected that provide water as free as possible from foreign materials.

- Volume and pressure: The water-supply flow rate and pressure should be capable of maintaining water discharge at the design rate and duration for all systems designed to operated simultaneously. For water-supply distribution systems, an allowance for the flow rate of hose streams or other fire-protection water requirements should be made in determining the maximum demand. Sectional control shut-off valves should be located with particular care so that they will be accessible during an emergency case. When only a limited water source is available, sufficient water for a second operation should be provided so that the protection can be re-established without waiting for the supply to be replenished.

- Sources: The water supply for water-spray systems should be from reliable fire-protection water supplies, such as:
Connections to waterworks systems
Gravity tanks (in special cases pressure tanks) or
Fire pumps with adequate water supply
- Fire-department connection: One or more fire-department connections should be provided in all cases where water supply is marginal and/or where auxiliary or primary water supplies may be augmented by the response of suitable pumper apparatus responding to the emergency. Fire-department connections are valuable only when fire-department pumping capacities are equal to maximum demand-flow rate. Careful consideration should be given to such factors as the purpose of the system, reliability, and capacity and pressure of the water system. The possibility of serious exposure fires and similar local conditions should be considered. A pipeline strainer in the fire-department connection should be provided if indicated by 13.8.5. Where a fire-department connection is required, suitable suction provisions for the responding pumper apparatus should be provided.

9.14   System design and installation

- Workmanship: Water-spray system design, layout, and installation should be entrusted to none but fully experienced and responsible parties. Water-spray system installation is a specialized field of sprinkler system installation which is a trade in itself. Before a water-spray system is installed or existing equipment remodeled, complete working plans, specifications and hydraulic calculations should be prepared and made available to interested parties.

9.14.1   Density and application

- Extinguishment: Extinguishment of fires by water spray may be accomplished by surface cooling, by smothering from steam produced, by emulsification, by dilution, or by various combinations thereof. Systems should be designed so that, within a reasonable period of time, extinguishment should be accomplished and all surfaces should be cooled sufficiently to prevent "flashback" occurring after the system is shut off. The design density for extinguishment should be based on test data or knowledge concerning conditions similar to those that will apply in the actual installation. A general range of water-spray application rates that will apply to most ordinary combustible solids or flammable liquids is from 6.1 $(L/min.)/m^2$ to 20.4 $(L/min.)/m^2$ of protected surface.
Note: There is some data available on water application rates needed for extinguishment of certain combustibles or flammables; however, much additional test work is needed before minimum rates can be established. Each of the following methods or a combination of them should be considered when designing a water-spray system for extinguishment purposes:
Surface cooling
Smothering by steam produced
Emulsification
Dilution
Other factors.

- Cable trays and cable runs: When insulated wire and cable or nonmetallic tubing is to be protected by an automatic water-spray (open nozzle) system maintained for extinguishment of fire that originates within the cable or tube (i.e., the insulation or tubing is subject to ignition and propagation of fire), the system should be hydraulically designed to impinge water directly on each tray or group of cables or tubes at the rate of 6.1 (L/min.)/m^2 on the horizontal or vertical plane containing the cable or tubing tray or run. Exception: Other water-spray densities and methods of application should be used if verified by tests and acceptable to the authority of N.I.O.C. Automatic detection devices should be sufficiently sensitive to rapidly detect smoldering or slow-to-develop flames. When there is concern that spills of flammable liquids or molten materials will expose cables, nonmetallic tubing, and tray supports, design of protection systems should be in accordance with that recommended for exposure protection. When electrical cables or tubing in open trays or runs are to be protected by water spray from fire or spill exposure, a basic rate of 12.2 (L/min.)/m^2 of projected horizontal or vertical plane area containing the cables or tubes should be provided. Water-spray nozzles should be arranged to supply water at this rate over and under or to the front and rear of cable or tubing runs and to the racks and supports. Where flame shields equivalent to 1.6 mm thick steel plate are mounted below cable or tubing runs, the water density requirements may be reduced to 6.1 (L/min.)/m^2 over the upper surface of the cable or rack. The steel plate or equivalent flame shield should be wide enough to extend at least 152 mm beyond the siderails of the tray or rack in order to deflect flames or heat emanating from spills below cable or conduit runs. Where other water-spray nozzles are arranged to extinguish, control, or cool exposing liquid surfaces, the water-spray density may be reduced to 6.1 (L/min.)/m^2 over the upper surface, front, or back of the cable or tubing tray or run. Fixed water-spray systems designed for protecting cable or tubing and their supports from heat of exposure from flammable or molten liquid spills should be automatically actuated.
- Control of burning: A system for the control of burning should function at full effectiveness until there has been time for the flammable materials to be consumed, for steps to be taken to shut off the flow of leaking material, for the assembly of repair forces, etc. System operation for hours may be required. Nozzles should be installed to impinge on the areas of the source of fire, and where spills may travel or accumulate. The water application rate on the probable surface of the spill should be at the rate of no less than 20.4 (L/min.)/m^2. Pumps or other devices that handle flammable liquids or gases should have the shafts, packing glands, connections, and other critical parts enveloped in directed water spray at a density of no less than 20.4 (L/min.)/m^2 of projected surface area.

9.14.2   Exposure protections

The system should be able to function effectively for the duration of the exposure fire that is estimated from knowledge of the nature and quantities of the combustibles and the probable effect of firefighting equipment and materials. System operation for hours should be required.

Automatic water-spray systems for exposure protection should be designed to operate before the formation of carbon deposits on the surfaces to be protected and before the possible failure of any containers of flammable liquids or gases because of the temperature rise. The system and water supplies should, therefore, be designed to discharge effective water spray from all nozzles within 30 sec following operation of the detection system.

The densities specified for exposure protection contemplate minimal wastage of 2.0 (L/min.)/m^2.

- Spray nozzles: Care should be taken in the application of nozzle types. Distance of "throw" or location of nozzle from surface should be limited by the nozzle's discharge characteristics. Care should also be taken in the selection of nozzles to obtain waterways, which are not easily obstructed by debris, sediment, sand, etc., in the water.

- Selection: The selection of the type and size of spray nozzles should be made with proper consideration given to such factors as physical character of the hazard involved, draft or wind conditions, material likely to be burning, and the general-purpose of the system.

- Position: Spray nozzles may be placed in any position necessary to obtain proper coverage of the protected area. Positioning of nozzles with respect to the surfaces to be protected or to fires to be controlled or extinguished should be guided by the particular nozzle design and the characteristics of water spray produced. The effect of wind and fire draft on very small drop sizes or on larger drop sizes with little initial nozzle velocity should be considered, since these factors will limit the distance between nozzle and surface, and will limit the effectiveness of exposure protection, fire control or extinguishment. Care should be taken in positioning nozzles that water spray does not miss the targeted surface and reduce the efficiency or calculated discharge rate $(L/- min.)/m^2$. Care should also be exercised in placement of spray nozzles protecting pipelines handling flammable liquids under pressure, where such protection is intended to extinguish or control fires resulting from leaks or ruptures.

- Strainers: Main pipeline strainers should be provided for all systems utilizing nozzles with waterways less than 9.5 mm and for any system where the water is likely to contain obstructive material. Mainline pipeline strainers should be installed so as to be accessible for flushing or cleaning during the emergency. Individual strainers should be provided at each nozzle where water passageways are smaller than 3.2 mm. Care should be taken in the selection of strainers, particularly where nozzle waterways are less than 6.5 mm in least dimension. Consideration should be given to the size of screen perforation, to volume available for accumulation without excessive friction loss, and to the facility for inspection and cleaning.

- Vessels: The rules for exposure protection should consider emergency relieving capacity for vessels, based on a maximum allowable heat input of $18\,930\,W/m^2$ of exposed surface area. The density should be increased to limit the heat absorption to a safe level in the event required emergency relieving capacity is not provided. Water should be applied to vertical or inclined vessel surfaces at a net rate of no less than 9.2 $(L/min.)/m^2$ of exposed uninsulated surface. Where run-down is considered for vertical or inclined surfaces the vertical distance between nozzles should not exceed 3.7 m. The horizontal extremities of spray patterns should at least meet. Spherical or horizontal cylindrical surfaces below the vessel equator cannot be considered wettable from run-down. Where projections (manhole flanges, pipe flanges, support brackets, etc.) will obstruct water-spray coverage, including run-down or slippage on vertical surfaces, additional nozzles should be installed around the projections to maintain the wetting pattern that otherwise would be seriously interrupted. Bottom and top surfaces of vertical vessels should be completely covered by directed water spray at an average rate of no less than 9.2 $(L/min.)/m^2$ of exposed uninsulated surface. Consideration should be given to slippage but on the bottom surfaces the horizontal extremities of spray patterns should at least meet. Special attention should be given to distribution of water spray around relief valves and around supply piping and valve connection projections. Uninsulated vessel skirts should have water spray applied on one exposed (uninsulated) side, either inside or outside, at a net rate of no less than 9.2 $(L/min.)/m^2$.

- Structures and miscellaneous equipment: Horizontal, stressed (primary), and structural steel members should be protected by nozzles spaced not greater than 3 m on centers

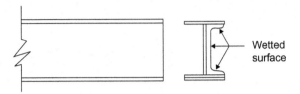

Figure 9.3 The wetted surface of structural member-beam or column.

(preferably on alternate sides) and of such size and arrangement as to discharge no less than 4.1 (L/min.)/m^2 over the wetted area.

The wetted surface of structural member-beam or column is defined as one side of the web and the inside surface of one side of the flanges as shown above.

Vertical structural steel members should be protected by nozzles spaced not greater than (3 m) on centers (preferably on alternate sides) and of such size and arrangement as to discharge no less than 9.2 (L/min.)/m^2 over the wetted area (see Figure 9.3).

Metal pipe, tubing, and conduit runs should be protected by water spray directed toward the horizontal plane surface projected by the bottom of the pipes or tubes.

Nozzles should be selected to provide essentially total impingement on the entire horizontal surface area within which pipes or tubes are or could be located.

For single-level pipe racks, water-spray nozzles should discharge onto the underside of the pipe at a plan view density of 9.2 (L/min.)/m^2.

For two-level pipe racks, water-spray nozzles should discharge onto the underside of the lower level at a plan view density of 8.2 (L/min.)/m^2 and additional spray nozzles should discharge onto the underside of the upper level at a plan view density of 6.1 (L/min.)/m^2.

For three-, four-, and five-level pipe racks, water-spray nozzles should discharge onto the underside of the lowest level at a plan view density of 8.2 (L/min.)/m^2 and additional spray nozzles should discharge onto the underside of alternate levels at a plan view density of 6.1 (L/min.)/m^2. Water spray should be applied to the underside of the top level even if immediately above a protected level.

For pipe racks of six or more levels, water-spray nozzles should discharge onto the underside of the lowest level at a plan view density of 8.2 (L/min.)/m^2 and additional spray nozzles should discharge onto the underside of alternate levels at a plan view density of 4.1 (L/min.)/m^2. Water spray should be applied to the underside of the top level even if immediately above a protected level.

Water-spray nozzles should be selected and located such that extremities of water-spray patterns should at least meet and the discharge should essentially be confined to the plan area of the pipe rack.

Spacing between nozzles should not exceed 3 m and nozzles should be no more than 0.8 m below the bottom of the pipe level being protected.

Consideration should be given to obstruction to the spray patterns presented by pipe supporting steel. Where such interferences exist, nozzles should be spaced within the bays.

- Exceptions: Water-spray protection with the same density as specified previously may be applied to the top of pipes on racks where water-spray piping cannot be installed below the rack due to possibility of physical damage or space is inadequate for proper installation. Vertically stacked piping may be protected by water spray directed at one side of the piping at a density of 6.1 (L/min.)/m^2. Table 9.1 provides more information.

Table 9.1 **Protection of metal pipe, tubing, and conduit**

Number of rack levels	Plan view density at lowest level		Plan view density at upper level(s)*		Levels requiring nozzles
	Gpm/ft^2	$(L/min)/m^2$	Gpm/ft^2	$(L/min)/m_2$	
1	0.25	9.2	N/A	N/A	All
2	0.20	8.2	0.15	6.1	All
3, 4 or 5	0.20	8.2	0.15	6.1	Alternate
6 or more	0.20	8.2	0.10	4.1	Alternate

*The table values contemplate exposure from a spill fire.

- Transformers: Transformer protection should contemplate essentially complete impingement on all exterior surfaces, except underneath surfaces that in lieu thereof may be protected by horizontal projection. The water should be applied at a rate no less than 9.2 (L/min.)/m² of projected area of rectangular prism envelope for the transformer and its appurtenances and no less than 6.1 (L/min.)/m² on the expected nonabsorbing ground surface area of exposure. Additional application is needed for special configurations, conservator tanks, pumps, etc. Spaces greater than 305 mm in width between radiators, etc., should be individually protected. Water-spray piping should not be carried across the top of the transformer tank, unless impingement cannot be accomplished with any other configuration and provided the required distance from the live electrical components is maintained. In order to prevent damage to energized bushings or lightning arrestors, water spray should not envelop this equipment by direct impingement, unless so authorized by the manufacturer or manufacturer's literature.
- Belt conveyers
- Drive unit: Water-spray system should be installed to protect the drive rolls, the take-up rolls, the power units, and the hydraulic-oil unit. The rate of water application should be 9.2 (L/min.)/m² of roll and belt. Nozzles should be located to direct water spray onto the surfaces to extinguish fire in hydraulic oil, belt, or contents on the belt. Water spray impingement on structural elements should be such as to provide protection against radiant heat or impinging flame.
- Conveyer belt: Water-spray system should be installed to automatically wet the top belt, its contents, and the bottom return belt. Discharge patterns of water-spray nozzles should envelop, at a rate of [9.2 (L/min.)/m²] of top and bottom belt area, the structural parts and the idler-rolls supporting the belt. Water-spray system protection should be extended onto transfer belts, transfer equipment and transfer buildings beyond each transfer point. Or, systems for the protection of adjacent belts or equipment should be interlocked in such a manner that the feeding belt water-spray system will automatically actuate the water-spray system protecting the first segment of the downstream equipment. Special consideration should be given to the interior protection of the building, gallery, or tunnel housing the belt conveyer equipment. Also, the exterior structural supports for galleries should be protected from exposure such as fires in flammables located adjacent to the galleries. The effectiveness of belt conveyer protection is dependent on rapid detection and appropriate interlocks between the detection system and the machinery.

9.15   Fire and explosion prevention

The system should be able to function effectively for a sufficient time to dissolve, dilute, disperse, or cool flammable or hazardous materials. The possible duration of release of the materials should be considered in the selection of duration times. The rate of application should be based on experience with the product or on testing.

9.15.1   Size of system

Separate fire areas should be protected by separate systems. Single systems should be kept as small as practicable, giving consideration to the water supplies and other factors affecting reliability of the protection. The hydraulically designed discharge rate for a single system or multiple systems designed to operate simultaneously should not exceed the available water supply.

9.15.2   Separation of fire areas

Separation of fire areas should be by space, fire barriers, diking, special drainage, or by combination of these. In the separation of fire areas consideration should be given to the possible flow of burning liquids before or during operation of the water-spray systems.

9.15.3   Area drainage

Adequate provisions should be made to promptly and effectively dispose of all liquids from the fire area during operation of all systems in the fire area. Such provisions should be adequate for:

1. Water discharged from fixed fire-protection systems at maximum flow conditions
2. Water likely to be discharged by hose streams
3. Surface water
4. Cooling water normally discharged to the system

There are four methods of disposal or containment:

1. Grading
2. Diking
3. Trenching
4. Underground or enclosed drains

The method used should be determined by:

1. Extent of the hazard
2. Clear space available
3. Protection required

Where the hazard is low, the clear space is adequate, and the degree of protection required is not great, grading is acceptable. Where these conditions are not present, consideration should be given to dikes, trenching, or underground or enclosed drains.

9.16   Valves

9.16.1   Shut-off valves

Each system should be provided with a shut-off valve so located as to be readily accessible during a fire in the area the system protects or adjacent areas, or, for systems installed for fire prevention, during the existence of the contingency for which the system is installed. Valves controlling water-spray systems, except underground gate valves with roadway boxes, should be supervised open by one of the following methods:

Central station, proprietary, or remote station alarm service
Local alarm service that will cause the sounding of an audible signal at a constantly attended point
Locking valves open
Sealing of valves and approved weekly recorded inspection when valves are located within fenced enclosures under the control of the owner

9.17   Automatically controlled valves

Automatically controlled valves should be as close to the hazard protected as accessibility during the emergency will permit, so that a minimum of piping is required between the automatic valve and the spray nozzles. Remote manual tripping devices, where required, should be conspicuously located where readily accessible during the emergency and adequately identified as to the system controlled.

9.17.1   Hydraulic calculation

Table 9.2 shows details of process unit fire-water requirements.

9.18   Fixed-roof tanks containing high-flash liquids

When the stored product has a closed-cup flash point of 65°C (150°F) or higher, a fixed-roof tank can be considered relatively safe. Then water for foam extinguishment is not required, provided the following conditions are met:

If the product is heated, there must be no possibility of the storage temperature exceeding either the flash point or 93°C (200°F).
There must be no possibility of hot oil streams entering the tank at temperatures above 93°C (200°F) or their flash point.
Cutter stock having a flash point below the storage temperature must never be pumped into the tank for blending purposes.
Sufficient fire water should be available to cool exposed adjacent tankage in the event of ignition. Then the tank should be pumped out or allowed to burn out.

The product should not be crude oil with boilover characteristics. If the product were crude, the fire would have to be extinguished before the heat wave reached water at the tank bottom.

Storage temperatures between 93°C (200°F) and 121°C (250°F) should be avoided, as water lenses or water at the tank bottom may reach boiling temperature at any time, resulting in a serious frothover.

If the product is heated above 121°C (250°F) foam extinguishment cannot be accomplished and slopover will occur if foam is applied.

9.19   Floating-roof tanks

Floating-roof tanks are considered virtually ignition-proof, except for rim fires. Thus, there should be sufficient fire water to cool the shell and extinguish a rim fire.

9.19.1   Pressure storage

The water requirement for cooling pressure storage spheres or drums may exceed the maximum cone roof tank fire-water requirement when spheres are of large diameter, or when a number of spheres or drums are closely spaced. However, when adjacent spheres are not over 15 m (50 ft.) in diameter and are at least 30 m (100 ft.) apart, shell to shell, cooling of these spheres may be disregarded.

- Low-pressure refrigerated storage: Water for monitor-cooling streams should be available to cool the tank shell if it is exposed to fire. Water must not be applied directly to refrigerated LP-gas or cryogenic flammable liquid spills or spill fires, since much more rapid vapor evolution or increased fire intensity will result.

Table 9.2 Process unit fire water requirements

Type of process unit	Minimum fire water rate	
	M³/H	US gpm
Atmospheric distillation, vacuum, or combination units with up to 15,900 M³/d (100,000 bbl/d) throughput; treating plants; asphalt stills; others	598	2500
Atmospheric distillation, vacuum, or combination units with 15,900 M³/d (100,000 bbl/d) or higher throughput; catalytic cracking units	900	4000
Light-end units containing volatile oils and hydrogen, such as reformers, catalytic desulfur-izers, and alkylation units	900	4000
Lube oil units and blending facilities	454	2000

9.19.2 Cooling water for exposed tankage

During a tank or sphere fire, cooling streams may be needed for adjacent tankage. However, this does not apply to process units where flammable liquids are not contained in sufficient volume to generate enough heat to require cooling adjacent tankage. Allow at least two 57 m³/h (250 US gpm) cooling streams, or a total of 114 m³/h (500 US gpm), for each adjacent, unshielded fixed, or floating-roof tank within the following limits:

> Within 15 m (50 ft.) of a burning tank or sphere of any size, regardless of wind direction
> Within one tank diameter and within a quadrant that will require the maximum amount of cooling water for tanks that fall within the quadrant
> Within 45 m (150 ft.) of a sphere and within the most congested quadrant

9.19.3 Fire-water capacity vs crude throughput

Figure 9.4 shows fire-water capacity, in cubic meters per hour (thousands of US gallons per minute), plotted against crude throughput, in cubic meters per day (thousands of barrels per calendar day). The curve represents an average of data collected from plants all over the world. This curve can be used as a guide to the required quantity when calculating refinery fire-water capacity by the methods described in this section.

Fire water should be obtained from an unlimited source, such as a natural body of water. When this is not possible, the supply should always be available in a storage tank or reservoir. Fire-resisting suits should have a high thermal insulating value, but practically it may not be possible to have sufficiently high insulating protection against high rate of heating.

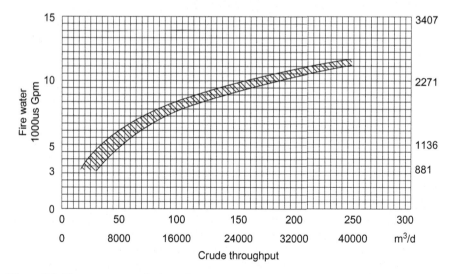

Figure 9.4 Fire water vs crude throughput.

Different types of clothing may be required for protection against heating by radiation and against heating caused by hot air and flame lick. Metalized reflecting fabrics provide effective protection against radiant heat. In standards the use of fire blanket, fire-resisting blanket, curtains, and shields will be discussed.

9.19.4   Specification for fire-resistant suits

- Structural (Firefighter Suit): Where people are working in extremely high temperatures up to 1000 to 1100°C, such as furnace and oven repair, cooking, slagging, firefighting and rescue work, the use of aluminized fabrics are essential. Combination of "Celanece pbi" fiber (25%) and "Conex" meta aramid fiber (75%). The flame-resistant outer shell should not break nor lose the inherent flexibility after exposure to 1200°C flame for period of more than 65 sec. For use by structural firefighters encountering dangerous radio-activity

Figure 9.5 Aluminum-coated, heat-protective suit is used in fighting fires without entering the burning area. Transparent faceshield is metal coated to offer increased heat protection. Head fitting includes chin strap.

pollution hazards and radio-active contamination during firefighting and related life saving operations at the "hot job" locations. These suits consist of:
Trousers
Coats
Gloves
Boots
Hoods
One piece from head to foot air-fed to reduce heat and increase comfort
• Aluminized clothing: Aluminized clothing with range of ceramic fiber (1450°C) this type of clothing falls into two classes:
Fire-proximity suits (Figure 9.5) Not to enter the flame area
Emergency suits (Figures. 9.6 and 9.7): For temperature exceeding 550°C

Notes:

1. Never use fire-proximity clothing where fire entry suits are required.
2. Clothing for protection of close approach and other emergencies is given in Table 9.3.

Figure 9.6 Fire entry suit for use in entering a burning area. Note the self-contained breathing apparatus.

Figure 9.7 Demonstration of fire-entry suit. Spun glass material of suit is chemical resistant and will not burn, even in pure oxygen atmosphere.

Ordinary clothing can be protected against flame or small sparks by flame-proofing. Flame-proofing will make material:

Highly flame-resistant
Effective water soluble for flame-proofing is 228.8 grams of borax and 113.4 grams of boric acid in 3.8 L of hot water
Flame-proofed clothing should be marked for distinction

Ordinary clothing can be protected against flame or small sparks by flame-proofing. Flame-proofing will make material:

Highly flame-resistant
An effective water soluble for flame-proofing is 226.8 grams of borax and 113.4 grams of boric acid in 3.8 L of hot water
Flame-proofed clothing should be marked for distinction

Table 9.3 Clothing for protection against intense heat

Type of hazard	Example of hazards	Flame resistance	Suggested method of protection	Fitting of suit	Head protection		Degree of ventilation
					Type		
Radiant heat	Close approach fire	Outer material should be "Flame Proof" and interlining should be of low flam inability	High reflective surfaces for high rate of heating thermal resistance.	Free ventilation desirable to allow evaporation and prevent local heating	Faceshield of wire gage or transparent material which may be reflective coated		Naturally ventilated
Radiant heat and occasional flame lick	Rescue work and fire fighting operation inproximity of flame	Outer materials and inter lining should be flame proof	Reflecting surfaces against radiant heat and thermal resistant as high as practicable *	As little entry of air and as much free circulation of air inside the suite	Helmet with visor to drape to enclose the mead and nock visor reflectively coated		Ventilation may be under control of weather but shouldbe closable
Radiant heat and pockets of flame	//	//	//	Negligible entry of air and much precirculation of air inside the suit	//		//
Radiant heat and complete static immersion	Oil fires fire entry work	Outer material should be non combustible. Properties, inter lining to be flame proof *	//	//	//		Should be as air tight as practicable

Aramid fiber/Conex (PBI Blended, AEX FIRE) Pre-carbon fiber/PYRO MEX/ Glass Fiber/ceramic fiber.
*With material characteristic.

9.19.5 Materials

The material of any articles of clothing used against heat and fire proximity should be flame-proof. Any lining material that because of the design of the clothing could come into contact with flame should be made of flame-proof material.

- Design and Make-Up: There should be no pockets external to the assembly. The trousers and the sleeves of the jacket should not have turn-ups. Wherever possible seams and sewing threads should be protected. The threads should be compatible with the body fabric and should not impair the effectiveness of the protection afforded by the garment.

9.19.6 Head wear

Helmets intended for use against fire proximity should be tested complete with visors, there should be no discoloration. The visor should show no sign of cracking or breakdown, and all seams should be substantially undamaged. Helmets required to provide protection against impact and should pass the test for shock absorption.

The amount of respirable air within the headwear and suit should be made clear to the purchaser by the manufacturer and should be consistent with the use to which the equipment is to be put and for the time for which it is to be used.

Fasteners should be so designed or protected that they cannot be damaged by heat or cause head injury to the user.

The field of vision should meet the requirements of the operations to be conducted by the user and should be agreed between the purchaser and the manufacturer.

The headwear should be designed so that the visor or faceshield does not mist up in use to an extent that reduces the visibility.

9.19.7 Visor/facepiece

The visor or faceshield should be constructed of at least two independent layer of material, and their edges should be effectively protected by suitable frames or by the design of the helmet itself.

The degree of transparency to light passing through a visor should be specified.
When the visor or faceshield is sprayed with water, it should not have no fragment and neither the field of vision of the user nor the transparency of the visor or faceshield should be reduced by more than 50%.
The visor of faceshield should not crack, fracture, or become detached from its frame when tested.
Acrylic facepiece containing lead-equivalent of.3 mm
Coating thickness, with heat-protective film, supported outside as giving a wide vision.

9.19.8 Hand wear

Gloves should be graded as light duty or heavy duty and should be designed so they will not slip off in use but should be easy to take off.

9.19.9 Footwear/heat-resistant boots

The trouser legs of the protective suit should fit snugly into or around the boot to prevent the entry of flame. The outer shell of heat-resistant boots should be of aluminized agamid fabric lined with felt or leather. The sole should be made of heat-resistant rubber that meets UL 96.VO class requirements. The toe should be protected with steel protector for impact and compression.

9.20 Instructions and markings

The manufacturer's instructions should be provided with each suit. These should give information on how the best results may be obtained in use and on the limitations of the clothing, in particular, full information should be provided concerning the undergarments used in assessing its performance, and it should be stressed that the protective clothing for proximity and fire entry should be used only by trained personnel. The instructions should also give information on the amount of respirable air contained in the suit in terms of "the time for which it can be safely worn."

9.20.1 Markings

Each separate article of protective clothing and each garment, except the visor and faceshield, should be permanently marked with the following:

Number of accepted standard
Warning must be adhered according to manufacturer's instructions
Type of heat against which clothing is designed to give protection, "flame," "radiation," or both
Each protective garment should bear a permanent label bearing the manufacturer's identification mark and drawing attention to the necessity of consulting the manufacturer's instructions regarding the use of undergarments.

9.20.2 Marking of visor and faceshield

Visors and faceshields should be marked with the following:

Number of accepted standard
Manufacturer's identification mark
· Testing: Flame-proof clothing should be tested in accordance with BS EN 367, BS EN ISO 6942, and certified to be flame-proof for class and types of hazards.

9.21 Fire-resistant blanket

The fire blanket should be made of woven glass fiber fabric with silicon rubber coating on both sides.

- Containers: The fire blanket should be packed and stored in a travel bag with handling loops and ready for use by its unique quick-release system. The bad should include the instructions for use in the language specified.
- Performance Requirements: All the test requirements laid down in BS 7944, BS EN 1869 should be carried out and certified by manufacturers.
- Sizes: The following sizes should be used:
  1200 mm × 1200 mm
  1800 mm × 1200 mm
  1800 mm × 1800 mm

9.21.1 Use

Fire blankets can be used for fire extinction in:

Catering establishment
Schools
Hospitals and nursing homes
Laboratories
Garage and workshops
Boats and caravans
Ships and galleys
Numerous industrial outlets
Extinction of fire on a people's clothing
Restaurants
Flammable liquid cans
Cinema projection rooms

9.22   Fire blanket

A fire blanket is a flexible sheet of material intended to be used for small fires by smothering or as a protection against radiant heat or small hot objects. Fire blankets are classified as:

Light duty: For extinguishing small fires in containers of cooking fat or oil and fires in clothing worn by people.
Heavy duty: For industrial applications with ability to resist penetration by molten metals ejected from cutting and similar processes and any conducted or radiant heat transfer when used for insulation purposes, in addition to the uses mentioned for the light-duty blankets.
- Size and shape: Fire blankets should be rectangular or square with no edge longer than 1800 mm. Light-duty fire blankets should have no edge less than 900 mm. Heavy duty fire blankets should have no edge less than 1200 mm. Fire blankets should have a maximum mass of 10 kg.
- Hand-holding devices: Hand-holding devices if provided should not comprise loops or pockets and should not become detached from the blanket during testing.
- Appearance and line lateral use: The two sides of firc blanket should be of similar appearance finish or color and should give the same result when tested.
- Flexibility: Fire blankets should be capable of being rolled without permanent deformation and along any axis completely around a 50 mm dia bar.

- Ease of removal and unpacking: Fire blankets should be stowed or packed in such a way that they can be taken from the storage position unfolded and held ready for use in not more than 4 S. The force required to remove the fire blanket from its carrying bag should not exceed 80 N.
- Resistance to fraying: The edge of fire blankets should not fray or tear during testing.

9.22.1 Performance tests

Fire blankets should be certified by vendor for the following tests in accordance with BS 7944, BS EN 1869:

Thermal insulation (heavy-duty fire blanket only)
Resistance to the effects of hot cutting (heavy-duty fire blanket only)
Electrical insulation (resistance 1 mega ohm)
Reusability
Fire performance test

9.22.2 Marking blankets

Each fire blanket/fire-resisting blanket should be marked with the following:

Word "fire blanket/fire-resisting blanket"
Word "heavy duty," "light duty," and "reusable" as appropriate
Manufacturer's name and address

9.22.3 Container

Each container should be marked with word "fire blanket" in white letters no less than 15 mm high on a rectangular background.

- Container or instruction sheet: Each container or instruction sheet for fixing near to the storage position of the fire blanket should be marked with following:
  Words "fire blanket/fire-resisting blanket" on the front
  Words "heavy duty," light duty" as appropriate
  Instructions for use
  Manufacturer's name and address
  Model or other identification of the fire blanket
  Size in meters
  Reusable or should be discarded after use
  Washing or cleaning instructions (reusable only)
  Checking and maintenance instructions including when to discard if damaged or contaminated

9.23 Proscenium fire-resisting curtain

The proscenium opening of every approved stage should be provided with a curtain made of approved materials constructed and mounted so as to intercept hot gases, flames and smoke to prevent a glow from a severe fire on the stage from showing on the auditorium side for a period of 5 min. The closing of the curtain from the full open position should be effected in less than 30 sec, but the last 2440 mm of travel should require no less than 5 sec.

The proscenium curtain should be constructed in accordance with standards listed in NFC 101 code 8.3.2.1.7 (1992).

The curtain should be automatic closing without the use of applied power. In addition to these protections, the following items should also be considered:

Noncombustible opaque fabric curtain arranged so that it closes automatically.

Automatic fixed water-spray deluge system should be located on the auditorium side of proscenium opening and be so arranged that the entire face of curtain will be wetted. The system should be activated by combination of rate-of-rise and fixed-temperature detectors located on the ceiling of the stage. Detectors should be spaced in accordance with their listing. The water supply should be controlled by a deluge valve and should be sufficient to keep the curtain completely wet for 30 min or until the valve is closed.

Curtain should be automatically operated in case of fire by a combination of rate-of-rise and fixed temperature detectors that also activates the deluge spray system. Stage sprinklers and vents should be automatically operated in case of fire by fusible elements.

Operation of the stage sprinkler system or spray deluge valve should automatically activate the emergency ventilating system and close the curtain.

Curtain vents and spray deluge system valve should also be capable of manual operation.

Every stage provided with fire-resisting curtain and larger than 45 sq. m in area should have a system of sprinkler at the ceiling and in usable spaces under stage.

- Flame retardant requirement: The material used for fire-resisting curtain should meet the requirement of NFPA 701, standard methods of fire tests for flame-resistant of textile. Foamed plastics may be used only by specific approval of experts in the oil, gas, and petrochemical industries. Scenery and stage properties on thrust stages should be either noncombustible or limited-combustible materials.
- Standpipes: Regular stages over 93 sq. m in area and stages approved by Government authority should be equipped with Standpipe located at each side of stage.

9.24 Fire-resisting shields

Fire-resisting shields covered by ceramic fiber or aluminized asbestos materials should be fabricated of metal (steel) frame with a ceramic fiber cover and wherever considered essential to be provided with water-spray protection. The shield may be of dolly type for ease of its movement. The shield should be fixed with two fires resisting glass windows and opening for firefighting nozzles.

9.24.1 Size

Portable fire-resisting shield should not be less than 1200 mm wide and 2000 mm in height. The fire-resisting shield should be made locally provided that material used should be flame and heat resistance of no less than half an hour.

9.25 Notes on design, maintenance, and operating instructions

9.25.1 Design

Clothing for protection against intense heat should be primarily designed to prevent heat reaching the user, and the entry of hot air and fumes. This can be done by allowing ambient air circulate freely under the protective clothing. In the design of helmet and suit, care should be taken to ensure that when in use there should be sufficient air trapped in the helmet and suit to meet the respiration requirements of the user for the exposure period. The suit should be as airtight as possible and the helmet or its visor or face shield has to be provided with ventilation opening, which should be easily closed to ensure a reserve of fresh air when it is in close position.

When protective clothing is used for short periods weight will not be an important factor, but where clothing is worn for long periods, the weight should be as low as possible. Care should be taken to ensure that body and limb movements are not hampered and the protective clothing should be proportioned. Either it should be made of flexible fabric or the clothing should be designed to give flexibility. Other important aspects of design considerations are:

1. Correct fitting
2. Ease and speed of donning and removal
3. Comfort in wear

The gripping power of the soles depend on the nature of the surface that user is walking, on namely the material, angle of slope and condition (e.g., wet, dry, oily). To resist slipping performances of the different designs using several different users on sloping surfaces should be achieved. Nonskid qualities should last throughout the life of the footwear.

9.26 Maintenance

Garments should be examined thoroughly at regular intervals as well as after each time of use, and all tears, broken or defective fasteners, etc., should be repaired before re-use. The materials used in the repaired portion should meet the requirements of standards.

If a garment is soiled, it should be cleaned as soon as possible since contamination by flammable substances such as oil or grease may impair the flame-proof properties of materials. Any cleaning process on garments should be of such a nature that the cleaning agents and treatment have no deleterious effect. If clothing is dry-cleaned, no residual solvents giving rise to toxic effects should be used. The Manufacturer should recommend methods of cleaning.

The uppers of leather boots, except those made of suede leather, should be dressed periodically to maintain suppleness and waterproof ness. Flammable oils and fats should not be used.

Metalized material garments can be used only if reflective surfaces are untarnished. These garments should be washed with soapy water and wiped with a very soft rag. Normal room temperature is suitable for storage of these garments, which should be hung so that unnecessary folding is avoided. Once the reflective surfaces are no longer bright the garment should be discarded.

9.26.1 Operating conditions

It is essential that users should know the limitations of the clothing and that they retreat from danger before failure is likely. These garments should only be used by trained users who have experience in wearing them. Users should be prepared for the moment when the protective clothing and the air within it becomes warmer than the skin. Training should be planned so that each user can recognize the approach of danger and the time to leave the danger zone. Prevention of mist on the inside of a visor or face shield is important. Various proprietary anti-mist compounds and devices, which will alleviate this trouble to a large extent, should be used. The wetting of hot dry assemblies should be avoided as this may cause scalds.

9.27 Clothing for protection against heat and fire; general recommendations for users and for those in charge of such users

Rules and instructions that are essential to know and observe with the use of clothing for protection against heat and fire should be in hand as a "checklist" of safety requirements for those responsible for checking. When new rulings specify regulations other than those given in standards, the stricter specifications should be applied.

It is essential to teach that no clothing for protection against heat and fire can offer unlimited protection. Variable and interdependent factors affect the time that such clothing can offer protection in an area of heat and fire. For one and the same garment, this period may vary enormously from one operator to another. It is also important to realize that, if the operator has an accident or feels unwell, the absence of movement on his part reduces the circulation of air inside the garment and may increase the effects of the external heat.

9.27.1 Operators

- State of health: Any person using the garment giving protection against heat and fire must be free from any physical or mental defects, especially if he is to wear a breathing apparatus.
- Training: Protective clothing against intense heat with or without fire should only be used by people who undergo systematic training in its use.

Regular training has several objectives, the most important of which are:

To acquire a routine to permit the reduction to a minimum of the time required to put on the clothing and special equipment

To keep the operator informed of the properties and limiting factors of the material he has to wear

To accustom the operator to move about in such clothing

To allow the operator to accustom his body to prolonged effort, while learning to recognize his physiological limit of endurance, and also to assess the approach of the moment when he is still able to retreat from the danger zone in total safety

Training of operators should be carried out with garments corresponding to those used in practical operations. The old clothing of the same type and style should be used and kept exclusively for training.

9.27.2 Materials

- Fusible materials: People likely to find themselves in an area where there is a risk of heat or fire should not wear clothing or underclothing made of fusible material next to the skin even if they are protected by special garments.
- Permeable and absorbent materials: People clothed in permeable garments or in garments of which the material of the outer layer absorb water or flammable products (liquids, dusts, gases or vapors) should be aware of the danger of entering an area of intense heat or fire when those garments have been in, or in contact with such products. Specific safety measures should be taken to prevent permeable or absorbent garments from coming into contact with liquid oxygen.

9.27.3 Electricity

- Static electricity: Certain garments may become charged with and discharge of static electricity. The use of such garments is dangerous in areas contaminated by explosive or flammable gases.
- Electric shock-electrocution: Before entering an area where there is an electrical hazard, the person in charge of rescue operations or firefighting should ensure that electricity-supply systems have been separated from the supply source.

9.28 Safety provisions

9.28.1 Operational groups

Any operation requiring special protective clothing or equipment should be carried out by a group of at least two workers who are in constant physical contact with each other and with a safety station situated outside the danger area. At this safety station, for each group taking part in the operation, a stand-by group of at least the same number of men, protected at least as effectively as the first group should be ready to take immediate action at the slightest alert.

9.28.2 Cooling by wetting

Unless the garment has been specially designed for it, it should never be cooled by wetting.

9.29   Inspection, storage, and maintenance

9.29.1   Inspection

Garments for protection against heat and fire should be checked at regular intervals and maintained in perfect condition. Particular attention should be paid to the fastening devices to make sure they are operating properly. Any defects that are discovered or suspected should be pointed out to the manufacturer or the certified representative who is responsible for declaring that the garment is capable of offering protection corresponding to its classification in accordance with the standards laid down.

Note: The inspection garments is especially responsible work; it requires special technical knowledge and oftentimes equipment.

9.29.2   Storage

The manufacturer's recommendations regarding the conditioning and storage of clothing should be strictly observed. Each type of garment should be arranged in a group for rapid identification of its classification. A check should be made at regular intervals to see that all these recommendations are observed.

Protective clothing, particularly if provided with a special surface to reflect heat, should be stored in such a way as to avoid folding the material and to protect it against dust and other dirt that may decrease its efficiency. Protective clothing made of woven, porous, or absorbent materials should be stored in such a way as to avoid its contamination by products likely to make its use dangerous.

9.29.3   Maintenance

The manufacturer's instructions regarding the maintenance, use, and cleaning of the garments should be strictly observed.

9.30   Used and reconditioned garments

9.30.1   Used, converted, or reconditioned garments

The classification of a garment that has been used, whether reconditioned or not, should be re-examined according to the standards drawn up for new garments without relying on its former classification. Any re-examination should ensure that the garment in question is then supplied with the symbol of its appropriate classification. Any reconditioning of a protective garment is likely to change its protective and other characteristics. This task should be entrusted to a highly qualified person who then has to re-examine the classification of the garment.

Note: "Reconditioning" means any work carried out on the original garment, for the purpose of restoring it to a suitable condition for use. Even the replacing of a defective fastening device of a new garment is an act of reconditioning from the perspective of the standard.

9.31 Glass-fiber fire blankets

Fire blankets are made of texturized woven glass fiber, which gives them a rough surface providing stability. Designed to enable simple storage of the blanket, the container is noncorrosive, rigid self-extinguishing white PVC. Blankets are available in the following sizes:

Blanket size (cm)	Container size (cm)
90 × 90	27 × 8 × 8
122 × 122	31 × 8 × 8
180 × 122	36 × 8 × 8
90 × 90	27 × 8 × 8
90 × 90	27 × 8 × 8

Based on BS 7944, BS EN 1869:

Light duty	(reusable) Fire blankets
that comply with the above packed in white rigid uPVC containers	
Blanket Size	**Container Size**
120 cm × 120 cm	8 cm × 8 cm × 0.5 cm
180 cm × 120 cm	8 cm × 8 cm × 35.5 cm

9.32 Delivery, inspection, quality control, and commissioning of firefighting pumps

Fire pumps of adequate capacity, water pressure with reliable power and water supplies are recognized as prime fire-protection equipment in the petroleum industries. Fire pumps should be of proper design, correctly installed, and subjected to periodic tests and if not conscientiously maintained will be found out of order in the event of fire and resulting serious consequences. This section covers the minimum requirements for acceptance tests done at the time of commissioning, tests and inspections done at regular intervals, and preventive maintenance of the pumping apparatus built for fire protection and used in the oil, gas, and petroleum industries.

9.32.1 Acceptance, operation, and maintenance of fire pumps (fixed installations)

• Field Acceptance Tests: Project engineer should notify the following responsible authorities of the time and program of acceptance fire-pump tests:
  Plant head of operation

Responsible maintenance engineer (mechanical and electrical)

Chief fire officer or head of safety and fire

Manufacturer/supplier representative

A copy of the manufacturer's certified pump test characteristic curve should be available for comparison of result of field acceptance test. The fire pump as installed should equal the performance as indicated on the manufacturer's certified shop test characteristic curve within the accuracy limits of the test equipment.

The fire pump should perform at minimum rated and peakloads without objectionable overheating of any components.

Vibration of the fire-pump assembly should not be of a magnitude to warrant potential damage to any fire-pump component.

- Test procedures
- Test equipment: Test equipment should be provided to determine net pump pressure, rate of flow through the pump, volts, and ampers for electric motor-driven pump and speed.
- Flow test: The minimum rated and peak load of the fire pump should be determined by controlling the quantity of water discharged through the approved test devices.
- Measurement procedure: The quantity of water discharging from the fire-pump assembly and pressure should be determined and stabilized. Immediately thereafter, the operating conditions of fire pump and driver should be checked.

For electric motors operating at rated voltage and frequency, the ampere demand should not exceed the product of a full load ampere rating times the allowable service factor as stamped on the motor nameplate.

For electric motors operating under varying voltage, the product of the actual voltage and current demand should not exceed the product of the rated voltage and rated full load current times the allowable service factor. The voltage at the motor should not vary more than 5% below or 10% above rated (nameplate) voltage during the test.

Engine-driven units should not show signs of overload or stress. The governor of such units should be set to properly regulate the minimum engine speed at rated pump speed at the maximum pump brake horsepower.

The steam turbine should maintain its speed within the limits as specified.

The gear-drive assembly should operate without excessive objectionable noise, vibration, or heating.

- Load start test: The fire-pump unit should be started and brought up to rated speed without interruption under the conditions of a discharge equal to peak load.
- Controller acceptance test: Fire-pump controller(s) should be tested in accordance with the manufacturer's recommended test procedure. As a minimum, no less than 10 automatic and 10 manual operations should be performed during the acceptance test. A fire-pump driver should be operated for a period of at least 5 min at full speed during each of the above operations. The automatic operation sequence of controller should start the pumps from all provided starting features. This should include pressure switches or remote starting signals. Tests of engine drive controllers should be divided between both sets of batteries.
- Alternative power supply: On installations with an emergency source(s) of power and an automatic transfer switch, loss of primary source should be simulated and transfer should occur while the pump is operating at peak load. Transfer from normal to emergency source and retransfer from emergency to normal source should not cause opening of overcurrent protection devices in either line. At least half of the manual and automatic operations should be performed with the fire pump connected to the alternative source.
- Emergency governor: Emergency governor valve for steam should be operated to demonstrate satisfactory performance of the assembly (hand tripping is acceptable).

Alarm conditions both local and remote should be simulated to demonstrate satisfactory operation.
- Test duration: The fire pump should be in operation for no less than 1 hour total during all of the foregoing tests.

9.32.2   Maintenance

Maintenance includes running the pump a few minutes each week. During such runs water is discharged through the relief valve or other opening. The run is carried up to nearly full speed and pressure. The condition of the pump and its associated equipment is least determined by an operating test. If a pump that is an important unit of fire protection shows more than 15% slip, a recommendation for repair is in order and the cause should be found and remedied at once. There are four main types of maintenance: predictive, preventive, corrective action, and improvement.

- Annual fire-pump tests: The annual flow test should be conducted to determine its ability to continue to attain satisfactory performance at shut-off, rated and peak loads. All alarms should operate satisfactorily. All valves in the suction line should be checked to assure that they are fully open. The pressure relief valve should be verified by actual test to be correctly adjusted and set to relieve at the appropriate pressure. The annual test should be performed by personnel trained in operation of fire pumps. Test results should be recorded. The speed of the pump driver should be determined and recorded. Any significant reduction in the operating characteristics of the fire-pump assembly should be reported and repair made immediately.

9.32.3   Fire-pump operation

The fire pump should be maintained in readiness for operation. After any test, the fire pump should be returned to automatic operation. All valves should be returned to normal operating positions. The fire-pump room should be kept clean, dry, orderly and free of miscellaneous storage. Access to this room should be restricted. In the event of fire, qualified personnel should be dispatched to the fire-pump room to determine that the fire pump is operating in a satisfactory manner.

The fire-pump unit should be operated weekly and at least one start should be accomplished by reducing the water pressure. This may be done with a test drain on a sensing line and with flow from the fire-protection system. Qualified operating personnel should be in attendance during the weekly pump operation. The satisfactory performance of the pump driver, controller, and alarms should be observed and noted.

9.32.4   Preventive maintenance and inspection

A preventive-maintenance program should be established in accordance with the pump manufacturer's recommendations. Records should be maintained on all work performed on the pump, driver, and controller.

- Diesel-engine operation and maintenance weekly run: Engines should be started no less than once a week and run for no less than 30 min to attain normal running temperature. They should run smoothly at rated speed.
- Engine maintenance; Engines should be kept clean, dry, and well lubricated. The proper oil level should be maintained in the crankcase. Oil should be changed in accordance with manufacturer's recommendations, but no less frequently than annually.
- Battery maintenance: Storage batteries should be kept charged at all times. They should be tested frequently to determine the condition of battery cells, and the amount of charge in the battery. The automatic feature of a battery charger is no substitute for proper maintenance of battery and charger. Periodic inspection should determine that the charger is operating correctly. The water level in the battery should be correct, and the battery is holding its proper charge. Only distilled water should be used in battery cell. The plates should be kept submerged at all times.
- Fuel-supply maintenance: The fuel storage tanks should be kept as full as possible at all times, but never less than 50% of tank capacity, they should always be filled by means that will ensure removal of all water and foreign material.
- Temperature maintenance: Temperature of the pump room, pump house or area where engines are installed should never be less than the minimum recommended by the engine manufacturer. The engine manufacturer's recommendations for water heater and oil heater should be followed.
- Emergency starting and stopping: The sequence for emergency manual operation, arranged in a step-by-step manner should be posted on the fire-pump engine. It should be the engine manufacturer's responsibility to list any specific instructions pertaining to the operation of equipment during the above-mentioned sequences.

9.33   Acceptance, service test, operation and maintenance of fire-service pumping units

Fire-service pumping units are classified as:

1. Pumping unit mounted on fire trucks
2. Trailer mounted
3. Portable

In this section operation, service test and maintenance of trailer mounted and portable lightweight pumping units are discussed.

9.33.1   Test site requirements

- Tests at draft: When tests are to be performed with the pump drafting, the test site should be adjacent to a supply of clear water at least 1.25 m deep with the water level of 3 m below the center of suction inlet. The suction strainer to be submerged at least 0.60 m below the surface of the water when connected to the pump by 6 m of suction hose.
- Other tests: When suitable site for drafting is not available, the site should provide a level area for stationing the pump, a source of water such as fire hydrant connected to a water-distribution system, and an area suitable for discharging the water.

9.33.2 Environmental conditions

Pump tests should be performed when conditions are as follows:

Air temperature	0 to 38°C
Water temperature	2 to 32°C
Barometric pressure corrected	(737 mm Hg) minimum at sea level

- Equipment
- Suction hose and strainer: When testing the pump at draft 6 m of suction hose of appropriate size for the rated capacity of the pump and suction strainer that well allow flow with total friction should be furnished. When testing a pump from a hydrant or other source of water at positive pressure, the suction hose may be of any convenient size and length that will permit the necessary amount of water to reach the pump with the minimum suction pressure of 69 kPa (0.7 bar) and only the strainer at the pump inlet should be required.
- Discharge system: Sufficient fire hose should be provided to allow discharge of rated capacity to the nozzles or other flow-measuring equipment without exceeding the flow velocity of 9.7 m/s(35 Ft/s). Approved flow gauge or pitot tube should be used and all gauges should have been calibrated within a week proceeding tests. Speed measuring equipment should consist of either a tachometer measuring revolutions per minute or a revolution counter and stopwatch.

9.33.3 Acceptance tests

Tests should be conducted by the assigned fire-protection engineer, mechanical engineer, and manufacturer/supplier representative. A copy of the manufacturer's certified pump test characteristic curve should be available for comparison of results of acceptance test. The fire pump should perform at minimum rated and peak loads without objectionable overheating of any component. Elements of the certification test should be duplicated in-so-far as practical. The pumping test, overload test, pressure control test, and pump vacuum test should be performed as a minimum. Result of all tests should be recorded and the testing authorities should decide if the specified criteria have been met. Where test results are not acceptable, the manufacturer should be notified in writing of the discrepancies and other matters to be remedied.

9.33.4 Service tests

- Quarterly output test: Pump should be subject to a pumping test from open water using one length of hose per delivery. The length of test should be at least 15 min and any pump found incapable of sustaining the pressure indicated below with the lift as near as possible to, but not exceeding 3 m from the surface of water to the pump inlet should be reported. The test should be recorded. See Table 9.4 for sample details.

Notes: Where the nominal output of a pump falls between any figures in Table 9.4, the number and/or size of the nozzle(s) should be adjusted accordingly. Pumps having capacity below that shown above should be tested to about 75% of the performance specified by the maker.

Table 9.4 **Pump output**

Normal output of pump at 7 bar L/min.	Pump test pressure minimum bar	Number of hoselines	Size of nozzles mm
4500	5.5	2	28
		2	25
4050	5.5	4	25
3600	5.5	3	28
3150	5.5	1	28
		2	25
2700	5.5	2	25
		1	20
2250	5.5	4	20
1800	5.5	3	20
1350	5.5	2	20
900	5.5	2	15
450	5.5	1	15

9.33.5 Quarterly vacuum test

This test should be carried out immediately after the output test given above. All length of suction should be coupled up to the suction inlet of the pump with the blank cap in position at the end of the last length but with the blank caps left off all delivers. The primer should be run at priming speed for not more than 45 sec. Priming should cease after obtaining 0.8 bar or more vacuum and the compound gauge needle should then be watched. If the needle falls back to 0.3 bar in less than 1 minute an excessive air leak is present. This may be due to a defective pump gland, to leakage at compound or pressure gauge connections, delivery valves, cooling water connections, or faults in the suction hose or couplings. Any leak should then be rectified. A leak in the suction hose may be found by the water-pressure test.

The ambient air temperature, water temperature, vertical lift, elevation of test site, and atmospheric pressure (corrected at sea level) should be determined and recorded. The engine pump and all parts should exhibit no undue heating, loss of power, over-speed, or other defects during the entire test. The capacity, discharge pressure, suction pressure, and engine speed should be recorded at the beginning and the end of each phase of pumping test.

9.33.6 Maintenance of portable and trailer fire pumps

On all occasions when a pump has been used other than from a hydrant on open water, the pump should be washed out from such a hydrant as soon as possible. The pump should not be allowed to run dry unless the pump bearings are of a type that can withstand such treatment.

Mobile pumps should be housed in a room in which temperature is not allowed to fall below 4°C but as an additional precaution during the winter antifreeze should be added to the engine cooling system if permitted by the manufacturer, who will of course advise against this if the engine has direct cooling system. It is also advisable to drain the water from the pump casing in the cold weather.

The engine should be regularly oiled in accordance with manufacturer's instruction and an occasional spot of oil given to the controls such as the throttle control. It is sound practice to grease the threads on the suction inlet and the suction couplings at the same time.

A suitable record book should be prepared showing the history of the pump and all maintenance and repair work done on it. The record should also contain the number of hours that the pump has been running.

The practice of starting and running the engine of a pump for only a few minutes every shift during the day and night is very damaging to the engine because it will be continually running on the chock and this procedure will cause sooted plugs, dilution of engine oil with petrol resulting in dry cylinder walls, and moisture condensation throughout the exhaust system, leading to general corrosion and rapid deterioration of the engine. The pump therefore should be tested by pumping clean water.

9.34 Installation, inspection, and testing of firefighting fixed systems

In the oil, gas, and petrochemical industries various oil product and chemical units are protected against the out break of fire with special type of firefighting fixed systems that are relevant to the nature of fire protection needed. In this section the construction requirement for fire systems is discussed. This section specifies the minimum requirements for the construction, installation, inspection, and testing of various types of firefighting fixed systems and is divided in five parts as follows:

- Part 1: CO_2 Fire-extinguishing Systems
- Part 2: Dry Chemical Powder Fire-extinguishing Systems
- Part 3: Sprinkler Systems
- Part 4: Water-supply Systems
- Part 5: Foam Systems

This section is prepared for the use and guidance of those charged with the installation, commissioning, testing and approving of extinguishing systems in order that installed equipment function as expected throughout their life. Only those skilled in the field of safety and fire are authorized to approve the installation testing and commissioning of the systems. It is necessary for those charged to consult with fire-protection engineers to be aware and experienced with design of fire-extinguishing systems, which would enable them to effectively discharge their respective duties.

9.34.1 Installation, testing, and quality control of CO_2 extinguishing systems

- Planning: Where a fixed carbon dioxide-extinguishing system is being considered for new or existing building the following authorities should be consulted during construction and installation of the system:
  Plant manager
  Operation/production superintendent
  Fire-protection engineer
  Authorities involved with the project
- System Layout Drawings: Prior to installation of system the layout drawing with full details and dimensions should be prepared to define clearly both the hazards and the proposed system. Detail of hazards should be included to show the materials involved and also the location and/or limits of the hazard and any other materials that are likely to become exposed to the hazard in the event of fire. The means of egress from the area to be protected (if it is an automatic total flooding system), for personnel who are likely to be present in that area should be indicated together with the number of occupants. The location and size of piping and nozzles should be clearly indicated together with the location of carbon dioxide-supply cylinder, fire detection devices, manual control locking devices, and all auxiliary equipment. Features such as dampers, means of escape, delay system, and doors related to the operation of the system should also be shown with all details of calculations used in assessing the quantity of carbon dioxide. Further information should be given separately indicating the equipment such as length of pipe and fittings, flow rates, and pressure drops throughout the system.
- Tests and Approval of Installations: The completed system should be inspected and tested by qualified personnel of the supplier before the approval of the owner is obtained. Only listed or approved equipment and devices should be used in the system. The supplier of the equipment or their agent should arrange tests of completed installation to the satisfaction of the owner and show that it complies the standards. The tests should include the following except that the discharge of CO_2 should not be carried out in the special cases:
  A thorough visual inspection of the installed system covering hazard area, the layout of piping and operational equipment should be carried out. Discharge nozzles should be inspected for proper size and location. The location of alarms and manual emergency releases should be checked. The configuration of hazard should be compared to the original specification. The area should be inspected closely for unclosable openings and sources of agent loss, which may have been overlooked in the original layout. The area should be inspected for ease of exit before system be operated.
  Warning: The CO_2 extinguisher system should not be tested and discharged into an area where employees are present.
  A check that all components of the system have been installed in a proper manner.
  A check that all nuts, bolts, and fittings have been correctly tightened.
  A check that all electrical connections are safe and in working order.
  A check of labeling of devices for proper designations and instructions. Nameplate data of the storage containers should be compared to the specifications.
  Carbon-dioxide gas tests to check the tightness of closed suction of pipe work. Separate gas discharges should be made into each space to ensure that the piping is continuous and that nozzles have not been blocked.
  Note: A minimum of 10% of the required quantity of gas should be discharged through the system pipe work into each space. Prior to testing proper safety procedures should be reviewed.

A full discharge test should be performed on all systems except when specially waived by relevant authority.

When multiple hazards are protected from a common supply, then a full discharge test should be performed for each hazard.

- Local application: Full discharge of entire design quantity of carbon dioxide through system piping to ensure carbon dioxide effectively covers the hazard for the full period of time required by the design specifications and all pressure-operated devices function as intended.
- Total flooding: Full discharge of entire design quantity of carbon dioxide into the hazard area to ensure that the concentration is achieved and maintenance in the period of time required by the design specifications and all pressure-operated devices function as intended.
- Handheld hose line: Full discharge test of handheld hose-line system require evidence of liquid flow from each nozzle with an adequate pattern of coverage.

Note: A partial discharge is appropriate for most installations, but for others a total discharge with measurement of carbon dioxide concentrations achieved is desirable.

- Checklist: The installer of equipment as supplier should provide a checklist to enable the owner to witness that the tests are being carried out in a satisfactory manner. The list should include the following:

Check that the system has been installed according to the relevant drawing and documents.

Check that all the following detection equipment functions correctly.

In fusible link systems, ensure that control cable lines are free and that operating control weights develop sufficient energy to operate container and/or direction valve control mechanisms.

In pneumatic rate of rise systems check with manometer to ensure correct breathing rate and leak free capillary lines. Also apply heat to detectors to ensure correct operation and subsequent activation of control mechanisms.

In electrical detector systems, check electrical circuitry and supply voltages for integrity. Apply heat flame and smoke to detectors to check operation of control mechanism.

Operate manual release devices to ensure correct functioning.

Check operation of all alarm devices.

Check correct operation of all safety devices.

Carry out a test of CO_2 gas discharge using an adequate percentage of the total CO_2 capacity to check:

That the direction valves when shut, hold leak gas

That feed pipes lead to the correct protected space

That no leak occur when equipment is fitted to pipe work and at pipe fillings

That pressure-operated devices function correctly and the items they control, such as shutters and alarms function correctly

That, where possible, discharge nozzles pass gas and are not blocked

Ensure test containers are replaced and that all containers are filled with the correct quantity of carbon dioxide.

Check that nameplate and construction plated are correctly worded and displayed. Ensure the containers have been mechanically tested.

When the installation has been completed and tested, a completion certificate should be signed by the vendor and owner's authorized personnel.

9.34.2 Installation, testing, and quality control of dry-chemical powder systems

- Planning: Where a dry powder-extinguishing system is being considered for a new or existing building the proper authorities should be consulted.

- Drawings: Prior to installation of system manufacturer's layout drawings should be prepared and studied. These should have sufficient details to define clearly the nature of hazard and proposed appropriate type of system. Details of the hazard and materials that are likely to become exposed to the fire should also be included. Considering the contents about the location, sizes of piping and nozzles should be clearly indicated, together with the location of the powder supply, fire-detection devices, manual control, all auxiliary equipment, doors related to the operation of the total flooding system and all details of calculation used in assessing the quantity of discharge rate of powder, and the type of system should be shown in the layout. Further information should be given separately indicating the equivalent length of pipe and fittings; flow rates and pressure drop throughout the system.
- Commissioning and Acceptance Tests: The supplier/vendor should arrange tests of the completed installation to the satisfaction of the owner's relevant authorities to show that the system complies with standards and functions as designed. Discharge of powder should be done through monitor and hose reel systems when testing. Application of other means will cause spread of powder and consequent expense of cleaning the powder from contaminated areas.
- Warning: The powder system should not be test discharged into areas where atmosphere is explosive. Electrostatic effects can induce sparking that may cause ignition of any flammable vapors or gases that may be present.
- Commissioning test program: The supplier should submit to the owner a test program, which should include instructions to:
  Check that all components of the system are installed according to the designed drawings and documents and in the correct manner
  Check that all nuts, bolts, and fittings are correctly tightened and that all pipework supports are correctly fitted
  Check that all electrical connections are safe and in working order
  Check that all pipe work and nozzles are the correct size
  Check that all equipment functions correctly
  Operate manual release devices to confirm correct functioning
  Check operation of all alarm devices
  Check correct operation of all safety devices
  Carry out a test discharge of propellant gas to check:
    That distribution valves when shut hold pressure
    That feed pipes lead to the correct protected space
    That leak occur at joints of fittings
    That pressure-operated devices function correctly and the items they control such as shutters and alarms function correctly
    That piping is continuous and nozzles are not blocked
  Train all personnel who will be authorized to use monitor and hose-reel systems.
- System restoration: After completion of the commissioning and acceptance tests the system should be restored to operational condition and all containers checked for correct fill. The pipe work should be blown down to remove any residual powder.
- Completion certificate: When the system has been completed, tested, and restored, the owner should be provided with two copies of completion certificate, a complete set of instructions and drawings showing the system as installed and the statement that the system complies with all appropriate requirements of standard and giving details of any departure from appropriate standard. The completion certificate should be signed by the supplier and the owner.

9.35 Installation, testing, and quality control of firefighting sprinkler systems

9.35.1 Installation planning

Prior to installation the system layout drawing should be prepared and the proper authorities consulted.

- Drawings, information, and documents: The information should be provided to the authorities concerned in a meeting to make sure that all requirements are met. The information provided should include the following:
  General specification of the system
  Block plan of the premises showing:
    Type(s) of installation(s) and the hazard class(es) and the occupancy in the various buildings
    Extent of the system with detail of any unprotected areas
    Construction and acceptance of the main building and any communicating and/or neighboring buildings
    Cross-section of the full height of the building(s) showing the height of the highest sprinkler above the stated datum level
    Statement that the installation will comply with standard including any deviation(s) with reasons for the deviation(s)
- Installation Layout Drawings: Layout drawings should include the following information:
  The class(es) of installation according to the hazard including stock category and design storage height
  Constructional detail of floors, ceilings, roofs, exterior walls separating sprinklered and nonsprinklered areas
  Sectional elevation of each floor of each building showing the distance of sprinklers from ceilings, structural features, etc., that affect the sprinkler layout of water distribution from the sprinklers
  Indication of trunking, staging, platform, machinery, fluorescent light fittings, heaters, suspend open cell ceilings etc., that may adversely affect the sprinkler distribution
  Sprinkler type(s) and temperature rating(s)
  Location and type of main control valves and location of alarm motor and gangs
  Location and details of any water flow and air or water pressure alarm switches
  Location and size of any tail-end air valves, subsidiary stop valves, and drain valves
  Drainage slope of the pipe work
  Location and specification of any orifice plate
  Schedule listing of number of sprinklers, medium and high velocity sprayers, etc., and the areas of protection
- Pre-calculated pipe work: For pre-calculated pipe work details should be given on or with the drawing:
  Identification of the pressure loss between the control valve and the design point at the following design rate of flow:
    1. In a high hazard installation: 225 L/min
    2. Ordinary-hazard installation: 100 L/min

9.35.2   Work on-Site

• Care of material on-site: Components should be properly stored on-site until required for installation. Unloading, stocking, and storage should be carried out with care to prevent damage to pipes, pipe thread, valves, sprinkler heads, gauges, and any pumps and power units used in the system. Site locations should be prepared in advance of delivery so that heavy items such as fire pumps, strainers, and pressure tanks can be transported directly to the final locations. Pipes should be protected against entry of foreign objects such as rubber cordage, etc., into the bore and they should be examined and cleaned prior to erection. Open ends of pipes should be capped as work on-site proceeds. Sprinklers, controls, and sprayers should preferably be fitted to pipes in situ. Where fabricated ranges are used the sprinklers can be fitted immediately before erection using pipe rack to hold the racks and the ranges off the ground.

9.35.3   Fire protection of buildings under construction or modification

Work on the system should proceed with the progress of building. Installations and zones should be made operational as soon as practical.

• Hot work: Suitable precautions should be taken when hot work is performed. Procedures for hot work should be observed.

9.35.4   Commissioning and acceptance tests

The selected authorities of the owner should be invited to witness the tests and inspect the system in conjunction with the supplier's engineers.

9.35.5   Commissioning tests

All installation pipe work should be pressure tested as follows:

• Dry pipe work, pneumatically, to no less than 2.5 bars for no less than 24 hours
• Wet pipe work, hydraulically, to no less than 15 bars or 1.5 times the working pressure, whichever is greater for no less than 1 hour

Any faults disclosed such as permanent distortion, rupture or leakage should be corrected and the test repeated.

Note: In water sensitive areas, it is advised to pneumatically test pipe work before any hydraulic tests.

9.35.6   Water supplies and alarms

Each water supply should be tested with each installation in the system. The pump(s), if fitted in the supply, should start automatically and the supply pressure at the appropriate flow rate should not be less than the appropriate value and corrective action should be taken if necessary.

Note: Adequate facilities should be provided for the disposal of test water. A test facility including a direct reading flow meter suitable for sprinkler service should be provided at the pump delivery branch down stream of each outlet check valve to permit a running pressure test of the pump at the full load condition or nominal rating as appropriate.

- Alarms: The equipment for automatic transmission of alarms to fire-service or remote manned center should be checked for:
  Continuity of the connections
  Continuity of the connection between the alarm switch and the control unit
- Checks: The following should be checked and corrective actions taken if necessary:
  All water and air-pressure gauge reading on trunk mains and pressure tanks
  All water levels in elevated reservoir water storage and pressure tanks
  Each water motor alarm should be sounded for no less than 30 sec
  Automatic starting of the pump when it starts should record the starting pressure and check that this is correct. For diesel engine, run the engine for 30 min. or for the time recommended by manufacturer. Shut down the engine and use the manual start test button and check that the engine restarts
  Mode-monitoring system for stop valve on life safety installations should be tested
  Pipe work should be checked for electrical earthing connections
  Secondary electrical supply from diesel generator should be checked for operation
  All stop valves controlling the flow of water to sprinklers should be manipulated to ensure that they are in working order
  Dry alarm valves and any accelerators in any dry-pipe installations and tail-end extensions should be exercised
  Test of drainage facilities should be done

9.35.7 Completion certificate and documents

The supplier should provide the following:

Completion certificate stating that the system complies with all appropriate requirements of standards, and giving details of any departure from appropriate standards
Copy of test reports of commissioning
Complete set of operating instructions and or installed drawing including identification of all valves and instruments used for testing and operation and user program for inspection and checking
Certificate giving the result of in-situ testing of pipe fasteners
Appropriate certificate stating that components used in the system are suitable for sprinkler service and passed quality inspections

9.36 Sign and notices

A location plate suitable for sprinkler service of weather-resistant material and lettering should be fixed on the outside of the external wall as close as practical to the entrance nearest to the installation main control valve set(s). The plate should bear the words "sprinkler stop valve inside." A sign should be fitted also to the main and any subsidiary stop valves bearing the words "sprinkler stop valve."

All valves and instruments used for testing and operation of the system should be appropriately labeled. Where the sprinkler system comprises more than one installation, each alarm valve should be permanently marked with the number identifying the installation in control:

Each water motor alarm gang should also be marked with the number of installation.

Where water flow into an installation initiate an automatic alarm to the fire station, notice to that effect should be fixed adjacent to the alarm test valve(s).

The alarm at both the diesel engine controller and the responsibly manned location should be marked as appropriate:

Diesel fire-pump failure to start and/or
Diesel fire-pump starter switched off and pump running manually operated shut down mechanism should be labeled "SPRINKLER PUMP SHUT-OFF"
Each switch in the dedicated power feed to an electric sprinkler fire-pump motor should be labeled: "sprinkler pump motor supply not to be switched off in the event of fire."

9.37   Installation, testing, and quality control of water-supply systems

Water-supply systems for firefighting include fire pumps, systems of piping, control valves, hydrants, and risers brought from tanks, rivers, or other sources such as reservoirs or almost unlimited natural sources, which can be used for firefighting. The magnitude of the water supply for firefighting depends on the size and number of streams likely to be required and the length of time such streams may have to be operated.

9.37.1   Planning for construction

- Drawings and documents: All drawings and documents should carry the following information:
  General specifications of the system
  Block plan of the area showing:
    Type of hazards and the intend of the system
    Particulars of the water supplies including flow data and pressure
    Statement that the installation will comply with standards including any deviation(s) with reasons
    List of materials including pump details (manufacturer's data sheets)
    Hydraulically calculated of water flow and pressure at each ring main

Any recommendations including planning stages of the installation should also be documented.

9.37.2   Installation

- Work on-site: Adequate provision should be made by responsible authorities to protect materials and equipment on-site from loss, deterioration, and damage. Unloading and stacking should be carried out with care to prevent damages to the hydrants, couplings, and other components used in the system.
- Water pipes: General practice in tropical should be to lay down fire-water mains as far as possible over ground, but in cold climate the fire-water main is to be laid underground. The

depth, the type of pipe, and protection required should be specified in design specifications. The depth of cover to provide protection against freezing will vary depending on lowest temperature. The pipe should be properly guided in places where the pipe is subject to shock or vibration. Special care is necessary when running pipe under roads carrying heavy trucking. In such areas pipes should be run in a covered pipe trench or be otherwise properly guarded.

• Laying of pipe: Pipe should be clean inside when put in trenches and open end should be plugged when work is stopped to prevent stones or dirt from entering.

• Back filling: Earth should be well tamped under and around pipes and should contain no ashes, cinders, or other corrosive materials.

• Flushing: After a system has been completed and before underground pipes is permanently filled with water, the entire inside system should be properly flushed out under pressure to move the larger obstructing materials from underground piping. Pipes should be securely anchored before any pressure or flow tests are carried out. Steel pipes when used underground or water supplies known to have unusual corrosive properties should be protected against corrosion.

9.37.3 Initial inspections and acceptance tests

• Fire-water mains and hydrants: Before final approval testing, the supplier/contractor should furnish a written statement to the effect that the work has been completed in accordance with the approved specification and plans. Inspection and, where practicable, a wet test should be done by supplier/contractor in conjunction with fire authorities. The test should include flushing out the outlet and checking the outlet connections. The flow and pressure at the outlet should also be measured and be satisfactory. The inspection should also verify that earthing requirements have been carried out satisfactorily or certified by selected electrical authorities. The system should then be completely charged with water to a pressure of 10 bar measured at the inlet for a period of 15 min. During this period an inspection of the system should be made to check that no leakage of water is taking place. All piping should be tested for no less than 2 hours at a pressure of 4 bars in excess of the maximum static pressure when this is 10 bars. After the initial inspection and test completed a flow test should be carried out. For this test water should be passed through the system under pressure and the flow gauge reading recorded. Inability to sustain an effective for firefighting jet, or any undue pressure loss should be investigated. If as a result of these tests any defects are found, these should be remedied as necessary and a retest of the system should be carried out. The water supply tested should be representative of the supply that may be available at the time of a fire. All isolating valves should be locked in the open position.

9.38 Installation, testing, and quality control of foam systems

9.38.1 Planning for construction

Drawings and documents and all the information specified in design of the systems should be provided and authorities should be consulted. Prior to installation of systems full consideration should be given to the following:

Purpose and function of the system
Application rate and the duration of discharge of the system, and the appropriate minimum values given in standards

Hydraulic calculation
Pipework including support details
Detection system layout (if specified) and method of operation
Type, location, and spacing of foam discharge devices
Type and location of foam proportioning devices
Source of water and quantity needed
Quantity and type of foam concentrate, its design concentration, the method of storage, and the quantity to be held in reserve

9.38.2 Extensions and alterations

Any extension or alteration to an existing system should comply with the appropriate requirements of standards. Any extension or alteration to the foam system should be carried out by the installer or his contractor. The organization that services the system and the relevant authorities should be notified promptly of any alteration. The effect on available water supply and minimum required quantity of foam concentrate should be considered at the design stage of extension or alteration to a system, and full hydraulic calculations should be carried out on the new system layout prior to commissioning.

9.38.3 System description

A system consists of an adequate water supply, a supply of foam concentrate, suitable proportioning equipment, a proper piping system, foam makers, and discharge devices designed to adequately distribute the foam over the hazard. Some systems may include detection devices. These systems are of the open-outlet type in which foam discharges from all outlets at the same time, covering the entire hazard within the confines of the system.

Self-contained systems are those in which all components and ingredients, including water, are contained within the system. Such systems usually have a water supply or premix solution supply tank pressurized by air or inert gas. The release of this pressure places the system into operation.

There are four basic types of systems:

1. Fixed
2. Semi-fixed
3. Mobile
4. Portable

9.38.4 Fixed systems

These are complete installations piped from a central foam station, discharging through fixed delivery outlets to the hazard to be protected. Any required pumps are permanently installed.

9.38.5 Semi-fixed systems

The type in which the hazard is equipped with fixed discharge outlets connected to piping that terminates at a safe distance. The fixed piping installation may or may not include a foam maker. Necessary foam-producing materials are transported to the scene after the fire starts and are connected to the piping.

The type in which foam solutions are piped through the area from a central foam station, the solution being delivered through hose lines to portable foam makers, such as monitors, foam towers, hose lines, etc.

9.38.6 Mobile systems

This includes any foam-producing unit that is mounted on wheels, and that may be self-propelled or towed by a vehicle. These units may be connected to a suitable water supply or may utilize a premixed foam solution.

9.38.7 Portable systems

These systems produce foam by equipment that can be hand carried and connected to pressurized water or premixed solution by fire hose. Generally the foam containers should be kept in proper place to protect against sun radiation.

9.38.8 Commissioning and acceptance tests

The supplier of the system or his supervising engineers should arrange for the completed system to be inspected and tested to determine that it is properly installed and that it will function as designed to the satisfaction of the user and the relevant authorities. A commissioning test program should be submitted by the installer to the user, and the test should be carried-out by competent people.

9.38.9 Inspection

A visual inspection should be conducted to ensure that the system has been installed correctly. All normally dry horizontal pipework should be inspected for drainage pitch. The inspector should check for conformity with design drawings and specifications, continuity of pipework, removal of temporary blinds, accessibility of valves, controls, and gauges and proper installation of foam makers, vapor seals, and proportioning devices. All equipment should be checked for correct identification and operating instructions.

Water-supply pipework, both underground and above ground, should be flushed thoroughly at the maximum practicable rate of flow before connection is made to system piping in order to remove foreign materials that may have entered during installation or that may have accumulated in the mains systems at lower rates of flow. The minimum rate of flow for flushing should not be less than the water demand rate of the system.

Foam concentrates have a lower surface tension than water, and they may cause internal pipe scale or sediment to loosen with the risk of blockage of sprayers, proportioning equipment, etc. Pipes and fittings should be carefully cleaned before assembly and any loose jointing material should be removed.

9.38.10   Flushing after installation

In order to remove foreign materials that may have entered during installation, water-supply mains, both underground and aboveground, should be flushed thoroughly at the maximum practicable rate of flow before connection is made to system piping. The minimum rate of flow for flushing should not be less than the water demand rate of the system, as determined by the system design. The flow should be continued for a sufficient time to ensure thorough cleaning.

Disposal of flushing water must be suitably arranged. All foam system piping should be flushed after installation, using its normal water supply with foam-forming materials shut off, unless the hazard cannot be subjected to water flow. Where flushing cannot be accomplished, pipe interiors should be carefully visually examined for cleanliness during installation.

9.38.11   Acceptance tests

The completed system should be tested by the supplier's qualified personnel for approval. These tests should be adequate to determine that the system has been properly installed and will function as intended.

9.38.12   Inspection and visual examination

Foam systems should be examined visually to determine that they have been properly installed. They should be inspected for such items as conformity with installation plans, continuity of piping, removal of temporary blinds, accessibility of valves, controls, and gauges, and proper installation of vapor seals, where applicable. Devices should be checked for proper identification and operating instructions.

9.38.13   Pressure tests

All piping, except that handling expanded foam for other than subsurface application, should be subjected to a 2 hours hydrostatic pressure test at 13.80 bars (200 psi) gauge or 3.45 bars (50 psi) in excess of the maximum pressure anticipated, whichever is greater. There should be no permanent distortion or rupture and no substantial leakage during this test.

A full-scale discharge test should be conducted to ensure that the system discharge at the design rate, function in accordance with all other design requirements and produces and maintains an even foam blanket over the surfaces to be protected.

9.38.14   Operating tests

Before acceptance, all operating devices and equipment should be tested for proper function. Where conditions permit, flow tests should be conducted to ensure that the hazard is fully protected in conformance with the design specification.

Static water pressure, residual water pressure at the control valve, and at a remote reference point in the system, actual discharge rate, consumption rate of foam-producing material, and the concentration of the foam solution should be considered. The foam discharged should be visually inspected to ensure that it is satisfactory for the purpose intended. Particular checks should be made during the discharge tests to ensure that these factors have been taken properly into account. Water may be used instead of foam solution for some tests to avoid the need of extensive cleaning of the system after tests.

The inspections and tests should cover:

Rate of application of foam solution
Foam properties
Foam distribution
Running pressures
Concentration of the foam solution
Manpower requirements

- System Restoration: After completion of the acceptance test, the pipework should be flushed, strainers inspected and cleaned, and the system restored to operational condition.
- Completion Certificate: The installer should provide to the user a completion certificate stating that the system complies with all the appropriate requirements of standards, and giving details of any departure from appropriate recommendations.

9.38.15   Operation

- Method: All operating devices whether manual or automatic should be suitable for the service conditions they will encounter. They should not be readily rendered inoperative, nor be susceptible to inadvertent operation, as a result of relevant environmental factors such as high or low temperature, atmospheric pollution, humidity, or marine environments. The choice of method of operation will be governed by the potential rate of fire development, the likelihood of spread to other risks, and the degree of life hazard.
- Operating instructions and training: Operating instructions for the system should be provided at the control equipment and also at the plant or fire-control center. People who are authorized to operate the system should be thoroughly trained in its function and method of operation.
- Manual controls: The location and purposes of the controls should be plainly indicated and should be related to the operating instructions. Manual controls for systems should be located in an accessible place sufficiently removed from the hazard to permit them to be safely operated in emergency, yet close enough for the operator to be aware of conditions at the hazard.
- Color coding of pipework: The pipes should be color-coded in accordance with any scheme for pipework.

9.38.16   Emergency plans and operation training

- Planning: Fire and emergency plans should be prepared and approved by responsible authorities. The plan should consist of the following items:
Classification of emergencies
Designation of responsibilities
Alarms and communications
Outside sources and helps
The emergency plans and instructions should be posted at prominent places.

- Instructions: Operating and maintenance instructions and layouts should be posted at control equipment with a second copy on file. All people who are expected to inspect, test, carry out maintenance, or operate foam-generating apparatus should be thoroughly trained and kept thoroughly trained in the functions they are assigned to perform.

9.39 Training facilities and firefighter qualifications

The major concern of a company's safety and fire-protection group is the protection of employees from threat of fire and industrial accident in their workplace. Fire-service personnel must deal and handle fire and emergency cases that may occur and take rescue measures and protect facilities against fires. It is the responsibility of fire authorities to provide relevant services required to all emergencies and ensure that firefighters have the required skill and knowledge of:

Nature of manufacturing, processing, handling, and storage of products
Hazards involved
Provision and use of fire equipment suitable for extinction of flammable and combustible materials

The safety and fire authorities should also provide means for prevention of accidents and safeguard employees against the risk of mishaps. This section specifies the minimum requirements for training facilities, organizing, operating an effective fire service, and qualifications required for firefighters and fire officers.

The standard is divided into two parts as follows:

- Part 1: Training Facilities
- Part 2: Firefighter Qualifications

9.39.1 Training and training facilities

The fire-service training should be carried out under strict discipline and at no time should be allowed to degenerate into anything but the serious undertaking. A training program should be prepared in advance covering the topics and period applicable to the requirement of individual and groups responsibilities. It should consist of instructions and practical exercises and drills and laboratory work. The schedule of instructions should mainly consist of talks or lectures and practices. Training techniques include explanation, demonstration, and participation.

- Training Facilities: The engineering standard for training center includes:
Layout of training ground
Equipment details
Installations

The training facilities given here are for safety and fire-service training center and are separate from a training and drill work done at each fire stations. There is an element of convenience, if the training center and fire station can be in the same general neighborhood. Training facilities should include adequate classrooms, lecture halls, and audio-visual for the needs of safety and the fire department. These space requirements may be used as joint facilities for conference, and assembly areas for other training and development or gathering, but a classroom of sufficient size or similar lecture facility intended for continuing educational training should be available for safety and fire purposes.

Adequate audio-visual, special apparatus, and reference materials should be available to support the departments training activities and cover all subjects later discussed in standards. In addition the department should have access to training facilities for live fire training to demonstrate flammable liquid fires pumping and drafting operations and driver training.

The training building should be supplemented by sufficient yard space to provide room for fuel pits, oil tanks and facilities of fire and emergency that can be created to make training realistic. The site should be sufficiently isolated from other properties to eliminate inconvenience to the public from training activities. The site should have proper drainage, adequate water supplies for hydrants, appropriate pollution abatement equipment and a pond or reservoir for pumping operation. Suitable lighting and loud hailer systems should be provided. Also adequate facilities to monitor operation and safeguard trainees must be provided. Finally extra site for parking of vehicles should be provided. The training facilities should be secured, the site should be fenced and lighted in accordance with standard. The training center should include an appropriate selection of built-in fire-protection equipment including, smoke detectors, automatic sprinklers, stand-pipe systems, fire pumps, storage tanks, and other equipment with which employees should become familiar.

9.39.2   Training program for plant personnel

Safety and fire authorities are responsible for developing a training program for all departments and should designate a qualified training officer to act in administrating the program. He should foresee that the budget required for training facilities, expendable supplies, training aids, and training staff including in-house and guest instructors are considered.

All heads of line organization should fully support training activities initiated by fire authorities and make sure that the program is performed as prescribed. Department heads should consider hazards involved at their work site and recommend special training needed for their personnel assigned to implement their jobs. Fire authorities should include this special training requirement in the program for action. To improve or maintain standards of proficiency, the training should be planned carefully and be followed by an evaluation discussion in which all personnel should be encouraged to participate.

• Training Officer: As a member of fire authority, the training officer should segregate all topics recommended by the department heads and should make an appropriate program suitable for each department personnel and should ensure that necessary facilities are in hand before the training course begins. The training officer should define the syllabus of the training course for inclusion in the program and develop schedules to assure that the respective department personnel have received instructions accordingly. Instructions can be given either at the department's own facilities or at the safety and fire training center. The effectiveness of training should be continuously evaluated using critiques as an aid to such evaluation.

At least annually the program should be reviewed for updating to include new techniques and equipment in the material of the course.

If the course is conducted by instructors the training officer should supervise and witness the course of lectures, etc., and the outdoor practical exercises to see that the training procedures are followed as directed.

Records should be kept of all personnel attending the course. Reports of instructors should be reviewed by the training officer to evaluate performance of personnel of the facts demonstrated during the operation of the facilities. He should supervise the work of instructors and other personnel assigned in conducting the course and should see that each training session can be measured against a planned program.

The training officer should inspect and find that the operation of training equipment, first aid, and other facilities are in good working order.

* Firefighter Training Requirements: Success in fighting fires will be achieved if the following points are fully considered and adhered to:

Discipline should be enforced when crew members attend the scene of a fire.

Firefighters should obey the instructions given by the fire master.

Every one of the crew of firefighters should carry out the special duty assigned to him as specified in the *Fire-service Drill Book*, published by Home Office Fire Department (England), can be referred to when required.

Firefighting is a teamwork. The crew members must always help each other during the operation as this is a recognized activities resulting in the extinction of fire.

Fire authorities should develop a program to include all essential drills and exercises for firefighters to practice.

Fire authorities should appoint the training officer or his assistant to supervise the daily drill practices and other activities in accordance with schedule of program made.

Performance of firefighters should be evaluated periodically to determine the effectiveness of drills and exercises and provide a base for improvement and upgrading the program.

Refresher course should be programmed to refresh the memory of firefighters to help smooth running of the operation when called to attend fires.

New techniques should be considered for inclusion in the program in order to promote the knowledge of firefighters. Any changes in design and operation of fire equipment must be discussed and demonstrated.

* Training for New Recruits for Fire Brigade: New recruits should be given comprehensive training at least for the period of 6 months as probationers. This period should be considered as a basic training course for subsequent in service training. The course of probational training should be consistent with the performance objective as outlined below:

Appropriate firefighting drills as outlined in the *Fire-service Drill Book* by Home Office Fire Department (England).

Firefighters professional qualification (see NFPA 1001 2002).

Some of the following subjects are extracted from practical firemanship:

* Elements of combustion and extinction
* Fire-service equipment
* Fire extinguishers and foam equipment
* Incidents involving air craft and shipping
* Special emergency and rescue appliances

- Breathing apparatus and resuscitation
- Pumps and pump operation
- Structural fire protection
- Fire protection in buildings
- Communications
- Practical firefighting
- Emergency plans
- Topography of the area
- Some information about fire and emergencies at oil wells
- Toxic gases and fumes and detection instruments
- Safety in oil, petrochemical, and gas industries including fire prevention
- Fire detectors
- First-aid and rescue operations

The course duration for new recruits training should be at least the minimum number of hours necessary to complete the topics given in the schedule of the program.

- Regular training: At least 2 hours of each shift on duty should be devoted to training activities. This activity should be in form of classroom instruction during the night, and fitness practice drill and familiarization inspection the during day.

9.40 Firefighter qualifications

A firefighter must be physically fit for work at a fire and emergency that will involve considerable physical exertion. He must be courageous and yet be calm, as his manner and conduct will depend on his reactions in an emergency. He must be patient when executing his duties. He must have initiative and must possess the will to keep going for long periods under adverse conditions. He must cultivate his power of observation to the utmost and must also possess an inquiring mind. He must have a keen sense of discipline for unless he himself is able to obey orders without question. His duty may be summed up as, firstly, to save life, secondly, to prevent the destruction of property and thirdly to render humanitarian service. He should endeavor to learn as much as possible of plant and processes, so that if he is called to a fire and emergency in the units and premises he will be aware of the conditions he will meet and of any precautions he must take.

9.40.1 Selection of firefighter

The selection, training, direction, and employment of fire-department personnel is a major phase of fire-department administration and operation. If these personnel functions are properly conducted, fire-prevention and firefighting procedures can be channeled and handled smoothly, but without properly selected and trained personnel, fire-protection efforts are likely to be spasmodic and inefficient with too many important details left to chances. Care should be taken to get the right type of workers, and character reliability must be unquestioned. His intelligence should be keen. Emergencies require quick decisions and good judgment.

It is an accepted practice to require at least satisfactory completion of a basic fireman-training course and a reasonable degree of health and physical fitness. Each firefighter must not only be physically capable of withstanding the hard work of firefighting over a period of years but must be an outstanding member of firefighting force.

It is the fire department's responsibility to select and nominate a high level of personnel from physical fitness point of view as well as mental ability of the applicants.

9.40.2 Firefighter requirements

- Medical requirement: Prior to being accepted for fire-service membership, new recruits should be examined and certified by physician as being medically and physically fit. The medical and physical fitness requirements should take account of the risks and tasks associated with the assigned duties. Firefighters should be re-examined annually by industrial medical doctor as being medically and physically fit. Temperament of the applicants is equally important because firefighting is primarily a team function in which every member plays a vital assigned role. Firefighters should be encouraged to maintain good medical and physical condition and should be required to report any changes in their physical and medical fitness that could impair their performance as a firefighter. A thorough background investigation of each applicant from security point of view will be required. The applicants should have educational background of no less than technical diploma. The applicants should be selected after passing the physical examinations. When they have been selected, they should be engaged on probationary basis for a period of no less than 6 months. The probationary period should not be merely a perfunctory period for basic or recruit training, but rather a period permitting a close supervision of each applicant to see that he is sincerely interested in becoming a professional firefighter. During the 6 months probation the individual recruit should be examined closely to ensure his willingness and capability and after the fire authorities have confirmed their satisfaction the applicant can be assigned to a firefighter's duty.
- Personal Record and Promotion: After firefighters have completed basic training at the training center and have finished their probationary period, the department should maintain constant interest in each individual firefighter, with a record of all assignments, accomplishments, and performance. In some cases these records should be taken into consideration in giving credit toward promotion. Credit should be given for certain amount of longevity and activity under theory that experience gained in years of actual firefighting, emergency services, and rescue operations have a value that cannot be measured in written examination alone. In general promotional examinations for each rank should be held at two years intervals or when vacancies arise. Examinations should be based on the type of work to be performed in the rank in question following a careful analysis of the duties of the rank involved.

9.40.3 Operation of fire-service apparatus

The fire-service apparatus should be operated only by members who have been qualified in its proper operation by formal training using performance based standards. Drivers of fire trucks should have valid drivers' licenses for the type of vehicle as specified in traffic regulations. Vehicles should be driven in compliance with all applicable traffic signs and regulations. Drivers of fire-service vehicles should be directly responsible for safe and prudent operation under all conditions. Drivers of fire trucks are also generally operators of fire-truck fixed equipment such as pumps, foam proportioners, chemical dry powders, firefighting booms, etc.

- Ranking: Emergency activities and firefighting operation should be carried out under-strict discipline and orders should not be degenerated into anything but the serious undertaking. Therefore, the uniform and ranking is an important factor and should be officially selected and recognized.
  Firefighter ranks:
  Firefighter grade 2
  Firefighter grade 1
  Fire truck driver and operator
  Crew leader controller, communication
  Section leader-maintenance mechanic
  Fire officer ranks:
  Station fire officer
  Firemaster
  Fire-protection engineer
  Training fire officer
  Deputy chief fire officer
  Chief fire officer

9.40.4 Responsibilities

Brief responsibilities of above ranking and job descriptions are as follows:

- Firefighter grade 2: The firefighter grade 2 (lowest ranking) has completed recruit training course and passed satisfactory performance of probationary period of 6 months and been assigned to firefighting duties (for more information see Chapters 5 and 6 of NFPA 1001–2002).
- Firefighter grade 1: A firefighter with 4 years of experience with a good record of background service, physically strong, and has passed an official examination with the ability to be assigned as a leader of 2–3 firefighters grade 2.
- Crew leader: Firefighter with minimum of 6 years of experience with an excellent background record with good discipline, physical fitness, and the ability to be a leader of a 5 to 6 men crew assigned to a fire truck and working as a team leader in assigned firefighting and emergency duties.
- Section leader: Firefighter with minimum of 10 years of experience with an excellent record and good discipline. Knowledgeable of direct firefighting team, all emergency conditions, and rescue operations. Acting as shift fire-station leader with team-work and decision-making abilities.
- Senior fire-control operator (section leader rank): With minimum of 10 years of experience with firefighting and as a crew leader. Knowledgeable of communication system and emergency call procedures. Responsible for all call out and communication of messages during emergency conditions. He is assigned in a central main control room.
- Maintenance mechanic (section leader rank): Mechanical foreman with 2 years of experience with all aspects of firefighting equipment and appliances. Responsible for testing, inspections and repair work of firefighting and emergency equipment. Section leader and a helper to be assigned to service and maintenance of all types of fire extinguishers.
- Station fire officer (officer's ranking): He is the officer in charge of the station. A comprehensive knowledge of station ground and detailed information of any manufacturing

processes and risks involved. On arrival at the fire and emergency ground he takes command before arrival of the district fire master or chief fire officer.

- Fire master: He is responsible for all fire-protection systems of an area or district, reporting to the chief officer. He inspects all plants and fire risk areas and advises on fire-prevention methods and requirements with close cooperation of plant managers, and organizes firefighting team and makes sure that they are regularly trained for firefighting and emergencies conditions.
- Fire-prevention engineer (divisional fire officer): Being as a qualified engineer he is responsible for advising on requirements of fire-protection systems. He is responsible for the testing and servicing of all fire appliances. He advises on fire-prevention methods and makes sure that fire-service standards are met. He has broad knowledge of all types of fire-protection equipment.
- Training fire officer (divisional fire officer): As a qualified engineer, he is responsible for the following training courses:
  1. Fire-service recruit training
  2. Plants fire team training
  3. Fire officers advance training
  4. Supervisors fire-prevention courses

The instructors who teach the subjects should be familiar with lessen plans. The following is a good example of the organization of a lesson plan:

1. Title
2. Must indicate clearly and concisely subject matter to be taught
3. Objective
   Should state what the trainee should know or be able to do at the end of training period
   Should limit the subject matter
   Should be specific
   May be divided into a major and several minor objectives for each session
4. Training aids
5. Should include such items as actual equipment or tools to be used and charts, slides, films, etc.
6. Presentation
   Should give the plan of action
   Should indicate the method of teaching to be used (lecture, demonstration, class discussion or combination of these)
   Should contain suggested directions for instructor activity (show chart, write keywords on blackboard)
7. Application
   Should indicate, by example, how trainees will apply this material immediately (problems may be worked a job may be performed, trainee may be questioned on understanding and procedures)
8. Summary
   Should restate main points
   Should tie up loose ends
   Should strengthen weak spot in instruction
9. Test
   Tests help to determine if objectives have been reached they should be announced to the class at the beginning of the session
   Assignment should give references to be checked or indicate materials to be prepared for future lessons

9.41   Safety boundary limit

- Design for Safety: Efficiency and Safety in industrial operations can be greatly increased by careful planning of the location, design and layout of a new plant or an existing one in which major alteration should be made. Numerous accidents, explosions, and fires are preventable if suitable measures are taken right from the earliest planning stages. The size, shape, type of buildings and structures, spacing, the nature of processes and materials, working conditions, chemical and physical properties of dangerous substances and their processing methods are the major factors to be considered. It is always preferable that high-hazard processes be located in small isolated buildings of limited occupancy or in areas away from hazard involved. Lower-hazard operations can justify larger unit.
- Safe Distance Limits: Selection of safe distances from possible hazards involves consideration of a number of factors: possible hazards to the community and their relationship to climate and other conditions, highly flammable materials (liquid and gases), amount of harmful substances, drainage, and waste disposal. Plan for safety boundary limits should include all necessary safety precautions, and each case should be carefully studied and planned by competent engineers. This section specifies minimum requirements for spacing of hydrocarbon production, gas and oil refineries, petrochemical complexes, and safe distance of oil and gas wells to other production facilities, high-tension electrical poles, roads, and residential areas. Standards are also guidelines for normal operations but each special case should be carefully studied considering all factors of possible hazards.

9.41.1   Hydrocarbon production and processing plants

- Layout and design: Spacing of equipment should be in accordance with oil insurance and Tables 9.5, 9.6, and 9.7. When the topography of the site is level, arrange drainage to minimize exposure of process areas to large spills. Otherwise, locate storage tanks at a lower elevation than process areas.
- Storage tanks: The selection, design, construction, installation and testing, as well as fire protection, of storage tanks should be in accordance with Tables 9.8, 9.9, 9.10, and 9.11.
- Emergency shutdown system
Gas and Product Line Control Valves: High-pressure gas lines should not pass through a process area or run within 30 m of important structures or equipment without shutdown valving to insure that portions of piping within the process area can be isolated from the main gas line and depressurized in the event of an emergency. However, extensive use of shutdown valves may not be needed, since the increased complexity of the system will require a greater degree of preventive maintenance if unwarranted shutdowns should be avoided. Shut-off valves, sometimes known as "station isolation valves," should be provided on all gas and product pipelines into and out of the plant. A bypass line with a normally shut valve may be required between plant inlet and discharge lines. All station isolation valves-and bypass valves, if any, should be located at a minimum distance of 75 m but not more than 150 m from any part of the plant operations. Care should be taken in locating these valves so they will not be exposed to damage by plant equipment or vehicular traffic.
Emergency Shutdown Stations: At least two remote emergency shutdown stations, located at a minimum distance of 75 m apart, should be provided. Locate actuating points at least 30 m from compressor buildings and high-pressure gas lines. More than two shutdown stations may be required, depending on the size and complexity of a given plant. One of the

Table 9.5 **Oia (oil insurance association) general recommendations for spacing in refineries**

Table 9.6 Oia general recommendations for spacing in petrochemical plants

Table 9.7 **General recommendations for spacing in gas plants**

MAXIMUM DISTANCE IN METERS	SERVICE BUILDINGS	GAS COMPRESSOR HOUSE	LARGE PROCESS OIL PUMP HOUSE	DISTILLATION AND FRACTINATION	UTILITIES	PRESSURE TANKS	ATMOSPHERIC TANKS	LOADING RACKS	FIRED HEATERS	COOLING TOWERS	SKID UNITS FOR PACKAGE PLANT	CONTROL HOUSES *
SERVICE BUILDINGS												
GAS COMPRESSOR HOUSE	SEE CHART											
LARGE PROCESS OIL PUMP HOUSE	30	15										
DISTILLATION AND FRACTINATION	30	15	9									
UTILITIES	30	30	30	30								
PRESSURE TANKS	15	60	60	60	two be of cement							
ATMOSPHERIC TANKS	45	60	60	60	15	15						
LOADING RACKS	30	30	30	30	30	30	45 TO 60					
FIRED HEATERS	30	30	30	30	30	68	15	15 TO 30				
COOLING TOWERS	30	15 TO 30	12	15 TO 30	15	30	30	30	60			
SKID UNITS FOR PACKAGE PLANT	15 TO 30	15	15	15	30	30	60	30	60	60		
CONTROL HOUSES *	15	30	30	30	60	60	60	60	60	60	60 TO 150	

Note: * Control houses serving unusually large or hazardous units and central control houses for multiple units of housing computer equipment, require greater spacing and may require blast resistant construction.

Table 9.8 **Proximity of refrigerated storage vessels to boundaries and other facilities**

Boundary lines or other facilities	Minimum spacing of dome roof tanks	Minimum spacing of spheres or spheroids
Property lines adjacent to land which s developed or could be built upon public highways main line railroads	60 m(1)	60 m(1)
Utility plants, buildings of high occupancy (offices, shops, labs, warehouses, etc.)	1-½ vessel diameter but not less than 45 m not exceed 60 m(1)	60 m(1)
Process equipment (or nearest process unit limits if firm layout not available)	1 vessel diameter but not less than 45 m need not exceed 60 m(1)	60 m(1)
Non-Refrigerated pressure storage facilities	1 vessel diameter but not less than 30 m need not exceed 60 m	¾ vessel diameter but not less than 30 m need not exceed 60 m
Atmospheric storage tanks (stock closed cup flash point 55°C)	1 vessel diameter but not less than 30 m need not exceed 60 m	1 vessel diameter but not less than 30 m need not exceed 60 m
Atmospheric storage tanks (stock closed cup flash point 55°C or higher)	½ vessel diameter but not less than 30 m need not exceed 45 m	½ vessel diameter but not less than 30 m need not exceed 45 m

Note:
1. Distance from boundary line or facility to centerline of peripheral dike wall surrounding the storage vessel should not be less than 30 m at any point.

Table 9.9 **Proximity of atmospheric storage tanks to boundaries and other facilities**

Boundary lines or other facilities	Minimum distance from			
	Low flash or crude stocks in floating roof tanks	Low flash stocks in fixed roof tanks	Crude stocks in fixed roof tanks	High flash stocks(1) in any type of tanK
Property lines adjacent to land whic is developed or could be built upon public highways,main line, railroads and manifolds located on marine piers. Building of high occupancy(offices, shop, labs, warehouses, etc.)	60 m	60 m	60 m	45 m (3)

(*Continued*)

Table 9.9 (Continued)

Boundary lines or other facilities	Minimum distance from			
	Low flash or crude stocks in floating roof tanks	Low flash stocks in fixed roof tanks	Crude stocks in fixed roof tanks	High flash stocks(1) in any type of tanK
Building of high occupancy (offices, shop,labs, warehouses, etc.)	1-½ tank diameter but not less than 45 m need not exceed 60 m	1-½ tank diameter but not less than 45 m need not exceed 60 m	60 m	1 tank diameter butnot less than 30 m need not exceed 45 m (3)
Nearest process equipment, or utility plant (or nearest unit limits if firm layout not available)	45 m	45 m	60 m	½ tank diameter but not less than 30 m need not exceed 45 m (3)(4)

Table 9.10 Proximity of atmospheric storage tanks to each others

Type of stocks and tankage	Minimum spacing between (1) (2)		
	Single or paired tanks	Grouped tanks	Adjacent rows of tanks inseparate groups (1)
Low flash or crude stocks in floating roof tanks	¾ tank diameter need not exceed 60 m	½ tank diameter need not exceed 60 m	¾ tank diameter not less than 25 m need not exceed 60 m
Low flash stocks in fixed roof tanks	1 tank diameter	½ tank diameter	1 tank diameter not less than 30 m
Crude oil stocks in floating roof tanks	¾ tank diameter need not exceed 60 m	Not permitted	
Crude oil stocks in fixed roof tanks	1-½ tank diameter (pairing not permitted)	Not permitted	
High flash stocks in any type tank	½ tank diameter need not exceed 60 m	½ tank diameter need not exceed 60 m (3) (4)	½ tank diameter not less than 15 m need not exceed 60 m

Table 9.11 **Proximity of non-refrigerated pressure storage vessels/ drums to boundaries and other facilities**

Boundary lines or other facilities	Minimum spacing to spheres, spheroids and drums
Property lines adjacent to land which is developed or could be built upon public highways, main railroads, and manifolds located on marine piers	60 m (1)
Building of high occupancy (offices, shop, lab, warehouses, etc.)	60 m (1)
Nearest process equipment, or utilities, point (or nearest unit admits if firm layout not available)	60 m (1)
Refrigerated storage facilities	¾ tank diameter, but not less than 30 m need not exceed 60 m
Atmospheric storage tanks (stock closed cup flash point of 55°C and below)	1 tank diameter, but not less than 30 m need not exceed 60 m
Atmospheric storage tanks (stock closed cup flash point above 55°C)	½ tank diameter, but not less than 30 m need not exceed 45 m

Note:
1. Distance from boundary line or facility to centreline of peripheral dike wall surrounding the storage vessel should not be less than 30 m at any point.

actuating stations should be located in the control room. It should be distinctively marked and equipped with signs stating the proper method of actuation in the event of an emergency.
- Wastewater separators: Wastewater separators handling hydrocarbons should be spaced at least 30 m from process unit equipment handling flammable liquids and 60 m from heaters or other continuous sources of ignition. Preferably, wastewater separators should be located downgrade of process equipment and tankage.
- TEL blending plants: Tetraethyl Lead (TEL) blending plants should be spaced 30 m from process equipment handling flammable liquids and 45 m from fired heaters or other continuously exposed sources of ignition. Arrange to reduce any possibility of flammable liquid draining near the TEL plant.
- Flares: Spacing of flares from process equipment depends on the flare stack height, flare load in pounds per hour and the allowable heat intensity at the equipment location. Flare locations should be at lower elevations than process areas, should be curbed to contain hydrocarbon carry-over, and should be at least 60 m from equipment containing hydrocarbons. Also, areas where personnel may be present and where the public has free access must be considered. For spacing requirements, refer to attached figures for oil and gas separation units (see also Tables 9.12 and 9.13).
- Blowdown drums: Blowdown drums are used for liquid relief in emergencies and are not usually installed when a suitable pressure relieving system and flares are provided. When used, blowdown drums should be 30 m minimum from process unit battery limits and 60 m from storage tanks and other refinery facilities.
- Fire-training areas: Fire-training areas are ignition sources when in use. Because of the smoke produced, they can also create a nuisance for the refinery and neighboring facilities. Fire-training areas should be 60 m from process unit battery limits, main control rooms, fired steam generators, fire pumps, cooling towers and all types of storage tanks. They

Table 9.12 The minimum distances of production units flares from public roads

Flares	Public main roads meters	Private or branch roads meters
Oil or gas burning	200	200
Pitsgroundlevel flares	200	150
High level flares	150	100
Units cold flares	100	50

Notes:
1. If the above figures can not be followed, the case should be thoroughly examined by committee of production engineers and authorities concerned.
The committee will prepare drawing of the area with detailed conditions stating why the above distances can not be observed and recommend the proposed distances.
2. Distances between flares should not be less than 100-meters.

Table 9.13 Minimum distances of oil/gas wells from other production facilities

Structures		Asmari (Meters)	Bangestan (Meters)
1	Gas pipelines laid on the ground	200	200
2	Gas pipelines buried	60	60
3	Oil pipelines laid on the ground level	200	200
4	Burried oil pipeline	60	60
5	High tension electrical pole	200	200
6	Telephone lines	200	200
7	Oil & gas production units and facilities	400	400
8	Burning pits of productions units	300	300
9	Ground level flares	300	300
10	Production units flare stacks	150	150
11	Cold flares	300	300
12	Residential areas	400	400
13	Public roads	300	300
14	Private and branch roads	200	200
15	Oil/Gas wells	200	200

should also be 75 m from property boundaries, administration, shops and similar buildings, and from the main substation.

1. Spacing may be reduced to 30 m for a tank or group of tanks meeting all of the following:
All tanks are an integral part of the given process operation.
Each tank is less than 15 m in diameter.
The total capacity of the group does not exceed 7950 m^3 (50,000 bbl).

2. Spacing need not exceed 30 m provided that all of the following requirements are met:

The stock is stored at ambient temperature and the closed cup flash point is above 93°C or if heated, not above 93°C and not within its flash point.

The stock is not received directly from a process unit where upset conditions could lower its flash point.

The total capacity of any tank does not exceed 31800 m3 (200,000 bbl) and the total capacity of any group of tanks does not exceed 79500 m3 (500,000 bbl).

There are not tanks storing low flash stocks within the same group.

3. Spacing need not exceed 15 m provided that all of the following requirements are met:

All tanks are an integral part of the given process operation.

Each tank is less than 25 m in diameter and the total capacity of a group of tanks does not exceed 7950 m³ (50,000 bbl).

There are not tanks storing low-flash stocks within the same group.

4. Spacing between high flash and low flash tank groups should be governed by the low-flash criteria.

5. A minimum spacing of 3 m should be provided between any tank shell and the peripheral dike or toe wall.

6. Finished stocks with a closed cup flash point above 93°C may be spaced a minimum of 2 m apart provided that all of the following requirements are met:

The stock is stored at ambient temperature: if heated, not above 93°C and not within 10°C of its flash point.

The stock is not received directly from a process unit where upset conditions could lower its flash point below the limits of subpara. above.

There are not tanks storing low-flash stocks within the same group.

7. Finished stocks with a closed cup flash point of 54°C or higher but less than 43°C may be spaced 1/6 of the rim of their diameters apart, except:

8. Where the diameter of one tank is less than one-half the diameter of the adjacent tank, the spacing between the tanks should not be less than one half the diameter of the smaller tank, provided that all of the following requirements are met:

The spacing between tanks is no less than 2 m.

The stock is not heated above 93°C and not within 10°C of its flash point.

Group Tanks do not exceed a total capacity of 15900 m³ (100,000 bbl) and there are no tanks storing low-flash stocks within the same group.

The stock is not received directly/from a process unit where upset conditions could lower its flash point below the limits of subpara b above.

Safety and firefighting equipment, part 2

10

This chapter is prepared for the use and guidance of authorities in charge of purchasing and operating firefighting equipment in order that such equipment will function as intended throughout its life.

Nothing in any standards is intended to prevent the use of new methods or equipment, provided sufficient technical data are submitted to the company's fire and safety departments to demonstrate that the new methods or equipment are equivalent in quality, effectiveness, and durability to those described by the standards. It may be necessary for those charged with the purchasing and approval of equipment to consult with an experienced fire-protection engineer competent in the field to select the best safety and firefighting equipment to meet the standards.

The standards cover the minimum requirements for physical properties and performance of firefighting equipment to be used and/or purchased in the oil, gas, and petrochemical industries. The standards only include the necessary essentials to make the standard workable in the hands of those skilled in this field.

10.1 Fire-service valves

Hydrants and isolating valves installed on the main water lines and other types of valves utilized in plant units and industrial areas are designed for firefighting and fire-protection systems specifically manufactured where quick operation and reliability are the main factors.

In this section the following types of fire-service valves are discussed:

Underground Hydrant Valve.
Aboveground Hydrant Hose Valve.
Aboveground Butterfly Valve (Rubber-lined).
Ball Valve Used on Fire Trucks and in Foam Systems.
Check Valve (Nonreturn) Used in Fire-protection Systems.

10.1.1 Underground fire hydrant

Underground fire hydrants should be of the following types:

Wedge Gate.
Screw-Down.
Butterfly Type (Figure 10.1).

Personnel Protection and Safety Equipment for the Oil and Gas Industries.
DOI: http://dx.doi.org/10.1016/B978-0-12-802814-8.00010-9
© 2015 Elsevier Inc. All rights reserved.

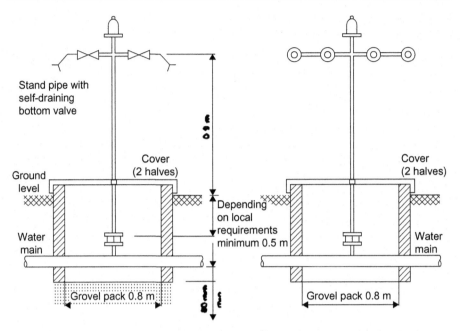

Figure 10.1 Two-way and four-way hydrants for underground main lines.

Wedge gate valves should comply with the requirement of BS 5163 for PN 16 and material for duckfoot bends should be chosen from grey cast-iron (CI) or spheroidal graphite cast-iron (SG). The flange thickness should not be less than 17mm at position of any bolt hole.

- Materials: The hydrant should be fitted with captive valve and the following materials should be used:
  For the threaded part that engages with the spindle: Gun metal or high tensile brass.
  For the body: Grey cast-iron (CI) or spheroidal graphite cast-iron (SG).
  For body seating: Gun metal or high tensile brass.
  For spindle: High-tensile brass or stainless steel.
  For screwed outlet: Gun metal, die cast brass, or high-tensile brass.
  For spindle cap: Cast-iron.
  For surface base and frame: Grey cast-iron or spheroidal graphite cast-iron.
- Manufacture and workmanship: All cast units should be cleanly cast and should be free from air holes, sand holes, cold shuts and chill. They should be neatly dressed and carefully fettled. All castings should be free from voids, whether due to shrinkage, gas inclusions, or other causes.
- General requirements:
  Screwed Outlet: Should be provided with the cap to cover the outlet threads. It should be securely attached to the hydrant outlet by a chain.
  Bolting: The dimensions of bolting should comply with the requirement of the appropriate ISO metric standard.

Corresponding part of hydrants of the same design and manufacture should be interchangeable.

Valve should be closed by turning the spindle in a clockwise direction when viewed from above.

All cast-iron parts should be thoroughly cleaned and coated before rusting.

Spindle sealing should be of the following types:

   Stuffing box and gland type.

   Toroidal sealing ring (O) ring.

Where spindle sealing is of toroidal type, two such seals should be used. The packing and seals of all types should be capable of being replaced with the valve under pressure.

When fitted with standard round-thread outlet the hydrant should deliver no less than 2000 L/min at constant pressure of 1.7 bar at the inlet to the hydrant.

- Test requirement: A single set of tests only is required to ascertain that the hydrant design meets the stated requirements of this clause:

   Hydrostatic pressure.

   Wedge gate.

- Hydrant valve and seating: The valve should comply with the requirements of BS 5163: 1986 when tested in accordance with the methods of that clause.

- Duckfoot bend: There should be no visible sign of leakage from the duckfoot bend when tested.

10.1.2   Screw-down type hydrants

- Hydrant seating: There should be no visible sign of leakage past the valve when the hydrant is tested.

- Complete hydrant: There should be no visible sign of leakage from the hydrant when tested.

- Screwed outlets: There should be no visible sign of leakage from the screwed outlet when tested. If the screwed outlet is supplied and attached to the hydrant then it should be tested either integrally with it or after removal from the hydrant.

- Hydrostatic test certificate: Provision should be made for a certificate to be made available that certifies that the hydrant has complied with the requirements.

- Markings: Each hydrant valve, duckfoot bend, and outlet should be clearly marked, either integrally with the stated components or on a plate of durable material securely fixed to that component, as follows:

   Direction of valve opening on the gland or upper part of the hydrant.

   Material designation for grey cast-iron "CI" or for spheroidal graphite cast-iron "SG".

   Manufacturer name and trademark.

- Coatings: Units should be thoroughly cleaned and dried before being given a short term coating of:

   Hot applied coal tar or bitumen based coating material.

   Cold applied black bitumen solution.

- Clear opening frame and cover: The minimum clear openings in surface box frames should be:

   For wedge gate 220–500 mm max

   For screw down 230–380 mm max

The depth of the frame should not be less than 100 mm for grade wedge gate and 75 mm for grade screw down. The minimum bedding width of the frame should be 50 mm. Surface box frames and covers should be designed so that the top of the cover is flush with the top of the frame. Surface box covers should be clearly marked by having the words "fire hydrant" in letters no less than 30 mm high, or the initials "FH" in letters no less than 75 mm high, cast into the cover.

10.1.3   Above-ground hydrant valve

Above-ground hydrant valves are generally come in two types: rubber-lined butterfly valve and hose valve.

10.1.4   Rubber-lined butterfly valve (Figure 10.8)

This type of valve is designed and used in cold climates.

• Valve design specification: The butterfly valves should be designed in accordance with British Standard BS 5155.

All valves should have a pressure rating of PN 16, which is the pressure rating for fire-water systems. The aboveground valves should be wafer type with wrench and the underground valves should be single-flange or lug type with gear. For a further description of the design and materials, see Figure 10.2.

• Additional design requirements.
Design should include a separate stem seal, even if the primary seal is accomplished by the lining.

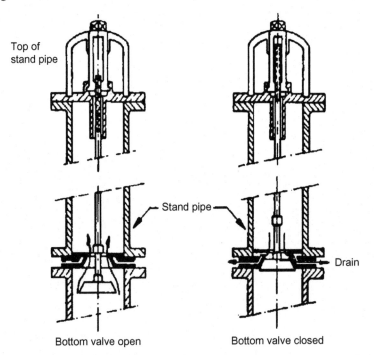

Figure 10.2 Typical detail of self-draining butterfly valve.

Valves should be designed such that they can be locked in the open position.

Weatherproof gearbox should be provided with an indicator to show the "open" and "close" position visible from all positions at a distance of 15 m.

Assembly of valve and gearbox should not allow of the indicator showing the position of the disc incorrectly.

Outer barrel of the extension for the underground valve should have a diameter of no less than 60 mm, and a thickness of no less than 5.5 mm and should be made from carbon steel pipe.

Shaft should be made from Monel alloy k-500 when exposed to the medium. When the shaft is isolated from the medium using a lining, it may be made from stainless steel material. The shaft extension may be made of carbon steel.

- Valve size and type selection: The nominal size and type of the valves to be tested should be determined after consultation between the buyer and the manufacturer. For the selected item(s), the following should be stated in writing:

  By the buyer.

    Valve size.

    Class rating.

    Valve type: wafer, single flange, or lug type.

  By the manufacturer.

    Drawing and specification numbers.

    Lining material.

    Body and extension material.

    Make of gear and material.

- Specific requirement of materials
- Lining: Rubber lining should not have surface defects such as blisters, cracks, porosity, or other imperfections.

  Repair of the lining is not acceptable.

  Lining should be classified in accordance with ASTM D 1418.

  To verify that the lining applied by the valve manufacturer is the type that has been specified, an identification test is required in accordance with the method described in ASTM D 3677 or BS 903.

  The thickness of the lining should conform within ±10% to the thickness that has been specified by the manufacturer.

  Minimum of three measurements should be taken.

  The hardness of the rubber should conform to the value specified for the type or grade applied.

  Minimum of three hardness tests should be performed on the lining. The hardness reading should be expressed in Durometer A or Durometer D in accordance with ASTM D 2240 and should be within ±5% of the specified value.

- Tests

  Adhesion test: To test the adhesion, samples should be prepared from the same material as the lining as well as the body material. The adhesion should be tested in accordance with the ASTM D 429, method B. The adhesion value can be calculated from the load at failure, the original bonded area and the values should not be less than those specified by the manufacturer.

- Other tests: The following tests should be carried out by manufacturer:

  Leakage test for seat and seal.

  Body strength test.

  Disc and shaft strength test.

  Cycling operation test.

Torque test.

Capacity valve test.

Torque test in dry condition.

- Certificates: Material certificates in accordance with DIN 50049-3.1 B with physical properties and chemical analyses should be provided for the valve body, shaft, and disc. A certificate with the physical properties of the rubber lining material should be provided by the rubber manufacturer and a statement from the valve manufacturer that the certified material has been applied for the valve.
- Marking: Valves should be marked with the following information shown on a corrosion-resistant nameplate permanently attached to the valve: Manufacturer's name, catalog reference number, size, class, differential pressure rating, and material identification, including body and times.
- Packaging: Valves package should be palletized, or packed in cartons, boxes, or crates.
- Shipment: Opening-valve ends should be fully blanked to protect the sealing surfaces and valve internals during shipment and storage.

10.1.5   Hose valve

These requirements cover angle-pattern and straightway-pattern hose valves intended for use on standpipes, fire pumps, and hydrants supplying water for fire-protection service.

- Type and size.
  Angle (90 degree) pattern type for use on stand-pipes having inlet and outlet opening of the same size or the inlet larger than the outlet (outlet coupling 65 mm (2½ inch)).
  Angle (90–120 degree) pattern type for use on inlet pipe sprinkler equipment having 40 to 65 mm nominal outlets.
  Straightway pattern type for use on fire pumps and hydrant having inlet and outlet opening of the same nominal 65 mm (2½ inch) size.
  Straightway-pattern type for use on stand-pipes having inlet and outlet opening of the same size or with the inlet larger than the outlet opening of 65 mm (2½ inch).
- Working pressure: Hose valves should be constructed for a minimum working pressure of 12 bar for all types.
- Materials
  Intention of use: Hose valve intended for use on stand-pipes and fire pumps should be made entirely of brass and bronze except for the hand wheel and for the valve seal. Hose valve intended for use on hydrants and assembled by bolting the valve to the outside of the hydrant barrel may have the cast-iron body and bonnet intended to be bolted together.
  The remaining valve parts should be made of brass, bronze, or other materials having equivalent corrosion-resistant properties.
  Casting: Casting should be smooth and free from scale, humps, cracks, blisters, holes, and defects that could impair its intended use.
  Direction to open: A hose valve should open by turning the handwheel to the left (counter clockwise) as viewed from the top.
  Seat ring: The seat ring should be made of brass, bronze, or other equivalent corrosion-resistant material.
  Outlet and attachment: A hose valve intended for use on fire pump and standpipe should be fitted at the outlet with female instantaneous coupling 65 mm (2½ inch) (BS 336).

Hose outlet blank cap should be made of brass or equivalent corrosion-resistant material. The blank cap should be of male instantaneous with brass chain attachment (BS 336).

Stuffing box and seals

- A valve should include a stuffing box, or other means for sealing, so that there should be no leakage at the valve stem. The bearing surface provided in a stuffing box gland or seal retainer for the stem should be made of material having corrosion-resistant equivalent to brass or bronze.
- A stuffing box should include a gland or follower with a packing nut. There should be no threads within the stuffing box.
- The stuffing box should have sufficient width to contain packing so that there is no leakage around the stem and should have sufficient space for entrance of packing removal tools.
- The stuffing box bottom and the end of the gland should be beveled.
- A rubber ring, such as an "O" ring, used to provide a stem seal should be made of vulcanized natural rubber or synthetic rubber compound having uniform dimensions. A ring should have the properties of A.B.C item 13.5 UL 668.
- A valve should be constructed to permit repacking of the stuffing box or replacement of at least one seal ring when the valve is fully open and under rated working pressure. A stem seal formed rubber rings should include at least two rings and stem seals using "O" rings should include at least one ring.
- In a cast-iron valve, the entire stuffing box should be made of brass or bronze, and the stem opening through the bonnet should be brass or bronze bushed.
- Handwheel: The gandwheel should not have a diameter of no less than 100 mm.
- Seat ring-angle pattern.
  - The seat ring when finished should not extend into the valve interior beyond the near side of the outlet opening.
  - The seat seal holder should be free to turn on its stem so that the seal may seat without any rotary as scraping action likely to damage the seal.
  - The means for securing a locknut used to secure a seal holder to its stem should give securement equivalent to that provided by the use of a pin.
  - A rubber part used to provide a seat seal should (1) be made of vulcanized natural rubber or a synthetic rubber compound, and (2) have uniform dimensions. A seat seal material should have the properties as item 15.6 A.B.C of UL 668.
  - A resilient seat seal should be made of a nonmetallic material firmly secured and assembled so that it may be easily replaced. The seal holder should enclose the outer edge of the seal for its entire thickness. The valve seal, seal holder, and seal clamping ring should have dimensions so that the seal face overhangs the body seat ring both inside and outside.
  - The clearance between the edge of the seal holder and the inside of the body should not be less than specified in standards. The clearance between the inside of the seat and the seal nut and clamping ring, should not be less than 1.6 mm.
  - The seal retainer nut or clamping ring should be pinned in place or restrained from movement by locking feature.
- Additional requirements for straight-pattern type valves:
  A straightway-pattern valve may be of the nonrising-stem or rising-stem construction.
  A straightway-pattern valve, when fully open, should have a straight through unobstructed waterway with a circular cross-section, whose area at any point should not be less than the cross-sectional area of the waterway of the size of pipe with which the valve is intended to be used.

The gate of straightway-pattern valve should be either of the following: solid-wedge, split-wedge, or parallel-seat type.

A straightway-pattern valve should have guides for the gate cast integral with the body.

A valve that seats tightly with the gate in one position only should have integrally cast guides of unequal widths, or other equivalent means, to provide for intended assembly.

A valve for use on hydrants may have a gate with a single seating face and with guides cast in the body.

A valve body should have a boss formed on the underside at the outlet end that may be drilled to receive a drip cock.

 • Performance tests: Representative samples of valves should be subject of tests specified in standards.
 • Manufacturing test and production tests: The manufacturer should provide the necessary production control, inspection, and tests. The program should include at least the following:

Each valve should be factory tested for body and seat leakage. Each test should be conducted at twice the rated working pressure for 1 minute. The seat-leakage test should be conducted hydrostatically between the inlet and the closed seat. The body-leakage test should be conducted hydrostatically with the valve partially open and pressure applied on all parts, including the bonnet and body joint, and the stuffing box or sealing device. The valve should show no seat leakage, no weepage, or leakage through body and bonnet castings or at the joint between the two castings, and no leakage past the stuffing box or sealing device.

Straightway-pattern valves having metal-to-metal seats with provisions for attachment to hydrants only should have no more than slight weeping past the seat. Slight leakage at the valve stem is permitted.

• Marking: A hose valve should be marked with the following:

Name or identifying symbol of the manufacturer or private labeler.

Nominal size of valve.

Distinctive model number, catalog designation, or equivalent.

Rated working pressure.

Markings of valves should be in raised cast letters; on permanent stamped metal nameplates; or applied by a method that affords equivalent permanence. Markings may be at any convenient location on the valve.

An arrow, 31.8 mm (1¼ inches) long, showing the direction to turn the handwheel to open the valve, with the word "Open" at the feather end, or in a break in the shaft, should be cast on the rim of the handwheel so as to be easily readable. If the shaft of the arrow is broken to admit the word "Open," the sum of the parts of the arrow should not be less than 24 mm. If a vendor produces valves at more than one factory, each valve should have a distinctive marking to identify it as the product of a particular factory.

10.1.6 Ball valves

Ball valves are generally used on fire-truck water delivery sections of the pump, foam system, fire-extinguishing system, and where quick action for opening of flow is intended. Ball valves are also used where simultaneous opening and closing of two or three valves by quick action is needed.

• Bodies should be of one piece or split construction: In case of split body valves, the minimum design strength of the split body. Joint(s) should be equivalent to that of the body end flange of a flange body, or the appropriate equivalent flange for a butt-weld-end, socket weld-end, or threaded end body. Bolted covers should be provided with no less than four bolts, stud bolts, stud or sockethead cap, or hexagon-headed screw (Figure 10.3).

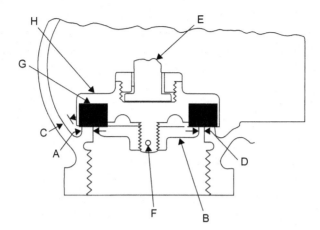

Figure 10.3 Angle pattern valve seat details. **A** Seat width. **B** Nut or clamping ring for Resilient Seat Seal. **C** Clearance between the edge of the Resilient Seat Seal holder and the inside of the body. **D** Clearance between inside of the seat and the nut or clamping ring for the Resilient Seat Seal. **E** Valve stem. **F** Locking pin for Retainer Nut or Clamping Ring. **G** Resilient Seat Seal. **H** Holder for Resilient Seat Seal.

- Flanged ends: End flanges should be cast or forged integral with the body or end piece of a split body design, or attached by butt-welding.
- Stems, ball shanks, stem extensions: Stems, ball shanks, stem extensions, stem mounted handwheels, or other attachments should be provided with permanent means of indicating port position and should be designed to prevent misorientation.
- Stem retention: The valve design should be such that the stem seal retaining fasteners, e.g., packing gland fasteners, alone do not retain the stem. The design should ensure that the stem should not be capable of ejection from the valve while the valve is under pressure by the removal of the stem seal retainer, e.g., gland, alone.
- Body seat rings: Body seat rings or seat ring assemblies should be designed so as to be renewable except for those valves having a one = piece sealed (welded) body construction.
- Ball: On full-bore valves the ball port should be cylindrical. Sealed-cavity balls should be designed to withstand the full-hydrostatic body test pressure. The typical types of ball construction are given in (Figure 10.4(b)).

Notes:

1. The buyer should state on his inquiry or order if reduced-bore valves are required with ball valves having cylindrical ports.
2. Solid, sealed-cavity, and two-piece ball valves are shown with cylindrical ports in Figure 10.4A.

- Wrenches and handwheels: When used, wrenches and handwheels should be designed to withstand a force no less than that given in BS 5351.
- Antistatic design: Valves should incorporate an antistatic feature that ensures electrical continuity between stem and body of valves DN 50 or smaller, or between ball, stem, and body of larger valves if specified. The use of conductive packing is permitted provided that the packing:
Forms part of the primary stem seal.
Is essential for the proper functioning of the valve.
Cannot be removed by removing the gland and gland packings alone.

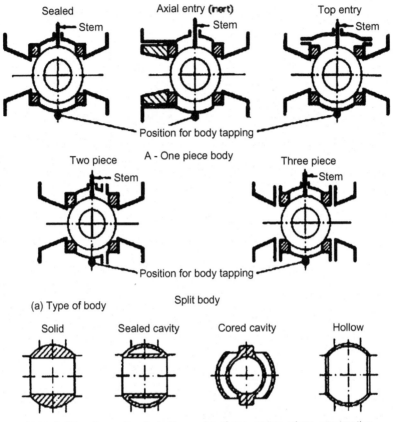

(a) Type of body

Note: Solid and cored cavity balls may be of one- or two -piece construction.

(b) Type of ball

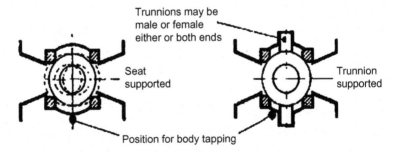

Figure 10.4 Typical variations of construction ball valve.

- Operation: Valves should be operated by a handwheel or wrench.
  Note: For manually operated valves, clockwise closing will always be supplied. Anticlockwise closing to be supplied by special request.
  The length of the wrench or diameter of the handwheel for direct-operated valves should (after opening and closing a new valve at least three times) be such that a force not

exceeding 350 N should be required to operate the ball from either the open or closed position under the maximum differential pressure recommended by the manufacturer.

Handwheels should be marked to indicate the direction of closing.

Handwheels and wrenches should be fitted in such a way that while held securely, and they may be capable of being removed and replaced where necessary.

All valves should be provided with an indicator to show the position of the ball port. When the wrench is the sole means of indicating port position, the design should not permit incorrect assembly and should then be arranged so that the wrench lies parallel to the line of flow in the open position.

Stops should be provided for both the fully open and fully closed positions of the valve.

All valves should be provided with same form of indicator for the position of the ball port.

- Materials: The body, body connector, insert, and cover materials should be selected to withstand at least twice the working pressure which is 12 bar (175 psi) for fire service. Valve used on foam-liquid system should be selected from materials to withstand corrosion effect. Body seat rings, stem seal, body seals, and gaskets should be suitable for use of foam-liquid concentrate. Valve used for salt water or extinguishing agents such as dry-chemical powder, should be specified in the purchasing order. Wrench and handwheel should be of steel, malleable cast iron, or spheroidal graphite cast-iron. Screwed body ends should have female threads complying with requirement of applicable ISO standard, either taper or parallel at the manufacturer's option unless the particular form is specified in the purchasing order. Flanged-end, butt-weld end, socket weld end, extended-weld end, and threaded-end valves should comply with BS 5351.
- Tests: All valves should be tested hydrostatically by the manufacturer before dispatch. Test should be carried-out with water. Testing requirement for the body and seat should be in accordance with BS 5146 Part 1.
- Pressure retention: The pressure retention of ball valve used should withstand without leakage of hydrostatic test pressure of 22.5 bar for 2 min.
- Test certificate: The manufacturer should issue a test certificate confirming that the valves have been tested in accordance with BS 5146 Part 1 and stating the actual pressures and medium used in the test.
- Preparation for dispatch: After testing, each valve should be drained, cleaned, prepared, and suitably protected for dispatch (painting of finish valve should be specified in purchase order) in such a way as to minimize the possibility of damage and deterioration during transit and storage. All ball valves should be in open position when dispatched. Body-end should be suitably sealed to exclude foreign matter during transit. Valves should have their jointing surfaces protected.
- Marking: Each valve should be marked clearly on the body or on a plate securely fixed to the body. Identification marking should be in accordance with Section 7 items 28 to 31 of BS 5159. Information should be specified by the buyer.

10.1.7   Gate valves

Gate valves covered by these requirements are of the outside screw and yoke type for nonrising stem type and flanged-end for installation either above ground or below ground. The gate valve covered by these requirements is intended for installation and use for:

Low-expansion foam combined agent and deluge foam-water-spray systems.
Installation of sprinkler systems.
Installation of standpipe, hose systems, and hose reels.
Water-spray fixed systems for fire protection.
Isolation of fire-water mains.

- Construction and design: A gate valve of the outside-screw-and-yoke type should be constructed for use with standard pipe thread size 12.5 mm or larger. A gate valve of the nonrising stem type should be constructed for use with standard pipe of thread size 65 mm or larger. Valve sizes refer to the nominal diameter of the waterway through the inlet and outlet connections and to the pipe size for which the connections are intended. Exception: A 12.5 mm size valve may consist of a 20 mm valve assembly having 12.5 mm pipe threads tapped in the metal of the body (see Figures 10.5 and 10.6).
- Material: Valves should be one of the following types:
Type A, which is for T-key operation only.
Type B, which is for key/bar operation, but which can be operated by a T-key. However, both types should be capable of operation by a handwheel.

Note: Type B valves are designed for heavier duty than type A valves and will withstand higher torque loads.

- Bodies and bonnets
The body of a valve should be of the straightway type and should provide, when the gate is fully open, a waterway diameter equal to or greater than the inside diameter of a mating pipe. The diameter measurement should be made at points away from projecting lugs used for the seat ring assembly. Exception: A gate valve providing a waterway having a diameter less than the diameter of the mating pipe is acceptable, if the valve incorporating such a waterway complies with the requirements of the friction loss test for valves having reduced waterways.
The body and bonnet of a 50 mm or smaller valve should be made of material having strength, rigidity, and resistance to corrosion at least equivalent to bronze.

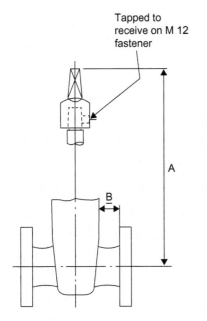
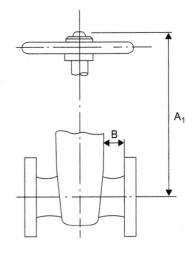

Figure 10.5 Typical gate valve.

The body and bonnet of a valve larger than 50 mm should be made of materials having strength, rigidity, and resistance to corrosion at least equivalent to cast iron or bronze.

A casting should be smooth and free from scale, lumps, cracks, blisters, sand holes, and defects of any nature that could make them unfit for the use for which they are intended. A casting should not be plugged or filled, but may be impregnated to remove porosity.

Guides should be cast integrally with the body. If the gate can be assembled in other than the intended manner, the guides should be of unequal width, or other equivalent means should be provided to facilitate correct assembly.

- Dimensions: Face-to face, body flange, maximum height, and flange-to body dimensions should be in accordance with BS 5163 Section 1.6.
- Gates.

The gate for a 50 mm or smaller valve should be of material having resistance to corrosion at least equivalent to bronze.

The gate for a valve larger than 50 mm should be of cast-iron or other material having at least equivalent corrosion resistance.

The central part of a gate for a valve larger than 25 mm should be recessed.

Any cast-iron surface of a gate should be so constructed as to clear the body seat ring in all positions.

For a cast-iron gate for an iron-bodied valve, guides or links should be provided to reduce the risk of the gatering seating surfaces rubbing on the body or bonnet during operation.

- Seating surfaces

For a valve having a metal-to-metal seating surface, all seating surfaces of the gate and body should be of bronze or material having at least equivalent corrosion resistance.

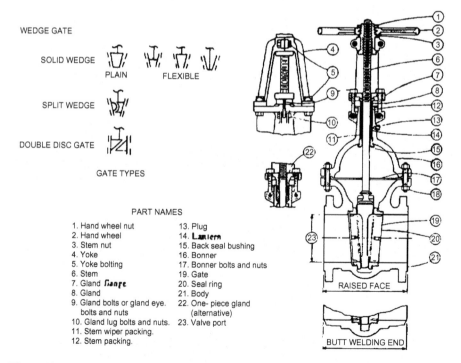

WEDGE GATE

SOLID WEDGE

PLAIN   FLEXIBLE

SPLIT WEDGE

DOUBLE DISC GATE

GATE TYPES

PART NAMES

1. Hand wheel nut
2. Hand wheel
3. Stem nut
4. Yoke
5. Yoke bolting
6. Stem
7. Gland flange
8. Gland
9. Gland bolts or gland eye. bolts and nuts
10. Gland lug bolts and nuts.
11. Stem wiper packing.
12. Stem packing.

13. Plug
14. Lantern
15. Back seal bushing
16. Bonner
17. Bonner bolts and nuts
19. Gate
20. Seal ring
21. Body
22. One- piece gland (alternative)
23. Valve port

RAISED FACE

BUTT WELDING END

Figure 10.6 Typical gate valve.

A seating surface that is constructed by a resilient material should (1) be made of bronze or other metal having at least equivalent corrosion resistance or (2) have a protective organic coating.

• Stem

A stem should be made of material having thread standard strength and resistance to corrosion at least equivalent to bronze.

Stem threads should be in accordance with acme, modified acme, half "V," or square.

The connection between a stem and its gate should be so aligned that the stem will not be bound when the gate is seated.

A stem nut should be of material having strength, wear resistant, and corrosion resistance at least equivalent to bronze.

The stem of a nonrising stem valve should, when the valve is closed, enter the stem nut a distance equal to at least 1¼ times the outside diameter of the stem.

A 125 mm or larger outside-screw-and-yoke valve should be provided with a bronze washer between the yoke and the handwheel, unless the construction of the stem nut does not permit the yoke and handwheel to come into contact.

The stem of a nonrising stem valve should be provided with a square tapered end to closely fit the wrench nut. The diagonal of the base of the square should be at least equal to the diameter of the stem.

• Stem sealing: The design of the stem seal valves should be one of the following:

Stuffing box and gland.

Injector packing form.

Toroidal sealing rings (o-rings).

Seals or packings should be capable of being replaced, with the valve under pressure and in the fully open position.

Note: The user is warned that there may be some leakage to atmosphere during this operation. When the seal is a toroidal sealing ring the following additional requirements should apply:

1. At least two such seals should be used.
2. Dust seal should be positioned above the seals to prevent ingress of foreign matter.

• Handwheel

A handwheel should be constructed to be readily grasped by the hands (Figure 10.7).

An arrow showing the direction to turn the handwheel to open the valve, with the word "open" at the feather end or in a break in the shaft, should be cast on the handwheel so as to be easily readable.

• Direction of closure: Manually-operated valves should be closed by turning the key or handwheel in a clockwise or anticlockwise direction when facing the top of the valve.

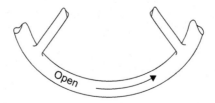

Figure. 10.7 Handwheel detail.

- Wrench nut: The wrench nut for a nonrising stem valve should be made of material having strength and resistance to corrosion at least equivalent to cast-iron (Figure 10.8). It should be fitted to the tapered square end of the stem and should be secured by a nut, pin, key, or cap screw.
- Tests: The manufacturer should conduct a type test on each type and size of valve. Type testing should consist of the following tests in the sequence specified in accordance with BS 5163.
Pressure testing.
Strength testing, followed by pressure testing.
Functional testing.

The results should be recorded and retained by the manufacturer and should include the results of a visual examination of the valve components after type testing.

Before commencing the tests, the number of turns of the stem to accomplish full obturator travel of the particular valve under test should be determined. Following the strength test, the valve should be required to operate through the same number of turns to verify that no damage to component parts has occurred.

- Performance tests: Representative of each-size gate valve should be subject of the following tests in accordance with either UL 262 or BS 5163 Section 4.
Metallic materials test.
Nonmetallic material tests.
Tensile strength and elongation tests.
Accelerated oxygen pressure-aging test.
Hardness tests.
Organic coating material for seating surface test.
Resilient seat material securement and cycling tests.
Stuffing box repacking test.
Leakage test.
Mechanical strength test.

To verify compliance with these requirements in production, the manufacturer should provide the necessary production control, inspection, and tests. The program should include at least factory testing of each valve for body and seat leakage. The

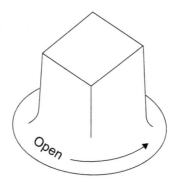

Figure 10.8 Wrench nut.

body leakage test should be conducted hydrostatically at twice rated working pressure applied to all internal parts with the valve open and pressure exerted on both sides of the gate. There should be no leakage through the body or distortion. The seat-leakage test should be conducted hydrostatically at twice rated working pressure or pneumatically at rated working pressure.

The pressure should be applied between one end and the closed gate or, for double-disc gate valves, between the valve discs. If tested pneumatically, the valve should be fully submerged in water. If tested hydrostatically, there should be no leakage past a metal-to-metal gate seat in excess of amount prescribed in standards.

- Markings: A gate valve should be marked with the following:
Nominal size.
Manufacturer's name or identifying symbols.
Rated pressure.
Distinctive model number or catalog designation.
Identifying symbols for material.

10.1.8 Check valves (nonreturn)

Check valves covered by these requirements are intended to be installed and used for:

Foam/water proportioning system.
Foam/water firefighting trucks.
Gravity water tanks.
Back flow of water to cross-connection system and multi units pumping systems.
- Materials: Check valve used for salt water should be made of materials suitable for that purpose and selected from materials specified in BS 1868 (Section 3). Check-valve used for foam water system should be made of corrosion-resistant materials equal to brass.
- Inspection and test: Check valves should be inspected and pressure tested in accordance with the requirements of BS 5146 and manufacturer should certify that the test has been carried out accordingly.
- Marking: Identification marking is required covering the following:
Nominal size DN.
Pressure rating PN.
Direction of flow (arrow) to be cast or embossed on the body of each valve.
Manufacturer's name or trademark.
Body material identification.
- Preparation for dispatch: All valves should be thoroughly cleaned and dried. Unmachined external surface of the valve should be painted in aluminum finish paint. Machine or threaded surfaces should be coated with an easy removable rust prevention material. Valve should be so packed to minimize the possibility of damage during storage and transit.

10.2 Hose reels (Figures 10.20 to 10.23)

The requirements of this section have been framed to ensure that the equipment can be manipulated by one person, while at the same time ensuring reasonable robustness for long life, efficient operation and avoidance of excessive maintenance (see Figures 10.9 to 10.12).

Dimensions are nominal

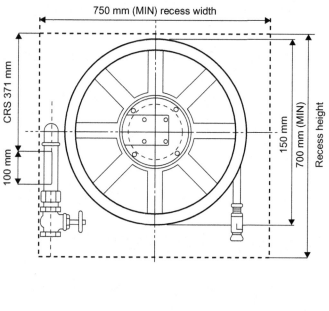

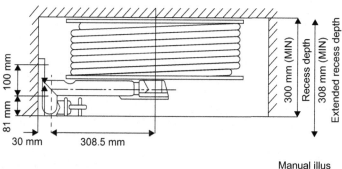

Manual illus

Figure 10.09 Swinging recess hose reels.

10.2.1 Classification

Hoses specified in standards should be classified as follows:

- Class A: Rubber covered for use on firefighting vehicles and for use on hose reels for fixed installations.
- Class B: Thermoplastics covered for use on hose reels for fixed installation.

  Each class of hose is further divided into:

- Type 1: Design working pressure 15 bar.
- Type 2: Design working pressure 40 bar.

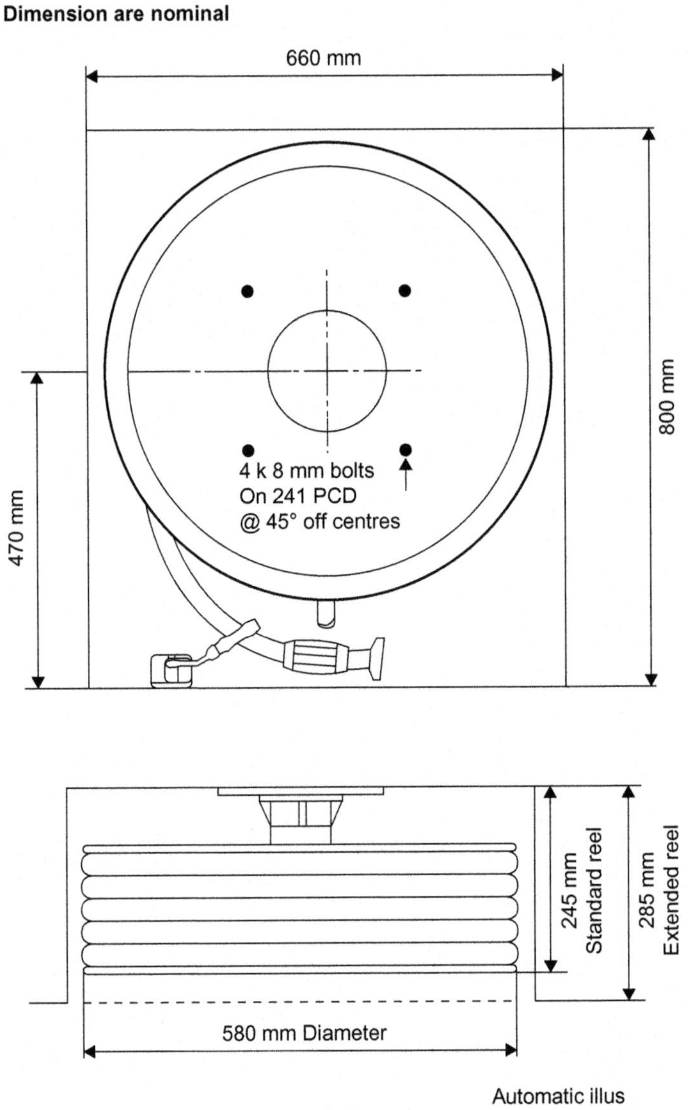

Dimension are nominal

Automatic illus

Figure 10.10 Fixed recess hose reels.

Notes:

1. The ambient temperature should be specified by buyer.
2. Class A hoses have better resistance when in contact with hot surfaces.
3. The hoses may be either mandrel or nonmandrel made, finished smooth, fluted or fabric marked.

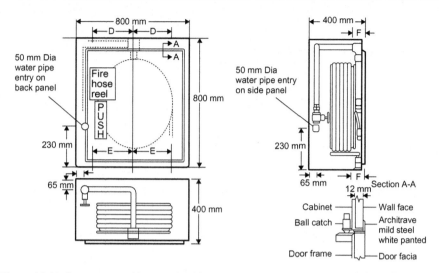

Figure 10.11 Door mounted hose reel cabinet.

10.2.2   Construction

Class A hoses should consist of:

Seamless rubber lining.
Textile reinforcement.
Rubber cover.

The cover should be black for firefighting vehicles and red for fixed installations. Class B hoses should consist of:

Seamless rubber lining.
Textile reinforcement.

The lining and cover of both classes of hose should be free from air holes, porosity, and other defects.

- Dimensions: The bore of the hose should be either 19 mm or 25 mm. The minimum thickness of the lining cover should be 1.5 mm. The variation from concentricity measured between inside and outside diameters should not exceed 2.0 mm. The maximum mass of the hose should be as follows: 19 mm bore (¾ in): 0.75 kg/m 25 mm bore (1 in): 0.90 kg/m.

10.2.3   Requirements for complete hose assembly for fixed installations

The complete assembly should be capable of operating at 10 bar maximum working pressure and of delivering water to any point within its specified range without leakage. When tested there should be no leakage from any part of the assembly.

Dimensions are nominal

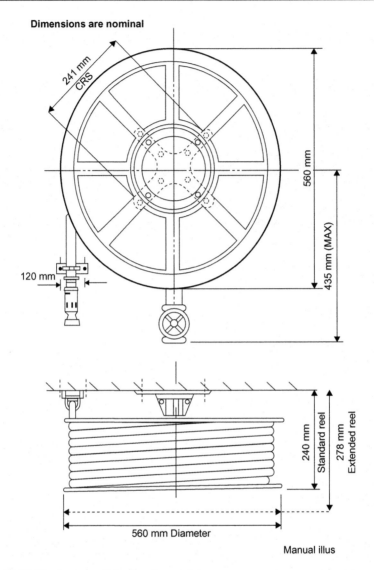

Figure 10.12 Fixed open wall hose reels.

The hose should be capable of being run out, through the hose guide if fitted, in any generally horizontal direction up to the limits of hose length. When tested the force required to start or restart rotation of the reel drum should not be more than 200 N.

The water should be turned on at the reel either:

Automatically when the reel is unwound, in which case the valve should be fully opened by not more than four complete revolutions of the reel, when the range of the jet should not be less than 6 m. or.

By manual operation of the inlet valve.

An interlocking device should be fitted whereby the nozzle cannot be withdrawn until the water supply has been turned on. When tested the water-flow rate should not be less than 24 L/min and the range of the jet should not be less than 6 m.

10.2.4 Design

The hose reel should rotate around a spindle so that the hose can be withdrawn freely.

The drum or hose support of the first coil of hose should not be less than 150 mm diameter. The fitting by which the hose is attached to the reel should be arranged in such a way that the hose is not restricted or flattened by additional layers of hose being placed upon it.

The inlet valve of manual reel should be a screw down stop valve or a gate valve. The valve should be closed by turning the handle in a clockwise direction. The direction of opening should be marked on the handle preferably by an engraved arrow and the word open. There should be no visible leakage or distortion.

Reels should be of sufficient size to carry the length of hose fitted, excluding the nozzle, within the space defined by the end plates. The length of hose fitted should not be more than 45 m for 19 mm internal diameter or 35 m for 25 mm internal diameter for fixed installation hose reels. The following types and models are generally used:

Swing and fixed open wall. The swing arm allows the hose to be pulled-off through 180° of movement (Figures 10.1 and 10.4).

Recess hose reel with omnidirectional hose guide to be pulled off through in any direction. Door-mounted hose-reel cabinet Figure 10.13. It may be also desirable to provide an anti-overrun device to prevent the hose from becoming entangled when run-out.

- Material and finish: The hose reels should be finished in signal fire-service red. Ferrous materials should not be used as a part of waterway. This does not include the connecting pipe and union between the valve and spindle. All components that would be adversely affected by external environmental condition should be treated to resist corrosion.
- Hose: The hose should comply with the requirements of standards. For fixed installation, and Class A Type 2 for fire-truck installation.
- Nozzle: Each hose should terminate in a shut-off nozzle to give either plain jet or jet/spray.
- Production Tests: Each reel and subassembly should be tested in accordance with BS 5274 for the following tests and certified by manufacturer:
Leakage test.
Strength test.
Import test.
Load test.
Unwinding test.
Water-flow and range test.
Operational test of automatic valve.
Production pressure test.
- Markings: Each reel and valve subassembly should be clearly marked with:
Name and address of manufacturer.
Date and the number of standard used.

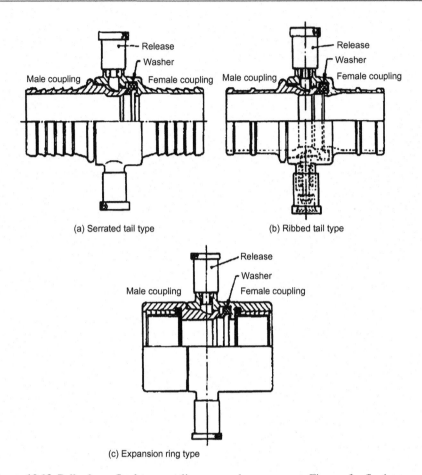

Figure 10.13 Pull release fire hose couplings general arrangement. Figures for fire hose couplings.

Notice and operating instructions:

Fire-hose reel assemblies should be provided with a notice bearing (fire-hose reel) in white letters on a red background or on adjacent to the hose reel.

Fire-hose reel assemblies should be provided with full operation instructions for display on or adjacent to the hose reel.

- Preparation for Dispatch: Hose-reel assembly should be dispatched in crate or in carton in such a way to minimize the possibility of damage during transit and storage.
- Hose Reels Mounted on Firefighting Truck
  Hose: The hose should be of class A. rubber covered type 2 for working pressure of 40 bar and the cover should be black.
  Dimension: The bore of the hose should be of 25 mm and the length 43 m or more as specified.

Valve: The hose-reel valve should be of quick opening ball type.

Nozzle: The hose reel should be terminated to a super fog spray nozzle with squeeze grip valve. The nozzle can be replaced with foam-branch nozzle with deflector when required.

10.3 Firefighting hoses and couplings

10.3.1 Firefighting hose

Fire hoses should deliver water effectively over a long period of time under a variety of conditions and hazards, including impact, abrasion, damage, contamination, burning, and weathering effects. The fire hose should be lightweight, flexible, and easy to handle wet or dry, highly resistant to chemical, heat, impact, and abrasion, easy to clean, and require minimum maintenance that can be carried out simply and effectively.

The hose should consist of the following:

Impermeable elastomeric lining.
Synthetic fiber reinforcement.
Externally applied coating of the reinforcement.

- Dimensions: The nominal length of fire hose should be of 25 m, the bore 45 mm, and 70 mm. The mass per unit length should not exceed 0.37 kg for 45 mm and 0.68 kg for 70 mm per meter.
- Construction: The fire hose should be made all synthetic. Nylon wrap and weft totally encapsulated in PVC to form a unified lining and cover. The hose should be made of first quality materials and to be protected against the effect of lining puncture. The hose should be easily cleaned and easily repaired.
- Hose assemblies: Hose assemblies should be fitted with delivery hose couplings tied in by binding with galvanized mild steel wire of diameter 1.6 mm, and applied over a hose guard of synthetic fiber (to protect the hose from wire). Multi-serrated type couplings should be secured by 20 continuous turns of wire and ribbed type couplings should be secured by ties of at least eight continuous turns on both sides of the rib. The end of the wire should be secured by twisting them together and embedding them in the hose guard.
- Requirement: The service test pressure for the fire hose should be specified by the user, but in any case each hose assemblies should be subject of proof test pressure of 22.5 bar.
- Tests: The following tests should be conducted by manufacturer in accordance with BS 5173: Hydrostatic tests.
Kink test piece under pressure.
Burst-pressure test.
Adhesion test.
Hot cub test.
Impact test.
Ozone test.
Abrasion test.
Oil test.
Acid test.

The test report should include:

1. Date of tests and tests results.
2. All details necessary for the complete identification of the hose under the tests.

10.3.2   Pump-suction hoses

This section covers physical properties and performance of three classes of suction hoses that are also divided into three types according to design working pressure.

*   Classification: Hoses and hose assemblies should be classified as follows:
    Class A: Smooth-bore rubber hose.
    Class B: Semi-embedded rubber hose.
    Class C: Smooth-bore polymer-reinforced thermoplastic hose.

Classes A and B are further divided into:

Type 1: Medium-duty hose with a design working pressure up to and including 5 bar.
Type 2: Heavy-duty hose with a design working pressure up to and including 7.5 bar.

Note: Ambient temperature to be specified.
Class C is further described as:

Type 3: Medium-duty hose with a design working pressure up to and including 5 bar (ambient temperature to be specified)

*   Construction: Class A and B hoses should consist of:
    Rubber lining.
    Textile reinforcement.
    Embedded single- or twin-wire helix.
    Rubber cover.
*   Lining and cover: The lining and cover should be concentric and should be free from holes, porosity, and other defects.
*   Wire helixes: All helix wire should be galvanized as specified in BS 443. The wire used for the internal helix in Class B hoses and the embedded helix in Class A hoses should have a minimum tensile strength of $1250 \, \text{N/mm}^2$. The wire used for the embedded helix of Class B hoses should have a minimum tensile strength of $650 \, \text{N/mm}^2$.
*   Hose ends: Hose ends should be compatible with suction-hose couplings, and soft ends should have an additional rubberized textile reinforcement applied as cuff over the soft.
*   Finishing: The hose should be consolidated and uniformly vulcanized.
*   Dimensions: The nominal bore of the suction hose should be of 75–100 and 140 mm. The mass of the hose excluding couplings should not exceed the values in Table 10.1.

Table 10.1 **The mass of the hose excluding couplings shall not exceed the following values**

Nominal	Maximum mass	
Bore mm	Type 1 kg/m	Type 2 kg/m
75	3.7	4.1
100	6.0	6.7
140	8.0	8.9

- Tests: The following tests should be carried out in accordance with BS 5173 by manufacturers and the test certificates issued:
  Hydrostatic test.
  Vacuum test.
  Vacuum test with flexing.
  Adhesion test.
  Flexibility test.
  Hydrostatic test of suction hose assembled with couplings.
- Attachment of couplings: When the couplings are fitted they should be either:
  Bound-in using steel wire complying with the requirements of BS 3592: Part 1 having a minimum tensile strength of $380\,N/mm^2$ and galvanized as specified in BS 443 or, secured by the use of stainless steel band and buckle-type clips.
- Polymer-reinforced thermoplastic hoses (Class C, Type 3): The hose should be uniform in color, opacity, and other physical properties. It should consist of a flexible thermoplastic material supported in its mass by a helix of polymer material of a similar molecular structure. The reinforcing and flexible components of the wall should be fused and free from visible cracks, porosity of foreign inclusions.
- Dimensions and tolerances: When measured in accordance with the method described in BS 5173: Section 101.1 the bore of the hoses should be compatible with suction-hose couplings. The bore of the hose compatible with suction couplings and mass of the hose excluding couplings should not exceed the values in Table 10.2.
- Hydrostatic-pressure test: At proof pressure the hose should exhibit no evidence of leakage, cracking or abrupt distortion indicating irregularity in materials manufactured. The tests as specified in BS 3165 should be carried out by manufacturers.
- Markings: For rubber hoses and assemblies a rubber label giving the following information should be vulcanized to the hose approximately 0.5 m from one end. For thermoplastic hoses the information should be marked with:
  Hose manufacturer's name or identification.
  Number of standard used.
  Nominal bore.
  Month and the year of manufacture.
  Design working pressure.

Table 10.2 The bore of the hose compatible with suction couplings and mass of the hose excluding couplings shall not exceed the values

Nominal bore mm	Maximum mass kg/m
75	3.0
100	4.5
140	6.0

10.3.3   Fire-hose couplings

Fire-hose couplings used in the oil, gas, and petrochemical industries are of two types:

Delivery hose instantaneous couplings.
Pump suction hose round threaded couplings.

Both types should be made in accordance with BS 336 (1989). The release mechanism of delivery hose couplings should be of pull-release type and delivery connector female coupling should be of single lug twist type.

- Delivery-hose couplings: The tail design should be of either multiserrated or ribbed type of the sizes given in Table 10.3.
  For delivery-hose connector sizes reference should be made to BS 336.

- Locking of plungers: Nuts on plungers should be of a self-locking type.
  Note: Peening over the end of the shank is not regarded as an efficient means of locking the plunger.
- Plunger springs: Plunger springs metal delivery hose coupling should be of such strength that they can be compressed to a length sufficient to free the plunger from engagement by a force of no less than 55 N or more than 110 N and for plastic couplings no less than 45 N and not more than 65 N.
- Washers: Coupling washers should be of natural rubber (Grades Y40, Z40) to BS 1154 or chloroprene (Grade C40) to BS 2752. Expansion ring washers should be of an elastomer of maximum shore hardness 70. The tolerances on the stated dimensions should be in accordance with Class M 3 of BS 3734.
- Materials: A copper alloy part containing more than 15% zinc in the construction of any part used should have corrosion-resistant properties equivalent to those of high-strength yellow brass or bronze (BS 2874). Materials may be used in accordance with BS 336 CZ 112-121-122-DCB3-LG2 PB 102 PB 103. High-strength aluminum alloy and plastic material may be used if specified by buyer. Any plastic materials complying with BS 336 may be used for the body but the release plunger should be made of metal.
  Note: On diecast fittings and plastic injection moldings where a bore is shown as parallel, a reasonable taper should be allowed to facilitate core withdrawal, but this taper should be kept to the minimum required for this purpose.
- Castings and moldings: Castings should be clean, sound and free from gross porosity, cracks, and other surface imperfections. No filling or similar after treatment of castings should be carried out without the approval of the buyer.
- Finishing: Metal couplings should be smooth and polished. Delivery connectors may be chrome-plated, if specified by buyer.

Table 10.3   The tail design shall be of either multiserrated or ribbed type of the following sizes

Nominal Size Inch	Nominal Hose dia. mm	Tail ext. Diameter mm	External dia. Between ribs mm	Internal diameter mm	
1¾	45	44.5	42	32.9	33
2¾	70	69.9	65.9	56.8	58

- Tests: The manufacturer should certify that the fire-hose couplings and rubber-sealing materials are tested in accordance with UL 236 for:
  Hydrostatic-pressure test.
  Creep test.
  Pull test.
  Crushing test.
  Rough usage test.
  Mercurous nitrate immersion test.
  Salt-spray corrosion test.
  Test for rubber-sealing materials.
- Suction-hose couplings: Couplings should be made of rigid materials of copper alloy or die-cast brass. At one end female swivel and other end male. Couplings at each end should have two lugs for suction spanner. Both screws should be of round thread to meet BS 336 as shown in Figure 10.14.
- Sizes: Nominal bore should be for the following:
  75 mm (3 inch).
  100 mm (4 inch).
  140 mm (5½ inch).

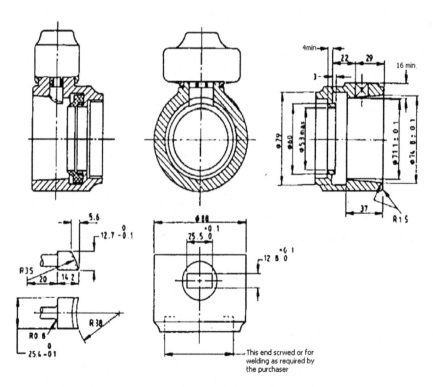

Figure 10.14 Single lug twist type delivery hose connector. All dimensions are in millimeters.

- Diameter of suction hose and the length should be specified by buyer: Couplings should be tested for the following and certified by the manufacturer:
  Crashing test.
  Rough usage test.
  Mercurous nitrate immersion test.
  Salt-spray test.
  Hydrostatic-pressure test.
  Washer test.

10.4 Firefighting nozzles

Firefighting nozzles are generally used for:

- Water spray.
- Foam application.
- Other extinguishing agents.

10.4.1 Water-spray nozzles

The essential feature of water spray that distinguishes it from the use of water in other forms is that water is applied in small drops using special types of nozzles. The discharge both from hose-line nozzles and sprinklers is simply water spray in a particular form. A great variety of water-spray patterns are now employed both for manually directed hose and fixed spray nozzles and for sprinkler system. Water-spray protection may be engineered to provide specialized protection for various industrial applications by using special type nozzles designed to achieve a better distribution of water at a given pressure.

Water spray may be used for any one or combination of the following purposes:

Extinguishment of fire.
Control of fire, by application of spray, to produce controlled burning where materials burning are not susceptible to complete extinguishment, or where the extinguishment is not desirable, as in the case of explosive gases.
Exposure protection by applying spray to exposed structures wetting of exposed surfaces.
Prevention of fire by use of spray to dissolve, disperse dilute, or cool flammable materials.
- Types-hose nozzles: Hose nozzles are of three types:
  1. Open nozzles (nonadjustable) that provide solid streams.
  2. Adjustable fog nozzles that provide variable discharge and pattern from shut off to solid stream and from narrow to wide angel spray.
  3. Combination nozzle in that solid stream, fixed, or adjustable spray and shut off are selected usually by a two- or three-way control valve.
- Construction: Branch-pipe nozzle should be made of brass aluminum alloy and plastic as specified by the buyer. Integral branch pipe nozzle units should be made of metal and/or plastics. Handheld controllable branch pipes should be of metal and/or plastics. The means of controlling the flow of water should provide a shut-off/jet facility and additionally provide a spray (Figure 10.30(b)). A nozzle intended and marked for fire-service use may be with dual handholds or a single hand grip. The outer-end coupling of the nozzle-hose connection should be of instantaneous male in accordance with BS 336.
Note: There are a variety of hose nozzles as illustrated in Figure 10.30.

- Hose-reel nozzles: Hose-reel nozzles subject of standards are of two types:
  1. Super-fog gun 19–25 mm hose-reel hermaphrodite coupling adjustable to straight stream to extra-fine fog for fire-truck hose reel. Suitable pressure 7 to 10 bar or more with shut-off grip 60 to 100 LPM at 10 bar.
  2. Fine-fog nozzle 19 mm (¾ inch) hose reel with hermaphrodite or round thread standard connection adjustable to straight stream or spray with shut off lever suitable for 4 to 10 bar water-pressure standard hose reels.
- Water-fog nozzles: A great variety of fog nozzles are made by manufacturers but the most suitable handheld are those of constant/select flow feature that allows on site manual adjustment. The pattern selection is from straight to wide spray, 100 to 350 LPM at 10 bar.
- Fixed-spray nozzles: Fixed-spray system require proper engineering practice considering many factors including the specific characteristics of the hazard, the size and velocities of the dispersed water particles, location of nozzles and volume of water. For the selection of suitable spray nozzles, high-velocity spray nozzles that deliver their discharge in the form of spray-filled cone or low velocity spray nozzles that usually deliver a much finer spray in form either of a spheroid or of well filled cone should be used. In general, the higher the velocity and finer the courser and size of water droplets, the greater the effective range of spray.
- Materials: A metal used in construction of any part should have corrosion-resistant properties equivalent to those of high-strength yellow brass.
- Rubber seals:
  Should be made of vulcanized natural rubber of synthetic compound.
  Should have uniform dimension.
  Should be of such size, shape, and resiliency as to withstand ordinary usage and foreign matter carried by water.
- Tests: Each spray nozzle used for sprinkler or any types of water protection should be subject of the following tests:
  Discharge calibration.
  Hydrostatic test (2 times working pressure).
  Discharge flow and pattern.
  Salt-spray corrosion test.

Portable fire-service hose nozzles straight stream or fog nozzles should be subject to the following tests:

Discharge-flow and pattern.
Hydrostatic test (3 times working pressure).
Nonmetallic high temperature.
Leakage.
Rough usage.
Salt-spray corrosion.
Nonmetallic salt water.
Tensile strength.

The manufacturer should certify that the above tests have been carried out.

- Marking: Water-protection spray nozzles should be marked with the following information cast on them:
  Name of manufacturer or identifying symbol.
  Distinctive catalog designation.
  Rate of flow at designed pressure.

10.4.2 Foam-branch nozzles

A foam-branch nozzle is a portable device intended to discharge foam and may be either aspirating or nonaspirating. Portable foam branches may pick-up foam-liquid concentrate directly from a container (self-inducing) or utilize foam solution produced at some point before introduction into the nozzle.

Foam branches, portable and handheld, can be used for a wide range of flammable liquid fires. They also provide a rapid method of securing a flammable liquid spill with a vapor-suppressing foam blanket before ignition occurs. Portable, low-expansion foam-branch nozzles capacity varies from 220 to 900 L solution with expansion ratio of 8:1 to 10:1 at 5.5 to 8-bar.

Note: The expansion ratio of medium-expansion foam-branch nozzles is 50–150 to 1.

- Type: Foam-branch nozzles should be designed to produce fully expanded foam with all types foam concentrate of low expansion. Foam-branch nozzles should operate as low as 3.4 bar (50 psi), although the best performance is achieved at 5.4 to 8.6 bar (80 to 125 psi). The foam branch should be provided with pick up tube proportioning for either 3% or 6% with control lever as specified. In these nozzles proper proportioning of foam concentrate is accomplished with the suction created by water passing through the nozzles foam maker's venturi. The nozzle should be capable of working with any remote proportioning system. Foam-branch nozzles used by fire-truck hose reels should be of hermaphrodite coupling end of 25 mm and squeeze grip valve. Foam-branch nozzles should be of stream/spray deflector type allowing the operator to switch from long-range straight stream to spray stream pattern without shutting down the nozzles.

10.4.3 Construction

Foam-branch nozzles should usually be made of brass or anodized aluminum alloy suitable for salt water. Nonmetallic handheld foam-branch nozzle may be acceptable for the purpose and should be tested in accordance with standards. A metal used in construction of any part should have corrosion-resistant properties equivalent to those of high-strength yellow brass.

A foam-branch nozzle should consist of the following parts:

Water inlet male coupling instantaneous (BS 336).
Internal water head.
Foam compound pick-up connection.
Air inlet.
Internal forcing tube.
Deflector and swing handle.
Female instantaneous coupling for foam discharge if specified by buyer.

- Performance test: Foam-branch nozzles should be tested to determine the discharge flow of foam with predetermined expansion ratio.
  Nozzle with deflector should be measured for the range of flow in still atmosphere and spray pattern.
  Foam-branch nozzles should be measured for the amount of foam-liquid concentrate-flow rate, from pick-up tube at the predetermined pressure.

- Miscellaneous tests: Representative foam-branch nozzle of each size should be subject of the following tests:
  Discharge calibration.
  Discharge flow and pattern.
  Hydraulic operation.
  Nonmetallic high temperature.
  Rough usage.
  Hydrostatic pressure.
  Salt-spray corrosion.
  Leakage.
  Air oven.
  Nonmetallic water exposure and hot water.
  Tensile strength, ultimate elongation.
- Marking: Each nozzle should be marked with the following information, using stamped or cast figures and letters no less than 4 mm in height:
  Name of manufacturer or its identifying symbol.
  Distinctive catalog designation.
  Date of manufacture.
  Rate of flow at position of straight stream and full spray.

10.5   Firefighting monitors

Firefighting monitors are generally classified as two types:

1. Water-water/foam monitors.
2. Dry-powder monitor.

10.5.1   Portable water/foam monitor

Monitors should be supplied with portable base stabilizer assembly. The support legs of the stabilizer assembly should be removable for convenient storage and for mounting base onto fire-truck or trailer unit. The minimum rotation should not be less than 130° and horizontal no less than +45° (Figures 10.32 and 10.33). The monitor should have the following design advantages:

Internally cast-in water way should vary as such that reduces turbulence and friction loss.
Half-turn vertical and horizontal travel locks.
Stainless or anodized aluminum alloy construction for rugged serviceability and light weight.
Counterbalance for ease of operation.
Portable monitor should be capable of discharging 800 to 1500 LPM of foam solution and water at 10 bar with straight stream range of no less than 45 m and 18 to 20 m height.
Monitor should be provided with manual spray-control actuator.
Portable foam monitor should be fitted with pick up tube and metering valve, so foam concentrate may be supplied through the pick-up tube from a container.
Inlet connections to be of 2 × 65 mm instantaneous male connections (BS 336).
Center of gravity dimensions should be based on nozzle in horizontal position without pick-up tube installed.

10.5.2 Trailer-mounted water/foam monitors

Water-foam monitors of this type may be designed for flow rate of 1000 to 2000 LPM or more water or foam solution with straight stream range of 60 meter with 25° elevated nozzle at 10 bar pressure. The trailer may be of double axel with hose bin and a foam concentrate tank up to 2000 liters. The monitor may be of self-oscillating type if specified. The monitor may be provided with spray/stream actuator arm that can adjust the spray-patterns variations ranging from full spray to straight stream.

10.5.3 Foam/water monitors mounted on fire trucks

Fire-truck monitors should be designed as specified in standards. Fire-truck monitors may also be designed to be hydraulically operated from the cab if specified.

10.5.4 Elevated fixed water/foam monitors

Elevated fixed monitors with larger capacity are required for high-risk areas such as near process and production units, in petrochemical complexes, refineries, loading terminals, on fire boats, tugs, and on tankers for cargo-space protection. The following types depending on hazard factors should be considered:

- Unmanned sweep protection water powered oscillating should be with the following features (Figure 10.34(b)):
  Automatic horizontal oscillation sweep with manually adjustable elevation that can be locked prior to operation with override mechanism.
  Provided with test connection for oscillating mechanism setting.
  Nozzle should be of aspirating type suitable for foam solution of 3 to 6% foam-liquid concentrate.
  Speed of oscillation to be adjustable from 0–30° per second.
  Arc of oscillation to be adjustable from 10° to 180°
  Angle of elevation adjustable from 45 below horizontal to above 60°.
  Water demand for oscillator 4 LPM at 7 bar.
  Monitor inlet pressure to be of 5 to 10 bars.
  Foam aspirating nozzle inlet flow to be between 1200 to 4000 LPM or more if specified.
- Self-oscillating hydraulic remote-controlled monitor with the following features:
  This monitor is normally elevated on suitable tower to provide a maximum area of coverage.
  Monitor is equipped with hydraulic motor to provide remote control of the monitor through 340° traverse arc and 125° of elevation arc (−45° to +80°).
  Monitor movement to be accomplished by selective rotational hydraulic motor driving through direct wormgear trains attached to the vertical and horizontal joints on the monitor.

10.5.5 Electric remote-control monitors

Using an electric remote control incorporated into a console control center, an individual section should operate a series of monitors protecting any potential hazard within the field of view. Foam aspirating nozzle used for hydraulic electric remote-control monitors can be of different flow rates from 700 to 4000 LPM of 3 to 6% foam-liquid concentrate solution or water.

10.5.6 Dry-powder extinguishing monitor

Dry-powder monitor is generally installed on truck or in locations in high-risk areas. It should be installed in such a way that can be easily operated. Its capacity should be selected to combat the large size of fire without flash back risk. The dry-powder monitor of the fire truck should have the capacity of no less than 20 kg/s (1200 kg/min) with the throw between 30 to 50 meters.

The monitor should be horizontally adjustable over 140° on each side from the straightforward position. The vertical elevation should be at least 90° from the most downward position. An operating handle should be installed at the monitor to open and close. The monitor should be equipped with a locking device and a cover on the barrel to prevent water entry.

10.5.7 Materials and construction

The monitor and the bearings should be made of corrosion-resistant materials of stainless, aluminum alloy or bronze see ASTM A 276, for barrel and deflector see ASTM B 179. The choice of material should also be made with due regard to possible metallic corrosion when different metals are used and in contact of moisture.

Oscillator components should be made of cast brass and stainless steel and the enclosure of steel. If the monitor is electric drive the motor should be of explosion proof or totally enclosed. If series of monitors are operated from a remote location, the console control should include the following standard features:

Start switch for hydraulic pump.
Run light to indicate hydraulic pump energized.
Shrouded push button for spray or straight stream selection.
Operating mechanism for horizontal and vertical of nozzle.

Welds should be free from lack of fusion, cracks, nonmetallic inclusion, porosity, and cavities.

• Finishing: Machined surface should be smooth and should be of grade N7 in accordance with BS 1134. When aluminum used for exposed surface, they should have a sealed anodized finish of thickness and less than grade specified in accordance with methods described in BS 1615 or BS 5599. Parts not machined should be finished clean as cast. Waterway should have a smooth finish. The exterior of all components should be sufficiently rounded and smoothed. Monitor nozzle to be of aluminum alloy or stainless steel finish, the other parts painted red enamel.
Performance Tests: Tests should be in accordance with UL 162 as follows:
Pressure-retaining parts should withstand without leakage by hydrostatic test pressure of no less than 2 times highest working pressure.
Flow rate as specified by the user in LPM should be maintained.
Expansion ratio of foam-liquid concentrate and drainage time should be indicated.
Straight stream range in meters in still air should be specified.
Speed of oscillation and water demand for oscillation should be specified.
Electric-powered oscillating motor should be also tested and certified, if oscillating system is electric motor.
Remote-control monitors should also be tested and certified in writing.

- Marking: Each foam/water monitor should be marked with the following information using stamped and cast figures or metal nameplate and letters no less than 8 mm in height:
  1. Name of the manufacturer or its identifying symbols.
  2. Distinctive catalog designation.
  3. Date of manufacture.
  4. Flow rate at 10 bar pressure and expansion ratio.
  5. Minimum working pressure.
- Shipping: Monitor units and related equipment should be properly prepared for transit to prevent damage from handling warehousing or shipping and should be labeled to ensure that it is not lost on transit and the following measures should be taken:
  1. All external connections should be protected by temporary closures.
  2. One package list should be included inside every package.
  3. Adequate shipping supports and packing should be provided in order to prevent internal damage during transit.
  4. For ocean transport, the equipment should be crated in heavy-duty container sealed with strong tape or metal bands.
- Guarantee: Manufacturer should guarantee by letter of acceptance the satisfactory performance of the equipment mentioned in all sections in standards and replace without charge any or all parts defective due to faulty material, design, or poor workmanship for period of 18 months after the date of shipment.

10.6   Butterfly-valve specifications

10.6.1   Carbon steel, lined body, above-ground application for cold climates

Design, dimensions, and marking generally follow BS 5155.

Type:	Wafer, tight shut-off, PN 16, suitable for mounting between raised-face flanges ANSI class 150 with serrated spiral finish.
Body:	ASTM A 216 WCC or WCB with maximum carbon content of 0.25% or ASTM A 105 normalized or Nodular cast-iron GGG 40.3.
Lining:	Polychloroprene or nitrile rubber.
Disc and bearings:	Aluminum Bronze Alloy ASTM B 148 - C 95800 or BS 1400: AB2.
Shaft:	Stainless steel AISI 316, or SAF 2205 or equivalent, provided that the steel is isolated from the medium by the lining. Otherwise it should be made as given in BS 3076 NA 18 (Monel alloy K-500).
Pins:	BS 3076 NA 18 (Monel alloy K-500).
Surface protection:	The outside surface protection of valve and barrel should be of rust-proof undercoat, covered by fire-service red.

Size (in)	DN	Face to face (mm)	Operating mechanism
2	50	43	Wrench
3	80	46	Wrench
4	100	52	Wrench
6	150	56	Wrench

10.7 Foam-making branch nozzles

The aspirating devices that are used to produce foam can be divided into three basic categories:

1. Foam-making branches (FMB) for LX or MX foam.
2. Generators for LX foam.
3. Generators for HX foam.

The above equipment is available in various sizes requiring from under 100 L/min to over 6000 L/min of foam solution. It is obvious that pumps supplying water for foam- making must have the capacity to meet the needs of the particular type and amount of foam-making equipment in use.

Some aspirating devices are fitted with means of picking up concentrate and are known as self-inducing. With other types, the concentrate has to be introduced into the water stream at an earlier stage by some form of induction equipment.

- LX FOAM-MAKING BRANCHES: For LX FMB designs will vary and will incorporate some or all of these features. The strainer is frequently omitted, as often is the on/off control. In the diagram are two orifice plates. The upstream orifice is the larger of the two and its function is to create turbulence in the space between the two orifice plates so that when the jet issues from the downstream orifice it rapidly breaks up into a dense spray. This fills the narrow inlet section of the foam-making tube and entrains the maximum quantity of air through the air inlet holes. Some FMBs have the upstream orifice plate fitted with disturbance notches. The downstream orifice is smaller and is precisely calibrated to give the designed flow rate. Some branch pipes have a swivel device in place of this orifice; others have several converging orifices. Most FMBs have a narrow section at the inlet end in which the air entrainment takes place, and then a wider section in which the foam forms. The wider section of the foam-making tube frequently contains "improvers" that are designed to enhance the foam quality, e.g., semi-circular baffles or a cone. At the outlet, the branch pipe is reduced in diameter to increase the exit velocity, thus helping the foam stream to be projected an effective distance. The design is crucial: too narrow an outlet produces back-pressure with less air entrainment and a lower-expansion (sloppy) foam. If the outlet is too large, the expansion is higher but the throw is reduced. Some branches are fitted with a dispersal mechanism, e.g., adjustable blades within the nozzle that enable a hollow conical spray to be produced. This overcomes the foam's tendency to remain in a coherent "rope" and allows the foam to fall more gently onto the fuel.
  Note:
  FMB: Foam-Making Branch LX: Low-expansion
  MX: Medium-expansion
  HX: High-expansion
- Types: In order to distinguish between capacities of FMBs, it is necessary to use a common factor. As the same FMB can produce different amounts of foam from different concentrates, classification by foam production is useless. Therefore, the classification used is by the nominal-flow requirement of foam solution in liters/min. This figure corresponds in each case to the nominal operating pressure for the particular branch. Some common models of LX FMB and their performance characteristics are listed in Table 10.4. The B225 FMB is specially designed for use with AFFF or FFFP, although it can be used with synthetic foam. Note the adjustable jaws giving the option of a cohesive jet or a spray, and the on/off trigger mechanism controlling the release of the foam. Table 10.4 gives the performance data for this branch. Special types of branches are available for use with hose-reel equipment.

Table 10.4 Lx foam-making branches: performance data

Branch	Nominal Flow requirement (litres/min)	Nominal Operating Pressure (bar)	Maximum Operating Pressure (bar)	Throw at Nominal Pressure (metres)	Throw at maximum Pressure (metres)	Expansions at Nominal Pressure (approx.)	Self-Inducing Capability	Remarks
FB 5 X MK II	230	5.5	19.5	20	74	10:1	Yes	Can very consentration from 3% to 6 % when sprated in self – inducing mode.
F 225	225	7	10	12* 20†	14* 23†	81* 10:1†	Yes	
B 225	225	7	8.8	13‡ 7§	14 8§	10:1	No	Design for use with film Forming foams.
FB 10/10	455	7	10.5	21	25	10:1	No	Can change from staight-Forward jet to conical spray.
F 450	450	7	10	18 21†	20 23†	8:1 10:1†	Yes	

FB 20X	970	7	10.5	25	27	10:1	No	Requires 2 men to manocurve It. Often adapted as a monitor Either free-standing or fited On an appliance.
F 900	900	7	10	21 24†	23 26†	8:1 10:1†	No	Not illustrated, but similar in Appearance to the f 450.

*Basic model, giving cohesive 'rope' type foam jet.
† Alternative version giving non-cohesive foam stream.
‡Jaws open, i.e. jet mode.
§Jaws closed, i.e. spray mode.

- Medium-expansion Foam Branch: Medium-expansion foam branches are designed to be used with synthetic foam concentrate and will produce foam at expansions usually ranging from 50:1 to 150:1. The greater expansion is due to ratio of medium-expansion foam is due to projection distance and is less than low expansion. The branch defuses and aerates the stream of foam solution and projects it through a gauge mesh to produce bubbles of uniform size.

10.8   Material and equipment standard for firefighting vessels

Tugs intended for berthing and unberthing the oil tankers are generally designed and built for firefighting and emergency rescue operations. The requirement in this material standard apply to the tugs and vessels primarily used for fighting fires and rescue operations on offshore and onshore structures. This specification covers, the initial consideration and planning, quotation requirement, inspections quality control, certifications, and all other purchase formalities of equipment intended for firefighting and rescue operations on offshore and onshore oil-loading terminals of the petroleum and petrochemical industries.

10.8.1   Service conditions

Firefighting tugs and vessels are generally stationed at oil loading terminals while the tugs' normal duties are berthing and unberthing the tankers. They should participate in firefighting and rescue operations in case of fire and emergency conditions. Firefighting systems and rescue equipment should be installed on the tugs and vessels in accordance with classifications and class notation specified in standards.

- Site Conditions: The vessels should be designed and fabricated for service at loading terminals, the temperature, humidity, dust, gaseous atmosphere, and wind velocity should be considered in accordance with local conditions. The seawater specific gravity is 1.03 with salt content of 35–40 thousands part per million and temperature of approx. 32 °C, can be used for making foam.
- Operational Conditions: One of the main duties of tugs is to tow or push off the tanker on fire or unberth the neighboring tankers to the safe locations and attend firefighting and rescue operations. When directed, the tugs also will attend the ships on fire and other emergencies. The tugs may also participate in sea oil pollution control.

10.9   Initial consultations and planning before purchasing

The owner should state his requirements giving as full description as possible that the supplier can prepare his proposal. Then both parties should review the relevant information to enable them to prepare a suitable material specifications and design of equipment. Selected authorities that are experts in the design and on materials should be consulted.

- Planning: In planning and layout design of firefighting equipment for vessels, particular consideration should be given to the following:
  Fire and emergency conditions giving the class notation for classification of firefighter vessel.
  Saving regard to the potential fire hazard in value, the requirement of fire-protection and emergency equipment should be considered.
  Consideration should also be given to the latest international convention for the safety of life at sea and the by-laws and regulations.

10.10 Quotation and technical information

- General technical information of the vessel including:
  Vessel size, propulsion, side thrusters, power, stability, and control system.
  Communication systems.
  Firefighter class notation.
  Number of crew and accommodation layout.
  Fuel-storage capacity.
- General technical information of firefighting systems.
  Number of firefighting monitors, type, and design.
  Length and height of throw water/foam.
  Type of monitor control and design support.
  Number of water pumps with detail of specifications including data sheet and flow charts.
  Fire-water piping, foam-generating, and proportioning system.
  Foam concentrate tank capacity, type of foam concentrate, and spare containers (the type of foam agent should be specified in purchase order).
  Fire-alarm systems.
- General technical information of self-fire protection of the vessel.
  Fire water-spray system.
  Pipeline and nozzles.
  Pumping system, capacity, and data.
  Fire-hose stations.
  High-expansion foam generator and capacity.
  Fire extinguishers.
- Miscellaneous
  Flood and search lights.
  Seawater inlets and sea chest.
  Firemen outfits.
  Breathing apparatus and compressor air supply.
  Corrosion protection and paints.
  Ventilation and air conditioning systems.
  Salvage operation system.
  Sign boards, markings, and notices.
  Power-generator and electrical-system layout.
- Emergency and rescue equipment.
  Life safety and rescue equipment.
  First-aid and resuscitation equipment.

10.11   Quality control

10.11.1   Quality control tests

The manufacturer should certify in writing that all quality control tests including welding have been carried-out in accordance with DNV-Part 2 Chapter 3, Material and Welding standards. The manufacturer also should quote all normal tests required as well as hydrostatic and operation simulation tests specified in the following.

All pressure-containing parts including fire-water piping system should be subjected to hydrostatic tests at a pressure no less than one and a half times the designed pressure.

10.11.2   Upon completion

The vessel should be given an operational test simulating actual design operating conditions as closely as possible with firefighting equipment in use. Tests should be made to verify that the vessel with firefighting systems and equipment is able to operate as intended and has the required capabilities. The manufacturer should conduct this test in presence of selected owner's representative unless a written waiver is given.

Manufacturer should furnish all equipment, material, and personnel required for the test. Defective parts, if present should be replaced with new parts and system retested until completely reliable and accepted.

10.12   Inspections

The owner's representative should witness the fabrication manufacturing, testing, and assembly of any part of manufacture's work that concerns the vessel. The supplier should agree, by his acceptance of the purchase order, to any inspection and rejection in accordance with practices and codes specified. Any inspection and testing in no way relieve the manufacturer of any responsibility for the vessel meeting all requirements of applicable codes.

The manufacturer should issue instructions for the proper quality inspections according to relevant standards. Weldings should be inspected in accordance with ISO group 0520 welding (brazing, soldering) equipment ISO codes and practices by an authorized marine-welding inspector. The official certificates should be issued and furnished by vendor before official deliver at the manufacturing berth in accordance with standard.

- Firefighting Monitors: Foam/water monitors and foam/water elevated fixed monitors should be of fixed self-oscillating mechanism setting. The maneuvering of the monitors should be remotely controllable by hydraulic and electric or hydraulic and pneumatic units and should be duplicated (two systems). The remote control should be arranged from a protected control station with a good general view. Foam monitors should be of capacity no less than 5000 liters/min with foam expansion of 15 to 1. The height of through 50 m above sea level and straight stream range approx. 70 meters in still air at 10 bar. The hydraulic cylinder to be for spray and straight stream functions.

- Foam-Generating and Proportioning System: Foam-proportioning systems should be of the by-pass variable inductor type, giving flexibility of use of foam and water from a single pump. It is also used in conjunction with large model foam/water monitors or deck hydrants. As small quantity of water is by-passed through a venturi it induces the foam compound at approximately the same rate. The foam compound/ water solution is conveyed at low pressure (1.5 bar) to the base of the monitor or headers that are fitted with special induction orifice. A negative pressure condition induces the solution into the water stream and quantity of foam liquid can be preset to the type of foam compound concentrate ratio used (3 to 6%). Foam injection is another method used to produce foam that should be installed on vessels.
- Foam-liquid Concentrate: To facilitate replenishment of the vessel foam compound storage tank from local stock should be used. The type of foam compound should be the same type as, used in oil terminals. Presently the foam compound solution type, Fluro-Protein is used and kept in storage stock at terminal (3% concentrate).
- Portable Fire Extinguishers: Portable fire extinguishers should also be the same standard as used in terminals. To facilitate recharging and replacement the buyer should specify the types and supplier's name.
- Firefighting Nozzles: Firefighting nozzles should be of 6 Nos. handheld of constant/select flow feature that allows on site manual adjustment. The pattern selection is from straight to wide spray 300 to 600 LPM.
- Fixed water-spray deluge nozzles used for self-protection of the vessel should be designed and installed in accordance with standard.
- Firefighting Valves: Vessel hydrant should be of four-way with hose delivery valve to be of straightway pattern having the inlet and outlet opening of the same 65 mm size. The valve should be made of bronze or other material having corrosion-resistant properties of seawater. The hose valve should be fitted at the outlet with female instantaneous coupling (2½ in) 65 mm. The hose valve should be provided with outlet blank cap made of corrosion-resistant material. The blank cap should be of male instantaneous with brass chain attachment. The hose valve should be designed and specifically manufactured for quick operational reliability.
- Firefighting Hoses: To be the same standard used in oil terminals fire service and of the same materials. The following are minimum requirements:

| 75 mm bore | 15 meter | length | 5 Nos. |
| 75 mm bore | 25 meter | length | 10 Nos. |

Delivery hose couplings should be of instantaneous pattern made. The brass or corrosion-resistant material. The release mechanism to be of pull-release type.

- Masks and Breathing Apparatus: Breathing apparatus should be the same as used in terminal fire service, and as approved by classification society. Self contained breathing apparatus, open circuit demand compressed air type. The air cylinder should be of minimum 1200 L capacity for minimum of ½ hour duration. The apparatus should have warning audible device that operates when the pressure of cylinder drops to warn the wearer:

| Number required | six sets MESC |
| Spare cylinders | six No. MESC |

- Personnel Safety and Firefighter Protective Clothing: Firefighters outfit (protective clothing) 12 sets of appropriate sizes. Protective clothing should be of garment, trousers, gloves, helmet, and footwear. The garment should consist of a composite of outer shell moisture barrier and thermal barrier. The garment trim to be utilized to meet visibility of 50 mm retroreflective and fluorescent surface. The garment configured to provide protection to upper torso, arms, and legs excluding head, hands, and feet. Helmet should essentially consist of shell, an energy absorbing and retention systems, faceshield, ear cover, and retroreflective marking. Footwear should protect firefighter's foot and ankle from adverse environmental effect. The footwear should consist of a nonslippery sole with heel upper with lining and insole with puncture-resistant device, an impact and compression-resistant toecap permanently attached. Gloves should protect adverse environmental effect to the firefighters hand, wrists, and should minimize the effect of heat sharp objects and other hazards protection and secure thermal and moisture.

10.13 Bollard pull-testing procedure

10.13.1 Testing procedure

The following testing procedures should be adhered to:

1. Proposed test program should be submitted prior to the testing.
2. During testing of continuous bollard pull BPcont the main engine(s) should be run at the manufacturer's recommended maximum continuous rating (MCR).
3. During testing of overload pull, the main engines should be run at the manufacturer's recommended maximum rating that can be maintained for minimum 1 hour. The overload test may be omitted.
4. The propeller(s) fitted when performing the test should be the propeller(s) used when the vessel is in normal operation.
5. All auxiliary equipment such as pumps, generators and other equipment that are driven from the main engine(s) or propeller shaft(s) in normal operation of the vessel should be connected during the test.
6. The length of the towline should not be less than 300 m, measured between the stern of the vessel and the shore.
7. The water depth at the test location should not be less than 20 m within a radius of 100 m of the vessel.
8. The test should be carried out with the vessel's displacement corresponding to full ballast and half fuel capacity.
9. The vessel should be trimmed at even keel or at a trim by stern not exceeding 2% of the vessel's length.
10. The vessel should be able to maintain a fixed course for no less than 10 minutes while pulling as specified in standards.
11. The test should be performed with a wind speed not exceeding 5 m/sec.
12. The current at the test location should not exceed 1 knot in any direction.
13. The load cell used for the test should be approved by Det norske Veritas and be calibrated at least once a year.
14. The accuracy of the load cell should be ±2% within a temperature range of −10°C and +40°C and within the range of 25–200 tons tension.

15. An instrument giving a continuous read-out and also a recording instrument recording the bollard pull graphically as a function of the time should both be connected to the load cell. The instruments should be placed and monitored ashore.
16. The load cell should be fitted between the eye of the towline and the bollard.
17. The figure certified as the vessel's continuous bollard the pull should be towing force recorded as being maintained without any tendency to decline for a duration of no less than 10 minutes
18. Certification of bollard pull figures recorded when running the engine(s) at overload, reduced RPM or with a reduced number of main engines or propellers operating can be given and noted on the certificate.
19. A communication system should be established between the vessel and the person(s) monitoring the load cell and the recording instrument ashore, using VHF or telephone connection, for the duration of the test.
20. The test results should be made available to the VERITAS Surveyor immediately upon conclusion of the test program.
21. For mean breaking strength of the towline.

10.14 Foam-liquid concentrate (FLC) proportioners, generators, and twin agents

Depending on concentration of foam liquid, foam generators and proportioners are usually designed to mix certain concentrated liquid foam with water and then mixed with air to produce the finished foam. There are four methods of applying foam into fire:

Nonaspirated	0–2 expansion ratio
Low expansion	2–20 expansion ratio
Medium expansion	20–200 expansion ratio

High-expansion 201 and greater expansion ratio: Finished foam is a mixture of foam-liquid concentrate with water and air or foam is an aggregation of air-filled bubbles of lower specific gravity than flammable liquid or water. Low-expansion foam extinguishes fires by resisting flame and heat attack in the process of falling from an overhead application and where it is formed initially, to a burning flammable or combustible liquid surface, where it flows freely, progressively and removing heat, forming an air-excluding continuous blanket or film over the fuel, thus sealing volatile combustible vapor from access to air. The foam produced by these systems possesses qualities of lower expansion, higher fluidity, and more rapid foam solution drainage than foams generated in other foam systems. Medium- and high-expansion foam may be used on solid fuel and liquid fuel fires but in depth coverage. High-expansion foam is an agent for control and extinguishment of Class A and Class B fires and particularly suited as flooding agent in confined spaces.

This section specifies the minimum requirements for performance of foam-liquid concentrate, proper testing, generating equipment including material specifications for purchasing of foam-liquid concentrate and equipment used, and methods of application of low, medium- and high-expansion foam systems. Application and material specification for twin-agent "foam/dry-chemical extinguisher" is also covered in standards. The standard is prepared in three parts as follows:

- PART I "Operational Methods – Foam Liquid and Proportioners"
- PART II "Materials Specification"
- PART III "Twin-agent Foam/Dry-powder Extinguisher"

10.14.1   Foam-liquid concentrate (FLC)

A liquid concentrate should be formulated so that it may be introduced into water flowing under pressure in pipe lines, by pressure induction, vacuum induction, or pump and motor (combined with balancing valves) induction methods. Foams are arbitrarily subdivided into three ranges of expansion:

- Low-expansion foam (LX) expansion 2 to 20.
- Medium-expansion foam (MX) expansion from 21 to 200.
- High-expansion foam (HX) expansion from 201 to greater expansion ratio.

10.14.2   Application of low-expansion foam

Foam system should include provision to minimize the danger when foam is applied to the liquids above 100°C, energized electrical equipment or reactive materials. Since all foams are aqueous solutions, where liquid fuel temperatures exceed 100°C they may be ineffective and, particularly where the fuel depth is considerable (e.g., tanks) may be dangerous in use. The foam and drainage of the water from the foam can cool the flammable liquid but boiling of this water may cause frothing or slop-over of the burning liquid particularly crude oil. Boil-Over, which may occur even where foam is not applied, is a more severe and hazardous event. Large-scale expulsion of the burning contents of a tank is caused by the sudden and rapid boiling of water in the base of the tank or suspended in the fuel. It is caused by the eventual contact of the upper layer of liquid fuel in the tank, heated to above 100°C by the fire, with the water layer.

Particular care should be taken when applying foam to high viscosity liquids, such as burning asphalt or heavy oil, above 100°C. Because foams are made from aqueous solutions they may be dangerous to use on materials that react violently with water, such as sodium or potassium, and should not be used where they are present. A similar danger is presented by some other metals, such as zirconium or magnesium, but only when they are burning. Low-expansion foam is a conductor and should not be used on energized electrical equipment; in this situation, it would be a danger to personnel.

10.14.3   Compatibility with other extinguishing media

The foam produced by the system should be compatible with any extinguishing media provided for application at or about the same time. Certain wetting agents and some extinguishing powders may be incompatible with foams, causing a rapid breakdown of the latter. Only media that are substantially compatible with a particular foam should be used in conjunction with it. Use of water jets or sprays may adversely affect a foam blanket. They should not be used in conjunction with foam unless account is taken of any such effects.

- Compatibility of foam concentrates: Foam concentrate (or solution) added or put into a system should be suitable for use and compatible with any concentrate (or solution) already present, in the system. Foam concentrates or foam solutions, even of the same class, are not necessarily compatible, and it is essential that compatibility be checked before mixing two concentrates or premixed solutions.
- Uses: Low-expansion foam systems are suitable for extinguishing fires on a generally horizontal flammable liquid surface. Extinction is achieved by the formation of a blanket of foam over the surface of the burning liquid. This provides a barrier between the fuel and air, reducing the rate of emission of flammable vapors to the combustion zone, and cooling the liquid. Low-expansion foam is not generally suitable for the extinction of running fuel fires, e.g., fuel running from a leaking container or from damaged pipework or pipe joints. However, low-expansion foam can control any pool fire beneath the running fire that may then be extinguished by other means. Low-expansion foam is not suitable for use on fires involving gases or liquefiable gases with boiling points below 0°C.

10.14.4   Medium- and high-expansion foam and alcohol-resistant foam liquid

Medium-expansion foam (expansion ratio 21 to 200) is generally used for protection against fires in:

Flammable liquid as spills of average depth not more than 25 mm.
Flammable liquids in defined areas such as bunds and heat treatment baths.
Combustible solids where up to about 3 m foam build-up is necessary to cover the hazard, e.g., engine-test cells and generating sets.

High-expansion foam (expansion ratio of 201 or greater expansion). This liquid concentrate is applicable in total flooding systems, local application system, portable and mobile systems. High-expansion foam is generally used in total flooding of warehouses, aircraft hangers, furniture stores, and other similar premises. High-expansion foam can also be used in situations where it would be hazardous to send personnel into in underground enclosures where smoke logging could occur and in consequence exit routes will be difficult to find. In local application smaller enclosures within larger areas such as pits, basements, etc., are places where filling the space is an effective means of dealing with an inaccessible fires. This system can be used both indoors and outdoors provided there is a means of shielding the foam from the effects of wind.

- Alcohol-resistant: Alcohol-resistant (AR) foam concentrates are formulated for use on foam destructive liquids, the foams produced are more resistant than ordinary foams to breakdown by the liquid. They may be of any of the classes and may be used on fires of hydrocarbon liquids with a fire performance generally corresponding to that of the parent type. Film-Forming foams do not form films on water miscible liquids. Alcohol-resistant foam concentrates are generally used at 6% concentration on water miscible fuels.

10.14.5 Foam-liquid proportioners and generators

- Low-expansion foam-liquid proportioners: For low-expansion foam liquid, proportioning, and mixing with water may be achieved by one or more of the following methods:
- Air foam nozzle with built-in eductor: In this type of proportioner the jet in the foam maker is utilized to draft the foam liquid. The length and size of pick-up tube and foam-liquid container and the foam maker should conform to the recommendation of the manufacturers and the bottom of the foam-liquid container should not be more than 1.8 meters below the foam nozzle.
- In line inductor: This unit is used to introduce foam concentrate into the water supply to produce a solution by way of venture system. This inductor is for installation in a hose line usually some distance from the foam maker. It must be designed for the flow rate of particular foam maker with which it is to be used. The device is very sensitive to down stream pressure and is accordingly designed for use with specified length of hose and pipe between inductor and the foam maker. The pressure drop of approximately 35% (not more than 40%) and the rate of the induction can be varied from 2 to 6%. A mobile unit consisting of hose, fixed inductor, branch pipe, and FLC (foam liquid concentrate) tank is also used. The FLC container can be refilled during firefighting operations.
- Mobile unit: The mobile unit consists of a fiberglass foam storage tank, in-line inductor, inlet and outlet hose connected to this foam making branch pipe (Figure 10.15). The unit can be used by one or two people.
- Primary-Secondary eduction method: This method of introducing air foam concentrate into the water stream en route to a fixed foam maker is illustrated in Figure 10.16. The unit consists of two eductor designated as the primary eductor and the secondary eductor. The primary eductor is located outside the firewall enclosure and is installed in a bypass line connected to and in parallel with the main water supply line to the foam maker. A portion of the water flows through the primary eductor and draws the concentrate from a container using pickup tube. The main water line discharges through the jet of a secondary eductor located at the foam maker proper, the mixture of water and concentrate from the primary eductor being delivered to the suction side of the secondary eductor.
- Limitations: The primary eductor may be installed as much as 150 m from the secondary eductor. The size of piping used, both in the water and the solution lines, should be as

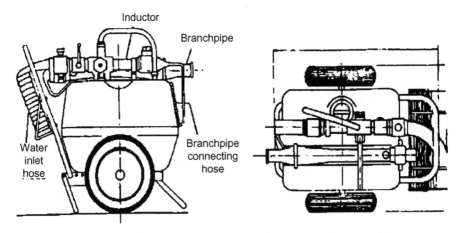

Figure 10.15 Mobile unit.

specified by the manufacturer. The elevation of the bottom of the concentrate container should not be more than 1.8 m below the primary eductor.

• Mechanical foam generator

• This method involves foam-liquid pick-up, aeration, and foam generation in one unit. It is mostly used by portable equipment where "rope" jet is required (Figure 10.17) There is a considerable pressure loss across the generator and for this reason, the pressure at the water head on the inlet side should not be less than 10 bar.

• High back-pressure foam generator: The use of high back-pressure foam generators is required for semisubsurface injection of fixed roof oil tank fires. When using a water pressure of 10 bar to the foam generators, the typical system will function in tank with the height of up to 18 meters. Water supply pressure should be determined for each individual installation or tank grouping and will depend on the requirements of the foam generators, injection devices, and the tank heights (Figure 10.18).

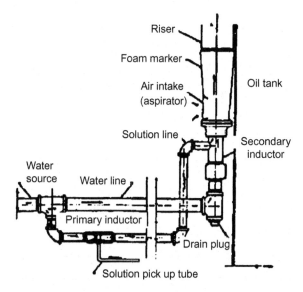

Figure 10.16 Introducing air foam concentrate into the water stream en route to a fixed foam maker. • Mechanical foam generator.

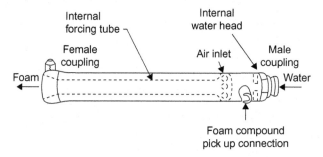

Figure 10.17 Foam liquid pick-up, aeration and foam generation in one unit.

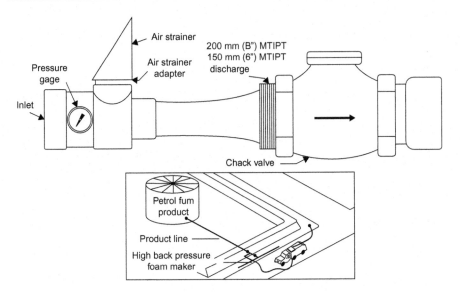

Figure 10.18 Fixed high back-pressure foam maker for fixed systems.

10.14.6   By-pass variable inductor

This is a preferred method for fire boats or tugs as it gives flexibility of use of foam and water from a single pump. It is also used in conjunction with certain large model foam/water monitors or deck hydrants. A small quantity of water is by-passed through a venture that induces the foam compound at approximately the same rate. The resultant 50/50 foam compound/water solution is conveyed at low-pressure 1.5 bar to the base of the monitor or headers that are fitted with manually operated water/foam valves and special induction orifice. When these valves are in the foam positions, a negative pressure condition exists on the outlet side of the valve that induces the solution into the water stream. Quantity of foam liquid can be adjusted by a lever from 0–360 LPM or more.

10.14.7   The pump proportioner

This proportioner may be applied with either fixed or variable inductor. The system may be used with advantage on fire tugs and boats fitted with large foam/water monitors that are not suitable for use with by-pass induction method. It may also be used for fixed equipment where water/foam compound solution is pumped through pipelines. This system also can be installed on fire trucks with water tank or where water from a hydrant is available to fill the water tank. Foam/water are mixed by the fire-truck pump. A portable type with foam concentrate tank or a pick-up tube called (M-J Inductor) can be supplied (Figure 10.19).

The system by-passes a small quantity of water from the delivery side of the pump. This induces foam-liquid concentrate via a fixed or variable inductor and return the

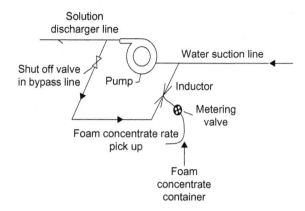

Figure 10.19 Around -the -pump proportioning system typical arrangement.

foam liquid/water solution to the low pressure side (suction) of the pump. The solution is then discharged from the pump deliveries. A disadvantage with this system is its lack of flexibility, in as much that water and foam cannot be used at the same time from single pump.

10.14.8   Automatic foam proportioner

There are several methods, adopted, one of which can be employed in mobile fire trucks. The system automatically proportions the correct percentage of foam liquid irrespective of water pressure and volume to the full pump capacity. The injector is attached to the water inlet of the pump. Operation of this method is by balancing hydraulic forces that act on the moving member that in turn controls several parts in the compound supply stream. One force is derived from the kinetic energy of the water stream striking a disk attached to the upstream end of the moving member and the opposing force from the jet reaction of the foam compound issuing from a nozzle attached to the downstream end. The forces are arranged to balance when the flow ratio of both liquids meet the required concentration (Figure 10.20).

10.14.9   Pressure-proportioning tank

This method employs water pressure as the source of power. With this device, the water supply pressurizes the foam concentrate storage tank. At the same time, water flowing through an adjacent venture or orifice creates a pressure differential.

The low-pressure area of the venturi is connected to the foam concentrate tank, so that the difference between the water supply pressure and this low- pressure area forces the foam concentrate through a metering orifice and into the venturi. Also, the differential across the venturi varies in proportion to the flow, so one venturi will proportion properly over a wide flow range. The pressure drop through this unit is relatively low.

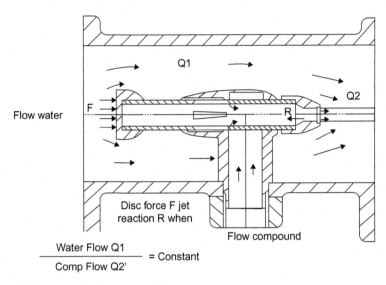

$$\frac{\text{Water Flow Q1}}{\text{Comp Flow Q2'}} = \text{Constant}$$

Figure 10.20 Automatic foam proportioner.

The system may be designed of twin tanks one tank may be replenished while the second tank is in operation. A special test procedure is available to permit the use of a minimum amount of concentrate when testing the pressure proportioner system (Figure 10.21A and 10.21B).

- Limitations
  1. Foam concentrate with specific gravities similar to water may create a problem by mixing.
  2. The capacity of these proportioners may be varied from approximately 50 to 200% of the rated capacity of the device.
  3. The pressure drop across the proportioner ranges from (1/3 to 2 bar) depending on the volume of water flowing within the capacity limits given above.
  4. When the concentrate is exhausted, the system must be turned off, and the tank drained of water and refilled with foam concentrate.
  5. Since water enters the tank as the foam concentrate is discharged, the concentrate supply cannot be replenished during operation, as with other methods.
  6. This system will proportion at a significantly reduced percentage at low-flow rates and should not be used below minimum design flow.

10.14.10   Diaphragm (bladder) pressure-proportioning tank

This method also uses water pressure as a source of power. This device incorporates all the advantages of the pressure-proportioning tank with the added advantage of a collapsible diaphragm that physically separates the foam concentrate from the water supply. Diaphragm pressure-proportioning tanks operate through a similar range of water flows and according to the same principles as pressure-proportioning tanks. The added design feature is a reinforced elastomeric diaphragm (bladder) that can be used with all concentrates listed for use with that particular diaphragm (bladder) material.

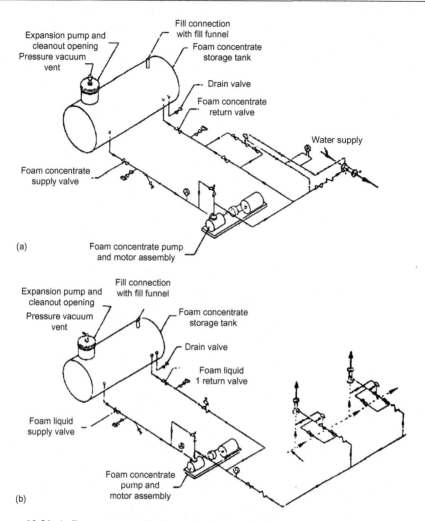

Figure 10.21 A. Pressure proportioning tank (horizontal). B. Balanced pressure proportioning with multiple injection points (metered proportioning) (vertical).

The proportioner is a modified venturi device with a foam concentrate feed line from the diaphragm tank connected to the low-pressure area of the venturi. Water under pressure passes through the controller and part of this flow is diverted into the water feed line to the diaphragm tank. This water pressurizes the tank, forcing the diaphragm filled with foam concentrate to slowly collapse. This forces the foam concentrate out through the foam concentrate feed line and into the low-pressure area of the proportioner controller. The concentrate is metered by use of an orifice or metering valve and joins in the proper proportion with the main water supply, sending the correct foam solution to the foam makers downstream (Figures 10.22(a) and 10.23(b)).

• Limitations

The limitations are the same as those listed in the Pressure-Proportioning Tank section, except the system can be used for all Type F concentrates.

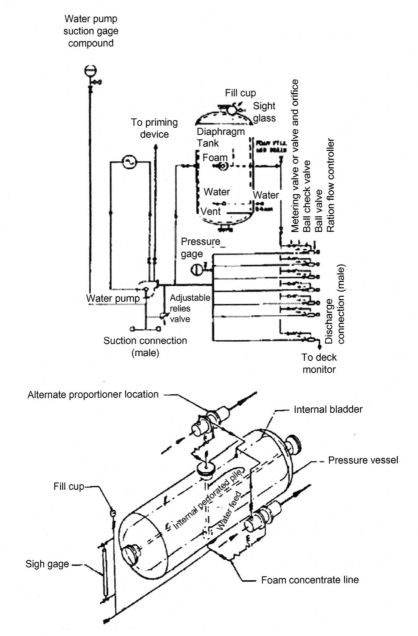

Figure 10.22 A. Pressure proportioning diaphragm vertical tank method. B. Pressure proportioning diaphragm horizontal tank method.

10.14.11 Wheeled diaphragm proportioner

The unit consists of a foam tank, venturi type proportioner with integral 1 to 6% foam concentrate metering orifice with length of fire hose and foam-branch nozzle. The fixed type of this unit is provided with a hose reel mounted on the ank. This unit is suitable for refinery area, offshore platforms, truck-loading racks, and industrial process areas.

10.14.12 Typical balanced pressure-proportioning system

In-line balanced proportioning system utilizing a foam concentrate pump discharging through a pressure regulating balancing valve and a metering orifice into a proportioning controller. A pressure regulating valve placed in the pump return line maintains constant pressure in the foam concentrate supply line at all design-flow rates. This constant pressure must be greater than the maximum water pressure under all conditions. This type of design is suitable when using multiple proportioning controllers located away from the central foam concentrate supply. A common foam concentrate supply line carries concentrate to each proportioning controller.

10.15 Medium- and high-expansion foam generators

Medium- and high-expansion foams are aggregations of bubbles mechanically generated by the passage of air or other gases through a net, screen, or other porous medium that is wetted by an aqueous solution of surface-active foaming agents. Under proper conditions, firefighting foams of expansions from 20:1 to greater expansion can be generated. Such foams provide a unique agent for transporting water to inaccessible places for total flooding of confined spaces; and for volumetric displacement of vapor, heat, and smoke.

Tests have shown that, under certain circumstances, high-expansion foam, when used in conjunction with water sprinklers, will provide more positive control and extinguishment than either extinguishment system by itself. High-piled storage of rolled paper stock is an example. Optimum efficiency in any one type of hazard is dependent to some extent on the rate of application and also the foam expansion and stability.

Medium- and high-expansion foams, which are generally made from the same type of concentrate, differ mainly in their expansion characteristics. Medium-expansion foam may be used on solid fuel and liquid fuel fires where some degree of in-depth coverage is necessary, e.g., for the total flooding of small enclosed or partially enclosed volumes such as engine test cells, transformer rooms, etc. It can provide quick and effective coverage of flammable liquid spill fires or some toxic liquid spills where rapid vapor suppression is essential. It is effective both indoors and outdoors.

High-expansion foam may also be used on solid and liquid fuel fires but in-depth coverage it can give greater than for medium-expansion foam. It is therefore most suitable for filling volumes in which fires exist at various levels. For example, experiments have shown that high-expansion foam can be used effectively against high rack storage fires provided that the foam application is started early and the depth of foam is rapidly increased. It can also be used for the extinction of fires in enclosures where it may be dangerous to send personnel, e.g., in basement and underground passages. It may be used to control fires involving liquefied natural gases and LPG and to provide vapor dispersion control for LNG and ammonia spills. High-expansion foam is particularly suited for indoor fires in confined spaces. Its use in outdoors may be limited because of the effects of wind and lack of confinement.

Medium- and high-expansion foam have several effects on fires:

When generated in sufficient volume, they can prevent free movement of air, necessary for continued combustion.

When forced into the heat of a fire, the water in the foam is converted to steam, reducing the oxygen concentration by dilution of the air.

The conversion of the water to steam absorbs heat from the burning fuel. Any hot object exposed to the foam will continue the process of breaking the foam, converting the water to steam, and of being cooled.

Because of their relatively low surface tension, solution from the foams that is not converted to steam will tend to penetrate Class A materials. However, deep seated fires may require overhaul.

When accumulated in-depth, medium- and high-expansion foam can provide an insulating barrier for protection of exposed materials or structures not involved in a fire and can thus prevent fire spread.

For liquefied natural gas (LNG) fires, high-expansion foam will not normally extinguish a fire but it reduces the fire intensity by blocking radiation feed back to the fuel.

Class A fires are controlled when the foam completely covers the fire and burning material. If the foam is sufficiently wet and is maintained long enough, the fire may be extinguished. Class B fires involving high flash point liquids can be extinguished when the surface is cooled below the flash point. Class B fires involving low flash point liquids can be extinguished when a foam blanket of sufficient depth is established over the liquid surface.

10.15.1   Mechanisms of extinguishment

Medium- and high-expansion foam extinguishes fire by reducing the concentration of oxygen at the seat of the fire, by cooling, by halting convection and radiation, by excluding additional air, and by retarding flammable vapor release.

10.15.2   Use and limitations

While medium- and high-expansion foams are finding application for a broad range of firefighting problems, each type of hazard should be specifically evaluated to verify the applicability of medium- or high-expansion foam as a fire control agent. Some important types of hazards that medium- and high-expansion foam systems may satisfactorily protect include:

Ordinary combustibles.
Flammable and combustible liquids.
Combinations of (a) and (b).
Liquefied natural gas (high-expansion foam only).

- Operating devices: A block diagram of a typical automatic medium- or high-expansion foam system is shown in Figure 10.11.
- Foam generators: At the present time, foam generators for medium- and high-expansion foam are of two types depending on the means for introducing air, namely, by aspirator or blower. In either case, the properly proportioned foam solution is made to impinge at appropriate velocity on a screen or porous or perforated membrane or series of screens in a moving air stream (Figure 10.23).The liquid films formed on the screen are distended by the moving air stream to form a mass of bubbles or medium- or high-expansion foam. The foam volume varies from about 20 to 1,000 times the liquid volume depending on the design of the generator. The capacity of foam generators is generally determined by the time required to fill an enclosure of known volume by top application within 1 to 5 minutes.
- Foam-generator aspirator type: These may be fixed or portable. Jet streams of foam solution aspirate sufficient amounts of air that is then entrained on the screens to produce foam. These usually produce foam with expansion ratios not over 250:1 (Figure 10.24).

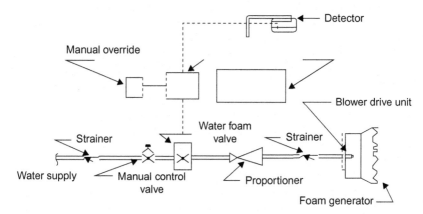

Figure 10.23 Block diagram of automatic medium-or high-expansion foam system.

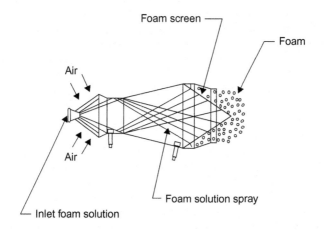

Figure 10.24 Aspirating-type foam generator.

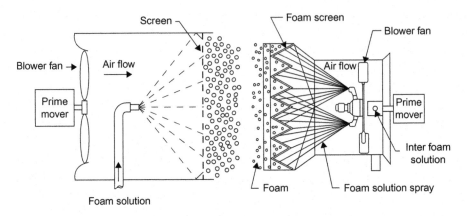

Figure 10.25 Blower-type foam generators.

- Foam-generator blower type: These may be fixed or portable. The foam solution is discharged as a spray onto screens through which an air stream developed by a fan or blower is passing. The blower may be powered by electric motors, internal combustion engines, air, gas, or hydraulic motors or water motors. The water motors are usually powered by foam solution (Figure 10.25).

10.16   Material specifications

10.16.1   Foam-liquid concentrate

Foams has for many years been recognized as an effective medium for extinction of flammable liquid fires. Foam concentrates are classified by composition as:

- Protein (old type compound) (P).
- Fluoro Protein (FP).
- Film Forming Fluoro Protein (FFFP) Synthetic (S).
- Aqueous Film Forming (AFFF) Alcohol-resistant or Universal (AR).

Because foams are made from aqueous solutions, they are dangerous to be used on materials that react violently with water such as sodium or potassium. This section specifies the requirements of foam-liquid types that has expansion rates depending on its range of application that are used for fire extinguishment of liquid hydrocarbon and special foam concentrates for Alcohol-resistant (AR) Fires. The requirements for medium- and high-expansion foam concentrates are also included.

10.16.2   Low-expansion foam liquid

- Grades: The foam concentrate should be graded for:
  Extinguishing performance as grades I, II, and II.
  Burn-back resistant as levels A, B, C, and D.

Typical anticipated extinguishing performance grades and burn-back levels are as given in Table 10.5 and should be tested in accordance with ISO 7203-1.

- Use with seawater: If a foam concentrate is marked as suitable for use with seawater the concentration for use with fresh water and seawater should be identical.
- Tolerance of the foam concentrate to freezing and thawing: The foam concentrate should be tested and graded with this requirement.
- Sediment in the foam concentrate: Percentage volume of sediment should not be more than 0.25% by volume as received (before aging) and not more than 1.0% after aging when tested.
- Comparative fluidity: The flow rate of the concentrate should not be less than kinetic viscosity of 200 mm I/S when tested before and after temperature conditioning in accordance with Annex D ISO 7203-1.
- PH: The PH of the foam concentrate before and after temperature conditioning should not be less than 6.0 and not more than 9.5 at 20° (±2)°C. If there is a difference of more than 0.5 PH between the two values the foam concentrate should be designated temperature sensitive. Surface tension of the foam solution, interfacial tension between the foam solution, and cyclohexane and spreading coefficient of foam solution on cyclohexane should be tested and its temperature sensibility determined in accordance with ISO 7203-1.

10.16.3   Expansion and drainage of foam

- Expansion: The foam produced from the foam concentrate with potable water should have the expansion within either $\pm20\%$ of the characteristic value or ±1.0 of the characteristic value, whichever is greater when tested in accordance with annex H.1 of ISO 7203-1. If any values for expansion obtained after temperature conditioning is less than 0.85 times or more than 1.15 times, the corresponding value obtained before temperature conditioning the foam concentrate should be designated temperature sensitive.
- Drainage: The foam produced from the foam concentrate with potable water and if appropriate with the synthetic seawater should have a 25% drainage time within $\pm20\%$ of the characteristic value when tested in accordance with F2 of ISO 7203-1. If any of the value for 25% drainage time obtained after temperature conditioning is less than 0.8 times or more than 1.2 times, the corresponding value obtained before temperature conditioning, the foam concentrate should be designated temperature sensitive.
  Note: For sampling and temperature conditioning see ISO 7203-1.

Table 10.5 Extinguishing performance and burnback resistance annex g

Type	Extin. Performance	Burnback Resistance	Type	Extin. Performance	Burnback Resistance
AFFF	1	A	FP (AR)	II	A
AFFF	1	A	P (AR)	III	B
FFFP	1	A	P (AR)	III	B
FFFP	1	A	S (AR)	III	D
FP	11	A/B	S (AR)	III	C

Note: For extinguishing performance Class I is the highest class and Class III is the lowest class for burnback resistance level, A, is the highest level and level "D" is the lowest class.

- Compatibility: The foam concentrate should be compatible with dry-chemical extinguishing powder when used simultaneously or successively and the user should be ensured by the manufacturer that any unfavorable interaction will not cause an unacceptable loss of efficiency. Foam concentrates of different manufacture, grade, or class are frequently incompatible and should not be mixed unless it has first been established that an unacceptable loss of efficiency will not result.
- Performance and quality assurance: The manufacturer should ensure the buyer that the foam-liquid concentrate have been sample tested in accordance with ISO 7203-1 for low expansion and ISO 7203-2 for medium and high expansion.
- Marking: The following information should be marked on shipping containers of low expansion: Designation (identifying name) of concentrate and the word "Low-expansion Foam Concentrate".
  Grade (I, II, or III) and level (A, B, C, or D) and if it complies the words "Film Forming".
  Recommended usage concentration (most commonly 1%, 3%, or 6%).
  Any tendency of foam concentrate to cause harmful physiological effects; the methods needed to avoid them and the first-aid treatment if they should occur.
  Recommended storage temperature and shelf-life.
  Nominal quantity in container.
  Supplier's name and address.
  Batch number.
  Suitable or not suitable with salt water.
  Any corrosiveness of the concentrate, both in storage and in use with seawater, as appropriate.

Notes:

1. It is extremely important that the foam concentrate, after dilution with water to the recommended concentration should not, in normal usage, present a significant toxic hazard to life in relation to the environment.
2. The packaging of the foam concentrate should ensure that the essential characteristics of the concentrate are preserved when stored and handled in accordance with supplier's recommendations.
3. Marking on shipping containers should be permanent and legible.
4. Foam concentrate of "medium and high expansion" should also bear the identification marks.
5. The supplier should provide a list of the characteristic data sheet at quotation stage.

10.16.4   Medium- and high-expansion foam concentrate

- Classification: The foam concentrate should be classified as medium- and/or high expansion and should comply with the specifications for the following:
  Use with seawater.
  Tolerance of foam concentrate to freezing and thawing.
  Sediment in the foam concentrate.
  Comparative fluidity.
  Ph of foam concentrate.
  Surface tension.
  Interfacial tension between the foam solution and cyclohexane.
  Spreading coefficient of the foam solution and cyclohexane.

- Expansion and drainage: The foam produced from the foam concentrate with potable water should have an expansion of no less than 50 and 25 to 50% drainage time within 20% of the characteristic value for medium expansion and have an expansion of no less than 201 and a 50% drainage time of no less than 10 min, for high expansion when tested.
  If the foam concentrate is marked as suitable for use with seawater, the foam produced from the foam concentrate with synthetic seawater should have expansion as:
  For medium expansion, the foam should have expansion value no less than 0.9 times and not more than 1.1 times the expansion value obtained from the same sample of foam concentrate tested with potable water.
  For high expansion, the foam should have an expansion no less than 0.0 times and not more than 1.1 times the expansion value obtained from the same sample of foam concentrate tested with potable water.
- Temperature sensitivity.
  1. Medium expansion: If the value for expansion and/or 25% or 50% drainage times obtained after temperature conditioning is less than 0.8 times or more than 1.2 times the corresponding value obtained before temperature conditioning the foam concentrate should be designated temperature sensitive.
  2. High expansion: If the value for expansion, and/or 50% drainage time obtained by using temperature conditioned foam concentrate and found it is less than 0.8 times or more than 1.2 times the corresponding value obtained by using foam concentrate not temperature conditioned, the foam concentrate should be designated temperature sensitive.
- Procedures for measuring expansion and drainage rates of foams.
  Foam sampling: The objective of foam sampling is to obtain a sample of foam typical of that to be applied to the burning surface under anticipated fire conditions. Inasmuch as foam properties are readily susceptible to modification through the use of improper techniques, it is extremely important that the prescribed procedures be followed.
  A collector (slider) has been designed to facilitate the rapid collection of foam from low-density patterns; it is also used for all sampling except where pressure -produced foam samples are being drawn from a line tap. A backboard is inclined at a 45-degree angle suitable for use with vertical streams falling from overhead applicators as well as horizontally directed streams (see Figure 10.26).
  The standard container is 20 cm deep and 10 cm inside diameter (1600 mL) preferably made of 1.6 mm thick aluminum or brass. The bottom is sloped to the center where a 6.4 mm drain fitted with a valve is provided to draw off the foam solution (see Figure 10.27).
  It is important that the foam samples taken for analysis represent as nearly as possible the foam reaching the burning surface in a normal firefighting procedure. With adjustable stream devices, it is usually desirable to sample both from the straight stream position, and the fully dispersed position, and possibly other intermediate positions.
  The collector should be placed at the proper distance from the nozzle so as to be the center of the ground pattern. The nozzle should be placed in operation while it is directed off to one side of the collector. After the pressure and operation have become stabilized, the stream is swung over to center on the collector. When a sufficient foam volume has accumulated to fill the sample containers, usually in a matter of only a few seconds, a stopwatch is started for each of the two samples in order to provide the "zero" time for the drainage test described later. Immediately, the nozzle is turned away from the collector, the sample containers are removed, and the top struck off with a straight edge. After all foam has been wiped off from the outside of the container, the sample is ready for analysis.

Note: For subsurface injection equipment, the foam sample should be obtained from a valved test connection on the discharge side of the foam maker.

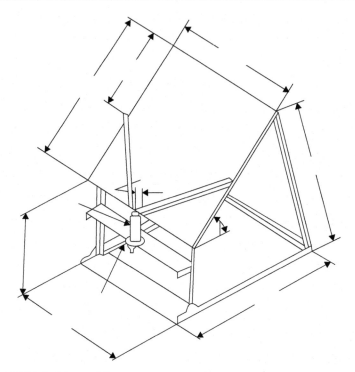

Figure 10.26 Typical foam slider.

10.16.5   Foam expansion

- Film forming: In this test a quantity of foam is placed on the surface of cyclohexane. The foam is swept from the surface by insertion of a conical screen and the exposed fuel surface is tested for the presence of an aqueous film by probing with a flame. If the film is present, the fuel will not ignite. In the absence of film, ignition will occur.

10.16.6   Fire-performance test

If the foam produced from the foam concentrate before and designated temperature sensitive after temperature conditioning and tested with potable water and if found appropriate with synthetic seawater in accordance with Annex "G" ISO 7203-2 should have an extinction time not more than the value given in Table 10.6 and burnback time no less than the medium expansion as stated in the Table 10.6.

10.16.7   Foam-quality laboratory-testing procedure

With the greatly expanding use of foam in hydrocarbon liquid firefighting, the need for a standardized laboratory procedure for analyzing and expressing the significant physical properties of the mechanical foam as related to firefighting capabilities has

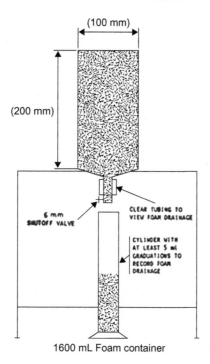

(100 mm)

(200 mm)

6 mm
SHUTOFF VALVE

CLEAR TUBING TO
VIEW FOAM DRAINAGE

CYLINDER WITH
AT LEAST 5 ml
GRADUATIONS TO
RECORD FOAM
DRAINAGE

1600 mL Foam container

Figure 10.27 1600 mL foam container.

Table 10.6 **Fire performance test**

Time	Medium expansion	Medium expansion
Extinction time 1% burnback time	Not more than 120 s Not less than 30 s	Not more than 150 s Not applicable

arisen. The numerical values obtained by tests generally enable characterization of the foam. Only by describing results obtained on standardized basis it will be possible to describe the optimum foam for the various operating conditions.

Considering the amount of foam liquid used and stored in numerical conditions in the Oil, Gas and Petrochemical Industries a central laboratory testing equipment should be available to test the foam-liquid concentrates and to advise that the foam liquid purchased or stored in unfavorable conditions for a long period of time is acceptable.

The equipment and procedure for testing should be in accordance with ISO 7203 Part (1) for low expansion and Part (2) for medium- and high-expansion foam concentrate (Annex A to J).

• Inspection of storage and simple quality control testing procedure: To determine the condition of foam liquid, at least annual inspection should be made of foam concentrate tanks or storage containers for evidence of excessive sludging or deterioration. Sample of concentrates should be either referred to qualified central laboratory for quality condition testing or at least conduct a quality control test.

- Foam-quality control test: A foam system will extinguish a flammable liquid fire if operated within the proper ranges of solution pressure and concentration and at sufficient discharge density per sq ft of protected surface. The acceptance test of a foam system should ascertain: All foam-producing devices are operating at "system design" pressure and at "system design" foam solution concentration.

The laboratory-type tests have been conducted, where necessary, to determine that water quality and foam liquid are compatible.

The following data are considered essential to the evaluation of foam-system performance:

Static water pressure.

Stabilized flowing water pressure at both the control valve and a remote reference point in the system.

- Rate of consumption of foam concentrate: The concentration of foam solution should be determined. The rate of solution discharge may be computed from hydraulic calculations utilizing recorded inlet or end-of-system operating pressure or both. The foam-liquid concentrate consumption rate may be calculated by timing a given displacement from the storage tank or by refractometric means. The calculated concentration and the foam solution pressure should be within the operating limit recommended by the manufacturer.

- Marking and packaging: The following information should be marked on the shipping containers by the supplier:

Designation (identifying name) of concentrate and as appropriate the word "medium" or "high" expansion.

If it complies, the words "film forming".

10.17 Materials specification

10.17.1 Foam proportioning and generating system

The system should be designed for low-expansion foam system utilizing 3 to 6% of foam-liquid concentrate. Equipment should be made of corrosion-resistant material suitable for the type of foam and operation with salt water. Material used for construction of equipment should resist galvanic corrosion and corrosion caused by atmospheric condition as determined by mercurous nitrate and salt-spray tests. All component parts such as check, flow control, by-pass drain, flash valves etc. and pressure gauges should be made in accordance with relevant standards. Depending on the system design, the equipment should be suitable for water pressure of up to 15 bar.

Before making a purchase request, all information related to the foam equipment and type of FLC should be sent from manufacturer or vendor along with relevant data sheets.

Hose couplings used should be of instantaneous male for inlets and female for outlets (BS336).

Any-gasket "O" rings and nonmetallic components should meet the requirement of ANSI 1474-UL199 standards. An internal operating part whose removal may become necessary during anticipated maintenance or repair should be accessible, removable, and replaceable without damage to the equipment. A tank that may be subjected to air, gas, or water pressure, or a combination thereof should be designed, constructed, tested, inspected and marked in accordance with section VIII of the ASME Boiler and Pressure Vessel Code.

10.17.2 Portable foam generators or proportioners

Portable foam proprtioners should be designed and calibrated for the type of FLC, foam making and aspirating units required to be used. Portable proportioners and branch pipes should be made of brass or anodized aluminum alloy. Pick-up tubes should be made of synthetic rubber tested to 20 bar with threaded or quick release couplings. Where back-pressure may cause contamination of FLC and water a check valve should be used.

Mobile self-contained units should be so designed and constructed to make it convenient to be carried by one man and used for rapid intervention. The foam tank should be constructed of high strength chemical resistant suitable for tropical climate. A flow regulator should be provided for mobile proportioners to regulate proportion of foam-liquid injection into water stream from 1 to 10%. The material of construction and parts should be so chosen to minimize maintenance requirement.

10.17.3 Fixed proportioners

All control valves used should be of a type that open and close smoothly and readily under all rated pressure, should effectively shut off the position of the system they control and should be sized to compensate the maximum flow and pressure required by the position of the system they control.

The function and operation of controls, operating devices, gauges, and drains should be clearly identified and should be accessible. The diaphragm or bladder should be made of material that will resist corrosion, breakdown, or loss of flexibility under condition of prolonged contact with the foam concentrate.

10.17.4 Test and quality inspection

Manufacturer should certify in writing that the appropriate tests and quality inspection have been carried-out in accordance with UL 162 for foam equipment. It is the responsibility of vendor to make sure that all components of foam generators and proportioners have been tested.

10.17.5 Test types

- Accuracy of proportioning: The foam system should proportion foam concentrate into water within $\pm 10\%$ of the recommended concentration range of design flows. There are two acceptable testing methods:
  With the foam system in operation at a given flow, a solution sample is collected from each outlet and the concentration measured by refractometer as described in NFPA 11 (standard for low-expansion foam and combined agent systems.)

With the foam system in operation at a given flow, using water as a substitute for foam concentrate, the water is drawn from a calibrated tank instead of foam concentrate. The volume of water drawn from the calibrated tank indicates the percentage of foam concentrate used by the system.

· PREPARATION FOR SHIPMENT: Each package unit or piece of equipment should be properly prepared for shipment to prevent damage by handling, and shipping and should be labeled to insure that they are not lost in transit. In addition the following measures should be taken:
All external connections should be protected.
Package list to be included inside every package and one attached to the package.
Shipping supports should be provided if considered necessary.

10.18   Twin-agent dry-chemical powder and foam system

In this system foam is applied to a hazard simultaneously or sequentially with dry-chemical powder. Systems of this type combine the rapid fire extinguishing capabilities of dry-chemical powders (as well as their ability to extinguish three-dimensional fires) with the sealing and securing capabilities of foam, and are of particular importance for protection of flammable liquid hydrocarbon hazards.

· Twin-agent system: These systems may be self-contained, and the application of each agent is separately controlled so that the agents may be used individually, simultaneously, or sequentially as the situation requires.
· Limitations: The manufacturers of the dry chemical and foam concentrate supplying the system should confirm that their products are mutually compatible and satisfactory for this purpose. Limitations imposed on either of the agents used in the system for the use of that agent alone should also be applied to the twin-agent system.
· Application rates: Minimum delivery rates for protection of a hazard, based on the assumption that all of the agent reaches the protected area, should be as follows:
AFFF solution should be delivered at a rate of 4.1 (L/min)/m2 (0.10 GPM/Sq. Ft.) of area to be protected. The ratio of dry-chemical discharge rate to premix AFFF discharge rate (kg dry chemical and kg AFFF solution per second) should be in the range of 0.6:1 to 5:1. This type of extinguisher is designed to cover the risk of the area to be protected and the size and specification of the equipment should meet the requirements of the user. The unit may be skid, trailer, or fire-truck mounted, and all appropriate component parts should comply with NFPA Standard Section 11 Chapter 4 on combined agent systems. Any exception to this specification should be stated in writing for the attention of manufacturers to include them in their quotations.

10.18.1   Extinguisher

· Twin agent: The twin-agent unit should contain dry-chemical and premixed aqueous film forming (AFFF) or (FFFP) tanks. Both agents should discharge independently through a twinned-hose reels terminated to manually triggered discharge nozzles for both agents.
· Nitrogen gas: Compressed nitrogen gas regulated to pressure of about 15 bar should be utilized for pressurization of storage containers and for the controlled discharge of agents from the pressurized storage containers separately or together through hose reels or/and paralleled fixed piping system, if provided.

• Dry-chemical containers: Dry-chemical container should have the following specifications. The container should be either of spherical or cylinderical shape and should have necessary gaseous provisions to issue complete fluidization of dry-chemical powder in the container at its maximum compact state and to maintain a uniform dry-chemical discharge rate throughout, no less than 95% of the specified discharged time period. The nitrogen gas should be used as the driving force. Each storage container should have the specified capacity of dry chemical with 15 bar (approximate working pressure). Containers should be designed and fabricated in accordance with the requirements of ASME section VIII boiler and pressure vessel code. The tank should be welded steel construction. Each tank should be equipped with level indication and pressure indicator. A manual means for depressurization of partially depleted dry-chemical container should be provided. Containers should also be provided with manual provision for completely drainage and flushing of the container assembly. Each container should be provided with one fill cap. The fillcap should consist of cast aluminum body equipped with two handles extending from opposite sides of the cap to permit hand tightening so that it is free from leakage under normal operating pressure without the use of tools. A safety vent hole should be located in the fill cap. A pressure-relief valve should be furnished to prevent the pressure in the tank from exceeding by 10% the maximum working pressure of the tank.

10.18.2   Foam-solution tank

The premixed foam-solution tank should be either of spherical or cylindrical shape and should have the necessary provisions for pressurization and expulsion of all the stored solution using a gaseous nitrogen source as driving energy. Each tank should have the specified capacity of 600 to 800 liters with 15 bar pressure (approximate) working pressure. The tank should be designed and fabricated in accordance with the requirements of ASME VIII boiler and pressure vessel code for the specified working pressure. The tank should be of stainless-steel welded construction.

10.18.3   Nitrogen cylinders

A nitrogen gas system should be integral part of twin-agent units with related nitrogen storage cylinders. Standard nitrogen cylinders with adequate capacity and 197 bar design pressure should be provided as driving force for both agents. The quantity of nitrogen gas should be adequate to expel the whole foam solution and dry chemical as well as the flush-out of the system. The nitrogen cylinders should be securely placed in a horizontal or vertical position. The method of placement should be in such a way to permit easy access for operation and replacement of the cylinders.

The nitrogen cylinders should be manifolded and connected to the agent tank. Each cylinder should have a minimum of one regulator for dry-chemical nitrogen supply each regulator should be designed for an inlet pressure of 197 bar and should be set to deliver nitrogen at reduced pressure of about 15 bar. Each set of regulators should be equipped with a spring-loaded pressure-relief valve and should be connected to the nitrogen cylinders. Each valve should be provided at the end of nitrogen supply line to each agent. The system should be designed in a manner that the agents can be delivered individually or together.

10.18.4 Hose and hose reels

Single-length twinned hose mounted on a suitable location should be provided. Hose material should be nonkink rubber type suitable for working pressure of at least 17 bar. The minimum hose diameters should be 25 mm diameter for foam and 20 mm diameter for dry chemical. Metal hose reels with manual rewind and straight through internal fittings designed for minimum pressure drop should be provided.

10.18.5 Nozzles

Manually triggered (pistol grip type) physically linked, liquid agent and dry-chemical discharge nozzle for use by single operator should be supplied. Nozzle should be provided with minimum effective stream ranges and discharge rate in accordance with technical data. Nozzles should be provided with an integral shut-off valve.

10.18.6 Type of agent

Aqueous film forming or film forming fluoro protein of foam concentrate known as AFFF (or FFFP) should be used with 3 to 6% concentration. The manufacturer should supply the quantity of foam required if so stated in purchasing order. In addition if so stated, the manufacturer should indicate the quantity of reserve supply and the ways and conditions required for storage of foam.

10.18.7 Dry-chemical powder

While dry powder of potassium bicarbonate, or Monnex is preferred, any sodium bicarbonate or potassium-sulphate base compatible with AFFF foam and suitable for class "B" and "C" if specified by buyer could be used. The dry chemical supplied should be guaranteed not to accelerate the break down or interact with the foam supplied under this specification.

10.19 System operation and control

10.19.1 Actuation

The system should be actuated manually by pull box at local and remote (hose-reel station) by hand manual. The valving and piping should be installed so that for normal operation the nitrogen from the cylinders passes through the regulator, manifolds, and piping into the agent tank to adequately fluidize and pressurize the tank. The flow of the agent from the tank into the hose should be controlled by a ball-type valve. The action system should be designed so that both agents can be discharged simultaneously by the operator.

10.19.2 Operating devices

Operating devices should include nitrogen pressure-regulator discharge control with quick-opening ball valves, shut-down equipment, actuation valves, hose reel, and manual overrides and agents nozzles with shut-off valves. All operation devices considered as integral parts of the system should function with system operation.

10.19.3 Alarms

A visual and an audible alarm should be provided if desired by the buyer to summon aid.

10.20 Paint and finishing

Paint and finishes should be desirable on manufacturer's standards and should adequately protect all pieces from their environment. However, corrosive, humid, and chemical conditions should be considered and the equipment should be painted with proper corrosion-resistant primer and final coating for hot, humid, corrosive, and unshaded conditions. The company's color codes should be used.

- Marking: Each twin-agent extinguishing system should be identified with permanently attached corrosion-resistant nameplate. The nameplate should be located where it can be easily visible after installation. The nameplate should contain the following information: Manufacturer's name or private labeler or its identifying symbol.
  Purchase order, item number and tag number.
  Capacity of each agent's tank and operating pressure.
  Nitrogen cylinders capacity and pressure.
  A marking plate should be permanently attached to the most suitable location easily visible to show "how to operate the extinguisher".

10.21 Items to be furnished by manufacturer

The following should be furnished by the manufacturer as the minimum requirements:

Units including foam solution tank, dry-chemical tank, nitrogen propellant system, manual actuation devices, extinguishing agents (if specified by buyer) and regulators.
Two 20 meters twinned hose lines with related hose reels, one for unit and one for extra hose-reel station if specified.
Both agents discharge nozzles or twin discharge nozzles as specified.
Fabrication drawing including piping layout supports, installation details, and wiring diagram (if any).
All operating devices that are necessary to match with alarms, if the system includes them.
All connections that are necessary to connect the system as a unit.
Operation, maintenance instructions, and spare parts lists.
System test procedures, initial and periodic, including special tools for this purpose.
Bolts, nuts, washers, clamps, gaskets, etc., required for assembling and mounting. The supplied materials should be suitable for the environment conditions stated in the specification.

- Testing: The manufacturer should quote all the normal tests required as well as hydrostatic and simulation tests as specified below. All pressure-containing parts of twin-agent fire extinguisher system should be subject to hydrostatic test at a pressure of no less than 1½ times the design pressure of that part. The system should be given prior to shipment an operational test giving actual design operating conditions as closely as possible, with extinguishing discharge. Manufacturer should conduct such tests in the presence of buyer or representative unless a written waiver is given. The buyer should have in hand from the manufacturer notification of test 30 days prior to date:
  Defective parts, if present, should be replaced with new parts and system retested, until completely reliable and accepted.

Manufacturer should furnish all equipment, materials, and manpower required for the test. The manufacturer should issue instruction for the proper initial and periodical testing of installed system without loss of extinguishing agent and should provide any special equipment required for calibration and testing during operation of the system.

Test information and result as specified below should be provided in a letter certifying that the system was tested and met all requirements specified. It should include:

1. Date of test.
2. Purchase order, item number, tag number.
3. Shipping destination.
4. Equipment serial number.
5. Test procedure(s).

Officially certified summary of test observation results and conclusions, any mulfunctioning and, or system connections should be reported. In addition photographs of the system should be furnished.

- Inspections: If so desired the buyer's representative should be offered the opportunity to witness the mulfunctioning, testing, assembly, or any part of the manufacturer's work that concerns the system ordered. The manufacturer should agree, by his acceptance of the purchase order, to carry any inspection and rejection stipulations in accordance with standard practices and codes specified herein. Any inspection and testing in no way relieve the manufacturer of any responsibility for the system meeting all requirements of this specification and applicable codes. The manufacturer should issue instructions for the proper inspection of the system, according to acceptable international and NFPA standards.

- Information to be Furnished by Manufacturer: The manufacturer should furnish with his quotation at least the following information:

Manufacturer names and model numbers.

Comprehensive catalogs, technical data, and descriptive literature of the equipment offered.

An explicit statement of any deviation from this specification.

List of spare parts for commissioning and two years operation with prices.

Preliminary dimensional drawing and description of operation.

List of all necessary tests with price including those specified herein.

List of recommended special tools for installation and future maintenance and their price.

- At ordering stage: The manufacturer should furnish the buyer within 6 weeks after receipt of purchasing order, the following information:

Five sets of drawing of the system and its components. The fabrication should not start until after manufacturer receipt of approved drawings. Vendor should supply one set of corrected drawings within weeks after receipt of drawings that have been approved or marked (approved).

Manufacturer should furnish buyer the following information prior to the shipment:

1. Ten copies of test certification. This will be prerequisite for final acceptance and invoice approval.
2. Five sets of recommended spare parts list for commissioning, 2 years of operation, and a list of special tools for stock.
3. Five sets of installation, maintenance, and operating instructions including comprehensive troubleshooting instructions.
4. Five copies of certified outline drawing.

- Shipment: Each package unit and related equipment should be properly prepared for transit to prevent damage from handling, warehousing or shipping and should be labeled to insure that it is not lost in transit. In addition the following measures should be taken:

  All external connections should be protected by temporary closures to exclude dirt and other foreign matter.

  One packing list to be included inside every package and one packing list to be in metal enclosure attached to the package.

  Adequate shipping supports and packing should be provided in order to prevent internal damage during transit.

  For ocean transport, the equipment should be crated in heavy-duty container sealed with strong tape or metal bands. Also provision should be taken to protect the equipment from possible marine exposure.

Glossary of terms

3.8 Chemical Protective Clothing   For the purpose of this standard the following definitions shall apply:

Acceptance Tests   Tests made at the time of apparatus commissioning to assure the Purchaser that the pump meets the performance requirement of the purchase contract.

Acoustic Test Fixture (ATF)   A device that approximates certain dimensions of an average adult human head and is used for measuring the insertion loss of ear muffs. For this purpose it includes a microphone arrangement for measuring sound pressure levels.

Air-Impermeable materials   Materials through which permanent gases cannot pass except by undergoing a process of solution.

Air-Supplied clothing   Clothing that is fitted with facilities for the entry of air that may provide for respiration and/or for thermal conditioning of the user. Air supplied clothing may provide complete cover either to the whole or to part of the body according to the circumstances of use.

Airway   Passageway for gas into and out of the lungs.

Aluminized Clothing   Aluminized Clothing with range of Ceramic Fiber (1450° C) clothing made of aluminized coated flameproof fabric that reflects and insulates heat and fire for short period of time and are of two types:

Fire proximity or reflective suit used in proximity of high temperature where flame is not entered or is designed to provide protection against conductive, convective and radiant heat.

Entry clothing protective clothing that is designed to provide protection from conductive, convective and radiant heat and permit entry into flame.

Anchorage Line   A rigid or flexible line secured to a structure to which a positioning device may be secured.

Apparatus for Gamma Radiography   An apparatus including an exposure container and accessories designed to enable radiation emitted by a sealed source to be used for industrial radiography.

Aqueous Film Forming Foam (AFFF)   The characteristic of a foam or foam solution forming an aqueous film on some hydrocarbon liquids.

Aspirated Foam   Foam produced by the mixing of air and foam solution within the equipment.

Assemblies forming a modular system   In this case the parts are not necessarily put together by the manufacturer of the assembly and placed on the market as a single functional unit. The manufacturer is responsible for the compliance of the assembly with the directive as long as the parts are chosen from the defined range and selected and combined according to his instructions.

Assemblies with a fully specified configuration of parts   These are put together and placed on the market as a single functional unit by the manufacturer of the assembly. The manufacturer assumes responsibility for compliance of the integral assembly with the directive and must therefore provide clear instructions for assembly/installation/ operation/maintenance, and so forth. The EC declaration of conformity as well as the instructions for use must refer to the assembly as a whole. It must be clear which is/are the combination(s) that form(s) the assemblies.

Assembly   A combination of two or more pieces of equipment, with components if necessary, placed on the market and/or put into service as a single functional unit.

Associated apparatus   Electrical apparatus that contains both intrinsically safe and nonintrinsically safe circuits and is constructed so that the nonintrinsically safe circuits cannot adversely affect the intrinsically safe circuits.

Assumed maximum area of operation (AMAO)   The maximum area over which it is assumed, for design purposes, that sprinklers will operate in a fire.

Assumed maximum area of operation, hydraulically most favorable location   The location in a sprinkler array of an AMAO of specified shape at which the water flow is the maximum for a specific pressure.

Assumed maximum area of operation, hydraulically most unfavorable location   The location in a sprinkler array of an AMAO of specified shape at which the water supply pressure is the maximum needed to give the specified design density.

Attenuation   The algebraic difference in db between the 1/3 octave band pressure level, as perceived by a real ear at threshold in a specified sound field under specified conditions, with the hearing protector absent and the sound pressure level with the hearing protector being worn, with other conditions identical.

Automatic fire hose reel assembly   A firefighting appliance consisting essentially of a reel, inlet pipe and automatic valve, hose, shut-off nozzle and where required a hose guide.

Automatic Resuscitator   Resuscitator in which the cyclic flow of gas for inflation of the lungs is independent of any inspiratory effort of the patient or repetitive action of the operator.

Automatic/manual or manual-only changeover device   A device that can be operated before a person enters a space protected by a fire-extinguishing system preventing the fire detection system from activating the automatic release of carbon dioxide.

Auxiliary equipment   Listed equipment used in conjunction with the dry chemical systems (i.e., to shut down powder, fuel, or ventilation to the hazard being protected or to initiate signaling devices).

Back Leak   Volume of expired gas that does not pass through the expiratory port but returns to the resuscitator.

Back tacking   Sewing and reverse sewing at the beginning or end of a seam to secure the stitching.

Backing lens   A transparent plate used between the eye and the welding filter.

Bag Inlet Valve   Valve activated by the sub-atmospheric pressure in the compressible unit of the resuscitator to refill the compressible unit with gas at ambient pressure.

Bag Refill Valve   Valve, with no manual trigger, activated by the sub-atmospheric pressure in the compressible unit of the resuscitator to refill the compressible unit from a compressed gas source.

Balanced system   A powder fire extinguishing system, with more than one discharge nozzle, in which the powder flow divides equally at each junction in the pipework.

Bar tacking   Reinforcement by means of stitching at point of stress e.g., button holes, pocket corners, seam ends and loops.

Bark Pocket   An opening between annual growth rings that contains bark. Bark pockets appear as dark streaks on radial surfaces and as rounded areas on tangential surfaces.

Basic eye-Protector   An eye-protector that satisfies the minimum requirements, but does not give the additional protection detailed in the following 5 Clauses 10

   1) Impact eye-protector   An eye-protector able to withstand the impact test to grade 1 or grade 2 and providing lateral protection to the orbital cavities. Grade 1 impact eye-protectors are able to withstand a velocity of impact of 120 m/s, and grade 2 are able to withstand 45 m/s.

   2) Molten metals eye-protector   An eye-protector that provides protection against molten metal splash and hot solids.

   3) Gases eye-protector   An eye-protector that provides protection against gases and vapors.

   4) Dusts eye-protector   An eye-protector that provides protection against dusts.

   5) Liquids eye-protector   An eye-protector that provides protection against splashes or droplets of liquids.

Bell-nipple   A short piece of pipe at the entry to a well that is belled at the top to guide tools into the hole. Usually has side connections for the fill-up and mud return lines.

Blending Tanks   A tank used for any mixture prepared for the special purpose "e.g." the product of a refinery are blended for marketing.

Blow down Drums   A stock into which the contents of a unit are emptied in an emergency.

Blowout   An uncontrolled and often violent escape of reservoir fluids from a drilling well when a high-pressure reservoir has been encountered and efforts to prevent or control the escape have failed. Production wells can also blow out due to surface equipment failure or if well servicing operations get out of control.

Boiling point   The temperature of a liquid at which the vapor pressure of the liquid equals the atmospheric pressure.

Booster pump   An automatic pump supplying water to a sprinkler system from an elevated private reservoir or a town main.

Bound seam   A seam having its material edges bound with a strip of additional material.

Boundary   Boundary of the equipment is the term used in a processing facility by an imaginary line that completely encompassed the defined site. The term distinguishes areas of responsibility and defines the processing facility for the required scope of work.

Bounding area   The area of the real or notional surface (sides, bottom, and top) of an enclosure around a hazard protected by a total flooding system.

Breathing Apparatus   An apparatus that enable the user to breath independently the immediate atmosphere with limits set out in this standard.

Breathing Tube   A flexible tube through that air or oxygen flows to the facepiece of a respirator or breathing apparatus.

Breeching Attachment   An attachment permitting a pole belt to be worn so that the load on the user is taken by the buttocks. It consists of a waist belt with two or more droppers to which the pole belt is attached.

Brim   An integral part of the shell extending outward over the entire circumference.

Canister   A container of materials that will remove certain contaminants in the air passing through them.

Capsule   Protective envelope used to avoid any damage to actual source and for easy handling.

Cargo   In this standard refers to liquids having flash point below 60°C.

Cartridge   A small, sealed, replaceable unit containing materials that will remove certain contaminants in the air passing through them.

Cement process   Such process that the periphery of upper is lasted to the insole, adhesive is coated on the periphery of upper and periphery of outsole, and then the outsole is bottomed using a sole press machine.

Certification Tests   Test made at the manufacturers plant of pumps and witnessed by the representative of a testing organization and approved by the Iranian Oil, Gas and Petrochemical Industries.

Chain stitch   A stitch formed with one or more needle threads and characterized by intralooping.

Check   A separation of the wood along the fiber direction that usually extends across the rings of annual growth, commonly resulting from stresses set up in the wood during seasoning.

Chemical hazard   The potential of a chemical, derived from the intrinsic properties of the chemical, to cause harm to the human body by contact with the skin.

Chin strap   An adjustable strap that fits under the chin to secure the helmet to the head.

Class C nozzle   A nozzle that produces a spray having a minimum cone angle of 30 degrees.

Closed-circuit Apparatus   An apparatus in which the exhaled air is rebreathed by the user after the carbon dioxide has been removed and a suitable oxygen concentration restored.

Clute patterns   A four-finger and thumb design, having one-piece palm, including the fronts of all four fingers and a separate cuff (see Fig. 1(a)). The back comprises three of four separate pieces of material.

Collimator   A device used for restricting the useful radiation beam to a specific direction and size.

Combustible dust Finely divided solid particles, 500 μm or less in nominal size, which may be suspended in air, settle out of the atmosphere under their own weight, burn or glow in air, and form explosive mixtures with air at atmospheric pressure and normal temperatures.

Commissoning Test Tests made at the time that apparatus is installed at its location and fixed with fire water piping and hydrant system.
Portable pumping unit is tested immediately after receiving at the site, before putting it into service.

Competent Authority An authority appointed by the Atomic Energy Organization of Iran.

Components Any item essential to the safe functioning of equipment and protective systems but with no autonomous function.

Compound Gauge A gauge that indicates pressure both above and below atmospheric pressure.

Compressed Air Line Apparatus An apparatus by which the user is supplied from a source of compressed air.

Compression failure A deformation (buckling) of the fibers due to excessive compression along the grain. This deformation may appear as a wrinkle across the surface. In some cases, compression failures may be present but not visible as wrinkles; in such cases they are often indicated by "fiber breakage" on end grain surfaces.

Compression wood An aberrant (abnormal) and highly variable type of wood structure occurring in softwood species.

Concentration The percent of foam concentrate contained in a foam solution. The type of foam concentrate being used determines the percentage of concentration required. A 3% foam concentrate is mixed in a ratio of 97 parts water to 3 parts foam concentrate to make foam solution. A 6% concentrate is mixed with 94 parts water to 6 parts foam concentrate.

Concentration The percent of foam concentrate contained in a foam solution.

Conductive dust Dust with electrical resistivity equal to or less than 103 Ωm.

Connector Any single item or arrangement of items that connects the safety belt or harness to the appropriate connecting feature on the positioning device or anchorage line including any tail on the harness or belt.

Constant flow (gallonage) spray nozzle An adjustable pattern nozzle in which the flow is delivered at a designed nozzle pressure. At the rated pressure the nozzle will deliver a constant gallonage from straight stream through a wide spread pattern. This is accomplished by maintaining a constant orifice size during flow pattern adjustment.

Constant pressure (automatic) spray nozzle An adjustable pattern nozzle in which the pressure remains constant through a range of flows. The constant pressure provides the velocity for an effective stream to reach at various flow rates. This is accomplished by means of a pressure activated self-adjusting orifice.

Constant/select flow (gallonage) feature A feature of a nozzle that allows on-site manual adjustment of the orifice to change the flow rate to a predetermined flow. The flow remains constant throughout the pattern range of pattern selection from straight stream to wide spray.

Construction classification numbers   A series of numbers from 0.5 to 1.50 that are mathematical factors used in a formula to determine the total water supply requirement of this book.

Container   General term designating any enclosure that may surround a sealed source.

Contaminant   Harmful or nuisance dusts and gases.

Contamination, Radio-active   The presence of a radio-active substance or substances in or on a material or in a place, where they are undesirable or could be harmful.

Continuous grade source of release   A source that will release continuously or is expected to release for long periods or for short periods that occur frequently.

Control Console   Panel having controls and indications for the potential, current, exposure time of the X-ray tube and any other parameter.

Coupling   A device for connecting lengths of hose so as to secure continuity from the source of a water supply to the delivery point.

Cover lens   A transparent cover used in front of the welding filter as a protection against welding splatter, etc.

Coverall   A one-piece type of legged workwear often capable of being fastened at wrist and ankle.

Crew Leader   Head of a crew assigned to a fire truck.

Cross grain (slope of grain)   A deviation of the fiber direction from a line parallel to the sides of the piece. Cross grain may be diagonal or spiral, or both.

Crown straps   The part of the suspension that passes over the head.

Cuff   The extension on a glove or mitt that covers the wrist (examples are shown in Fig. 2).

Cup   A hollow, approximately hemispherically shaped component that is mounted on the headband and to which a cushion and a liner are usually fitted.

Note: In this context, a cup is sometimes referred to as a shell.

Cushion   A deformable cover, usually foam plastics or liquid filled, fitted to the rim of the cup to improve the comfort and fit of the ear muffs on the head.

Note: In this context, a cushion is sometimes referred to as a seal.

Danger   The conceptual combination of the chemical hazard with its associated risk, taking into account the quantity of the chemical that may be released during an undesired event.

Dark shade   Shade number corresponding to the minimum value of luminous transmittance ôd. (see BS. 679).

Dead man control   A control that shut-off or significantly reduces water flow when force is released from it.

Decay   The disintegration of wood due to the action of wood-destroying fungi; also known as dote and rot.

Decibel (db)   One-tenth of a bel, a scale unit used in comparison of the magnitude of powers. The number of bels, expressing the relative magnitudes of two powers, is the logarithm to the base 10 of the ratio of the powers.

Deck   A platform in a ship.

Deck House   A super structure (as a cabin) built on the upper deck of a ship but not extending to the sides.

Deck Lights   A piece of heavy glass set in a ship deck or hull to admit light.

Degrees of flammability hazards   The degrees of hazards are ranked according to the susceptibility of materials to burning and are numbered from 4 to 0. Number 4 has a severe hazard that materials will burn readily and number 0 indicates that materials will not burn.

Degrees of health hazards   Degrees of hazards are ranked according to the probable severity of hazard to personnel and are numbered from 4 to 0. Number 4 has a severe hazard degree and number 0 offers no hazard degree.

Degrees of reactivity hazards   The degrees of hazards are ranked accord-ing to ease, rate, and quantity of energy released, and are numbered from 4 to 0. Number 4 is for materials that readily detonate or explode at nor-mal temperature and pressure and number 0 is for materials that are normally stable and not reactive with water.

Delivered Oxygen Concentration   Average concentration of oxygen in the gas delivered from the resuscitator.

Deluge installation   An installation or tail-end extension fitted with open sprayers and either a deluge valve or a multiple control arrangement so that an entire area is sprayed with water on operation of the installation.

Deluge valve   A valve suitable for use in a deluge installation. Note that the valve is operated manually and usually also automatically by a fire detection system.

Demand Valve   A device fitted in a breathing apparatus whereby the user receives air on demand from an air supply.

Design density   The minimum density of discharge, in mm/min of water, for which a sprinkler installation is designed, determined from the discharge of specified group of sprinklers, in L/min, divided by the area covered, in m^2.

Design point   A point on a distribution pipe of a precalculated installation downstream of which pipework is sized from table and upstream of which pipework is sized by hydraulic calculation.

Detector sprinkler   A sealed sprinkler mounted on a pressurized pipeline used to control a deluge valve. Operation of the detector sprinkler causes loss of air pressure to open the valve.

Device   Any piece of equipment designated to utilize sealed source(s).

Diffusion apparatus   An apparatus in which the transfer of gas from the atmosphere to the gas sensing element takes place by diffusion (i.e., there is no aspirated flow).

Diffusion instrument   An instrument in which the transfer of gas from the atmosphere to gas sensor take place by diffusion. There is no aspirated flow.

Direct vulcanizing process   Such process that the periphery of upper is lasted to the insole, then this assembly and the shoe bottom are set in a vulcanizing press machine, un-vulcanized rubber is introduced into the machine and the shoe bottom is fixed to the upper by vulcanizing the introduced rubber by heating while the said components are pressed.

Note: The shoe bottom means outsole and heel.

Discharge Device   Fixed, semifixed or portable devices, such as foam chamber, fixed foammakers, monitors nozzles, spray nozzles and sprinklers that direct the flow to the fire or flammable liquid surface.

Discharge Pressure   Suction pressure plus differential pressure that pump is able to develop when operating and is determined by gauge.

District Firemaster   Heads of a fire services in an area or district.

Drainage Time   The time for defined percentage of the liquid content of a foam to drain out under specified conditions.

Droppers   That part of the breeching attachment providing the connecting links between the waist belt and the pole belt.

Dual shade filter   A type of welding filter, part of which is made in a lighter shade and allows the welder to set up the work with a helmet or headshield in position before starting the welding operation; during welding the welder views the process through the darker part of the filter.

Dummy Sealed Source   Facsimile of a radio-active sealed source the capsule of which has the same construction and is made with exactly the same materials as those of the sealed source that it represents but containing, in place of the radio-active material, a substance resembling it as closely as practical in physical and chemical properties.

Dynamic Suction Lift   The sum of the vertical lift and the friction and entrance loss due to the flow through the suction strainer and hose.

Ear Muffs   A hearing protector, either fitting over and enclosing the pinna and sealing against the side of the head (circumaural) or sealing against the pinna (supraaural). Over-the-head, and ear muffs are designed to be worn with the headband passing over the top of the head and behind the head respectively. A head strap supports behind-the head ear muffs by being in contact with the top of the head. Universal ear muffs can be worn in either mode.

Ear Plugs   A hearing protector inserted and worn in the ear canal or in the ear cavity.

- Disposable   Intended for one fitting only.

- Reusable   Intended for more than one fitting.

- Sonic ear plug   Insert tipe ear protector utilizing a moving diaphragm, that attenuate harmful high level noises without blocking normal background sound.

Effective Duration   The time for which the apparatus can be expected to function satisfactorily. This time will be equal to the working duration plus a reserve period of at last 10 minutes for apparatus of less than 45 minutes working duration and 15 minutes for working duration between 45 and 75 minutes.

Escape Breathing Apparatus   The apparatus that is intended for escape purpose only from irrespirable atmosphere.

Exhalation Valve   A nonreturn valve to release exhaled air.

Expansion   The ratio of air to water in foam. A measure of the volume of foam produced for each volume of foam solution used.

Expansion Ratio   The ratio of the volume of foam to the volume of foam solution from that it was made.

Expellant gas   The medium used to discharge dry chemical from its container (i.e., CO_2 and nitrogen).

Expiratory Port   Opening through which gases and/or vapors pass from the patient during expiration.

Explosion-proofing Any electrical equipment used in hazardous area including gas detection equipment must be tested and approved to ensure that even under fault condition it cannot initiate an explosion.

Explosion-protected apparatus Any form of apparatus with a recognized type of protection.

Explosive gas atmosphere A mixture with air under normal atmospheric conditions of flammable materials in the form of gas, vapor, or mist, in which, after ignition, combustion spreads throughout the unconsumed mixture.

Explosive limit Lower explosive limit (LEL) The concentration of flammable gas, vapor, or mist in air, below which an explosive gas atmosphere will not be formed

Upper explosive limit (UEL) The concentration of flammable gas, vapor, or mist in air, above which an explosive gas atmosphere will not be formed

Explosive range The range of gas or vapor mixture with air between the explosive (flammable) limits over which the gas mixture is explosive.

Exposure Container A shield in the form of a container designed to allow the controlled use of gamma radiation and employing one or more gamma radiography sealed sources. For the purpose of this I.P.S, an apparatus for gamma radiography is classified according to the mobility of the exposure container:

Class P A portable exposure container, designed to be carried by one man alone.

Class M A mobile but not portable exposure container, designed to be moved easily by a suitable means provided for the purpose.

Class F A fixed installed exposure container or one with mobility restricted to the confines of a particular working area.

Exposure Head A device that locates the gamma radiography sealed source in the selected working position.

Extending Ladder A leaning ladder consisting of two or three sections constructed so that the height can be varied, in increments of one rung spacing, by sliding the sections relative to each other.

Extl Ex testing laboratory

Eye Protection For the purpose of this section of standard the following definitions apply.

Eyepiece A gas-tight, transparent window(s) or lens(es) in a full facepiece through which the user can see.

Eye-protector Any form of eye-protective equipment covering at least the region of the eyes.

Face screen An eye-protector covering all or a substantial part of the face.

Face shield A device worn in front of the face to give protection to the eyes, face and throat. It is either made of the material of the filter itself or is fitted with the filter(s) and, when provided, the filter cover(s).

Facepiece A mask fitting to the face covering the nose and mouth. There are two types:

Half-mask (or nosal facepiece) covering the nose and mouth.

Full facepiece covering the eyes nose and mouth.

Faceseal A flexible lip or pneumatic cushion sealing the facepiece to the face.

Fastener   A device for attaching pipe hanger components to a building structure or racking.

Fault signal   An audible, visible, or other indication that instrument is not working satisfactorily.

Filling density   The ratio of mass of carbon dioxide charged in a container to the container volume.

Film Forming Fluoroprotein (FFFP)   A liquid concentrate that has both a hydrolyzed protein in fluorinated surfactant base plus stabilizing additive.

Filter   The part of an eye protector through which a user sees and that is designed to reduce the intensity of incident radiation.

Fire   Process of combustion characterized by the emission of heat accompanied by smoke or flame or both
Combustion spreading uncontrolled in time and space

Fire and gas detection system (FGDS)   The combination of a fire and gas detection system connected to emergency shutdown system and also activating automatic extinguishing systems.

Fire detection system   Fire detectors and associated control panel to detect and alarm to personnel for evacuation of the plant area and building as well as to indicate the location of the incident to fire brigade to proceed to the scene of the incident (if available).

Fire hazard   Any situation process, material, or condition that on the basis of applicable data may cause a fire or explosion or provide a ready fuel supply to augment the spread or intensity of the fire or explosion and that poses a threat to life or property.

Fire hazard properties   Properties measured under laboratory conditions may be used as elements of fire risk assessment only when such assessment takes into account all of the factors that are pertinent to the evaluation of the fire hazard of a given situation.

Fire hose   A woven jacketed lined flexible conduit for conveying water for firefighting.

Fire Hydrant (Underground Fire Hydrant)   An assembly contained in a pit or box below ground level and comprising a valve and outlet connection from a water supply main.

Fire Hydrant Pillar   A fire hydrant whose outlet connection is fitted to a vertical component projecting above ground level.

Fire point   The lowest temperature at which a liquid in an open container will give off sufficient vapors to burn when once ignited. It generally is slightly above the flash point.

Fire Resisting Curtain   A fixed wall type curtain fixed above the proscenium opening that in case of stage fire automatically closes without the use of applied power.

Fire Resisting Shield   A local made shield to be used by fire fighters to combat intense flame and heat such as oil well fire.

Fire water hydrant   A device with suitable valves by which water is discharged from a water main.

Firedamp A combustible gas formed in coal mines.

Fire-resistive Fire resistance rating, as the time in minutes or hours, that materials or assemblies have to withstand a fire exposure as established in accordance with the test of NFPA 251.

First Aid Immediate treatment by a first aider of an ill or injured person in an emergency before arrival of a physician/surgeon, such as artificial respiration, bandaging, massaging, and use of slings, splints, tourniquets, stretchers, antiseptics, emetics etc.

Flame resistance The property of a material whereby flaming combustion is prevented, terminated, or inhibited following application of a flaming or nonflaming sources of ignition, with or without subsequent removal of the ignition sources. Flame resistance can be an inherent property of the textile material, or it may be imparted by specific treatment.

Flame spread Low flame spread means that the surface thus described will adequately restrict the spread of flame with regard to the risk of fire in the spaces concerned, which is determined by an acceptable test procedure.

Flameproof (EU) A type of protection in which the parts that can ignite an explosive atmosphere are placed in an enclosure that can withstand the pressure developed during an internal explosion of an explosive mixture and that prevents the transmission of the explosion to the explosive atmosphere surrounding the enclosure flash point. The low-est liquid temperature at which a liquid gives off vapors in a quantity such as to be capable of forming an ignitable vapor/air mixture.

Flammability hazard It is the degree of susceptibility of materials to burning. Many materials that will burn under one set of conditions will not burn under others. The form or condition of the material as well as its inherent properties affects the hazard.

Flammable (explosive) limits All combustible gases and vapors are characterized by flammable limits between which the gas or vapor mixed with air is capable of sustaining the propagation of flame. Lower limit and upper limit are usually expressed as percentage of the material mixed with air by volume.

Flammable (explosive) range The range of flammable vapor or gas air mixture between the upper and lower flammable limits is known as the flammable range or explosive range.

Flammable gas or vapour Gas or vapor that, when mixed with air in certain proportions, will form an explosive gas atmosphere.

Flammable liquid A liquid capable of producing a flammable vapor or mist under any foreseeable operating conditions.

Flammable material Material consisting of flammable gas, vapor, liquid, and/or mist (see also Appendix C).

Flammable mist Droplets of flammable liquid dispersed in air so as to form an explosive atmosphere.

Flash point   Minimum temperature of liquid that it gives off sufficient vapor to form an ignitable mixture with the air near the surface of the liq-uid or within the vessel used.

Flock lined (rubber or PVC) gloves   Gloves that have their inner surface covered in a layer of pure cotton fibers, anchored into the rubber of PVC during manufacture. These absorb perspiration and help to keep the hands cool during use in a warm environment; conversely, they contribute to warmth when used in a cold application.

Fluoroprotein   Conventional protein foam modified by the addition of fluorocarbon surfactants.

Fluoroprotein (FP)   A liquid concentrate that is similar to protein, but with one or more fluorinated surfactant additive.

Foam   A firefighting agent made by mechanically mixing air with a solution consisting of fresh or salt water to which a foam liquid concentrate has been added.

Foam   A mass of bubbles formed by the mechanical agitation of foam and water solution.

Foam branches   Portable devices that are hand held during use. A wide variety of foam branches are made available in the market.

Foam Concentrate   The liquid foaming agent as received from the manufacturer and used for mixing with the recommended amount of water and air to produce foam. This term as used in this standard includes concentrates of the following types and film forming fluoroprotein (FFFP).:
Protein Foam, Fluoroprotein Foam, Aqueous Film Forming Foam (AFFF, and other Synthetic Foams).

Foam concentrate   A concentrated liquid foaming agent as received from the manufacturer.

Foam Concentrate Proportioner   A means for controlling the ratio of foam concentrate to the quantity of water.

Foam Inlet   Fixed equipment consisting of an inlet connection, fixed piping and a discharge assembly, enabling firemen to introduce foam into an enclosed compartment.

Foam Solution   A mixture of a proportioned of premixed foam liquid concentrate dissolved in either fresh or salt water.

Foam solution   A homogeneous mixture of water and foam concentrate in the proper proportions.

Foam-water spray system   A foam-water spray system is a special system pipe-connected to a source of foam concentrate and to a water sup-ply, and equipped with foam-water spray nozzles for extinguishing-agent discharge (foam or water sequentially in that order or in reverse order) and distribution over the area to be protected. System-operation arrangements parallel those for foam-water sprinkler systems as described in the foregoing paragraph.
A source of release may be one of the above three grades or may be a combination of two or three, in which case it is regarded as a multigrade source of release.

Foam-water sprinkler system A foam-water sprinkler system is a special system pipe-connected to a source of foam concentrates and to a water supply, and equipped with appropriate discharge devices for extinguishing agent discharge and for distribution over the area to be protected. The piping system is connected to the water supply through a control valve that is usually actuated by operation of automatic detection equipment installed in the same areas as the sprinklers. When this valve opens, water flows into the piping system, foam concentrate is injected into the water, and the resulting foam solution discharging through the discharge devices generates and distributes foam.
Upon exhaustion of the foam concentrate supply, water discharge will follow the foam and continue until shut off manually. Systems may be used for discharge of water first, followed by discharge of foam for a definite period, and this is followed by water until manually shut off. Existing deluge sprinkler systems that have been converted to the use of aqueous film forming foam are classed as foam-water sprinkler systems.

Folding Trestles An arrangement of two frames hinged together, each fitted with cross-bearers suitable for supporting a working platform (see Appendix M Fig. E).

Forward Leak Volume of gas produced by the resuscitator during the inspiratory phase that does not pass through the patient port to the patient but passes to the atmosphere.

Fresh Air Hose Apparatus Apparatus in which air is drawn from a fresh air source with or without the assistance of a blower.

Fume Airborne particles usually less than a micrometer in size and sometime visible as a cloud or smoke.

Garment An individual item of protective equipment, the wearing of which affords protection to the skin.

Gas-Tight suit A one-piece garment with hood, gloves and boots that, when worn with self-contained or compressed air-line breathing apparatus, affords the user a high degree of protection against harmful liquids, dusts and gaseous or vapor contaminants.

Gauntlet A type of glove that, relative to the wrist glove, provides additional protection for the wrist and part of or the whole of the arm.

Goggles An eye-protector fitted with a single or two separate oculars enclosing the orbital cavities.

Goodyear welt process Such process that the periphery of upper is lasted to the rib provided on the insole, stitched to the welt by a welt stitching machine, this assembly is set on the outsole, and then the welt is lock stitched to the periphery of outsole using an outsole stitching machine.

Gradient filter A filter used in sun glare spectacle in which luminous transmittance changes progressively in the vertical meridian, when the filter is mounted, over some or all of the filter.

Gridded configuration pipe array A pipe array in which water flows to each sprinkler by more than one route.

Group i instrument   Portable, transportable, and fixed instrument for sensing the presence of combustible gas concentration with air. The instrument or part thereof may be used or installed in mines susceptible to firedamp.

Group ii instrument   Apparatus for use in potentially explosive atmosphere other than mines susceptible to firedamp.

Gunn pattern   A four-finger and thumb design, having the face of the thumb, the palm, and first (index) and fourth (little) fingers made of one or two pieces of material. The back is of one piece up to the cuff and includes the back of the four fingers at least. The fronts of the second and third fingers may be one piece each, jointed to the palm at the base of the appropriate fingers (see Fig. 1(b)). The back of the glove may be jointed.

Half-Life Period   Due to the radio-active decay the activity of a source decreases according to specific physical laws. The time in which a source loses half of its original activity (Ao) is referred to as the half-life of the source (HLT).

Half-Life, Radio-active   The time required for the transformation of one-half of the atoms in a given radio-active decay process, following the exponential law (physical half-life).

Hand shield   A device held in the hand to give protection to the eyes, face and throat. It Is fitted with filter(s) and, where provided, filter cover(s).

Hanger   An assembly for suspending pipework from elements of building structure.

Hard suction hose   A rubber reinforcement contains a rigid helix to resist collapse under vacuum.

Harmonized standards   Standards developed specifically to allow a presumption of conformity with the EHSR of ATEX 95.

Harness   The complete assembly by means of which the helmet is maintained in position on the user's head.

Hazardous area   An area in which an explosive gas atmosphere is present, or may be expected to be present, in quantities such as to require special precautions for the construction, installation, and use of electrical apparatus.

Head band   A band, usually of metal or plastics, designed to enable the ear muffs to fit securely around the ears by exerting pressure through the cushions.

Head depth   Horizontal distance between tragus and vertical line through back of head when subject is sitting erect

Head Harness   An arrangement of straps for holding a facepiece or mouthpiece securely in place.

Head height   Vertical distance between tragus and top of head when subject is sitting erect (see Fig. 3).

Head strap   A flexible strap fitted to each cup, or the headband close to the cup. It can be adjusted to support the ear muffs, usually behind-the-head types, by fitting closely to the top of the head.

Head width   Maximum width of head when subject is sitting erect (see Fig. 3).

Headband   The part of the harness that encircles the head.

Health hazard   Any property of material that either directly or indirectly can cause injury or incapacitation, either temporary or permanent from exposure by contact, inhalation, or ingestion.

Helmet A device that shields the eyes, face, neck, and other parts of the head.

Helmet A device that is worn to provide protection for the head, or portions thereof, against impact, flying particles, electric shock, or any combination thereof; and that includes a suitable harness.

Helmet A device supported on the head to give protection to the face, ears and throat and part of the top of the head. It is fitted with filter(s) and, where provided, filter cover(s).

Helmet A device covering a substantial part of the head and generally having functions other than, or in addition to, hearing protection.

Hem Producing a folded edge by turning the edge of a material and securing it.

Hermetically sealed component A component that is sealed against entrance of an external atmosphere and in which the seal is made by fusion, such as soldering, brazing, welding, or the fusion of glass to metal.

High expansion Foam having an expansion ratio higher than 200 (generally about 500).

High-flash stock Those having a closed-up flash point of 55°C or over (such as heavy fuel oil, lubricating oils, transformer oils, etc.). This category does not include any stock that may be stored at temperatures above or within 8°C of its flash point.

Hood A device that completely covers the head, neck, and portions of the shoulders.

Hood combined with a cape A garment that completely covers the head, neck and portions of the shoulders or upper part of the body.

Hose Reel Fire-fighting equipment, consisting of a length of tubing fitted with a shut-off nozzle and attached to a reel, with a permanent connection to a pressurized water supply.

Hose reel system A system including a hose, stowed on a reel or a rack, with a discharge nozzle that is manually directed and operated.

Housing The part of the equipment that supports the filter(s), filter cover(s) and backing lens.

Hydrant Outlet The component of fire hydrant to which the standpipe is connected.

Ignition temperature Minimum temperature under prescribed test conditions at which the material will ignite and sustain combustion when mixed with air at normal pressure, without initiation of ignition by spark or flame.

Increased safety A type of protection in which additional measures are applied to give increased security against the possibility of excessive temperatures and the occurrence of arcs and sparks inside and

Induction Methods that uses the venturi principle to introduce a proportionate quantity of foam concentrate into a water stream. Induction methods are:

A) Pressure Induction This method employs the water supply to pressurize the foam concentrate storage tank. At the same time, water flowing through an adjacent venture or orifice creates a pressure differential. The difference between the water supply pressure and this lower pressure area forces the foam concentrate to flow through a fixed or metering orifice into the water stream.

B) Vacuum Induction   This method utilizes the negative pressure created by water passing through a venture to draw the liquid concentrate from the storage tank or container through a pick-up tube and mix it with the water stream. 6

C) Pump-and-Motor Induction   By means of an auxiliary pump-foam compound is injected into the water stream passing through an inductor.

The resulting foam solution is then delivered to a foam maker. The proportioner may be inserted in the line at any point between the water source and foam maker.

Infrared sensor   A sensor that the operation of which depends on the absorption of infrared radiation by the gas being detected.

Injection molding process   Such process that the periphery of upper is lasted to the insole, this assembly is set on an injection molding machine, and the shoe bottom is formed by injecting un-vulcanized rubber into the mold.

In-Line Inductor   A venturi eductor, located in the water supply line to the foam maker to create a reduced pressure in piping that leads from a supply of concentrate so that the concentrate is automatically mixed with water in the required proportion. It is precalibrated and it may be adjustable.

Insertion loss   The algebraic difference in db between the 1/3 octave band pressure level, measured by the microphone of the acoustic test fixture in a specified sound field under specified conditions, with the hearing protector absent and the sound pressure level with the hearing protector on, with other conditions identical.

Insole   The inner part of footwear upon which the foot rests and that conforms to the bottom of the last.

Installation   A combination of two or more pieces of equipment that were already placed on the market independently by one or more manufacturers.

Jacket   A short coat.

Jockey pump   A small pump used to replenish minor water loss to avoid starting an automatic suction or booster pump unnecessarily.

Kerma (K)   The sum of the initial kinetic energies of charged particles produced by the interaction of uncharged radiation (e.g., electromagnetic radiation or neutrons) per unit mass of the material in which the interaction takes place. Kerma is expressed in joules per kilogram or in rads. For electromagnetic radiation absorbed in air, the principal quantity in the relationship between kerma and exposure is the average energy required to produce an ion pair.

Knot   A portion of a branch or limb, embedded in the tree and cut through in the process of lumber manufacture. It is classified according to size, quality, occurrence, and location in the cross section of a piece. The size of the knot is determined by its average diameter on the surface of the piece.

Ladder Backed Steps   A standing step ladder in which the back is fitted with cross-bearers suitable for supporting a working platform [see Appendix M (d)].

Landing Valve   An assembly comprising a valve and outlet connection from a wet or dry riser.

Last   A solid form in the general shape of a foot around that footwear is constructed.

Leakage Transfer of radio-active material from the sealed source to the environment.

Leaning Ladder A ladder supported in use by a separate structure, e.g. A wall.

Level, Reference The value of a quantity that governs a particular course of action. Such levels may be established for any of the quantities determined in the practice of radiation protection; when they are reached or exceeded, all relevant information is considered and the appropriate action may be taken.

Licensee Licensee means a person who has been issued a license by (AEOI).

Life safety A term applied to sprinkler systems forming an integral part of measures required for the protection of life.

Light shade Shade number corresponding to the maximum value of luminous transmittance ôl. (see BS 679).

Lightweight Stagings A working platform constructed of stiles, cross-bearers and decking, to provide a flat working surface.

Ligne A units 0.635 mm (1/40 inch) for measuring the diameter of buttons.

Liner Material contained within the cup that can increase the attenuation of the ear muffs at certain frequencies.

Lining An all-inclusive term used to describe all of the various lining parts used for the inside of the upper of footwear.

Liquid droplets Very small mass particles or a liquid substance capable of remaining in suspension in gas.

Load-Bearing Component Any component of a safety belt, of a safety harness, or of a safety lanyard to which a load can be applied by the user's body while working or in the event of an arrested fall.

Local application system An automatic or manual fire extinguishing system in which a fixed supply of carbon dioxide is permanently connected to fixed piping with nozzles arranged to discharge the carbon diox-ide directly to a fire occurring in a defined area that has no enclosure surrounding it, or is only partially enclosed and that does not pro-duce an extinguishing concentration throughout the entire volume containing the protected hazard.

Local artificial ventilation Movement of air and its replacement with fresh air by artificial means (usually extraction) applied to a particular source of release or local area.

Lock stitch The "plain stitch" in which two separate threads are used in formation, one thread is passed through the material, forming a loop, while the second is passed through the loop on the underside of the material.

Looped configuration A pipe array in which there is more than one distribution pipe route along which water may flow to a range pipe.

Low expansion Foam having an expansion ratio up to 20 (generally about 10).

Low-Density wood Wood that is exceptionally light in weight and usually deficient in strength properties for the species. In softwood species, low density is frequently indicated by exceptionally wide, or some times by extremely narrow, rings and generally has a low proportion of latewood. On the other hand, low density hardwood, at least in ring-porous species, is most commonly indicated by excessively narrow annual rings in which the earlywood portion predominates.

Lower explosive limit (LEL)   The concentration of combustible (flammable) gas, vapor, or mist in air below that an explosive gas/atmosphere will not be formed.

Low-flash stocks   Those having a closed-up flash point under 55°C such as gasoline, kerosene, jet fuels, some heating oils, diesel fuels, and any other stock that may be stored at temperatures above or within 8°C of its flash point.

Low-pressure storage   Storage of carbon dioxide in pressure containers at a controlled low temperature of −18°C. Note that the pressure in this type of storage is approximately 21 bar.

Low-rise system   A sprinkler system in which the highest sprinkler is not more than 45 m above ground level or the sprinkler pumps.

Main distribution pipe   A pipe feeding a distribution pipe

Manipulator Rod   A rigid rod used for remote handling of a source pencil, normally of 2 m long.

Manual   Pertaining to a fire extinguishing system that under specified conditions functions by means of intervention of a human operator.

Manual fire hose reel assembly   Firefighting appliance consisting generally of a reel, inlet pipe and manual valve, hose, shut-off nozzle and where required a hose guide.

Manual hose reel system   A manual fire extinguishing system consisting of a hose, stowed on a reel or a rack, with a manually operated dis-charge nozzle assembly, all connected by a fixed pipe to a supply of carbon dioxide.

Material conversion factor (MCF)   A numerical factor that should be used when the minimum design concentration of carbon dioxide for the material at risk exceeds 34% to increase the basic quantity of carbon dioxide as obtained by application of the volume factor required for protection against surface fires.

Maximum Rating   The maximum activity, expressed in becquerels, followed by the value in curies in brackets, of a gamma-radiography sealed source specified for a given radionuclide by the manufacturer and marked on the exposure container, and not to be exceeded if the apparatus is to conform to this standard.

Mechanical pipe joint   A component part of pipework other than threaded tubulars, screwed fittings, lead or compound sealed spigots, and socket and flanged joints used to connect pipes and to produce a seal both against pressure and vacuum.

Medium expansion   Foam having an expansion ratio between 20 and 200 (generally about 100).

Melting point   The temperature at which a solid of a pure substance changes to a liquid.

MESG   Maximum experimental safety gap

MIC   Minimum ignition current

Mitt   A covering for the hand and wrist, having a separate thumb and a common covering for the fingers.

Mobile monitor   Monitor mounted on a trailer with 2 or more 65 mm hose instantaneous male connections. The unit with hose storage bin is towed and carried to the scene of fire.

Model   Descriptive term or number to identify a specific sealed source design.

Moisture barrier The component layer designed to prevent the transfer of liquid, water form the environment to the thermal Barrier.

Monitor Firefighting cannon that is capable of discharging large amount of water/foam, or dry chemical for cooling or extinguishing fires.

Monitor system A system of fixed piping with nozzles that can be manually directed and operated locally and/or remotely.

Montpelier pattern A four-finger and thumb design, having the palm and the fronts of all four fingers in one piece, and the back of the glove and the backs of all four fingers in one piece. This pattern has a fourchette between the fingers.

Multifunction A detecting instrument that detects 0%–100% LEL, 0%–25% oxygen, 00%–25% ppm hydrogen sulfide, and 0–50 ppm carbon monoxide.

Multigrade source of release A source of release that is a combination of two or three of the above-mentioned grades and

Is basically graded continuous or primary

Gives rise to a release under different conditions that create a larger zone but less frequently and/or for a shorter duration than as determined for the basic grade

Note that different conditions mean, for example, different release rate of flammable material but under the same ventilation conditions.

A source of release that is basically graded continuous may in addition be graded primary if the rate of release of flammable material, for the primary grade frequency and/or duration, exceeds that for the continuous grade.

It may, additionally or alternatively to the primary grade, also be graded secondary if the rate of release of flammable material, for the secondary grade frequency and/or duration, exceeds that for continuous and, if applicable, the primary grade.

Similarly, a source of release that that is basically graded primary may in addition be graded secondary if the rate of release of flammable material for the secondary grade frequency and/or duration exceeds that for the primary grade.

Multiple control A valve, normally held closed by a temperature-sensitive element, suitable for use in a deluge system or for the operation of a pressure switch.

Nape strap A strap that fits behind the head to secure the helmet to the head; it may be an integral part of the headband.

Natural ventilation Movement of air and its replacement with fresh air due to the effects of wind and/or temperature gradients

Neck shield An article of protective clothing that, when fitted to a helmet, affords protection from reflected radiation to the back and sides of the head and neck.

No ventilation No ventilation exists where no arrangements have been made to cause air replacement with fresh air.

Node A point in pipework at which pressure and flow(s) are calculated; each node is a data point for the purpose of hydraulic calculations in the installation.

Nonaspirated Foam Foam produced by the mixing of air and spray of foam solution, out-side the equipment.

Nonconductive dust Combustible dust with electrical resistivity greater than $10^3\,\Omega$m.

Nonhazardous area   An area in which an explosive gas atmosphere is not expected to be present in quantities such as to require special pre-cautions for the construction, installation, and use of electrical apparatus.

Nonincendive circuit   A circuit in which any arc or thermal effect produced under intended operating conditions of the equipment is not capable, under the test conditions specified, of igniting the specified flammable gas- or vapor-air mixture.

Nonincendive component   A component having contacts for making or breaking an incendive circuit and the contacting mechanism should be constructed so that the component is incapable of igniting the specified flammable gas- or air-air mixture. The housing of a nonincendive component is not intended to exclude the flammable atmosphere or contain an explosion.

Nonincendive field circuit   A circuit that enters or leaves the equipment enclosure and that under intended operating conditions is not capable, under the test conditions specified, of igniting the specified flammable gas- or air-air mixture or combustible dust.

Nonleachable   Term used to convey that the radio-active material in the form contained in the source is virtually insoluble in Water and is not convertible into dispersible products.

Nonsparking apparatus   Apparatus that has no normally arcing parts or thermal effects capable of ignition. Normal use excludes the removal or insertion of components with the circuit energized.

Nonstochastic Radiation Effects   Radiation effects for which a threshold exists above which the severity of the effect varies with the dose.

Normal operation   The situation when the plant equipment is operating within its design parameters.

Minor releases of flammable material may be part of normal operation. For example, releases from seals that rely on wetting by the fluid being pumped are considered to be minor releases.

Failures (such as the breakdown of pump seals, flange gaskets, or spillages caused by accidents) that involve repair or shutdown are not considered to be part of normal operation.

Nozzles discharge rating   A valve expressed as an observed flow rate at a preselected pressure for example 2250 LPM at 7 bar.

Nuisance Dust Course   Nontoxic particles.

Occupancy   Fraction of total time spent by radiation worker/members of the public in the radiation field.

Occupancy hazard classification number   A series of numbers from 3 through 7 that are mathematical factors to be used in calculating total water supplies for firefighting and fire protection.

NFC standards has allocated number 3, which is the lowest occupancy hazard number, as the highest hazard grouping and number 7, which is the highest occupancy hazard number, as the lowest hazard grouping.

Ocular   The transparent part of the eye-protector that permits vision, for example, lens, visor, screen.

Offshore installation   The term used to describe any offshore unit for the drilling or producing oil or gas.

One-Finger mitt   A covering for the hand and wrist, having a separate thumb and forefinger and a common covering for the remaining fingers.

Open Circuit   Compressed air carried in the cylinder is fed through a demand valve and breathing tube to a full facepiece. Inhaled air passes through a nonreturn valve to the atmosphere.

Open-ended pipework   Pipework between a valve (including a relief valve) and open nozzle that cannot be under a continuous pressure.

Open-Field Radiography   Radiography operations carried out on shop floors, erection sites or other such areas with provisions for adequate radiological safety for the radiography personnel and others including members of the public.

Open-path infrared sensor   A sensor that is capable of detecting gas at any location along an open path traversed by an infrared beam.

Orbital cavities   The apertures in the skull in which the eyes and their appendages sit.

Ordinary   General masonry walls with wood roof and/or wood floors; also all frame construction.

Oscillating monitor   A self contained sweep protection water powered or electric powered monitors.

Outsole and heel   The bottom surface of footwear that is exposed to wear.

Overall   (Work wear) usually designed to be worn over the everyday clothes to give protection to the body and part of leg.

Oxygen Deficiency   Air containing insufficient oxygen to support life.

Particulate   Occurring in the form of minute separate particles, such as dusts, fume and mist.

Particulate Matter   A suspension of fine solid or liquid particles in air, such as dust, fog, fume, mist, smoke, or sprays. Particulate matter suspended in air is commonly known as an aerosol.

Peak   An integral part of the shell extending forward over the eyes only.

Penetration   The passage of chemicals, in any physical form, from the outside of the clothing to the inside via essential openings, fastenings, seams, overlaps between items, pores and any imperfections in the materials of construction.

Permeation   A combined process of molecular diffusion of a chemical through a solid material forming the whole or part of clothing and its desorption into a specified medium.

Photochromic Filter   A filter used on sun glare spectacle that reversibly alter its luminous transmittance under the influence of sunlight.

Pin noise   Noise whose sound pressure spectral density is inversely proportional to frequency, i.e., equal energy in each 1/3 octave band.

Head dimensions

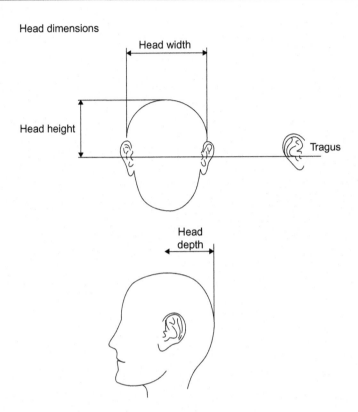

Pipe array   The pipes feeding a group of sprinklers. Note that pipe arrays may be looped, gridded, or branched.

Pipe rack   The pipe rack is the elevated supporting structure used to convey piping between equipment. This structure is also utilized for cable trays associated with electric-power distribution and for an instrument tray.

Pipe size   The international nomenclature diameter nominal-written as DN. 15-25-40-50-65-75-80 etc. Shall be used for pipe size (internal diameter).

Pitch pocket   An opening extending parallel to the annual growth rings that contains, or that has contained, either solid or liquid pitch.

Plot plan   The plot plan is the scaled plan drawing of the processing facility.

Polar Solvent Type Liquid Concentrate   A protein or synthetic based, low expansion liquid used in production of foam and intended to extinguish hydrocarbon and polar solvent (water miscible) fuel fire.

Polarizing filter   A filter used on sun glare spectacle in which the transmittance is dependent on the amount and orientation of the polarization of the incident radiation.

Pole Belt   The combination of waist belt and pole strap.

Pole Strap   The part of pole belt that is passed round a pole or similar structure.

Portable apparatus   Apparatus that is designed to be readily carried by the user from place to place as required.

Portable monitor   Portable monitors are fixed with portable base stabilizer assembly. The support legs of stabilizer assembly are remarkable for convenient storage and for mounting the base onto fire truck or trailer.

Positioning Device   A device that is normally locked onto an anchorage line and that requires manual release to permit free travel.

Premix Solution   A foam solution made by mixing foam concentrate and water in proper proportion and stored ready for use.

Premixed foam solution   Produced by introducing a measured amount of foam concentrate into a given amount of water in a storage tank.

Pressure rating   The international nomenclature pressure nominal-written as PN 20-50-68-100-150 etc. Shall be used for flange rating in this standard. (See Appendix B).

Pressure Reducer (Reducing Valve)   A device that reduces a high pressure to a normally constant low pressure.

Pressurization   The process of supplying an enclosure with a protective gas with or without continuous flow at sufficient pressure to prevent the entrance of a flammable gas or vapor, a combustible dust, or an ignit-able fiber.

Primary grade source of release   A source that can be expected to release periodically or occasionally during normal operation.

Primary Straps   Straps that take the direct load in the event of an arrested fall.

Projection Sheath   A flexible or rigid tube for guiding the source holder from the exposure container to the working position and comprising the necessary connection between the exposure container and the exposure head.

Proportioning   Is the continuous introduction of foam concentrate at the recommended ratio into the water stream to form foam solution.

Protective clothing   The combined assembly of those garments, the wearing of which affords protection to the skin.

Note: The primary function of individual garments may be to offer forms of protection other than protection of the skin as such.

Protective footwear   As used in this standard, footwear containing a protective toe box that is specially designed and manufactured to meet the performance requirements of this standard. However, protective footwear, many include, in addition many other types of protection for the user, such as metatarsal guards, antistatic properties, etc.

Protective padding   A material used to absorb the kinetic energy of impact.

Protective systems   Design units, that are intended to halt incipient explosions immediately and/or to limit the effective range of explosion flames and explosion pressures. Protective systems may be integrated into equipment or separately placed on the market for use as autonomous systems.

Protein foam (P)   A liquid concentrate that has a hydrolyzed protein base plus stabilizing additive.

Prototype Source   Original of a model of a sealed source that serves as a pattern for the manufacture of all sealed sources identified by the same model designation.

Prototype Testing   Performance testing of a new radio-active sealed source before sealed sources of such design are put into actual use.

Purging   The process of supplying an enclosure with a protective gas at a sufficient flow and positive pressure to reduce the concentration of any flammable gas or vapor initially present to an acceptable level.

Quality Control   Such tests and procedures as are necessary to establish the ability of the sealed sources to comply with the performance characteristics for that sealed source designed as defined in clause 7 of standard.

Quarter   The complete back part of the footwear upper.

Radiation Output   Number of particles and/or photons of ionizing radiation emitted per time unit from the sealed source in defined geometry. This is best expressed in terms of radiation flounce rate.

Radio toxicity   Of a radionuclide; the ability of a nuclide to produce injury by virtue of its emitted radiations when incorporated in the human body.

Radio-activity   The phenomenon exhibited by certain materials in which spontaneous emission of nuclear radiation occurs. Gamma radiation is the electromagnetic radiation emitted in such nuclear transformations.

Radiographer   A radiation worker who performs industrial radiography operations employing radiation sources and who possesses a valid certificate duly recognized or issued by the competent authority for this specific purpose.

Radiological Safety Officer (RSO)   A person who possesses valid RSO's certificate duly recognized or issued by the competent authority for this specific purpose.

Range pipe   A pipe feeding sprinklers directly or via arm pipes of restricted length.

Ranking   An official grade of position (in military).

Reactivity hazard   Susceptibility of materials to release energy either by themselves or in combination with water.

Recruit or Probational Training   Basic training.

Reel and valve subassembly   That part of the hose reel comprising the reel, inlet valve and the connection to the reel, but excluding the hose, shut-off nozzle and connectors or couplings.

Relative density of a gas or a vapour   The density of a gas or a vapor relative to the density of air at the same pressure and at the same temperature (air is equal to 1.0).

Relative vapor density   The mass of a given volume of the material in its gaseous or vapor form compared with the mass of an equal volume of dry air at the same temperature and pressure.

Remote Control   A device enabling the gamma radiography sealed sources(s) to be exposed by operation at a distance.

Remote control monitor   Monitor equipped with hydraulic motor to provide remote control of the monitor to traverse vertical and horizontal.

Respirator   A device designed to protect the user from inhalation of harmful atmospheres.

RHM/RMM   RHM is exposure rate in air expressed in rontgen per hour at 1 m from an unshielded gamma source of strength 1 curie. RMM is exposure rate in air expressed in rontgen per minute at 1 m from an unshielded gamma source of strength 1 curie.

Riser A vertical pipe feeding a distribution or range pipe above.

Rising Main, Dry (Dry Riser) A vertical pipe installed in a building for firefighting purposes, fitted with inlet connections at fire brigade access level and landing valves at specified point, which is normally dry but is capable of being charged with water usually by pumping from fire service appliances.

Rising Main, Wet (Wet Riser) A vertical pipe installed in a building for firefighting purposes and permanently charged with water from a pressurized supply, and fitted with landing valves at specified points.

Risk The probability of a specific undesired event occurring so that a chemical hazard will be realized (so as to cause harm to the unprotected user's body) during a stated period of time or in specified circumstances.

Rod Anode Extended anode of X-ray tube for intricate exposures.

Rosette (sprinkler rosette) A plate covering the gap between the shank or body of a sprinkler projecting through a suspended ceiling and the ceiling.

Rundown Tank One of the tanks in which are received the condensate from the still agitators or other refinery equipment and from which the distillates are pumped to larger tanks known as work tanks or storage tanks. Rundown tanks are also known as "pans" or receiving tanks. If the condensate were received directly into the larger storage tank, the lubing of a still would contaminate unnecessary perhaps thousands of liters or barrels of distillate.

Safety devices, controlling devices and regulating devices Devices intended for use outside potentially explosive atmospheres but required for or contributing to the safe functioning of equipment and protective systems with respect to the risks of explosion.

Safety Lanyard The line for connecting the safety belt or harness to an anchorage point.

Safety shoes Such shoes that principally protect the toes of user and also provided slip resistance.

Safety stitch A stitch formed by an overedge stitch reinforced by a chain stitch (or sometimes lock stitch) further in form the material edge.

Sampling probe A separate sample line that is attached to the instrument as required. It is usually short (1 m) and rigid but may be connected by a flexible tube to the instrument.

Sanitation Use of scientific knowledge in providing means to preserve health; use of things that contribute to hygiene and health cleanliness of working places and living quarters.

Sea Chests A casting connected to the side of a ship below the water line and to a valve for obtaining sea water.

Sealed device A device that is constructed so that it cannot be opened, has no external operating mechanisms, and is sealed to restrict entry of an external atmosphere without relying on gaskets. The device may contain arcing parts or internal hot surfaces.

Sealed Source   Radio-active source sealed in a capsule or having a bonded cover, the capsule or cover being strong enough to prevent contact with and dispersion of the radio-active material under the conditions of use and wear for which it was designed.

Seaming   Joining together the component part of a garment.

Secondary grade source of release   A source that is not expected to release in normal operation* and if it releases is likely to do so only infrequently and for short periods.

Secondary Straps   Straps used for connecting and positioning primary straps in assembly and in use.

Section   That part (which may be one or more zones) of an installation on a particular floor fed by a particular riser.

Secured Position   Condition of the exposure container and gamma radiography sealed source when the source is fully shielded and the exposure container is rendered inoperable by locking and/or other means.

Self-contained Breathing Apparatus (SCBA)   A portable device that includes the supply of respirable breathing gas for the fire fighters.

Semiconductor sensor   A sensor, the operation of which depends on changes of electrical conductance of a semiconductor due to chemical absorption of the gas being detected at its surface.

Sensing element   That part of a sensor that reacts in the presence of a flammable gas mixture to produce some physical change that can be used to activate a measuring or alarm function or both.

Service spaces   Those are used for galleys, pantries containing cooking appliances, lockers, and store rooms, workshops other than those forming part of the machinery spaces and similar spaces and trunks to such spaces.

Service Tests   Tests made occasionally (usually at least annually) after the pump has been put into service to determine if performance is still acceptable.

Shake   A separation along the grain, occurring most often between the rings of annual growth.

Shear Stress N/m2   Regular 3% and 6% -12 -13.

Sheathed incombustible or incombustible   Wood frame, incombustible sheathing.

Shelf Ladder   A single-section ladder fitted with treads that are intended to be horizontal in use.

Shell   A helmet without its harness, accessories, and fittings.

Short-term detector tubes   Tubes and associating aspirating pumps used for evaluating atmospheric contaminants at concentration in the range occupational exposure limit (OEL). It covers color tubes that are designed to give indication of concentration over a short period of time.

Shut-off Nozzle   A device that is coupled to the outlet end of hose reel tubing and by means of which the jet of water or spray is controlled.

Simulated Source   Facsimile of a radio-active sealed source the capsule of which has the same construction and is made with exactly the same materials as those of the sealed source that it represents but containing, in place of the radio-active material, a sub-stance with mechanical, physical and chemical properties as close as possible to those of the radio-active material and containing radio-active material of tracer quantity only. The tracer is in a form soluble in a solvent that does not attack the capsule and has the maximum activity compatible with its use in a glove box.

Single-Section Ladder   A leaning ladder constructed and used as a single unit.

Site-in Charge   A person who is so designated by the employer and who possesses a valid certificate for site-in-charge duly recognized or issued by the competent authority for this specific purpose.

Size   The length and breadth measurement of footwear based on the Iranian System of Grading.

Sling rod   A rod with a sling eye or screwed ends for supporting pipe clips, rings, band hangers, and so forth.

Soft suction   Collapsible hose used to supply water from hydrant to fire pump.

Source Drive System   Flexible cable system to drive the source pencil to the desired position.

Source Holder   Mechanical support for the sealed source. The following two terms apply to industrial radiography and gamma gauges and irradiation sources:

A) Source in device   Sealed source, which remains in a device giving mechanical protection from damage during use.

B) Unprotected source   Sealed source, which, for use, is removed from a device that would give mechanical protection from damage.

Source of release   A point or location from which a gas, vapor, mist, or liquid may be released into the atmosphere so that an explosive gas atmosphere could be formed.

Source Pencil   An assembly consisting of an encapsulated radio-active source and sometimes, a shield plug suitably encased with provision for attachment to a camera/flexible cable.

Spacing (of Rungs, Treads or Cross-Bearers)   The distance, measured along the longitudinal axis of the stiles, between the same relative positions of the members.

Specific gravity   The ratio of the weight of the substance to the weight of the same volume of water or air, whichever is applicable.

Spectacles   An eye-protector, the oculars of which are mounted in a spectacle-type frame, with or without side shields. Mounted oculars include lenses integral with the frame.

Spindle   The load-bearing axle on which the reel rotates.

Split   A separation of the wood parallel to the fiber direction due to tearing apart of the wood fibers, normally caused by external forces.

Spot-reading apparatus   Apparatus that is intended to be used for a short period of time as required.

Spray nozzles/spray nozzle assembly   A nozzle that has a water flow control that will provide a capability of full flow to completely shutting off the flow through the nozzle. This control device may be a permanently mounted valve or a break-apart shutoff butt assembly.

Spray tip   The primary adjustable pattern and flow appliance without a permanently attached shutoff butt. When used with fire hose mounted on standpipe systems, it may or may not have a shutoff capability.
Spray tip for fire department use operates from a wide spray pattern to a straight spray tip may or may not include a twist type pattern adjustment or shutoff.

Sprayer   A sprinkler that gives a downward conical pattern discharge.

Sprayer, high velocity   An open nozzle used to extinguish fires of high flash-Point liquids.

Sprayer, medium velocity   A sprayer of a sealed or open type used to control fires of lower flash-point liquids and gases or to cool surfaces.

Sprinkler, automatic   A temperature-sensitive sealing device that opens to discharge water for fire extinguishing. Note that the term "automatic sprinkler" is now rarely used. The term "sprinkler" does not include "open sprinkler."

Sprinkler, ceiling of flush pattern   A pendent sprinkler for fitting partly above but with the temperature-sensitive element below the lower plane of the ceiling.

Standing Step Ladder   A self supporting ladder consisting of a front and back hinged together and capable of being folded, the ascendable front being in the form of a shelf ladder.

Station Work Uniform   The definitions given in 3.9 shall apply for station firefighting uniform.

Stiles   The side members to which the rungs, treads or cross-bearers are fitted.

Stitch   Generally, the fundamental repeating unit produced by sewing material with one more sewing threads.

Stochastic-Radiation Effects   Radiation effects, the severity of which is independent of dose and the probability of which is assumed to be proportional to the dose without threshold at the low doses of interest in radiation protection.

Stored pressure system   A system in which the propellant gas is stored within and permanently pressurizes the powder container(s).

Structural Anchorage   A secure point of attachment on a structure to which an anchorage line may be secured.

Sub-Surface Injection   Discharge of foam into a storage tank below the liquid surface near the tank bottom.

Subsurface foam injection   Discharge of foam into a storage tank below the liquid surface near the tank bottom.

Suction Pressure   The pressure that pumps suction is subjected as determined by the gauge attached to suction side.

Suction pump   An automatic pump supplying water to a sprinkler system from a suction tank, river, lake, or canal.

Suit   A garment covering the upper part of the body from the head to the waist, and the arms to the wrist to which air suitable for respiration is supplied.

Suitable for sprinkler use   A term applied to equipment or components accepted by the authorities as for a particular application in a sprin-kler system, either by particular test or by compliance with specified general criteria.

Supply pipe   A pipe connecting a water supply to a trunk main or the instal-lation main control valve set(s), or a pipe supplying water to a private reservoir, suction tank, or gravity tank.

Surface fire   A fire involving flammable liquids, gases, or solids not subject to smoldering.

Surfactant   Also known as syndet or detergent foam.

Suspended open cell ceiling   A ceiling of regular open cell construction through which water from sprinklers can be discharged freely.

Suspension   The portion of the harness that is designed to act as an energy-absorbing mechanism. It may consist of crown straps, protective padding, or a similar mechanism.

Sweatband   The part of the headband, whether integral or replaceable, that comes in contact with at least the user's forehead.

Swing Back Steps   A standing step ladder in which the top is in the form of a tread and the back is merely a supporting frame.

Synthetic (S)   A liquid concentrate that has a base other than fluorinated surfactant or hydrolyzed protein.

Tail-end alternate (wet and dry pipe) extension   A part of a wet installation that is selectively charged with water or air according to ambient temperature conditions and which is controlled by a subsidiary dry or alternate alarm valve.

Tail-end dry extension   A part of a wet or alternate installation that is charged permanently with air under pressure.

Tank diameter   Where tank spacing is expressed in terms of tank diameter, the following criteria governs

If tanks are in different services, or different types of tanks are used, the diameter of the tank that requires the greater spacing is used

If tanks are in similar services, the diameter of the largest tank is used

Tank spacing   The unobstructed distance between tank shells or between tank shells and the nearest edge of adjacent equipment, property lines, or buildings.

Terminal main configuration   A pipe array with only one water supply route to each range pipe.

Terminal range configuration   A pipe array with only one water supply route from a distribution pipe.

Thermal conductivity sensor   A sensor that the operation of which depends on the changes of heat loss by conduction of an electrically heated element located in the gas to be measured compared with that of a similar element located in a reference gas cell.

Thermal semiconductivity sensor   A sensor that the operation of which depends on the condition of gases on an electrically heated catalytic element.

Thrusters   The force that is exerted endwise through a propeller shaft due to reaction of the water on the blades revolving.

Toe box   A stiffener designed to provide toe protection for the user as required by this standard.

Toe wall   A low earth, concrete, or masonry unit curb without capacity requirements for the retention of small leaks or spills.

Toggle support   A swivel device for securing hangers to hollow section ceilings or roofs.

Topside Application   A method of foam discharge wherein the foam is applied onto the top of a burning fuel surface.

Total flooding system   An automatic or manual fire extinguishing system in which a fixed supply of carbon dioxide is permanently connected to fixed piping with nozzles arranged to discharge the carbon dioxide into an enclosed space in order to produce a concentration sufficient to extinguish fire throughout the entire volume of the enclosed space.

Transport Index   A number expressing the maximum radiation level at 1 m from the surface of a package measured in mrem/h (1 mrem = 0.01 msv).

Transportable apparatus   Apparatus that is not intended to be portable, but which can be readily moved from one place to another.

Trim   Retroreflective and fluorescent material permanently attached to the outer shell for visibility enhancement.

Trunk main   A pipe connecting two or more water supply pipes to the installation main control valve set(s).

Tube Potential   Potential difference applied across the electrodes of an X-ray tube. This controls the energy of the X-rays generated. This is normally expressed in kilo Volts (kv) or mega Volts (mv).

Two-Piece suit   A suit that consists of a coat, jacket or other top garment and separate trousers and that covers at least the trunk, arms and legs but not the face, hands nor feet.

Upper   The upper parts of footwear including the outside and lining.

User   The person responsible for or having effective control over the fire safety provision adopted in or appropriate to the premises or the building.

Vamp   The complete fore part of the footwear upper back to the quarter.

Variable shade window   A device that enables observation of the workpiece before the welding arc is ignited and that automatically changes its shade number from a light shade to a dark shade when the welding arc is ignited.

Vertical Lift   The vertical distance from the surface of the water to the center of the pumping suction inlet.

Vessel   A craft used as a means of transportation on water.

Vessels Propulsion   The action driving forward or ahead.

Volume factor   A numerical factor that, when applied to the volume of an enclosure, indicates the basic quantity of carbon dioxide (subject to a minimum appropriate to the volume of the enclosure) required for protection against surface fires.

Wane   Bark, or lack of wood, on the corner of a piece.

Water solubility   The extent to which a substance mixes with pure water to form a molecular homogeneous system at a given temperature.

Welding goggles   Is a device enclosing 3 space in front of the eyes into which radiation arising from welding can penetrate only through filter(s) and, where provided, filter cover(s).

Wellhead   An assembly on top of the well casing strings with outlets and valves for controlling flow of production.

Wet water   Water to which a compatible wetting agent has been added.

Wet water foam   An admixture of wet water with air to form a cellular structure foam that breaks down rapidly into its original liquid state at temperatures below the boiling point of water at a rate directly related to the heat to which it is exposed in order to cool the combustible on which it is applied.

Wetting agent A chemical compound that when added to water in proper quantities, materially reduces its surface tension, increases its penetrating and spreading abilities, and can also provide emulsification and foaming characteristics.

Winter liner A sung-fitting cover worn under the helmet to protect the head, ears, and neck from cold.

Wood Characteristics Distinguishing features, the extent and number of which determine the quality of a piece of wood.

Wood Irregularities Natural characteristics in or on the wood that may lower its durability, strength, or utility.

Work Load Work load (known also as the weekly load) is expressed in terms of rontgen per week at 1 m from gamma ray sources and in ma min. Per week for X-ray sources.

Working Duration The maximum period of time for which the apparatus should be used.

Working Position Condition of the apparatus for gamma radiography when the beam is emitted for radiography.

Wrist glove A wrist length glove providing covering for the hand and wrist, having separate fingers.

   Wristing Additional fitment attached to the main body of the glove at the open cuff end to present a close fit to the wrist of the user.

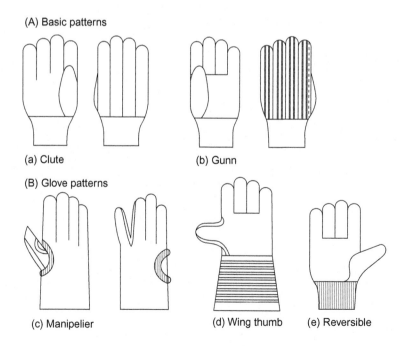

(A) Basic patterns

(a) Clute   (b) Gunn

(B) Glove patterns

(c) Manipelier   (d) Wing thumb   (e) Reversible

(C) Cuff patterns

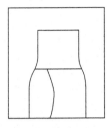 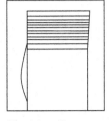 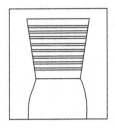

Knitted wrist   2" salety cult   Gaunner vasious lenghs

X-Ray Cable   The cable connecting the control console and X-ray tube.

X-Ray Tube   X-ray tube is a vacuum tube in which X-rays are produced by a cathode ray beam incident on the anode (target).

Yoke   The upper section of a garment covering the front and/or back from the shoulder seams usually the chest level.

Additional list of reading on retrofitting

2.2 Fire Hose and Couplings

BSI	**(BRITISH STANDARD INSTITUTION)**	
	BS 443 (1990)	"Specification for Testing Zinc Coating on Steel Wire and for Quality Requirement"
	BS 1154 (1986)	"Specification for Natural Rubber Compound"
	BS 1868	"Check Valve/Non-Return"
	BS 2752 (1990)	"Specification for Chloroprene Rubber Compound"
	BS 3592 (1986)	"Specification for Metallic Coated Steel Wire for the Bonded Reinforcement by Hydraulic Hoses"
	BS 3734 (1978)	"Specification for Dimensional Tolerances of Solid Molded and Extruded Rubber Product"
	BS 5163 (1986)	"Predominantly Key Operated Cast Iron Gate Valves For Water Works Purposes"
	BS 5173 (1977)	"Method of Test for Rubber and Plastic Hoses and Hose Assemblies"
UL	**(UNDERWRITER LABORATORIES)**	
	UL 162	"Hydrostatic Performance Test"
	UL 236 (1982)	"Couplings for Fire Hose"
	UL 668 Hose Valves For Fire Protection	

2.3 Fire Fighting Nozzles

ASME	**(AMERICAN SOCIETY MECHANICAL ENG.)**	
	Section VIII	"Boiler and Pressure Vessels Code"
ASTM	**(AMERICAN STANDARD FOR TESTING OF MATERIALS)**	
	ASTM-A 276 (1988)	"Specifications for Stainless and Heat-Resisting Steel Bar"
	ASTM-B 179 (1986)	"Specification for Aluminum Alloy in Ingot Form"
BSI	**(BRITISH STANDARD INSTITUTION)**	
	BS 1134 Part 1 (1988)	"Method for Assessment of Surface Texture"
	BS 1615 (1982)	"Method for Specifying Anodic Oxidation Coating on Aluminum and its Alloy"
	BS 5599 (1978)	"Specification for Hard Anodic Oxide Coating for Aluminum for Engineering Purpose"

NFPA   **(NATIONAL FIRE PROTECTION ASSOCIATION)**
            NFPA 11   "Low Expansions Foam/Combined Agent
                               Systems"

ISO   **(INTERNATIONAL ORGANIZATION FOR STANDARDIZATION)**
            ISO/DIS 7203 (1994)   "Fire Extinguishing Media Foam Concentrate"
            Part 1   "Specification for Low Expansion Foam
                               Concentrate for Application to Water-Immiscible
                               Liquids"
            Part 2   "Specification for Medium and High Expansion
                               Foam Concentrate"
            Part 3   "Specification for Low Expansion Foam
                               Concentrate for Top Application to Water
                               Miscible Liquids"

UL   **(UNDERWRITER LABORATORIES)**
            UL 401 (1989)   "Portable Hose Nozzles for Fire Protection"
            UL 262

ANSI   **(AMERICAN NATIONAL STANDARDS INSTITUTE)**
            ANSI # 89.1 (1986)   "Protective Headwear for Industrial Workers
                               Requirements"
            ANSI # 41.1 (1986)   "Personnel Protective Footwear"

API   **(AMERICAN PETROLEUM INSTITUTE)**
            API 2001(7th edition)   "Fire Protection in Refineries Seventh Edition"

ASTM   **(AMERICAN STANDARD FOR TEST OF MATERIAL)**
            ASTM D 2582-67   "Standard Test Method for Puncture Propagation
                               Tear Resistance of Plastic (1984) Films and Thin
                               Sheeting"
            DIN 4843 (1988)   "Safety Footwear, Safety Requirements Testing"

ASTM   **(AMERICAN STANDARD FOR TESTING OF MATERIALS)**
            ASTM-D 429 (1988)   "Rubber Property-Adhesion to Rigid Substracts"
            ASTM-D 2240 (1986)   "Rubber Property-Durometer Hardness"
            ASTM-A 105 (1978)   "Specification for Forging Carbon Steel for
                               Piping Components"
            ASTM-B 148 (1985)   "Specification for Aluminum Bronze Casting"
            ASTM-A 216 (1984)   "Specification for Steel Casting Carbon Suitable
                               for Fusion Welding for High Temperature
                               Service"
            ASTM-D 1418 (1985)   "Specification for Rubber Lattices
                               Nomenclatures"
            ASTM-D 3677 (1983)   "Rubber Identification by Infrared"

BSI   **(BRITISH STANDARD INSTITUTION)**
            BS 5306 Part 2 Section 5 Clause 24 (1990)

BSI   **(BRITISH STANDARD INSTITUTION)**
            BS 5566-1992   "Installed Dose Rate Meters, Warning
                               Assemblies and Monitors for Energy between
                               50 KeV and 7 MeV"

	BS 3783	"X-Ray, Lead - Rubber Protective Obsolescent Aprons for Personal Use"
	BS 5288	"Sealed Radio Active Sources"
BSI	**(BRITISH STANDARD INSTITUTION)**	
	BS 4667: (1974-1982), Parts 1, 2, 3, 4 and 5	"Specification for Breathing Apparatus"
	BS 7355 (1990) EN 136	"Full Face Mask for Respiratory Protective Devices"
	BS 7356 (1990) EN 140	"Half Masks and Quarter Mask"
	BS 6016 (1980)	"Specification for Filtering Face Piece Dust Respirators"
	BS 2091 (1969)	"Specification for Respirators for Protection Against Harmful Dusts, Gases and Scheduled Agricultural Chemicals"
	BS 4275 (1974)	"Recommendations for the Selection, Use and Maintenance of Respiratory Protective Equipment"
	BS 6850 (1987)	"Specification for Ventilatory Resuscitators"
BSI	**(BRITISH STANDARDS INSTITUTION)**	
	BS 5240 (1987)	"Industrial Safety Helmets"
	Part 1	"Specification for Construction and Performance"
	BS 6489	"Headforms for Use in the Testing of Protective Helmets"
	BS 679 (1977)	"Specification for Filters for Use During Welding and Similar Industrial Operations"
	B S 1542 (1982)	"Equipment for Face and Neck Protection Against Non Ionizing Radiation Arising During Welding and Similar Operations"
	BS 2092 (1987)	"Eye Protectors for Industrial and Non Industrial Uses"
	BS 2724 (1987)	"Sun Glare Eye Protectors for General Use"
	BS 2738 (1989)	"P.2 Spectacle Lenses" "Specification for Tolerances on Optical Properties of Uncut Finished Lenses"
	BS 3199 (1972)	"Method for Measurement of Spectacles Including a Glossary of Terms"
	BS 903	"Method of Testing Vulcanized Rubber Part A2. Determination of Tensile Stress-Strain Properties. Part A 19 Heat Resistance and Accelerated Air Aging Tests. Part A 38 Determination of Dimensions of Test Pieces and Products for Test Purposes"
	BS 1651 (1986)	"For Industrial Gloves"
	BS 2471 (1984)	"Methods of Test for Textiles-Woven Fabrics-Determination of Mass"

BS 3144	"Methods of Sampling and Physical Testing of Leather"
BS 5108 (1982)	"Method for Measurement of Sound Attenuation of Hearing Protectors"
(ISO 4869: 1981)	
BS 6344 (1988)	"Industrial Hearing Protectors"
(Parts 1 and 2)	
BS 5145 (1989)	"Lined Industrial Vulcanized Rubber Boots"
BS 5451 (1977)	"Electrically Conducting and Antistatic Rubber Footwear"
BS 2576 (1986)	"Method for Determination of Breaking Strength and Elongation (Strip Method) of Woven Fabrics"
BS 3870	"Stitches and Seams"
BS 3870	"Classification and Terminology of Stitch Types"
Part (1) (1991)	
BS 3870	"Classification and Terminology of Seam Types"
Part (2) (1991)	
BS 6629 (1985)	"Specification for Optical Performance of High Visibility Garments and Accessories for Use on the Highway"
BS 903 (1987)	"Methods of Testing Vulcanized Rubber"
Part A-16	"Determination of the Effects of Liquids"
BS 2576	"Method for Determination of Breaking Strength and Elongation (Strip Method) of Woven Fabrics"
BS 3084 (1981)	"Specification for Solid Fasteners"
BS 3424	"Testing Coated Fabrics"
Part 7, Method 9	"Method for Determination of Coating Adhesion Strength"
BS 3546	"Coated Fabrics for Water Resistant Clothing"
Part 1:	"Specification for Polyurethane and Silicone Elastomer Coated Fabrics"
Part 2:	"Specification for PVC Coated Fabrics"
Part 3:	"Specification for Natural Rubber and Synthetic Rubber Polymer Coated Fabrics"
BS 4724 (1986)	"Resistance of Clothing Materials to Permeation by Liquids"
Part 1 (1986)	"Method for the Assessment of Breakthrough Time"
Part 2 (1988)	"Method for the Determination of Liquid Permeating after Breakthrough"
BS 5438	"Methods of Test for Flammability of Vertically Oriented Textile Fabrics Assemblies Subjected to a Small Igniting Flame"
BS 6249	"Materials and Material Assemblies Used in Clothing for Protection Against Heat and Flame"

Part 1:	"Specification for Flammability Testing and Performance"	
BS 2092 (1987)	"Specification for Eye Protectors for Industrial and Non-Industrial Uses"	
BS 2723 (1988)	"Specification for Firements Leather Boots"	
BS 5145 (1984)	"Specification for Lined Industrial Vulcanized Rubber Boots"	

BSI   **(BRITISH STANDARD INSTITUTION)**

BS.AU 183 (1983)	"Specification for Passive Seat Belt Systems"
BS EN ISO 7500 (1999)	
BS 2087 (1981)	"Preservative Textile Treatments"
BS 3144 (1987)	"Methods of Sampling and Physical Testing of Leather"
BS 3146 (1984)	"Investment Castings in Metal"
BS 3382 (1968)	"Electroplated Coatings on Threaded Components"
BS EN 818 (1996)	
BS 7773 (1995)	
BS EN 696 (1995)	
BS EN 697 (1995)	
BS EN 699 (1995)	
BS EN 700 (1995)	
BS EN 701 (1995)	

BSI   **(BRITISH STANDARD INSTITUTION)**

BS 309	"Whiteheart Malleable Iron Castings"
BS 1203	"Specification for Synthetic Resin Adhesives (Phenolic and Aminoplastic) for Plywood"
BS 1204	"Synthetic Resin Adhesives (Phenolic and Aminoplastic) for Wood, Part 1 Specification for Gap-Filling Adhesives, Part 2 Specification for Close-Contact Adhesives"
BS 1210	"Specification for Wood Screws"
BS 1449	
BS 1470	"Wrought Aluminum and Aluminum Alloys for General Engineering Purposes-Plate, Sheet and Strip"
BS 1471	"Wrought Aluminum and Aluminum Alloys for General Engineering Purposes-Drawn Tube"
BS 1472	"Wrought Aluminum and Aluminum Alloys for General Engineering Purposes-Forging Stock and Forging"
BS 1474	"Wrought Aluminum and Aluminum Alloys for General Engineering Purposes-Bars, Extruded Round Tubes and Sections"

	BS 1490	"Aluminum and Aluminum Alloy Ingots and Castings"
	BS 4300	"Specification (Supplementary Series) for Wrought Aluminum and Aluminum Alloys for General Engineering Purposes"
	BS 4471	"Specification for Sizes of Sawn and Processed Softwood"
	BS 6125	"Specification for Natural Fiber Cords, Lines and Twines"
	BS 6681	"Specification for Malleable Cast Iron"

BSI   **(BRITISH STANDARDS INSTITUTION)**
BS 5306 Part 1   "System Design Water Supply"

BSI   **(BRITISH STANDARD INSTITUTION)**

BS 3120	"Performance Requirement of Materials for Flame Proofing"
BS Handbook NO 11	"Method of Tests for Textile"
BS EN 367	"Protective clothing heat transmission"
BS EN 180 6942	"Protective clothing method of test"
BS EN 531 (1995)	"Protective clothing for industrial workers exposed to heat"
BS 3119	"Method of Test Flameproof Materials"
BS 3791 EN 367 (92)	"Protective clothing – protection against heat and fire"
BS 7944	"Heavy Duty Fire Blanket"
BS EN 1869	"Fire blankets"
UL 96	"Standard for safety lightning Protection Components"

BSI   **(BRITISH STANDARD INSTITUTION)**

BS 903 Part A2 (1989)	"Determination of Tensile Stress-Strain Properties"
BS 1400 (1985)	"Copper Alloy Ingots and Copper Alloy Castings"
BS 2874	"Manganese Bronze"
BS 3076 (1989)	"Specification for Nickel and Nickel Alloy Bar"
BS 336 (1989)	"Specification for Fire Hose Couplings"
BS 5146 (1974)	"Inspection and Test of Valves"
BS 5155	"Valve Design Specification"
BS 5159 (1974)	"Cast Iron and Carbon Steel Ball Valves"

DIN   **(DEUTSCHES INSTITUTE FUR NORMUNG EV.)**
DIN 44425, DIN-6818
pt. 2

DIN   **(DEUTSCHE INDUSTRIE NORMEN)**
DIN 50049 (1986)   "Documents on Materials Testing"

IAEA   **(INTERNATIONAL ATOMIC ENERGY AGENCY)**
Safety Series Nos. 6 and 37

ISIRI	**(INSTITUTE OF STANDARDS AND INDUSTRIAL RESEARCH OF IRAN"**
	ISIRI UDC 614-891 "Specification for Industrial Safety Helmets (Heavy Duty)"
	No. 1375
	ISIRI 1944 (1992) "Cotton Fabrics, Specifications of Raw Wool in Packages for Wool Fibre Present"
ISO	**(INTERNATIONAL ORGANIZATION FOR STANDARDIZATION)**
	ISO-8382 (1988), "Resuscitators Intended for Use with Humans"
ISO	**(INTERNATIONAL ORGANIZATION FOR STANDARDIZATION)**
	ISO 361 "Basic Lionizing Radioactive Symbol First Edition 1975"
	ISO 3999 "Apparatus for Gamma Radiography Specification First Edition 1977"
	ISO TR 4826 "Sealed Radioactive Sources Leak Test Method 1979"
	ISO 2855 "Radioactive Materials Packaging Tests for Contents Leakage and Radiation Leakage"
ISO	**(INTERNATIONAL ORGANIZATION FOR STANDARDIZATION)**
	ISO 4850 (1979) "Personal Eye-Protectors for Welding and Related Techniques Filter-Utilization and Transmittance Requirement"
	ISO 4851 (1979) "Personal Eye Protectors-Ultra-Violet Filters-Utilization and Transmittance Requirement"
	ISO 4852 (1978) "Personal Eye Protectors-Infra-Red Utilization and Transmittance Requirement"
	ISO 4855 (1981) "Personal Eye Protectors-Non Optical Test Methods"
	ISO 2024 (1981) "Rubber Footwear, Lined Conducting Specification"
	ISO 2251 (1975) E "Lined Antistatic Rubber Footwear"
	ISO 6530 (1990) "Protective Clothing-Protection Against Liquid Chemical-Determination of Resistance of Materials to Penetration by Liquids"
JIS	**(JAPANESE STANDARD INSTITUTE)**
	JIS T 8103 (1983) "Anti-Electrostatic Footwears With/Without Safety Toes"
NFC	**(NATIONAL FIRE CODES) NFPA**
	NFC (1991), Chapter 1 to 4
NFC	**(NATIONAL FIRE CODES) – NFPA**
	NFC Code No. 1971 "Protective Clothing for Structural Fire Fighters"
	NFC Code No. 1972 "Helmets for Structural Fire Fighters"
	NFC Code No. 1973 "Gloves for Structural Fire Fighters"
	NFC Code No. 1974 "Station/Work Uniform"

NFC (NFPA) **(NATIONAL FIRE CODES)**

NFC Section 1231	"Water Supply"
NFC Section 15	"Water Spray System"
NFC Section 22	"Water Tanks"
NFC Section 24	"Mains Water Supplies"
NFPA-20	"Standard for the Installation Pumps for Fire Protection"
NFPA-15	"Standard for Water Spray Fixed systems for Fire Protection"

NFPA **(NATIONAL FIRE PROTECTION ASSOCIATION)**

NFPA 1001 – (2002)	"Standard for Fire Fighter Professional Qualifications"

NFPA **(NATIONAL FIRE PROTECTION ASSOCIATION)**

NFPA 101	"Life Safety Code"
NFPA 701	"Fire test for flame propagation of textiles"

National Safety Council Accident Manual for Industrial Operations Chapter 38 (6 th Edition)

NFPA 20	"Standard for the Installation of Stationary Pumps for Fire Protection 1999 Edition)
NFPA 1911	"Standard for Service Tests of Fire Pumps Systems on Fire Apparatus 2002 Edition"

References

Adams, N. J., & Kuhlman, L. G. (1993). Contingency planning for offshore blowouts. Paper # 7120. SPE.

Altena, J. W., & Zeckendorf, A. (January 16, 1995). Design, simulation creates low surge, low cost gas-injection compressor. *Oil and Gas Journal, 12.*

American Petroleum Institute. (1998). *Welded steel tanks for oil storage* (10th ed.) Standard 650. Washington, DC: API.

American Society of Mechanical Engineers Code Committee SC6000. (2000). *Hazardous release protection.* New York, NY: ASME.

Andrew, H. (1999). For whom does safety pay? The case of major accidents. *Safety Science, 32,* 143–153.

Apeland, S., & Scarf, P. A. (2003). A fully subjective approach to modelling inspection maintenance. *European Journal of Operational Research, 148,* 410–425.

API. (2006). Recommended practice for well control operations, Recommended practice 59 (2nd ed.).

API RP 2021 (R2006). (2001). *Management of atmospheric storage tank fires* (4th ed.). American Petroleum Institute.

Apostolakis, G. E., & Lemon, D. M. (2005). A screening methodology for the identification and ranking of infrastructure vulnerabilities due to terrorism. *Risk Analysis, 25*(2), 361–376.

Argyropoulos, C. D., Christolis, M. N., Nivolianitou, Z., & Markatos, N. C. (2008a). Assessment of acute effects for fire-fighters during a fuel-tank fire. In: *Proceedings of the 4th international conference on prevention of occupational accident in a changing work environment, WOS 2008.* Crete, Greece.

Argyropoulos, C. D., Christolis, M. N., Nivolianitou, Z., & Markatos, N. C. (2008b). Numerical simulation of the dispersion of toxic pollutants from large tank fire. In M. Papadakis, & B. H. V. Topping (Eds.). In: *Proceedings of the sixth international conference on engineering computational technology.* Stirlingshire, UK: Civil-Comp Press, Paper 49.

Argyropoulos, C. D., Sideris, G. M., Christolis, M. N., Nivolianitou, Z., & Markatos, N. C. (2010). Modelling pollutants dispersion and plume rise from large hydrocarbon tank fires in neutrally stratified atmosphere. *Atmospheric Environment, 44,* 803–813.

Arunraj, N. S., & Maiti, J. (2007). Risk-based maintenance – techniques and applications. *Journal of Hazardous Materials, 142*(3), 653–661.

ASME B96.1. Welded aluminium-alloy storage tanks, Publication date: Jan 1; 1999.

Attwood, D., Khan, F., & Veitch, B. (2005). Can we predict occupational accident frequency? *Process Safety and Environmental Protection, 84*(B2), 1–14.

Attwood, D., Khan, F., & Veitch, B. (2006). Occupational accident modelsdwhere have we been and where are we going? *Journal of Loss Prevention in the Process Industries, 19*(6), 664–682.

Aven, T., Sklet, S., & Vinnem, J. E. (2006). Barrier and operational risk analysis of hydrocarbon releases (BORA-Release) Part I. Method description. *Journal of Hazardous Materials, 137*(2), 692–708.

Baker, R. D., & Wang, W. (1992). Estimating the delay-time distribution of faults in repairable machinery from failure data. *IMA Journal of Mathematics Applied in Business and Industry, 3,* 259–281.

Bakke, J. R., Wingerden, K., Hoorelbeke, P., & Brewerton, B. (2010). A study on the effect of trees on gas explosions. *Journal of Loss Prevention in the Process Industries*, *23*, 878–884.

Balkey, R. K., & Art, J. R. (1998). ASME risk-based in service inspection and testing: An outlook to the future. *Society for Risk Analysis*, *18*(4).

Baron, M. M., & Cornell, M. E. P. (1999). Designing risk-management strategies for critical engineering systems. *IEEE Transactions on Engineering Management*, *46*, 87–100.

Basso, B., Carpegna, C., Dibitonto, C., Gaido, G., Robotto, A., & Zonato, C. (2004). Reviewing the safety management system by incident investigation and performance indicators. *Journal of Loss Prevention in the Process Industries*, *17*(3), 225–231.

Bauer, P. W. (1990). Recent developments in the econometric estimation of frontiers. *Journal of Econometrics*, *46*(1–2), 39–56.

Bedford, T. (2004). Assessing the impact of preventive maintenance based on censored data. *Quality and Reliability Engineering International*, *20*, 247–254.

Bernardo, M., Casadesus, M., Karapetrovic, S., & Heras, I. (2009). How integrated are environmental, quality and other standardized management systems: An empirical study. *Journal of Cleaner Production*, *17*, 742–750.

Bevilacqua, M., Braglia, M., & Gabbrielli, R. (2000). Monte Carlo simulation approach for a modified FMECA in a power plant. *Quality and Reliability Engineering International*, *16*, 313–324.

Biersack, W. M., Hyder, C. B., James, J. W., King, S. G., Kruse, E. M., & Veatch, J. D., et al. (2002). An infrastructure vulnerability assessment methodology for metropolitan areas. In *IEEE annual international Carnahan conference on security technology, proceedings* (pp. 29–34).

Bird, F. E., Germain, G. L., & Clark, M. D. (2003). *Practical loss control leader* (3rd ed.). Georgia: Det Norske Veritas (USA), Inc.

BP. (2010). Deepwater horizon accident investigation report.

BP, (March 11, 1993). BP proves compact oil dewatering. *The Chemical Engineer*, *1*.

Brown, K. A., Willis, P. G., & Prussia, G. E. (2000). Predicting safe employee behaviour in the steel industry: Development and test of a sociotechnical model. *Journal of Operations Management*, *18*, 445–465.

Buncefield Major Incident Investigation Board. (2008). The Buncefield incident 11 December 2005. Final report.

Burri, G. J., & Helander, M. G. (1991). A field study of productivity improvements in the manufacturing of circuit boards. *International Journal of Industrial Ergonomics*, *7*, 207–215.

Capelle-Blancard, G., & Laguna, M. (2010). How does the stock market respond to chemical disasters? *Journal of Environmental Economics and Management*, *59*, 192–205.

Carter, D. A., & Hirst, I. L. (2000). 'Worst case' methodology for the initial assessment of societal risk from proposed major accident installations. *Journal of Hazardous Materials*, *71*, 117–128.

CCPS. (1993). *Guidelines for engineering design for process safety*. New York: American Institute of Chemical Engineers.

Chang, J. I., & Lin, C. C. (2006). A study of storage tank accidents. *Journal of Loss Prevention in the Process Industries*, *19*, 51–59.

Changchit, C., & Holsapple, C. W. (2001). Supporting managers' internal control evaluations: An expert system and experimental results. *Decision Support Systems*, *30*, 437–449.

Chapman, H., Purnell, K., Law, R. J., & Kirby, M. F. (2007). The use of chemical dispersants to combat oil spills at sea: A review of practice and research needs in Europe. *Marine Pollution Bulletin*, *54*(7), 827–838.

Chen, C. Y., Wu, G. S., Chuang, K. J., & Mac, C. M. (2009). A comparative analysis of the factors affecting the implementation of occupational health and safety management systems in the printed circuit board industry in Taiwan. *Journal of Loss Prevention in the Process Industries, 22*, 210–215.

Cheyne, A., Tomas, J. M., Cox, S., & Oliver, A. (1999). Modelling employee attitudes to safety: A comparison across sectors. *European Psychologist, 4*(1).

Christer, A. H., Wang, W., Baker, R. D., & Sharp, J. (1995). Modelling maintenance practice of production plant using the delay-time concept. *IMA Journal of Mathematics Applied in Business and Industry, 6*, 67–83.

Christou, M., Papadakis, G., & Amendola, A. (2005). Guidance on the preparation of a safety report to meet the requirements of the directive 96/82/EC as amended by the directive 2003/105/EC 2002 (SEVESO II), EUR 22113 EN.

CNPC (China National Petroleum Corporation). (2006). *Blowout accidents in China national petroleum corporation*. Beijing, China: Petroleum Industry Press. (in Chinese).

Cohen, M., & Santhakumar, V. (2007). Information disclosure as environmental regulation: A theoretical analysis. *Environmental and Resource Economics, 37*, 599–620.

Comfort, L. K., Ko, K., & Zagorecki, A. (2004). Coordination in rapidly evolving disaster response systems: The role of information. *American Behavioral Scientist, 48*(3), 295–313.

Commission. (2011a). Chief Counsel's Report 2011. National Commission on the BP Deepwater Horizon Oil Spill and Offshore Drilling.

Commission. (2011b). Report to the President. National Commission on the BP Deepwater Horizon Oil Spill and Offshore Drilling.

ConocoPhillips. (2002). *Sustainable growth report*. ConocoPhillips.

Cook, W. D., & Zhu, J. (2005). *Modeling performance measurement: Applications and implementation issues in DEA* (1st ed.). Springer Science.

Cowing, M. M., Cornell, M. E. P., & Glynn, P. W. (2004). Dynamic modeling of the tradeoff between productivity and safety in critical engineering systems. *Reliability Engineering and System Safety, 86*, 269–284.

Cox, S. J., & Cheyne, A. J. T. (2000). Assessing safety culture in offshore environments. *Safety Science, 34*, 111–129.

Crawley, F. (July 13, 1995). Offshore loss prevention. *The Chemical Engineer*, 23–25.

Crippa, C., Fiorentini, L., Rossini, V., Stefanelli, R., Tafaro, S., & Marchi, M. (2009). Fire risk management system for safe operation of large atmospheric storage tanks. *Journal of Loss Prevention in the Process Industries, 22*, 574–581.

Danenberger, E. P. (1993). Outer Continental Shelf drilling blowouts, 1971e1991. Paper # 7248. SPE.

De Dianous, V., & Fievez, C. (2006). ARAMIS project: A more explicit demonstration of risk control through the use of bowetie diagrams and the evaluation of safety barrier performance. *Journal of Hazardous Materials, 130*(3), 220–233.

Dey, P. M. (2001). A risk-based model for inspection and maintenance of crosscountry petroleum pipeline. *Journal of Quality in Maintenance Engineering, 40*(4), 24–31.

DiMattia, D. G. (2003). *Human error probability index for offshore platform musters*, Ph.D. Thesis. Halifax, Nova Scotia: Chemical Engineering, Dalhousie University.

DHSG. (2011). Final report on the investigation of the Macondo well blowout.

DHSG (Deepwater Horizon Study Group). (2010). The Macondo blowout, 3rd Progress Report.

Dobson, J. D. (1999). Rig floor accidents: Who, when and why?—An analysis of UK offshore accident data. In Society of petroleum engineers/international association of drilling contractors conference, Amsterdam, The Netherlands.

Edwards, D. W., & Lawrence, D. (1993). Assessing the inherent safety of chemical process routes: Is there a relation between plant cost and inherent safety. *Trans IChemE (Process Safety and Environmental Protection), 71B*, 252–258.

Einarsson, S., & Brynjarsson, B. (2008). Improving human factors, incident and accident reporting and safety management systems in the Seveso industry. *Journal of Loss Prevention in the Process Industries, 21*, 550–554.

Embrey, D. E. (1992). Incorporating management and organisational factor into probabilistic safety assessment. *Reliability Engineering and System Safety, 38*.

Englund, S. M. (1991). Design and operate plants for inherent safety—part 1 and 2. *Chemical Engineering Progress 87*(3) and *87*(5), 85 and 79.

Etowa, C. B., Amyotte, P. R., & Pegg, M. J. (2001). Assessing the inherent safety of chemical processes: The Dow indices. In *51st chemical engineering conference*, Halifax, Canada.

Fabbrocino, G., Iervolino, I., Orlando, F., & Salzano, E. (2005). Quantitative risk analysis of oil storage facilities in seismic areas. *Journal of Hazardous Materials, A123*, 61–69.

Fernandez-Mu~niz, B., Montes-Peon, J. M., & Vazquez-Ordas, C. J. (2007). Safety management system: Development and validation of a multidimensional scale. *Journal of Loss Prevention in the Process Industries, 20*(1), 52–68.

Fitzgerald, M. K. (2005). Safety performance improvement through culture change. *Process Safety Environmental Protection, 83*, 324–330.

Flin, R. H., Mearns, K., Gordon, R. P. E., & Fleming, M. T. (1996). Risk perception in the UK offshore oil and gas industry. In *International conference on health, safety and environment*, New Orleans, Louisiana.

Geyer, T. A. W., & Bellamy, L. J. (1991). *Pipework failure, failure causes and the management factor*. London: The Institute of Mechanical Engineers.

Ghoniem, A. F., Zhang, X., Knio, O., Baum, H. R., & Rehm, R. G. (1993). Dispersion and deposition of smoke plumes generated in massive fires. *Journal of Hazardous Materials, 33*, 275–293.

Giannini, F. M., Monti, M. S., Ansaldi, S. P., & Bragatto, P. (2006). P.L.M., to support hazard identification in chemical plant design. In D. Brissaud (Ed.), *Innovation in life cycle engineering and sustainable development* (pp. 349–362). Springer.

Goins, W. C., & Sheffield, J. R. (1983). *Blowout prevention* (2nd ed.). Houstan, Texas: Gulf Publishing Company.

Gordon, J. E. (1949). The epidemiology of accidents. *The American Journal of Public Health.*

Gordon, R., Flin, R., & Mearns, K. (2001). Designing a human factors investigation tool to improve the quality of safety reporting. In *Proceedings of the 45th annual meeting of the human factors and ergonomics society.*

Gordon, R. P. E. (1998). The contribution of human factors to accidents in the offshore oil industry. *Reliability Engineering & System Safety, 61*(1–2), 95–108.

Haddon, W. (1973). Energy damage and the ten countermeasure strategies. *Human Factors.*

Halliburton. (2001). *HSE performance around the world; health, safety and environment annual report*. Halliburton.

Hanigan, N. (December 9, 1993). Solvent recovery: Try power fluidics. *The Chemical Engineer, 32.*

Hansen, M. D. (2001). Improving safety performance through rig mechanization. In *SPE/IADC drilling conference*, Amsterdam, The Netherlands.

Hansen, M. D., & Abrahamsen, E. (March, 2001). Engineering design for safety: Imagineering the rig floor. *Professional Safety*, 20–34.

Harnly, A. J. (1998). Risk based prioritization of maintenance repair work. *Process Safety Progress, 17*(1), 32–38.

Harstad, E. (1991). Safety as an integrated part of platform design. In: *Proceeding of 1st international conference on health, safety and environment*, The Hague, The Netherlands.

Haugen, S., Seljelid, J., & Nyheim, O. M. (2011). Major accident indicators for monitoring and predicting risk levels. Paper # 140428, SPE.

Hendershot, D. C. (1987). Safety considerations in the design of batch processing plants. *Preventing Major Chemical Accidents, 3*, 1.

Hendershot, D. C. (1995). Conflicts and decisions in the search for inherently safer process options. *Process Safety Progress, 14*(1), 52–56.

Hendershot, D. C. (1995). Some thoughts on the difference between inherent safety and safety. *Process Safety Progress, 14*(4), 227–228.

Hendershot, D. C. (1997). Measuring inherent safety, health and environmental characteristics early in process development. *Process Safety Progress, 16*(2), 78–79.

Herbert, I. (2010). The UK Buncefield incident – the view from a UK risk assessment engineer. *Journal of Loss Prevention in the Process Industries, 23*, 913–920.

Hibbert, L. (2008). Averting disaster. *Professional Engineering, 21*, 20.

Hill, T. G., & Bhavsar, R. (1996). Development of a self-equalizing surface controlled subsurface safety valve for reliability and design simplification. In *Offshore technology conference*, Houston, TX.

Hills, A. (2005). Insidious environments: Creeping dependencies and urban vulnerabilities. *Journal of Contingencies and Crisis Management, 13*(1), 12–20.

Holand, P. (1997). *Offshore blowouts: Causes and control*. Houston, Tex: Gulf Publ. Co.

Holand, P., & Skalle, P. (2001). *Deepwater kicks and BOP performance, unrestricted version*. Trondheim: SINTEF.

Hollnagel, E. (2004). *Barrier and accident prevention*. Hampshire, UK: Ashgate.

Houck, D. J., Kim, E., O'Reilly, G. P., Picklesimer, D. D., & Uzunalioglu, H. (2004). A network survivability model for critical national infrastructures. *Bell Labs Technical Journal, 8*(4), 153–172.

Hudson, P. T. W., Groeneweg, J., Reason, J. T., Wagenaar, W. A., Van der Meeren, R. J. W., & Visser, J. P. (1991). Application of TRIPOD to measure latent errors in north sea gas platforms: Validity of failure state profiles. Paper # 23293. SPE.

Hudson, P. T. W., Reason, J. T., Wagenaar, W. A., Bentley, P. D., Primrose, M., & Visser, J. P. (1994). Tripod delta: Proactive approach to enhanced safety. Paper # 27846. SPE.

Hurst, N. W. (1998). Risk assessment—The human dimension. *The Royal Society of Chemistry*.

IADC, (2002). Deepwater well control guidelines. *International Association of Drilling Contractors*.

IADC, (2010). Health, safety and environmental case guidelines for mobile offshore drilling units. *International Association of Drilling Contractors*.

IChem, E. (2008). BP process safety series, liquid hydrocarbon tank fires: Prevention and response (4th ed.). U.K.

IChemE, (July 25, 1991). Innovative technology: Multiphase pumping. *The Chemical Engineer*, 22.

ICivilE, (May, 1991). Revolutionary LEO prepares to clean up in oily water stakes. *Offshore Engineer*, 121.

ICivilE, (May, 1991). Spiral flow boosts shell and tube exchanger. *Offshore Engineer, 65*.

ICivilE, (August, 1992). Vortex choke cut erosion. *Offshore Engineer, 45*.

Iledare, O. O., Pulsipher, A. G., & Mesyazhinov, D. V. (1998). Safety and environmental performance measures in Offshore E&P operations: Empirical indicators for benchmarking. In *The society of petroleum engineers annual technical conference and exhibition*.

International Association of Oil and Gas Producers (OGP). (2002). OGP Safety performance of the global E&P industry 2001 Report no. 6.59/330. International Association of Oil and Gas Producers.

International Association of Oil and Gas Producers (OGP). (2004). OGP safety performance indicators 2003 Report no. 353. International Association of Oil and Gas Producers.

International Association of Oil and Gas Producers (OGP). (2005). OGP safety performance indicators 2004 Report no. 367. International Association of Oil and Gas Producers.

International Labour Organisation (ILO). (2003). *Safety in numbers, global safety culture at work*. Geneva: The International Labour Organisation.

Jahanshahloo, G. R., Memariani, A., Hosseinzadeh Lotfi, F., & Rezaei, H. Z. (2004). A note on some of DEA models and finding efficiency and complete ranking using common set of weights. *Applied Mathematics and Computation*, *166*(2), 265–281.

Johnson, D. M. (2010). The potential for vapour cloud explosions – lessons from the Buncefield accident. *Journal of Loss Prevention in the Process Industries*, *23*, 921–927.

Johnson, W. G. (1980). *MORT safety assurance systems*. Basel: Dekker.

Jones, K., & Rubin, P. H. (2001). Effects of harmful environmental events on reputations of firms. In *Advances in financial economics*, Vol. 6, pp. 161–182.

Jun,W., & Yue Jin, T. (2005). Finding the most vital node by node contraction in communication networks. In: *Proceeding of IEEE ICCCA S 2005* (pp. 1283–1286). Hong Kong.

Kaili, X., Liju, D., & Baozhi, C. (1998). Reliability study of man-machine monitoring system on major hazards. In: *Proceedings of the international symposium on safety science technology*.

Kao, C., & Liu, S. (2000). Fuzzy efficiency measures in data envelopment analysis. *Fuzzy Sets and Systems*, *113*(3), 427–437.

Khan, F. I., & Abbasi, S. A. (1998). *Risk assessment in chemical process industries:Advance techniques*. New Delhi: Discovery Publishing House. XC376

Khan, F. I., & Abbasi, S. A. (1999). Major accidents in process industries and an analysis of causes and consequences. *Journal of Loss Prevention in the Process Industries*, *12*, 361–378.

Khan, F. I., & Abbasi, S. A. (1999). Inherently safer design based on rapid risk analysis. *Journal of Loss Prevention in Process Industries*, *11*(3), 361–372.

Khan, F. I., & Abbasi, S. A. (2001). An assessment of the likelihood of occurrence, and the damage potential of domino effect (chain of accidents) in a typical cluster of industries. *Journal of Loss Prevention in the Process Industries*, *14*(4), 283–306.

Khan, F. I., & Amyotte, P. R. (2002). *Journal of Loss Prevention in the Process Industries*, *15*, 279–289.

Khan, F. I., & Haddara, M. R. (2004). Risk-based maintenance of ethylene oxide production facilities. *Journal of Hazardous Materials*, *A108*, 147–159.

Khan, K., Sadiq, R., & Haddara, M. (2004). Risk-based inspection and maintenance (RBIM): Multi-attribute decision making with aggregative risk analysis. *Process Safety and Environmental Protection*, *82*, 398–411.

Kjellen, U. (1995). Integrating analyses of the risk of occupational accidents into the design process. Part II: Method for predicting the LTI rate. *Safety Science*, *19*.

Kjellen, U., & Hovden, J. (1993). Reducing risks by deviation control a retrospection into a research strategy. *Safety Science*, *16*.

Kjellen, U., & Sklett, S. (1995). Integrating analyses of the risk of occupational accidents into the design process. Part I: A review of types of accident criteria and risk analysis methods. *Safety Science*, *18*.

Klassen, R. D., & McLaughlin, C. P. (1996). The impact of environmental management on firm performance. *Management Science*, *42*, 1199–1214.

Kletz, T. A. (1985). Inherently safer plants. *Plant/Operation Progress*, *4*, 164–166.

Kletz, T. A. (1991). Plant design for safety a user-friendly approach. New York: Taylor & Francis.

Kletz, T. A. (1991b). Process safetydan engineering achievement. In: *Proceedings of the Institution of Mechanical Engineers*. Part E: Journal of Process Mechanical Engineering, 1989–1996 (vols 203–210), 205, 11–15.

Kletz, T. A. (1998). *What went wrong? Case histories of process plant disasters* (4th ed.). Houston, Texas: Gulf Pub.

Kletz, T. A. (1998). *Process plants: A handbook of inherently safer design* (2nd ed.). Philadelphia, PA: Taylor & Francis.

Kletz, T. A. (2001). *Learning from accidents* (3rd ed.). Oxford; Boston: Gulf Professional.

Kletz, T. A. (2009). Accident reports may not tell us everything we need to know. *Journal of Loss Prevention in the Process Industries*, *22*, 753.

Knegtering, B., & Pasman, H. J. (2009). Safety of the process industries in the 21st century: a changing need of process safety management for a changing industry. *Journal of Loss Prevention in the Process Industries*, *22*, 162–168.

Koseki, H., Natsuma, Y., Iwata, Y., Takahashi, T., & Hirano, T. (2003). A study on largescale boilover using crude oil containing emulsified water. *Fire Safety Journal*, *39*, 143–155.

Kourniotis, S. P., Kiranoudis, C. T., & Markatos, N. C. (2000). Statistical analysis of domino chemical accidents. *Journal of Hazardous Materials*, *71*, 239–252.

Krishnasamy, L., Khan, F., & Haddara, M. (2005). Development of a risk-based maintenance (RBM) strategy for a power-generating plant. *Journal of Loss Prevention in the Process Industries*, *18*, 69–81.

Kujath, M. F., Amyotte, P. R., & Khan, F. I. (2009). A conceptual offshore oil and gas process accident model. *Journal of Loss Prevention in the Process Industries*, *23*(2), 323–330.

Kumar, U. (1998). Maintenance strategies for mechanized and automated mining systems: a reliability and risk analysis based approach. *Journal of Mines, Metals and Fuels*, *46*(11–12), 343–347.

Kumar, V. N. A., & Gandhi, O. (2011). Quantification of human error in maintenance using graph theory and matrix approach. *Quality and Reliability Engineering International*, *27*(8), 1145–1172.

Lawley, G. (1974). Operability studies and hazards analysis, loss prevention. CEP.

Le Bot, P. (2004). Human reliability data, human error and accident models—Illustration through the Three Mile Island accident analysis. *Reliability Engineering and System Safety*, *83*, 153–167.

Lees, F. P. (1996). *Loss prevention in the process industries* (2nd ed.). Oxford, U.K: Butterworth.

Lees, F. P. (1996). *Loss prevention in the process industries* (vol. 1). London: Butterworth Publications.

Lertworasirikul, S., Fang, S. -C., Joines, J. A., & Nuttle, H. L. W. (2003). Fuzzy data envelopment analysis (DEA): A possibility approach. *Journal of Fuzzy Sets and Systems*, *139*(2), 379–394.

Lessard, R. R., & DeMarco, G. (2000). The significance of oil spill dispersants. *Spill Science & Technology Bulletin*, *6*(1), 59–68.

Lewis, D. J. (1974). The Mond fire, explosions and toxicity index, applied plant lay-out and spacing, loss prevention symposium. CEP.

Li, Z., Shu-dong, H., & Xiang-rui, H. (2003). THERP + HCR-based model for human factor event analysis and its application. *Hedongli Gongcheng/Nuclear Power Engineering*, 24(3), 272–276.

Linstone, H. A., & Turoff, M. (1975). *The Delphi method techniques and application*. London: Addison-Wesley.

Little, R. G. (2004). A socio-technical systems approach to understanding and enhancing the reliability of interdependent infrastructure systems. *International Journal of Emergency Management*, 2(2), 98–110.

Liu, X., Li, W., Tu, Y. L., & Zhang, W. J. (2011). An expert system for an emergency response management in Networked SafeService Systems. *Expert Systems with Applications*, *38*, 11928–11938.

Lundberg, J., Rollenhagen, C., & Hollnagel, E. (2009). What-you-look-for-is-whatyou-find – the consequences of underlying accident models in eight accident investigation manuals. *Safety Science*, 47(10), 1297–1311.

Lutz, W. K. (1997). Advancing inherent safety into methodology. *Process Safety Progress*, *16*(2), 86–88.

Mannan, M. S. (2005). *Lees' loss prevention in the process industries hazard identification, assessment, and control* (3rd ed.). Amsterdam; Boston: Elsevier Butterworth-Heinemann.

Mannan, M. S., West, H. H., Krishna, K., Aldeeb, A. A., Keren, N., & Saraf, S. R., et al. (2005). The legacy of Bhopal: The impact over the last 20 years and future direction. *Journal of Loss Prevention in the Process Industries*, *18*, 218–224.

Mansfield, D. (1994). *Inherent safer approaches to plant design. HSE Research Project report 233*. London: HSEO Office.

Mansfield, D., & Cassidy, K. (1996). Inherent safer approaches to plant design. *IChemE symposium series* 134. Institution of Chemical Engineers, Rugby.

Mansfield, D., Poulter, L., & Kletz, T. A. (1996). *Improving inherent safety: A pilot study into the use of inherently safer designs in the UK Offshore oil and gas industry*. London: HSE Offshore Safety Division Research Project report, HSE Office.

Mansfield, D. P., Kletz, T. A., & Al-Hassn, T. (1996a). Optimizing safety by inherent offshore platform design. In *Proceeding of 1996 OMAE—Volume II (Safety and Reliability)*, Florence, Italy.

Markatos, N. C., Christolis, M., & Argyropoulos, C. (2009). Mathematical modeling of toxic pollutants dispersion from large tank fires and assessment of acute effects for fire fighters. *International Journal of Heat and Mass Transfer*, *52*, 4021–4030.

MARS, (1994). MARS: Diver assistance vehicle—ROV for IRM work. *Oil and Gas Technology*, *14*(October), 18.

Marsh, (2010). In: I. Clough (Ed.), *The 100 largest losses 1972–2009: Large property damage losses in the hydrocarbon-chemical industries*. London: Marsh Global Energy Risk Engineering.

McCauley-Bell, P., & Badiru, A. B. (1996). Fuzzy modelling and analytic hierarchy processing to quantify risk levels associated with occupational injuries—Part I: The development of fuzzy-linguistic risk levels. *IEEE Transactions on Fuzzy Systems*, *4*(2).

McEntire, D. A. (2001). Triggering agents, vulnerabilities and disaster reduction: Towards a holistic paradigm. *Disaster Prevention and Management*, *10*(3), 189–196.

McGrattan, K. B., Baum, H. R., & Rehm, R. G. (1996). Numerical simulation of smoke plumes from large oil fires. *Atmospheric Environment*, *30*, 4125–4136.

Medonos, S. (1994). Use of advanced methods in integrated safety engineering. In *1994 Offshore Mechanics and Arctic Engineering (OMAE) conference*, Houston, TX.

Meel, A., O'Neill, L. M., Levin, J. H., Seider, W. D., Oktem, U., & Keren, N. (2007). Operational risk assessment of chemical industries by exploiting accident databases. *Journal of Loss Prevention in the Process Industries, 20*, 113–127.

Mercan, M., Reisman, A., Yolalan, R., & Emel, A. B. (2003). The effect of scale and mode of ownership on the financial performance of the Turkish banking sector: Results of a DEA-based analysis. *Socio-Economic Planning Sciences, 37*(3), 185–202.

Mili, A., Bassetto, S., Siadat, A., & Tollenaere, M. (2009). Dynamic risk management unveil productivity improvements. *Journal of Loss Prevention in the Process Industries, 22*, 25–34.

Mohammad Fam, I., Nikoomaram, H., & Soltanian, A. (2012). Comparative analysis of creative and classic training methods in health, safety and environment (HSE) participation improvement. *Journal of Loss Prevention in the Process Industries, 25*(2), 250–253.

Mosleh, A., & Chang, Y. H. (2004). Model-based human reliability analysis: Prospects and requirements. *Reliability Engineering and System Safety, 83*, 241–253.

Munteanu, I., & Aldemir, T. (2003). A methodology for probabilistic accident management. *Nuclear Technology, 144*.

Nedic, D. P., Dobson, I., & Kirschen, D. S. (2006). Criticality in a cascading failure blackout model. *Electrical Power and Energy Systems, 28*, 627–633.

NFPA 11. (2002). National fire protection association standard: Standard for low-, medium-, and high-expansion foam.

NFPA 30. (1993). National fire protection association standard: Flammable and combustible liquids code.

NORSOK. (2004). *NORSOK standard: Well integrity in drilling and well operations, D-010.* Norwegian Technology Centre.

Norwegian Petroleum Directorate (NPD). (2004). Website, NPD.

Nuclear Regulatory Commission. (1983). USA, PRA procedure guide, NUREG/CR-2815.

Øien, K. (2001). A framework for the establishment of organizational risk indicators. *Reliability Engineering & System Safety, 74*(2), 147–167.

Olsen, E., Eikeland, T., & Tharaldsen, J. (2004). Development and validation of a safety culture inventory, to be presented at the EAOHP conference. Berlin, 20–21 November.

OSHA, (2012). Process safety management. *Occupational Safety and Health Administration (OSHA)* <http://www.osha.gov/> Accessed February 2012.

Owen, D., & Raeburn, G. (1991). Developing a safety, occupational health, and environmental (SOHE) program as part of a total quality program. In Offshore Technology Conference, Houston, Texas.

Papazoglou, I. A., Nivolianitou, Z., Aneziris, O., & Christou, M. (1992). Probabilistic safety analysis in chemical installations. *Journal of Loss Prevention in the Process Industries, 5*, 181–191.

Pate-Cornell, M. E., & Murphy, D. M. (1996). Human and management factors in probabilistic risk analysis: The SAM approach and observations from recent applications. *Reliability Engineering and System Safety, 53*, 115–126.

Patterson, S. A., & Apostolakis, G. E. (2007). Identification of critical locations across multiple infrastructures for terrorist actions. *Reliability Engineering and System Safety, 92*, 1183–1203.

Persson, H., & Lonnermark, A. (2004). Tank fires. SP Swedish National Testing and Research Institute. SP Report 2004:14, Boras, Sweden.

Pitblado, R. (2010). Global process industry initiatives to reduce major accident hazards. *Journal of Loss Prevention in the Process Industries, 24*, 57–62.

Pitblado, R. (2011). Global process industry initiatives to reduce major accident hazards. *Journal of Loss Prevention in the Process Industries, 24,* 57–62.

Pitblado, R., & Fisher, M. (2011). Novel investigation approach linking management system and barrier failure root causes. Paper # 22329. SPE.

Prem, K. P., Ng, D., Pasman, H. J., Sawyer, M., Guo, Y., & Mannan, M. S. (2010). Risk measures constituting a risk metrics which enables improved decision making: Value-at-risk. *Journal of Loss Prevention in the Process Industries, 23,* 211–219.

Rao, S. M. (1996). The effect of published report of environmental pollution on stock prices. *Journal of Financial and Strategic Decisions, 9,* 25–32.

Rathnayaka, S., Khan, F., & Amyotte, P. (2011). SHIPP methodology: Predictive accident modeling approach. Part I: methodology and model description. *Process Safety and Environmental Protection, 89*(2), 75–88.

Rausand, M. (2011). *Risk assessment: Theory, methods, and applications.* Hoboken, NJ: Wiley.

Rausand, M., & Høyland, A. (2004). *System reliability theory: Models, statistical methods, and applications* (2nd ed.). Hoboken, NJ: Wiley.

Reason, J. (1990). The contribution of latent human failures to the breakdown of complex systems. *Philosophical Transactions of the Royal Society of London Series B, Biological Sciences, 327*(1241), 475–484.

Reason, J., Hollnagel, E., & Paries, J. (2006). *Revisiting the «Swiss cheese» model of accidents.* France: EUROCONTROL Experimental Center.

Reason, J. T., Carthey, J., & De Leval, M. R. (2001). Diagnosing "vulnerable system syndrome": An essential prerequisite to effective risk management. *BMJ Quality & Safety, 10*(Suppl 2), ii21–ii25.

Rinaldi, S. A., Peerenboom, J. P., & Kelly, T. K. (2001). Identifying, understanding, and analyzing critical infrastructure interdependencies. *IEEE Control Systems Magazine, 21*(6), 11–25.

Robert, B. (2004). A method for the study of cascading effects within lifeline networks. *International Journal of Critical Infrastructures, 1*(1), 86–99.

Rundmo, T., Hestad, H., & Ulleberg, P. (1998). Organizational factors, safety attitudes and workload among offshore oil personnel. *Safety Science, 29,* 75–87.

Saati, M. S., Memariani, A., & Jahanshahloo, G. R. (2002). Efficiency analysis and ranking of DMUs with fuzzy data. *Journal of Fuzzy Optimization and Decision Making, 11*(3), 255–267.

Sanders, R. E. (1999). *Chemical process safety learning from case histories* (3rd ed.). Amsterdam; Boston: Elsevier Butterworth Heinemann.

Santos-Reyes, J., & Beard, A. N. (2008). A systemic approach to managing safety. *Journal of Loss Prevention in the Process Industries, 21,* 15–28.

Santos-Reyes, J., & Beard, A. N. (2009). A SSMS model with application to the oil and gas industry. *Journal of Loss Prevention in the Process Industries, 22*(6), 958–970.

Schönbeck, M., Rausand, M., & Rouvroye, J. (2010). Human and organisational factors in the operational phase of safety instrumented systems: A new approach. *Safety Science, 48*(3), 310–318.

Shaluf, I. M., & Abdullah, S. A. (2011). Floating roof storage tank boilover. *Journal of Loss Prevention in the Process Industries, 24,* 1–7.

Shapiro, S. (1990). *Piper alpha critique spurs together.* London: Business Insurance.

Shebeko, Y. N., Bolodian, I. A., Molchanov, V. P., Deshevih, Y. I., Gordienko, D. M., & Smolin, I. M., et al. (2007). Fire and exploS200sion risk assessment for large-scale oil export terminal. *Journal of Loss Prevention in the Process Industries, 20,* 651–658.

Shikdar, A. A., & Sawaqed, M. N. (2004). Sawaqed, ergonomics, occupational health and safety in the oil industry: A managers' response. *Computers & Industrial Engineering, 47,* 223–232.

Sklet, S. (2006). Safety barriers: Definition, classification, and performance. *Journal of Loss Prevention in the Process Industries, 19*(5), 494–506.

Sklet, S., Ringstad, A. J., Steen, A. S., Tronstad, L., Haugen, S., Seljelid, J., et al. (2010). Monitoring of human and organizational factors influencing the risk of major accidents. Paper # 126530. SPE.

Skogdalen, J. E., Khorsandi, J., & Vinnem, J. E. (2012). Evacuation, escape, and rescue experiences from offshore accidents including the Deepwater Horizon. *Journal of Loss Prevention in the Process Industries, 25*(1), 148–158.

Skogdalen, J. E., Utne, I. B., & Vinnem, J. E. (2011). Developing safety indicators for preventing offshore oil and gas deepwater drilling blowouts. *Safety Science, 49*(8e9), 1187–1199.

Smith, D. (2002). Health and safety performance of the global E&P industry 2000. In Offshore technology conference.

Sokovic, M., Pavletic, D., & Pipan, K. K. (2010). Quality improvement methodologies – PDCA cycle, RADAR matrix, DMAIC and DFSS. *Journal of Achievements in Materials and Manufacturing Engineering, 43*(1), 476–483.

S-RCM Training Guide. (2000). *Shell-reliability centered maintenance.* Shell Global Solution International.

Strutt, J. E., Wei-Whua, L., & Allsopp, K. (1998). Progress towards the development of a model for predicting human reliability. *Quality and Reliability Engineering International, 14,* 3–14.

Sutton, I. (2012). *Offshore safety management, implementing a SEMS program.* Waltham, MA: William Andrew.

Takala, J. (1999). Global estimates of fatal occupational accidents, epidemiology, September (vol. 10(5)), from the Occupational Safety and Health Branch. Working Conditions and Environment Department, International Labour Office, 4 Route des Morillons, CH-1211 Geneva 22, Switzerland.

Tanabe, M., & Miyake, A. (2010). Safety design approach for onshore modularized LNG liquefaction plant. *Journal of Loss Prevention in the Process Industries, 23,* 507–514.

Taylor, M. (April 26, 1990). Plate–fin exchangers offshore—the background. 23.

Tharaldsen, J., Olsen, E., & Eikeland, T. (2004). *A comparative study of safety culture, ABB offshore systems compared with other companies on the Norwegian Continental Shelf.* Norway: Work Life and Business Development, Rogaland Research.

Thompson, R. C., Hilton, T. F., & Witt, L. A. (1998). Where the safety rubber meets the shop floor: A confirmatory model of management influence on workplace safety. *Journal of Safety Research, 29*(1), 15–24.

Todinov, M. T. (2003). Setting reliability requirements based on minimum failurefree operating periods. *Quality and Reliability Engineering International, 20,* 273–287.

Tomas, J. M., Melia, J. L., & Oliver, A. (1999). A cross-validation of a structural equation model of accidents: Organisational and psychological variables as predictors of work safety. *Work and Stress, 13*(1), 49–58.

TotalElfFina. (2001). *Environment and safety report.* TotalElfFina.

UK HSE. (1992). Offshore installations (safety case) regulations 1992 (no. 2885, Health and Safety), ISBN 011025869X.

UK HSE. (1996). *Preventing slips, trips and falls at work.* HSE.

UK HSE. (2001a). Multivariate analysis of injuries data. Offshore technology report 2000/108, Prepared by the University of Liverpool for the HSE, Copyright.

UK HSE. (2001b). Safety culture maturity model. Offshore technology report 2000/049, Prepared by Dr Mark Fleming of The Keil Centre for the HSE, Copyright.

UK HSE. (2002b). Slips, trips and falls from height offshore. Offshore technology report 2002/001, Prepared by BOMEL Ltd. for the HSE, Copyright.

Van Heel, K. A. L., Knegtering, B., & Brombacher, A. C. (1999). Safety lifecycle management. A flowchart presentation of the IEC 61508 overall safety lifecycle model. *Quality and Reliability Engineering International, 15*, 493–500.

Vautard, R., Ciaisa, P., Fisher, R., Lowry, D., Breon, F. M., & Vogel, F., et al. (2007). The dispersion of the Buncefield oil fire plume. *Atmospheric Environment, 41*, 9506–9517.

Vesely, W. E., Belhadj, M., & Rezos, J. T. (1993). PRA importance measures for maintenance prioritization applications. *Reliability Engineering and System Safety, 43*, 307–318.

Vrijling, J. K., & van Gelder, P. H. A. J. M. (1997). Societal risk and the concept of risk aversion. In C. G. Soares (Ed.), *European safety and reliability conference Oxford, England.* New York, Lisbon, Portugal: Pergamon.

Wang, C. -H., Chuang, C. -CH, & Tsai, C. -CH. (2009). A fuzzy DEAeneural approach to measuring design service performance in PCM projects. *Automation in Construction, 18*(5), 702–713.

Wang, W., & Christer, A. H. (1998). A modelling procedure to optimize component safety inspection over a finite time horizon. *Quality and Reliability Engineering International, 13*(4), 217–224.

Warwick, A. R. (1998). Inherent safe design of floating production, storage and offloading vessels (FPSOs). In *Proceeding of 1998 offshore mechanics and arctic engineering conference*, Lisbon, Portugal.

WCTMWG (Well Control Training Material Writing Group of Sinopec Group). (2008). *Drilling well control technology.* Shandong, Dongying: Press of University of Petroleum, China.

Wen, M., & Li, H. (2009). Fuzzy data envelopment analysis (DEA): model and ranking method. *Journal of Computational and Applied Mathematics, 223*(2), 872–878.

Woo, D. M., & Vicente, K. J. (2003). Socio-technical systems, risk management, and public health: Comparing the North Battleford and Walkerton Outbreaks. *Reliability Engineering and System Safety, 80*, 253–269.

Wu, D., Yang, Z., & Liang, L. (2006). Efficiency analysis of cross-region bank branches using fuzzy data envelopment analysis. *Applied Mathematics and Computation, 181*(1), 271–281.

Wu, T. H., Chen, M. S., & Yeh, J. Y. (2010). Measuring the performance of police forces in Taiwan using data envelopment analysis. *Evaluation and Program Planning, 33*(3), 246–254.

Zhuang, J., & Bier, V. M. (2007). Balancing terrorism and natural disastersddefensive strategy with endogenous attacker effort. *Operations Research, 55*(5), 976–991.

Zimmerman, R. (2004). Decision-making and the vulnerability of interdependent critical infrastructure. *IEEE International Conference on Systems, Man and Cybernetics, 5*, 4059–4063.

Zutschi, A., & Sohal, A. (2003). Integrated management system: The experience of three Australian organizations. *Journal of Manufacturing Technology Management, 16*(2), 211–232.

Index

Note: Page numbers followed by "*f*" and "*t*" refer to figures and tables, respectively.

Printed in the United States
By Bookmasters